Photoorganocatalysis in Organic Synthesis

CATALYTIC SCIENCE SERIES

ISSN 1793-1398 (Print)
ISSN 2399-4495 (Online)

Series Editor: Graham J. Hutchings *(Cardiff University)*

Catalysis is at the forefront of the chemical industry and is essential to many fields in the chemical sciences. This series explores all aspects of catalysis in authored and edited volumes drawing on expertise from around the globe in a focussed manner. Volumes are accessible by postgraduate students and professionals in academia and industry.

Published

Vol. 18 *Photoorganocatalysis in Organic Synthesis*
edited by Maurizio Fagnoni, Stefano Protti and Davide Ravelli

Vol. 17 *Hydroprocessing Catalysts and Processes: The Challenges for Biofuels Production*
edited by Bo Zhang and Duncan Seddon

Vol. 16 *Electro-Catalysis at Chemically Modified Solid Surfaces*
by Jacques Simonet

Vol. 15 *Noble Metal Noble Value: Ru-, Rh-, Pd-catalyzed Heterocycle Synthesis*
edited by Xiao-Feng Wu

Vol. 14 *Enantioselective Titanium-Catalysed Transformations*
by Hélène Pellissier

Vol. 13 *Gold Catalysis: An Homogeneous Approach*
edited by F. Dean Toste and Véronique Michelet

Vol. 12 *Catalysis by Ceria and Related Materials (Second Edition)*
edited by A. Trovarelli and P. Fornasiero

Vol. 11 *Supported Metals in Catalysis (Second Edition)*
by J. A. Anderson

Vol. 10 *Concepts in Syngas Manufacture*
by J. Rostrup-Nielsen and L. J. Christiansen

Vol. 9 *Deactivation and Regeneration of Zeolite Catalysts*
edited by M. Guisnet and F. R. Ribeiro

More information on this series can be found at http://www.worldscientific.com/series/css

(Continued at end of book)

CATALYTIC SCIENCE SERIES — VOL. 18

Series Editor: Graham J. Hutchings

Photoorganocatalysis in Organic Synthesis

edited by

Maurizio Fagnoni
Stefano Protti
Davide Ravelli

University of Pavia, Italy

World Scientific

NEW JERSEY • LONDON • SINGAPORE • BEIJING • SHANGHAI • HONG KONG • TAIPEI • CHENNAI • TOKYO

Published by

World Scientific Publishing Europe Ltd.
57 Shelton Street, Covent Garden, London WC2H 9HE
Head office: 5 Toh Tuck Link, Singapore 596224
USA office: 27 Warren Street, Suite 401-402, Hackensack, NJ 07601

Library of Congress Cataloging-in-Publication Data
Names: Fagnoni, Maurizio, editor. | Protti, Stefano, 1979– editor. |
Ravelli, Davide, 1984– editor.
Title: Photoorganocatalysis in organic synthesis / edited by Maurizio Fagnoni (University of Pavia, Italy), Stefano Protti (University of Pavia, Italy) and Davide Ravelli (University of Pavia, Italy).
Description: New Jersey : World Scientific, 2018. | Series: Catalytic science series ; volume 18 | Includes bibliographical references.
Identifiers: LCCN 2018040371 | ISBN 9781786346049 (hc : alk. paper)
Subjects: LCSH: Organic compounds--Synthesis. | Photocatalysis.
Classification: LCC QD262 .P46 2018 | DDC 547/.215--dc23
LC record available at https://lccn.loc.gov/2018040371

British Library Cataloguing-in-Publication Data
A catalogue record for this book is available from the British Library.

For any available supplementary material, please visit
https://www.worldscientific.com/worldscibooks/10.1142/Q0180#t=suppl

Desk Editors: Herbert Moses/Jennifer Brough/Shi Ying Koe

Typeset by Stallion Press
Email: enquiries@stallionpress.com

Preface

The use of catalysts in organic synthesis is mandatory for the development of frontier research in academy and industry. Nowadays, catalysis has been actually monopolized by the use of metal catalysts able to activate several *X–Y* bonds in organic molecules [1–4]. The main drawbacks in the use of these metals is their high cost and their toxicity, with their removal from the desired compounds being a critical issue, if a pharmaceutical derivative is finally prepared. A way to circumvent such problem is to avoid the use of metal derivatives and promote the reaction by a substoichiometric quantity of an organic compound. As a matter of fact, organocatalysis has been growing at a regular pace in the last 10 years, in turn allowing the development of enantioselective processes. It is rather difficult to summarize the role that organocatalysts, such as amines, alkaloids, peptides, NHC carbenes, ureas and thioureas, alcohols, phenols, carboxylic and chiral Brønsted acids, have in synthetic planning [5–12]. The main drawback in this case, however, is the huge amount of the organic compound required (compared to commonly used metal complexes), that limited its application in industry.

A different catalytic approach is that of photocatalysis. The meaning and the use of photocatalysis dramatically changed in the last years. More than 10 years ago, a photocatalytic reaction was strictly related to a handful of processes, notably water splitting, photodepollution, or artificial photosynthesis. Indeed, only inorganic derivatives were previously dubbed as photocatalysts, titanium dioxide being the most famous and most employed one. The definition of photocatalyst, however, dramatically changed when photocatalytic processes were applied to organic synthesis. Actually, according to the IUPAC definition, the photocatalyst is a: "*Catalyst able to produce, upon absorption of light, chemical transformations of the reaction partners. The excited state of the photocatalyst repeatedly interacts with the reaction partners forming reaction intermediates and regenerates itself after each cycle of such interactions*" [13].

This definition, however, fits well even for organic molecules, acquiring the role of photoorganocatalysts (**POCs**). Ten years ago, the use of **POCs** was quite limited and mainly involved aromatic ketones, some quinones, and cyanoaromatics [14]. Nevertheless, with the dramatic advancement of photocatalysis in recent years, the use of **POCs** became more and more important and new organic derivatives were designed for this purpose, or other common derivatives (e.g. dyes) acquired a new role in photochemical reactions [14–26].

The recent impressive development in the use of **POCs** in organic synthesis well justify the publication of this book. The aim of this volume is to assimilate the photocatalyzed organic processes having a synthetic significance, where an organic molecule acts as the photocatalyst.

The chapters are organized according to the class of the **POC** used. Each chapter will start with a brief summary on the photophysical characteristics of the analyzed **POCs**, and selected representative synthetic applications will ensue. The recent publications in the field (mainly in the last 10 years) have been collected, though some "old" seminal papers have been quoted as well.

The classes of **POCs** considered are: simple (aromatic) ketones and aldehydes (Chapter 1), quinones (Chapter 2), along with unfunctionalized aromatics and cyano(hetero)aromatics (Chapter 3).

As pointed out in the literature, most of the **POCs** employed are heteroaromatics. Thus, the use of sulfur (methylene blue (**MB**) or thiapyrylium salts, Chapter 4), oxygen (pyrylium salts, Eosin Y, fluorescein, rhodamine, and Rose Bengal (**RB**), Chapters 5–7), and nitrogen (porphyrins, acridinium salts, imides, BODIPY, Chapters 8–10) heterocycles will come next. The last two chapters will be devoted to the applications of photoorganocatalysis in polymerizations (Chapter 11) and under flow conditions (Chapter 12). The structure of the main organic compounds acting as **POCs** is depicted in Chart 1.

In the majority of cases, photoorganocatalyzed reactions were quoted when the **POC** was present in a substoichiometric amount (usually in less than 30% mol quantity). Reactions carried out in the presence of a metal-based or a further organocatalyst (dual catalysis) [27] have been quoted as well, provided that the **POC** was uniquely responsible for light absorption.

On the other hand, photochemical processes where the organic molecule plays a different role were not included in this book. Accordingly, the following reactions will be not treated here:

— when the organic molecule forms an electron donor-acceptor (EDA) complex that is responsible for light absorption, but then the reaction causes the consumption of the starting organic compound (see Ref. [28] for example).

— when the (chiral) organic molecule is able to complex any photoreactive species, but not able to absorb the light itself (see Ref. [29] for example).

— when the organic molecule allowed for a photochemical reaction to occur and was involved in a catalytic cycle, but did not absorb light (e.g. thiols or disulfides, see Ref. [30]).

As for the above and according to the IUPAC definition, the type of reactions that can be promoted by a **POC** are gathered in Scheme 1.

The first process is a direct hydrogen atom transfer, where the excited **POC** (**POC***) is able to cleave homolytically a C–H bond from a hydrogen donor (R–H) and a carbon-centered radical chemistry results (path a). This reaction is photoorganocatalyzed by

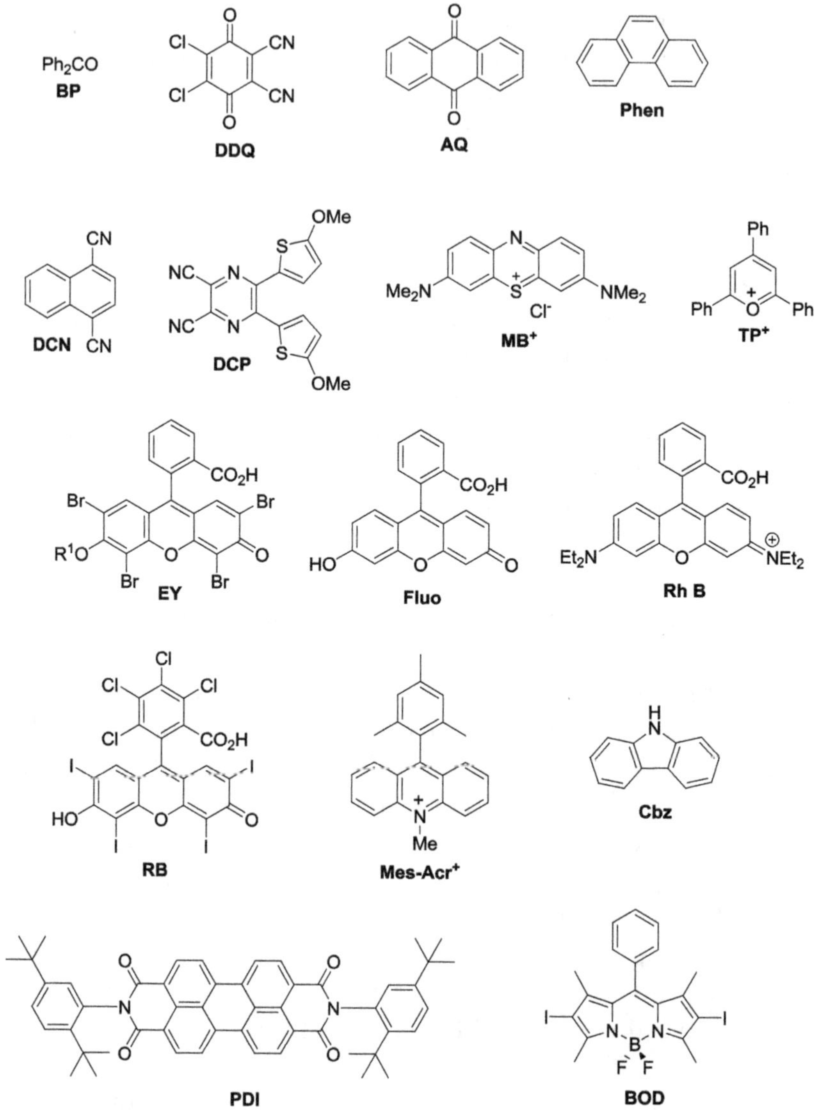

Chart 1. Main **POCs** used in organic synthesis and discussed in this book.

BP = Benzophenone; **DDQ** = 2,3-Dichloro-5,6-dicyano-1,4-benzoquinone; **AQ** = Anthraquinone; **Phen** = Phenanthrene; **DCN** = 1,4-Dicyanonaphthalene; **DCP** = 5,6-Bis(5-methoxythiophen-2-yl) pyrazine-2,3-dicarbonitrile; **MB**$^+$ = Methylene blue; **TP**$^+$ = 2,4,6-triphenylpyrylium salt; **EY** = Eosyn Y; **Fluo** = Fluorescein; **Rh B** = Rhodamine B; **RB** = Rose Bengal; **Mes-Acr**$^+$ = 9-mesityl-10-methylacridinium ion; **Cbz** = Carbazole; **PDI** = *N,N*-bis(2,5-di-*tert*-butylphenyl)perylene-3,4:9,10-tetracarboxydiimide; **BOD** = 2,6-Diiodo-1,3,5,7-tetramethyl-8-phenyl-4,4-difluoroboradiazaindacene.

Table 1. Selected photophysical and electrochemical data of **POCs** depicted in Chart 1.

	λ_{abs} [a] (ε [b])	Φ_F [c]	τ_S ns [d]	Φ_T [e]	τ_T μs [f]	E^0_{OX} **(POC$^+$/POC)** [g]	E^0_{RED} **(POC/POC$^-$)** [h]	E^*_{OX} **(POC$^+$/POC*)** [i]	E^*_{RED} **(POC*/POC$^-$)** [j]
BP	335	—	0.008	1	≤1	+2.39	−1.72	−0.61	+1.28
DDQ	330	—	—	1	—	—	+0.49	—	+3.18
AQ	326	—	—	1	3200	—	−0.96	—	+1.77
Phen	300	0.16	—	0.83	—	+1.50	−2.44	−2.1	+1.14
DCN	<350	—	10.3	—	—	—	−1.27	—	+2.30
DCP	<550	—	—	—	—	—	−1.45	—	+1.99
MB$^+$	610 (90000)	—	<1	0.52	32	+1.13	−0.30	−0.68	+1.60
TP$^+$	368	0.58	4.38	0.42	—	—	−0.32	—	+2.02
EY	539 (60800)	0.48	2.1	0.32	25	+0.76	−1.08	−1.11	+0.83
Fluo	491	0.93	4.73	0.03	—	+0.83	−1.22	−1.55	+1.25
Rh B	550	0.58	2.45	—	—	+0.91	−0.96	−1.31	+1.25
RB	549	0.09	0.50	0.77	—	+0.84	−0.96	−0.96	+0.81
Mes-Acr$^+$	425	<0.08	6	0.38	—	—	−0.49	—	+2.06
Cbz	330 (4250)	0.38	16.1	0.36	170	+1.16	—	—	—
PDI	526 (80000)	0.97	3.8	$<10^{-3}$	140	+1.65	−0.53	—	—
BOD	529 (39000)	0.03	0.13	0.97	220	+1.13	—	—	—

Notes: [a] Maximum of the lowest energy absorption band (nm). [b] Molar absorption coefficient (M^{-1} cm^{-1}). [c] Fluorescence quantum yield. [d] Lifetime of lowest singlet excited state S_1. [e] Quantum yield for triplet formation. [f] Lifetime of the lowest triplet excited state T_1. [g] Ground state oxidation potential (V vs SCE). [h] Ground state reduction potential (V vs SCE). [i] Oxidation potential in the excited state (V vs SCE). [j] Reduction potential in the excited state (V vs SCE).

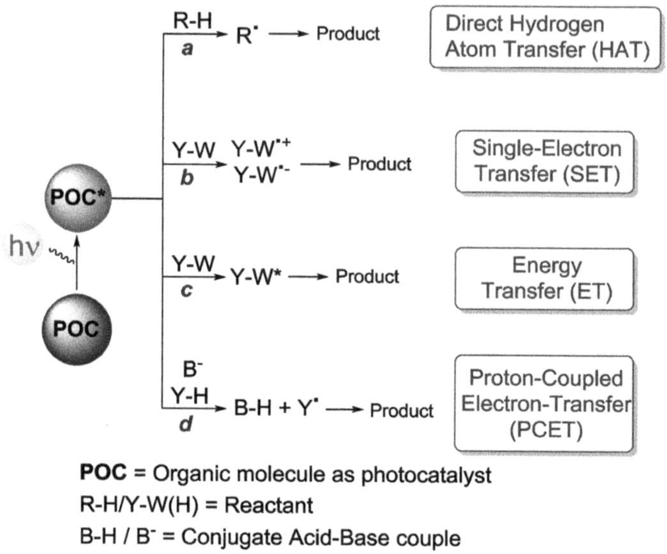

Scheme 1. Types of photoorganocatalyzed reactions discussed in this book.

(aromatic) ketones and quinones, since they have an excited state that resembles an alkoxy radical. A back-hydrogen transfer from the catalysts to a reactive species is mandatory in order to close the catalytic cycle. In most cases, the process is not efficient and requires a huge amount (up to equimolar) of the **POC** to take place. The preferred pathway is, however, that of single electron transfer (path b), nowadays known as photoredox catalysis. Here **POC*** may donate or accept an electron from an organic compound Y–W to generate the corresponding radical anion (Y–W$^{•-}$) or cation (Y–W$^{•+}$), respectively. A radical ion chemistry may then occur, albeit, in most cases, the fragmentation of the radical ion generates again a reactive radical. Most of the **POC**s are able to promote this reaction and a back-electron transfer is mandatory for **POC** regeneration. Depending on the oxidation (or the reduction) potential of Y–W, it is possible to choose a suitable **POC** that matches the redox properties of **POC*** (see Table 1), rendering the electron transfer feasible. A particular case is that of energy transfer processes (path c), where the

POC has the role of a photosensitizer, and this is the only case where the reaction involves the excited state of the organic substrate. In some instances, a radical may be formed by a proton-coupled electron transfer (PCET), and as a result a formal hydrogen atom abstraction from a Y–H bond may take place [31]. For the simplicity of the reader we collected in Table 1 some photophysical and electrochemical data of the **POCs** depicted in Chart 1.

We hope that this book will inspire and guide your future research!

Maurizio Fagnoni, Stefano Protti and Davide Ravelli
PhotoGreen Lab, University of Pavia
April 18, 2018

References

[1] Brandsma L.; Vasilevsky, S. F.; Verkruijsse, H. D. *Application of Transition Metal Catalysts in Organic Synthesis*, Springer-Verlag, Berlin, Heidelberg, **1999**.

[2] Tsuji, J. *Transition Metal Reagents and Catalysts: Innovations in Organic Synthesis*, John Wiley & Sons, Ltd., Chichester, **2000**.

[3] Suzuki, A. *Angew. Chem. Int. Ed.* **2011**, *50*, 6723–6737.

[4] Negishi, E. *Angew. Chem. Int. Ed.* **2011**, *50*, 6738–6764.

[5] Dalko, P. I.; Moisan, L. *Angew. Chem. Int. Ed.* **2004**, *43*, 5138–5175.

[6] Berkessel, A.; Groeger, H. *Asymmetric Organocatalysis*, Wiley-VCH, Weinheim, 2005.

[7] MacMillan, D. W. C. *Nature* **2008**, *455*, 304–308.

[8] Melchiorre, P.; Marigo, M.; Carlone, A.; Bartoli, G. *Angew. Chem. Int. Ed.* **2008**, *47*, 6138–6171.

[9] Bertelsen, S.; Jørgensen, K. A. *Chem. Soc. Rev.* **2009**, *38*, 2178–2189.

[10] Alemán, J.; Cabrera, S. *Chem. Soc. Rev.* **2013**, *42*, 774–793.

[11] Ed., North, M. *Sustainable Catalysis: Without Metals or Other Endangered Elements, Parts 1 and 2*, Vol. 41, RSC Green Chemistry Series, Royal Society of Chemistry, Cambridge, 2016.

[12] Qin, Y.; Zhu, L.; Luo, S. *Chem. Rev.* **2017**, *117*, 9433–9520.

[13] https://goldbook.iupac.org/html/P/PT07446.html.

[14] Fagnoni, M.; Dondi, D.; Ravelli, D.; Albini, A. *Chem. Rev.* **2007**, *107*, 2725–2756.

[15] Ravelli, D.; Dondi, D.; Fagnoni, M.; Albini, A. *Chem. Soc. Rev.* **2009**, *38*, 1999–2011.

[16] Ravelli, D.; Fagnoni, M.; Albini, A. *Chem. Soc. Rev.* **2013**, *42*, 97–113.

[17] Nicewicz, D. A.; Hamilton, D. S. *Synlett* **2014**, *25*, 1191–1196.

[18] Fukuzumi, S.; Ohkubo, K. *Org. Biomol. Chem.* **2014**, *12*, 6059–6071.

[19] Hari, D. P.; Koenig, B. *Chem. Commun.* **2014**, *50*, 6688–6699.

[20] Pérez-Ruiz, R.; Jiménez, M. C.; Miranda, M. A. *Acc. Chem. Res.* **2014**, *47*, 1359–1368.

[21] Nicewicz, D. A.; Nguyen, T. M. *ACS Catal.* **2014**, *4*, 355–360.

[22] Lerch, S.; Unkel, L.-N.; Wienefeld, P.; Brasholz, M. *Synlett* **2014**, *25*, 2673–2680.

[23] (a) Romero, N. A.; Nicewicz, D. A. *Chem. Rev.* **2016**, *116*, 10075–10166; (b) Zeitler, K. in *Visible Light Photocatalysis in Organic Chemistry*, Ed., Stephenson, C. R. J., Yoon, T. P., MacMillan, D. W. C. Wiley-VCH, Weinheim, **2018**, pp 159–232; (c) Sideri, I. K.; Voutyritsa, E.; Kokotos, C. G. *Org. Biomol. Chem.* **2018**, *16*, 4596–4614; (d) Li, R.; Byun, J.; Huang, W.; Ayed, C.; Wang, L.; Zhang, K. A. I. *ACS Catal.* **2018**, *8*, 4735–4750; (e) Bogdos, M. K.; Pinard, E.; Murphy, J. A. *Beilstein J. Org. Chem.* **2018**, *14*, 2035–2064; (f) Wang, Y.; Li, Y.; Jiang, X. *Chem. Asian J.* **2018**, *13*, 2208–2242.

[24] Montagnon, T.; Kalaitzakis, D.; Sofiadis, M.; Vassilikogiannakis, G. *Org. Biomol. Chem.* **2016**, *14*, 8636–8640.

[25] Majek, M.; von Wangelin, A. J. *Acc. Chem. Res.* **2016**, *49*, 2316–2327.

[26] Ravelli, D.; Protti, S.; Fagnoni, M. *Chem. Rev.* **2016**, *116*, 9850–9913.

[27] Skubi, K. L.; Blum, T. R.; Yoon, T. P. *Chem. Rev.* **2016**, *116*, 10035–10074.

[28] Arceo, E.; Bahamonde, A.; Bergonzini, G.; Melchiorre, P. *Chem. Sci.* **2014**, *5*, 2438–2442.

[29] Vallavoju, N.; Selvakumar, S.; Pemberton, B. C.; Jockusch, S.; Sibi, M. P.; Sivaguru, J. *Angew. Chem. Int. Ed.* **2016**, *55*, 5446–5451.

[30] Hager, D.; MacMillan, D. W. C. *J. Am. Chem. Soc.* **2014**, *136*, 16986–16989.

[31] Gentry, E. C.; Knowles, R. R. *Acc. Chem. Res.* **2016**, *49*, 1546–1556.

About the Editors

Stefano Protti (born in 1979) completed his Ph.D. in Pavia (2007) focusing on photochemical arylations via phenyl cations. He was a postdoctoral fellow at the LASIR laboratory (Lille, France) and at the iBitTec-S Laboratory (CEA Saclay, France). Since 2015, he has been Senior Researcher at the University of Pavia, Italy. His research is mainly focused on the development and the optimization of photochemical arylation processes under ecosustainable conditions. Stefano Protti is currently a co-author of more than 90 articles and reviews, 11 chapters in multiauthored books and the book *Paradigms in Green Chemistry and Technology* (2016, Springer IK, with Angelo Albini).

Davide Ravelli is currently a researcher at the PhotoGreen Lab of the University of Pavia. His main research interests are in the frame of photochemical reactions and attending applications in various fields, particularly in organic synthesis. He is highly interested in the discovery of methods for the facile generation of valuable intermediates (mainly radicals). In the last years, he has been deeply involved in the study of decatungstate-mediated photocatalyzed reactions (particularly, those involving a Hydrogen Atom Transfer — HAT — step). His experimental activity is supplemented by theoretical work (exploiting quantum chemical calculations), aimed at the rationalization of the processes under investigation.

Maurizio Fagnoni graduated at Pavia (Italy) under the supervision of Prof. Angelo Albini. He is currently an Associate Professor at the Department of Chemistry of the University of Pavia (PhotoGreen Lab). He is interested on the development of eco-sustainable processes induced by irradiation of UV/visible light. In particular, his research is mainly focused on the discovery of methods for the facile generation of valuable intermediates (radicals, diradicals, radical ions, and cations) to be used in synthesis or for technological application (photolithography as well as biological application). He is recently a co-author of the book *Photochemically-Generated Intermediates in Synthesis.*

About the Contributors

Manuel Anselmo, born in Genova on 17/02/1990, achieved his Bachelor in Chemistry (110/110 cum laude) in 2012 (Prof. Renata Riva). Then, he achieved his Master (110/110) in 2015 (Prof. Renata Riva) after an Internship in Germany, Düsseldorf (Prof. Thomas J. J. Müller). He is currently a Ph.D. student in University of Genova under the supervision of Prof. Andrea Basso, having recently returned in Italy after an internship in Germany, Regensburg (Prof. Burkhard König).

Andrea Basso graduated from the University of Genova in 1995 (Prof. Giuseppe Guanti), and then received his Ph.D. in 2000 working on Solid Phase Chemistry under the supervision of Prof. Mark Bradley (University of Southampton, UK). In 2001, he worked as a Postdoctoral-Fellow at the University of Basel in the group of Prof. Beat Ernst, and then moved back to the University of Genova, where he became Assistant Professor in 2005 and Associate Professor in 2015. His scientific interests are focused mainly on novel synthetic methodologies based on multicomponent reactions and photochemistry.

Aude-Héloïse Bonardi is a Graduated Engineer (equivalent to a M.S. degree) in Formulation Science from the Ecole Nationale Supérieure de Chimie de Mulhouse (ENSCMu, France), and also received a master degree in Material Science for the University of Haute-Alsace (Mulhouse, France). She is now a Ph.D. student in the group of Prof. Jacques Lalevée at the Institut de Science des Matériaux de Mulhouse (IS2M, France). Her research interests focus on the design of new photoinitiators for radical photopolymerizations under soft irradiation conditions and applications in 3D printing. She already co-authored 2 publications (international journals).

Sergio M. Bonesi was born in Argentina and obtained his M.Sc. degree from the University of Buenos Aires in 1988. He obtained his Ph.D. degree from the University of Buenos Aires in 1996, under the supervision of Prof. Dr. Rosa Erra Balsells. After a Postdoctoral stage under the supervision of Prof. Angelo Albini at the University of Pavia, he returned to Argentina and worked with Prof. Dr. R. Erra Balsells in the area of photochemistry of heteroaromatic compounds. He is currently Associate Professor of Organic Chemistry at the Department of Organic Chemistry, FCEyN, University of Buenos Aires. His interests are in the areas of photoinduced rearrangement reactions, synthetic photochemistry, and steady-state and time-resolved spectroscopies. He is devoted to studying photoreactions in confined media.

Paola Ceroni is Full Professor at the University of Bologna. In 1998, she obtained her Ph.D. degree in Chemical Sciences at the University of Bologna, after a period in the United States (Prof. Allen J. Bard's laboratory). Current research is focused on photochemistry and electrochemistry of supramolecular systems and nanocrystals. Her research on luminescent silicon nanocrystals was funded by an ERC Starting Grant PhotoSi and an ERC Proof of Concept SiNBiosys. She is fellow of the Royal Society of Chemistry. She is co-author of 190 scientific papers and of a book entitled *Photochemistry and Photophysics: Concepts, Research, Applications.*

Jia-Rong Chen completed his Ph.D. studies under the supervision of Prof. Wen-Jing Xiao at the Central China Normal University (CCNU) (2009). He has worked as a Humboldt Postdoctoral fellow with Prof. Carsten Bolm at the RWTH Aachen University (2011–2012). He is presently a Full Professor at CCNU. His research interests include the development of new catalysts for asymmetric catalysis and photoredox catalyzed radical chemistry.

Pier Giorgio Cozzi studied chemistry at the University of Milan, where he received his Laurea degree in 1989 with Prof. C. Gennari. After spending four years as a Research Associate in Lausanne (CH), he was appointed as Assistant, then Associate Professor and Full Professor at the University of Bologna. The development of new, enantioselective catalytic reactions, the design of new chiral ligands, and the introduction of new concepts in photoredox catalysis are his preeminent interests. P.G.C. has participated in several European projects and he was the coordinator of CATAFLU.OR Project (Fp7 European Program). P.G.C. is on the International Advisory Board of ChemPlusChem and AsianJOC. P.G.C. has received the CINMPIS award (2007) and the JSPS Fellowship from the Japanese Chemical Society (2014). In 2015, he received the Pino Gold Medal from Italian Chemical Society and, in 2018, the González-Ciamician Award from Spanish Royal Society of Chemistry.

Frédéric Dumur received his Ph.D. in Chemistry in 2002 from the University of Angers under the supervision of Prof. Pietrick Hudhomme. Following completion of his Ph.D., he was successively a Postdoctoral Research fellow at the University of Groningen (The Netherlands), Reims Champagne-Ardennes (France), and Versailles Saint-Quentin en Yvelines (France). In 2008, he joined the Faculty of Sciences at the Aix-Marseille University, where he is currently an Associate Professor. His research interests include the synthesis of dopants for OLEDs and photoinitiators of polymerization.

Peter Fodran studied chemistry at the Comenius University in Bratislava (Slovakia). His Master Research, under the guidance of Prof. R. Šebesta, focused on Cu-catalyzed tandem reactions. For his Ph.D. studies, he moved to the University of Groningen (The Netherlands), where he developed methods for stereoselective synthesis of glycerol-based lipids under the supervision of Prof. A. J. Minnaard. After successful defence of his Ph.D. in 2015, he stayed in the group of Prof. Minnaard as a Postdoctoral Researcher working on the total synthesis of methoxy-Mycolic acid. After completion of his project, he worked as a Research Scientist at the contract-research organization Syncom (Groningen, The Netherlands). Since 2017, he has been a Marie-Curie Individual Fellow in the group of Dr. C. J. Wallentin at the University of Gothenburg (Sweden), where he develops catalytic photoredox processes.

Hermenegildo García is a Full Professor since 1996, at the Instituto de Tecnología Química of the Technical University of Valencia. He made a Postdoctoral stay at the University of Reading with Prof. Andrew Gilbert in 1987 working on preparative organic photochemistry and several sabbatical leaves, and in the group of Prof. J. C. Scaiano at the University of Ottawa working on laser flash photolysis and time-resolved spectroscopy. Prof. García has been active in the field of heterogeneous catalysis working with porous catalysts and nanoparticles. This research fronts are related to the use of gold nanoparticles as heterogeneous catalysts, metal organic frameworks in catalysis, carbocatalysis, layered double hydroxides as photocatalysts, plasmon photocatalysis, and solar fuels production.

Didier Gigmes completed his Ph.D. in Organic Chemistry in 1998 at the University Paul Cézanne in Marseille (France). Then he moved to Elf Atochem North America in Pennsylvania (USA) to work as a Postdoctoral. In 2001, he was recruited as Researcher at CNRS and in October 2010, he was appointed as Research Director of the CNRS working at Aix-Marseille University. Currently, his research activities concern the synthesis of advanced polymer materials and the development of versatile photoinitiators. He is (co)author of 310 peer review articles in international journals, 17 patents, 10 book chapters, and editor of one book.

Katarzyna Goliszewska was born in Warsaw, Poland, in 1991. She joined the research group of Prof. Dorota Gryko in 2014 where her initial research interests included synthesis of B12 derivatives and their potential applications as catalysts. She graduated from the Warsaw University of Technology with a Masters in Engineering in 2015, and, in 2016, she started Ph.D. at the Institute of Organic Chemistry of the Polish Academy of Sciences. Her current research interests are focused on photoorganocatalysis with particular emphasis on porphyrins as photoredox catalysts.

Dorota Gryko obtained a Ph.D. from the Institute of Organic Chemistry at the Polish Academy of Sciences in 1997, under the supervision of Prof. J. Jurczak. After a Postdoctoral stay with Prof. J. Lindsey in North Carolina State University (1998–2000), she started an independent career in Poland. In 2009 and 2018, she received the prestigious TEAM grants from the Foundation for Polish Science. Her current research interests are focused on light-induced processes, with particular attention being paid to porphyrinoid catalysis, as well as on vitamin B12 chemistry, and its applications as a delivery vehicle.

Andrea Gualandi was born in Bologna, Italy, in 1978, and he studied Chemistry at the University of Bologna where, in 2003, he had his diploma. In 2008, he received his Ph.D. in Chemistry on the diastereoselective synthesis of amines, under the supervision of Prof. D. Savoia. In the same year, he joined the group of Prof. R. Neier at the University of Neuchâtel. Since 2010, he has been working at the University of Bologna in the group of Prof. P. G. Cozzi as a Postdoctoral Associate and as a Junior Assistant Professor. His main interests are the development of new catalytic stereoselective reactions.

Xiao-Qiang Hu received his B.S. from the Wuhan Polytechnic University in 2011. Subsequently, he began his Ph.D. studies under the supervision of Prof. Jia-Rong Chen and Wen-Jing Xiao at the Central China Normal University. Currently, he is working as a RESOLV Postdoctoral Researcher with Prof. Lukas J. Gooßen at the Ruhr-Universität Bochum. His research interests are transition-metal catalyzed C-H activation and radical cascade reaction.

Akichika Itoh obtained his Doctorate in Pharmacy in 1992 from Kyoto University under Prof. Kaoru Fuji, and completed his Postdoctorate in Michigan State University in 1994, under Prof. John W. Frost). He worked as an Assistant Professor in Gifu Pharmaceutical University in 1988, worked there as a Lecturer in 2003, and as a Full Professor from 2007 to present. His field of research covers Aerobic Photooxidative Reaction, Green Catalysis, Halogen Catalysis, Porous Materials, Methodology Development.

Fabio Juliá was born in Cieza, Spain. In 2016, he obtained his Ph.D. degree in the study of photophysical and photochemical properties of cyclometalated Pt(II) and Pt(IV) complexes at the University of Murcia (Spain) under the supervision of Prof. Pablo González-Herrero. In 2014, he spent 3 months as a visiting research student in the lab of Prof. Corey Stephenson at the University of Michigan, working in photoredox catalysis. After 6 months in industry, he joined in 2017 the group of Daniele Leonori (University of Manchester) as a Postdoctoral Researcher where he has been working since then.

Shin Kamijo received his B.Sc. degree from Tohoku University in 1997 and obtained his Ph.D. degree from the same University, under the supervision of Prof. Y. Yamamoto, in 2002. After working as a Research Associate in the same group (2002–2004), as a JSPS Postdoctoral Fellow with Prof. G. B. Dudley at Florida State University (2004–2006), and as a Postdoctral Associate with Profs. M. Shibasaki and M. Kanai at the University of Tokyo (2006–2007), he joined the research group of Prof. M. Inoue at the University of Tokyo as an Assistant Professor in 2007. In 2012, he moved to Yamaguchi University as an Associate Professor. He has been honored with the Incentive Award in Synthetic Organic Chemistry, Japan (2011) and the Thieme Chemistry Journal Award 2015. His research interests are focused on the development of chemical transformations involving activation and functionalization of unreactive bonds and their use in organic synthesis.

Jacques Lalevée, Ph.D., is Professor of Physical Chemistry at the University of Haute-Alsace, Mulhouse (France) and Institut Universitaire de France. His research interests are focused on the design of efficient photosensitive systems for photopolymerization processes. He co-authored about 315 publications (international journals), (co-) authored two books and edited two books.

Daniele Leonori was born and raised in Italy, and studied Medicinal Chemistry at the Universita' degli Studi di Perugia. He completed his Ph.D. in 2010 at the University of Sheffield under the supervision of Prof. Iain Coldham. After Postdoctoral studies with Prof. Magnus Rueping (RWTH Aachen University) and with Prof. Peter H. Seeberger (Max Planck Institute of Colloids and Interfaces), he joined the group of Prof. Varinder K. Aggarwal FRS as a Research Officer (University of Bristol). In 2014, he commenced his independent career as a Lecturer in Organic Chemistry at the University of Manchester. He was the recipient of an EPSRC Early Career Fellowship (2016) and the ERC Starting Grant (2017).

Marianna Marchini was born in Senigallia, Italy, in 1989. She got a bachelor degree in Chemistry in 2011 and a master degree in Photochemistry and Molecular Materials in 2013 at the University of Bologna. She completed Ph.D. in Chemistry (2016) in the group of Prof. Paola Ceroni with a thesis named "Photoactive Molecules in Supramolecular Architectures and Photoredox Reactions". During her Ph.D., she joined the group of Prof. A. D. Schlüter at the ETH in Zurich, where she worked in the field of 2D-polymers. Since 2017, she is a

Postdoctorate in the group of Prof. Paola Ceroni, and her research is focused on the photophysical and electrochemical study of supramolecular systems and photoactive molecules in organic reactions.

Lisa Moni graduated in 2005 at the University of Florence in Chemistry and Pharmacological Technologies with the highest mark (cum laude). In March 2009, she has completed the Ph.D. course in Chemical Sciences at the University of Ferrara under the guidance of Profs. A. Marra and A. Dondoni. Her Ph.D. research program dealt with the synthesis and biological applications of multivalent glycoside ligands. In 2010, she was Postdoctoral fellow at the University Joseph Fourier (Grenoble, FR) where she had developed diagnostic kits for Alzheimer's disease in collaboration with the Department of Biophysics and Nuclear Medicine. Since 2011, she has been working at the University of Genova, where she became Assistant Researcher in 2015. Her present research interests regard the development of original methodologies involving multicomponent reactions in combination with organocatalytic and enzymatic processes, and metal-catalyzed post-cyclizations.

Leticia Monjas obtained her M.Sc. degree in Organic Chemistry in 2012 from the Complutense University of Madrid (Spain). She performed her Master Thesis Research under the supervision of Dr. M. I. Rodriguez-Franco and Dr. S. Conde at the Medicinal Chemistry Institute of CSIC on multifunctional compounds for the treatment of Alzheimer's disease. For her Ph.D., she moved to the group of Prof. A. K. H. Hirsch at the University of Groningen (The Netherlands), to develop small-molecule modulators of vitamin transporters in bacteria. Thereafter, she moved to the University of Gothenburg (Sweden) where she worked first as a Postdoctoral Researcher in the group of Dr. C. J. Wallentin, and later joined the group of Prof. M. Grøtli, where she currently works on the development of kinase modulators for metabolic diseases.

Jason C. Morris completed his Bachelor of Science at Griffith University (Australia) with Honours, where he was involved in the synthesis of carbonic anhydrase inhibitors as potential cancer therapeutics. He then completed his Master of Applied Science at Queensland University of Technology (QUT), synthesizing novel profluorescent nitroxides for the detection of radical- and redox-related processes. Continuing in the thematic of photochemistry and nitroxides, he then completed a joint Ph.D. Program between QUT and Aix-Marseille University (AMU), investigating novel photosensitive alkoxyamines for application in nitroxide-mediated photopolymerization (NMP2). Since this time, his Postdoctoral focus has remained in the area of NMP2 at AMU.

Katarzyna Orłowska graduated from the Warsaw University of Technology with a Masters in Engineering in 2017. She joined the research group of Prof. D. Gryko in 2016. She is currently undertaking a Ph.D. at the Institute of Organic Chemistry of the Polish Academy of Sciences. Her research interests include photoredox catalysis, especially application of organic dyes as photocatalysts.

Al Postigo was born in Argentina and obtained his M.Sc. degree from the University of Buenos Aires in 1986. He moved to Canada in 1990, and obtained his Ph.D. degree from McMaster University in 1994, under the direction of Prof. Dr. William J. Leigh. After Postdoctoral positions in Canada, he returned to Argentina and worked with Prof. Dr. Roberto Rossi at the University of Córdoba in the area of radical ion reactions. He is currently Full Professor of Organic Chemistry at the Department of Organic Chemistry, Faculty of Pharmacy and Biochemistry, University of Buenos Aires. His interests are in the areas of radical chemistry, both carbon-centered radicals and metallic-centered radicals. He is devoted to studying radical reactions of these species in water and non-conventional media.

Sonia Remiro-Buenamañana obtained her Ph.D. from the University of East Anglia in 2015 under the supervision of Profs. Andrew N. Cammidge and Manfred Bochmann; her Doctoral thesis was focused on the synthesis and characterization of novel compounds related to porphyrin and phthalocyanine family and their subsequent use as encapsulants for opto-electronic applications, she also demonstrated the first syntheses of a completely new class of macrocyclic hybrid (SubTBDAPs). She is currently working as Marie Skłodowska-Curie Postdoctoral Fellow at the Instituto de Tecnología Química of the Technical University of Valencia in the group of Prof. Hermenegildo García. Her research interests include the synthesis and photocatalytic studies of a wide range of photocatalysts for solar fuel production.

Ségolène Villotte received a master degree in Organic Chemistry from the University Paris Sud (France) in 2015, after which she joined the laboratory of Dr. Didier Gigmes at the Faculty of Sciences of Aix-Marseille University (France). Her research interests focus on the design of new photoinitiators able to initiate photopolymerization reaction under soft irradiation conditions.

Carl-Johan Wallentin initiated his academic career in 2004 as a Ph.D. student at the Lund University (Sweden) under the directions of Prof. Kenneth Wärnmark. The main efforts during his time as a Ph.D. student centered on various aspects of supramolecular chemistry, with particular emphasis on hydrogen bonding and its utilization in the construction of discrete molecular aggregates. In 2011, he joined the research group of Prof. Corey Stephenson, then at Boston University (USA), to explore the upcoming field of photoredox catalysis. Thereafter, he joined the Department of Chemistry and Molecular Biology at Gothenburg University to established his own group active within the fields of method development and sustainable chemistry.

Wen-Jing Xiao completed his M.Sc. in 1990 under the supervision of Prof. Wen-Fang Huang from Central China Normal University. In 2000, he received his Ph.D. under the direction of Prof. Howard Alper at the University of Ottawa, Canada. After Postdoctoral studies with Prof. David W. C. MacMillan (2001–2002) at the California Institute of Technology in the States in 2003, Dr. Xiao became a Full Professor at the College of Chemistry at Central China Normal University. His current research interests include the development of new synthetic methodologies and the synthesis of biologically active compounds.

Nicolas Zivic received his Ph.D. degree in Chemistry in 2017 from the University of Aix-Marseille (Marseille, France) under the supervision of Dr. F. Dumur and Dr. Didier Gigmes. His work of thesis was focused on the synthesis of new photoinitiators active under soft irradiation conditions for cationic and radical polymerization. Since October 2017, he has joined Polymat (San Sebastian, Spain), working as Postdoctoral Researcher in the group of Prof. David Mecerreyes. Within Dr. Haritz Sardon's team, he carries on his research studying photobase generators able to catalyze the polymerization of biodegradable polymers.

Contents

Chapter 1

Ketones and Aldehydes

Shin Kamijo

Graduate School of Sciences and Technology for Innovation, Yamaguchi University, Yamaguchi, Japan
kamijo@yamaguchi-u.ac.jp

1.1 Introduction

Extensive studies have been carried out in the photochemistry of carbonyl compounds, especially ketones, since photoexcited ketones are known to induce various types of transformations [1–4]. Among them, one of the most significant features of aryl ketones, such as benzophenone (Ph_2CO) and its derivatives, is their ability to act as photoorganocatalysts (**POCs**) (Scheme 1.1).

For instance, Ph_2CO is readily excited to a singlet state with light irradiation [>350 nm for *n*–π*] and is then rapidly converted to a triplet state through intersystem crossing. The relatively longer lifetimes of the triplet state of aryl ketones [10 ns–1 μs for Ph_2CO depending on the solvent used] have allowed the promotion of photochemical transformations *via* energy transfer (EnT), hydrogen-atom transfer, and electron transfer (ET) reactions [redox potentials for triplet Ph_2CO: E^{T1}_{RED} + 1.28 V vs SCE, E^{T1}_{OX} – 0.61 V vs

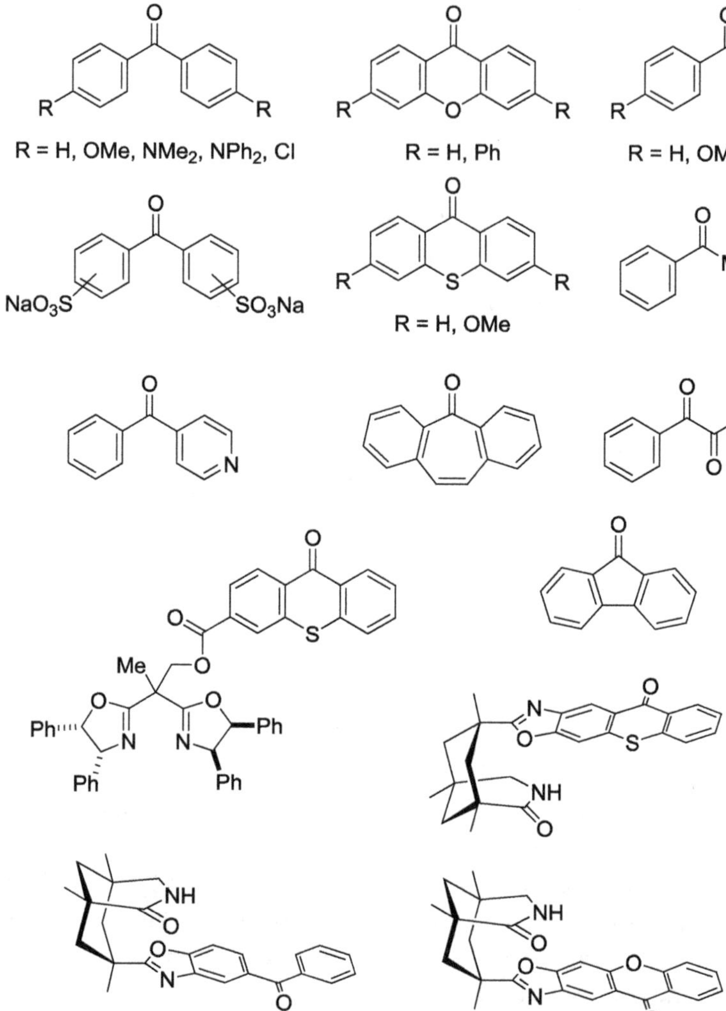

Scheme 1.1. Representative ketone and aldehyde **POCs**.

SCE; and energy for triplet Ph_2CO: $E_{0,0}^{T1}$ 3.0 eV]. This chapter mainly covers recent representative catalytic transformations involving carbonyl compounds, including ketones and aldehydes, under photoirradiation conditions.

1.2 Functionalizations of Unsaturated Bonds

1.2.1 *Alkylation of olefins*

One of the pioneering investigations on the alkylation of oxygen containing compounds, such as alcohols and acetals, *via* the photoinduced conjugate addition to carbohydrate-derived enones in the presence of a catalytic amount of benzophenone (Ph_2CO) was carried out by Fraser-Reid and co-workers [5–7]. As an example, irradiation of the enone derived from acetylated glucose in MeOH containing a catalytic amount of Ph_2CO (16 mol%) resulted in the regioselective formation of the 1,4-adduct in 50% yield (Scheme 1.2) [5]. The reaction was proposed to proceed with hydrogen-atom

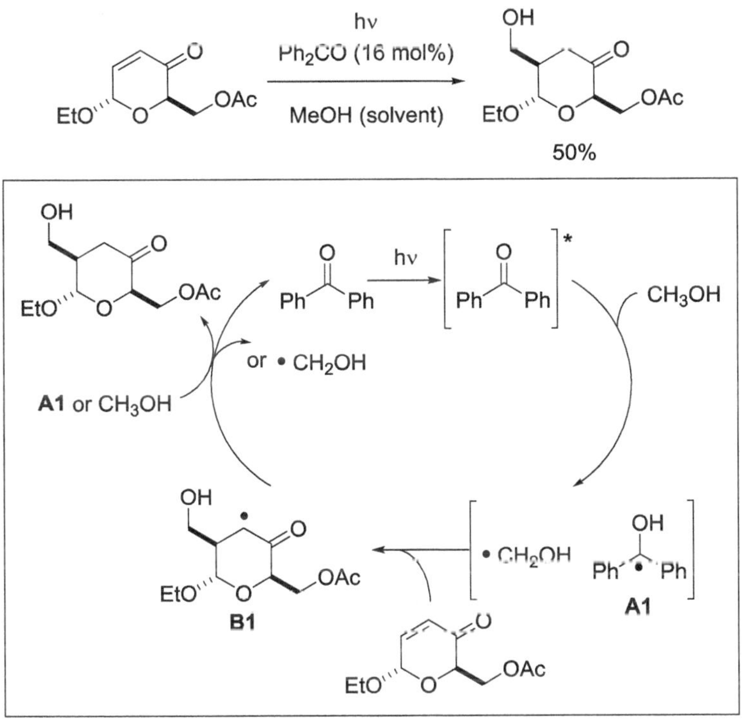

Scheme 1.2. Addition of alcohols to cyclic enones.

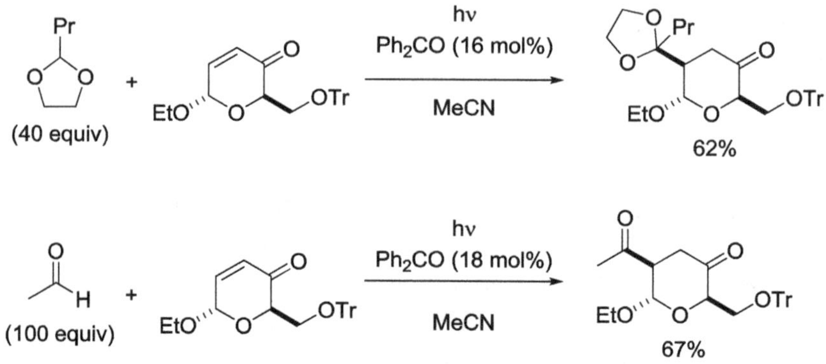

Scheme 1.3. Addition of acetals or aldehydes to cyclic enones.

abstraction from MeOH by the photoexcited Ph_2CO. The derived hydroxymethyl radical then added to the enone in 1,4-fashion from the less-hindered face. This addition step should be favored due to the nucleophilic nature of the hydroxymethyl radical. Subsequently, the hydrogen-atom transfer between the generated radical **B1** and the ketyl radical **A1** or MeOH took place to form the adduct [6]. In addition to MeOH, the applicable substances were expanded to include diols, acetals, and aldehydes (Scheme 1.3) [7]. A similar type of diastereoselective alkylation of isopropanol, 1,3-dioxolane, and tetrahydrofuran (THF) was also investigated by Mattay and co-workers [8].

The research group of Fagnoni and Albini achieved the catalytic photoinduced alkylation of 1,3-dioxolane with α,β-unsaturated ketones (aliphatic, both acyclic and cyclic, as well as aryl-substituted) in the presence of a ketone as **POC** [9]. The irradiation of the alkyl ketone in dioxolane with Ph_2CO (40 mol%) led to the formation of the addition product with a yield of 92% (Scheme 1.4(a)). On the other hand, the reaction employing the aryl ketone substrate was efficiently catalyzed by anthraquinone (**AQ**, 2 mol%) rather than Ph_2CO, furnishing the expected adduct in 85% yield (Scheme 1.4(b)). Further investigations enabled the alkylation of dioxolane using α,β-unsaturated aldehydes in aqueous solution [10], as well as by designing the water-soluble photocatalyst disodium benzophenonedisulfonate (**BPSS**) (Scheme 1.5) [11].

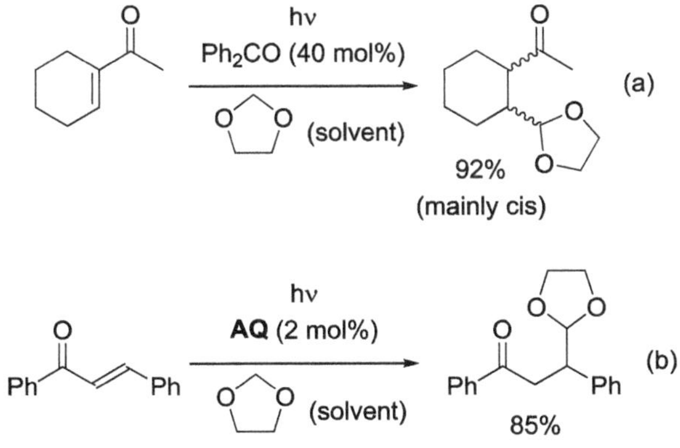

Scheme 1.4. Addition of 1,3-dioxolane to enones.

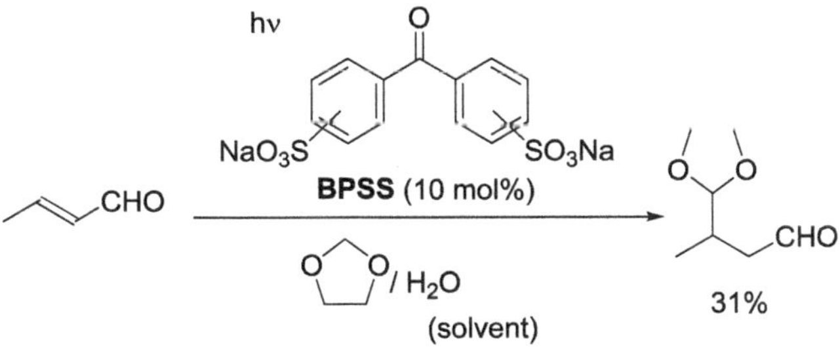

Scheme 1.5. Addition of 1,3-dioxolane to enals.

Hoffmann and co-workers investigated intensively the diastereoselective alkylation of amine derivatives by a chiral furanone *via* a photoinduced conjugated addition in the presence of a catalytic amount of aryl ketones [12–16]. In the initial stage of the investigation, the alkylation of *N*-methylpyrrolidine was attained (Scheme 1.6) [12].

The treatment of the pyrrolidine (20 equiv) and the chiral furanone bearing a menthyloxy group in MeCN with a catalytic amount of 4,4′-dimethoxybenzophenone (10 mol%) under irradiation produced the addition products in 85% combined yield. A complete facial selectivity at the furanone ring was observed; however, the other asymmetric center adjacent to the nitrogen atom could not be controlled. The current strategy for diastereoselective alkylation of

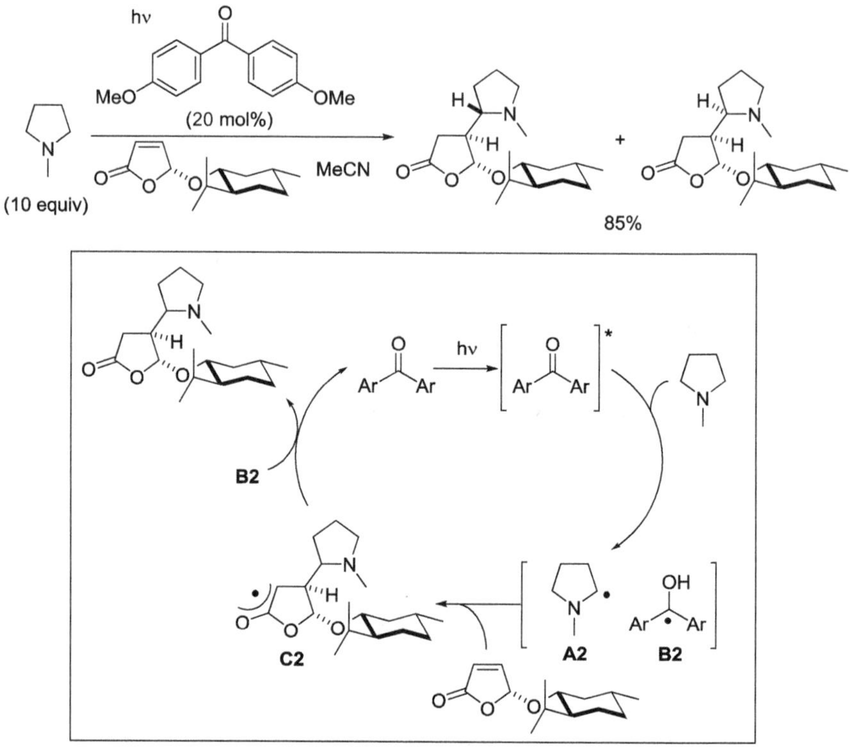

Scheme 1.6. Diastereoselective addition of pyrrolidine to a furanone.

pyrrolidine derivatives was applied to the total synthesis of two pyrrolizidines alkaloids, *viz.* (–)-isoretronecanol and (+)-lauburine [13]. In the proposed catalytic cycle, the reaction was initiated by photoinduced electron transfer (PET) between the photoexcited aryl ketone and pyrrolidine followed by proton transfer to generate the α-aminoalkyl radical **A2** and the ketyl radical **B2** [14]. The derived radical **A2** was added to the electron-deficient double bond of the furanone from the less-hindered face to form the oxoally radical **C2**. A hydrogen-atom transfer from the ketyl radical **B2** to the radical **C2** accounted for the recovery of the aryl ketone after the reaction. When dimethylaniline was employed as a starting amine in the presence of 4,4′-bis(*N*,*N*-dimethylamino)benzophenone (8 mol%), a radical tandem reaction took place to

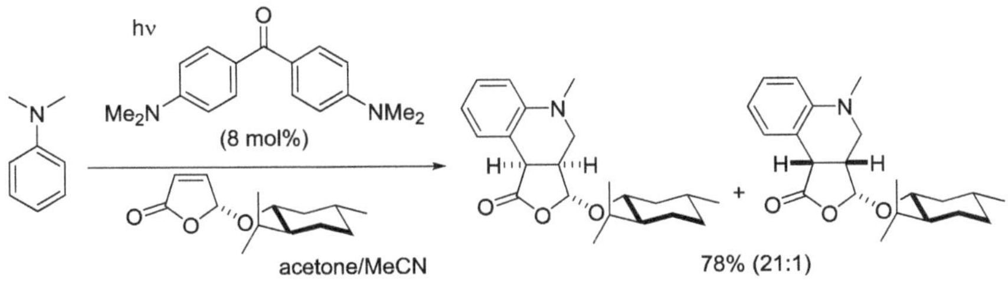

Scheme 1.7. Diastereoselective addition of aniline to a furanone.

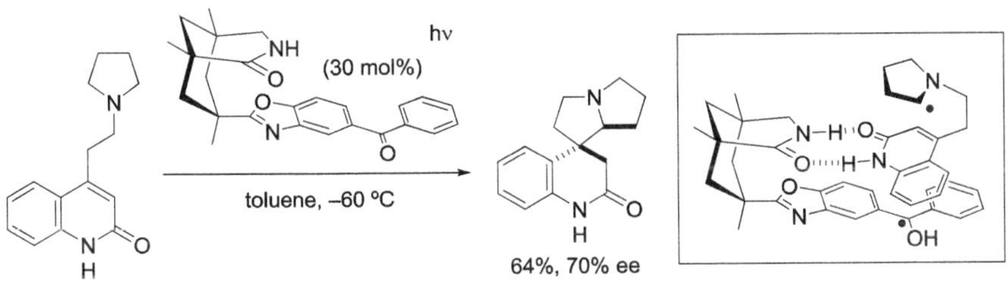

Scheme 1.8. Enantioselective intramolecular addition of pyrrolidine to a quinolone.

form tetrahydroquinolines in a highly diastereoselective manner (Scheme 1.7) [15]. Further investigations enabled the alkylation of aliphatic amines, including Et_3N and six-membered azacycles with electron-deficient olefins [16].

The enantioselective intramolecular alkylation of an amine was achieved by Bach and co-workers using a chiral hybrid catalyst (Scheme 1.8) [17]. The hybrid catalyst contained a benzophenone unit and a rigid bicyclic lactam part to bind the starting substances through two hydrogen bonds. The reaction of the pyrrolidine moiety tethered at the C-4 position of the quinolone in toluene in the presence of the chiral catalyst (30 mol%) under irradiation furnished the cyclized product in 64% yield with 70% ee.

Geraghty and co-workers investigated a photoinduced addition starting from cycloalkanes and electron-deficient alkynes (Scheme 1.9) [18]. The irradiation of methyl propiolate in cyclopentane in the presence of a catalytic amount of Ph_2CO (15 mol%) afforded the corresponding vinylated adduct in high yield. Five- to

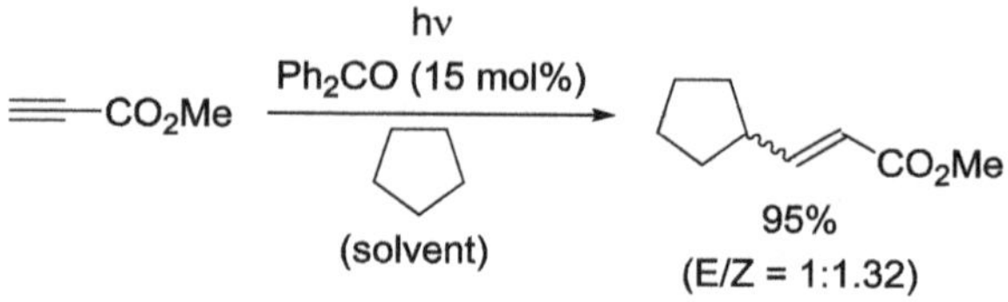

Scheme 1.9. Photoorganocatalyzed addition of alkanes to an alkynoate.

eight-membered cyclic alkanes were successfully added to terminal alkynes substituted with an electron-withdrawing functionality, such as ester, cyano, and sulfonyl groups.

1.2.2 *Trifluoromethylation of olefins*

Hoffmann and co-workers developed the catalytic trifluoromethylation of olefins with sodium triflinate (CF_3SO_2Na) as a CF_3 source and a benzophenone derivative as the **POC** (Scheme 1.10) [19]. The present protocol allowed the functionalization of electron deficient maleimide derivatives, which is characteristic, since the CF_3 radical is generally considered an electrophilic species. As an example, the mixing of equimolar amounts of *N*-phenylmaleimide and CF_3SO_2Na in MeCN, in the presence of 4,4′-dimethoxybenzophenone (10 mol%) and hexafluoroisopropanol (HFIP), under irradiation, provided the expected trifluoromethylated adduct in good yield. In the initial stage of the proposed catalytic cycle, the redox reaction between the photoexcited aryl ketone and CF_3SO_2Na generated the CF_3 radical with release of SO_2. The derived CF_3 radical added to the maleimide and the ketyl radical anion was protonated by HFIP. Then, a hydrogen-atom transfer regenerated the ketone and provided the trifluoromethylated adduct.

1.2.3 *Thioetherification of olefins*

Kokotos and co-workers succeeded in the construction of C–S bonds through the photoinduced addition of thiols to olefins utilizing a glyoxylic acid derivative as a catalyst (Scheme 1.11) [20].

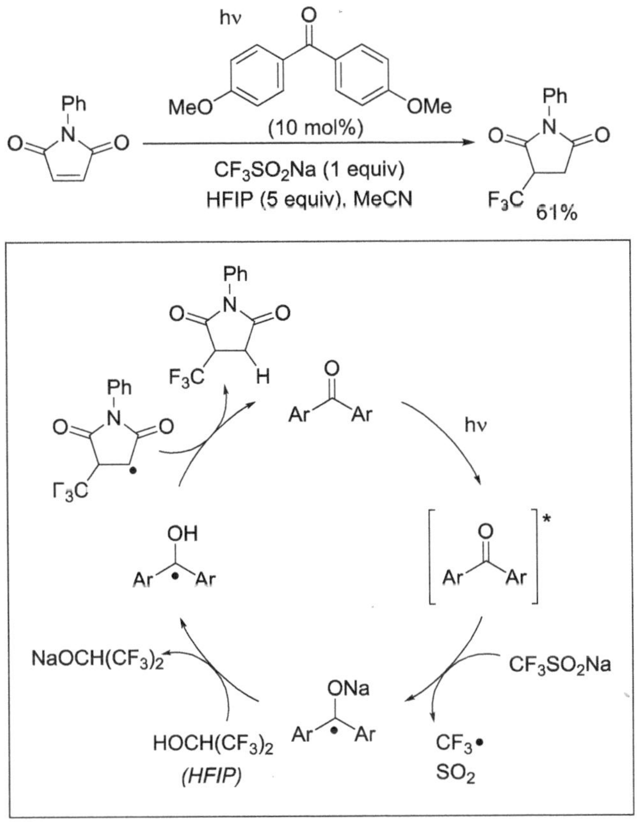

Scheme 1.10. Trifluoromethylation of maleimides.

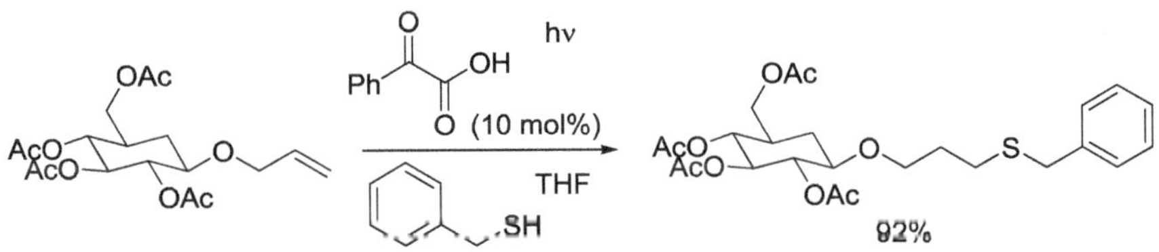

Scheme 1.11. Thioetherification of an olefin with a thiol.

The screening of the catalytic ability of carbonyl compounds disclosed the higher reactivity of phenylglyoxylic acid, than that of Ph_2CO, since both the aromatic substituent and the carboxylic moiety stabilized the excited state formed upon irradiation. Even

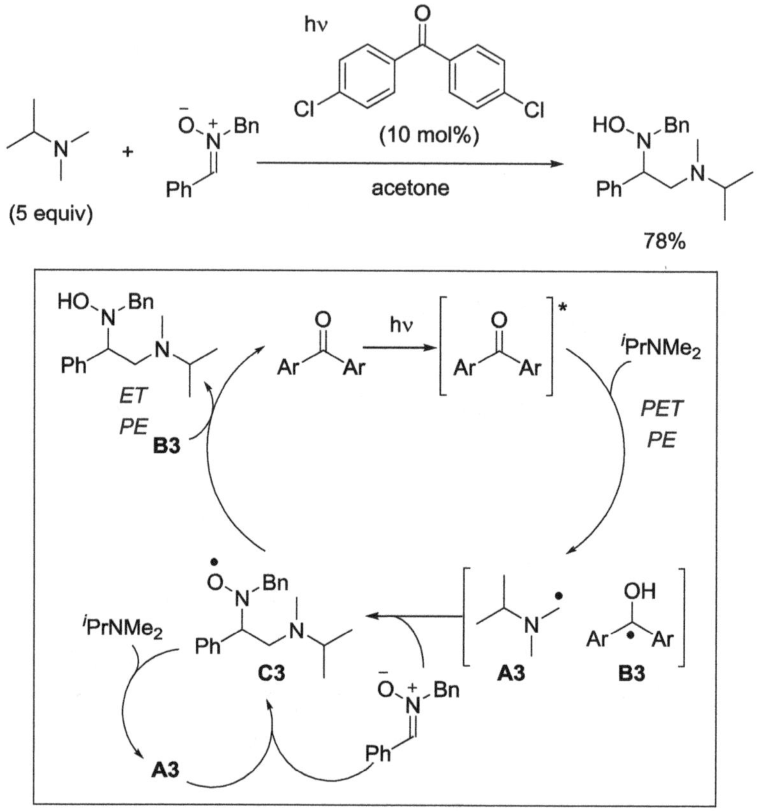

Scheme 1.12. Photoorganocatalyzed alkylation of nitrones.

when a fully acylated glucopyranoside served as the starting olefin, the corresponding thioether was obtained in a high yield.

1.2.4 *Alkylation of nitrones*

The research group of Suga and Fujii achieved the photoinduced synthesis of *β*-amino hydroxylamines from tertiary amines and nitrones by making use of the excellent radical accepting ability of nitrones (Scheme 1.12) [21]. The treatment of the nitrone with $^{i}PrNMe_2$ (5 equiv) and a catalytic amount of 4,4′-dichlorobenzophenone (10 mol%) in acetone under irradiation promoted the formation of the β-amino hydroxylamine in a selective manner. In the proposed mechanism, the catalytic cycle started with the generation of α-aminoalkyl radical **A3** and ketyl radical **B3** through PET and proton exchange (PE) between the starting amine and the photoexcited ketone. Then, the α-aminoalkyl radical **A3** was added to the nitrone to give the oxyl radical **C3**. The latter was reduced through either ET and PE with the ketyl radical **B3** or through hydrogen-atom abstraction from the starting amine.

1.2.5 *Hydroacylation of azodicarboxylates*

Kokotos and co-workers investigated the photoinduced addition of aldehydes to azodicarboxylates leading to acylated hydrazines (Scheme 1.13) [22–24]. Among the aryl ketones examined, phenylglyoxylic acid and its esters showed a high catalytic activity. The irradiation of a petroleum ether solution of heptanal (1.5 equiv) and diisopropyl azodicarboxylate (1 equiv) in the presence of phenylglyoxylic acid (10 mol%) led to the hydroacylation of the N=N bond, giving the acylated hydrazine in an excellent yield. It should be noted that the present transformation proceeded even with water as the solvent. The derived products, acyl hydrazines, served as activated acyl compounds and they could be readily converted to amides and hydroxamic acids by treatment with amines and hydroxylamines, respectively.

1.2.6 *Double functionalization of olefins*

Melchiorre and co-workers reported that the intermolecular atom transfer radical addition (ATRA) of a variety of haloalkanes to

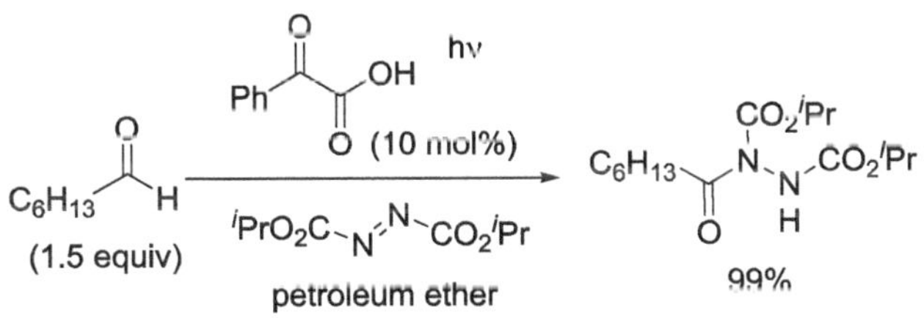

Scheme 1.13. Hydroacylation of an azodicarboxylate with an aldehyde.

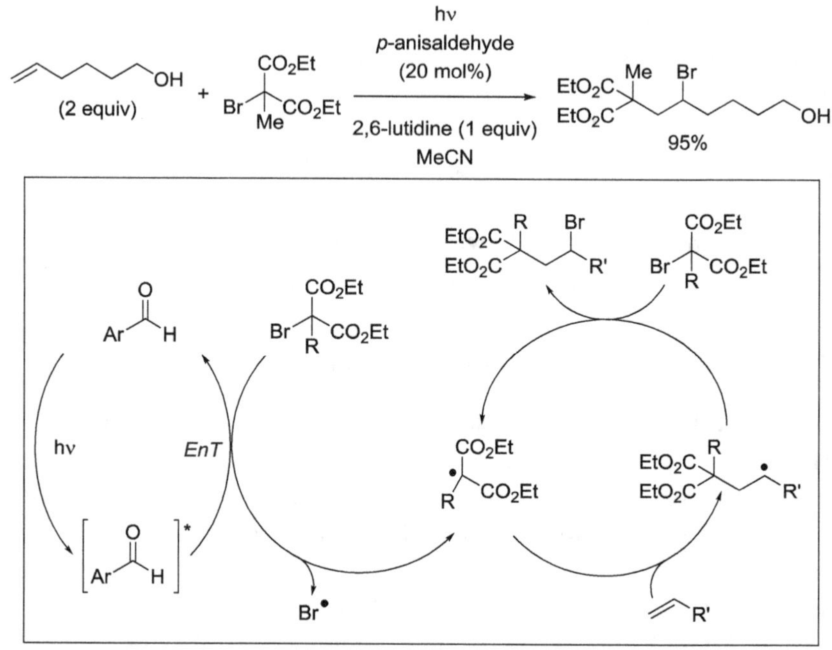

Scheme 1.14. Difunctionalization of an olefin with a bromomalonate.

olefins was efficiently catalyzed by *p*-anisaldehyde under irradiation (Scheme 1.14) [25]. For instance, the reaction of 2-bromo-2-methylmalonate (1 equiv) and 5-hexenol (2 equiv) in MeCN, with a catalytic amount of *p*-anisaldehyde (20 mol%) and 2,6-lutidine (1 equiv), under irradiation at ambient temperature produced the desired addition product in high yield. In most cases, terminal olefins and haloalkanes bearing at least one electron-withdrawing substituent were employed as starting materials. Polar functionalities, such as hydroxy, siloxy, and acetoxy groups, as well as carbamate, bromine, and oxirane units, were compatible. The addition of 2,6-lutidine was essential to promote the reaction by scavenging trace amounts of HBr generated under irradiation. An ET between the photoexcited *p*-anisaldehyde and the haloalkane was suggested to cause the homolytic cleavage of the C–Br bond.

Vincent and co-workers achieved the iodoperfluoroalkylation of unsaturated C–C bonds utilizing a Ph_2CO-catalyzed photochemical

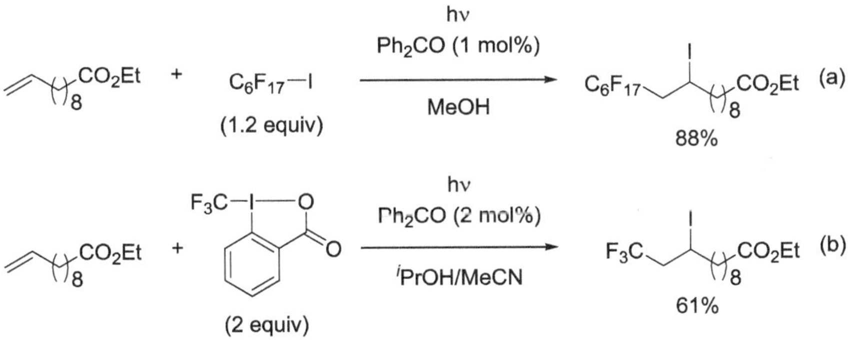

Scheme 1.15. Iodoperfluoroalkylation of olefins.

ATRA approach (Scheme 1.15(a)) [26]. The irradiation of a MeOH solution of undecenoate (1 equiv) and $C_6F_{17}I$ (1.2 equiv) proceeded in the presence of Ph_2CO (1 mol%), giving the expected addition product in high yield. The iodoperfluorination could be conducted not only with terminal alkenes but also with terminal alkynes. The same group later reported the iodotrifluoromethylation of alkenes using Togni's reagent, a benziodoxolone, as the source of both CF_3 and I units under similar reaction conditions (Scheme 1.15(b)) [27].

Xia and co-workers elaborated on the formation of phenanthrene and anthrone derivatives through a radical tandem cyclization involving the double functionalization of alkenes with formation of C–CF_3 and C–Ar bonds [28]. When a mixture of an *α,β*-unsaturated aromatic ketone and CF_3SO_2Na (2 equiv) in MeCN was irradiated in the presence of Ph_2CO (5 mol%), a phenanthrene derivative was obtained (Scheme 1.16). The reaction was proposed to proceed through the sequential generation of the CF_3 radical from CF_3SO_2Na, its addition to the terminal olefinic carbon, and cyclization.

Zhang and co-workers investigated the generation of phthalimide-*N*-oxyl radical (PINO) from *N*-hydroxyphthalimide (NHPI) *via* a photoinduced proton-coupled electron transfer (PCET) process using a photoexcited aryl ketone [29]. The effectiveness of aryl ketones was evaluated by conducting the dioxygenation of styrene, and 4,4′-bis(*N,N*-diphenylamino)benzophenone (**DPA-BP**) appeared

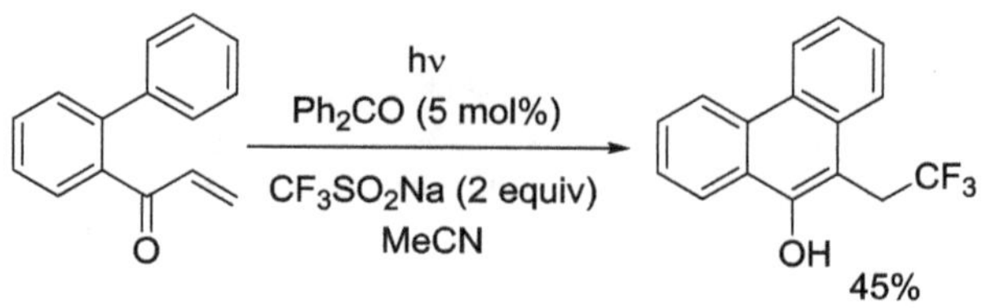

Scheme 1.16. Sequential olefin trifluoromethylation and cyclization.

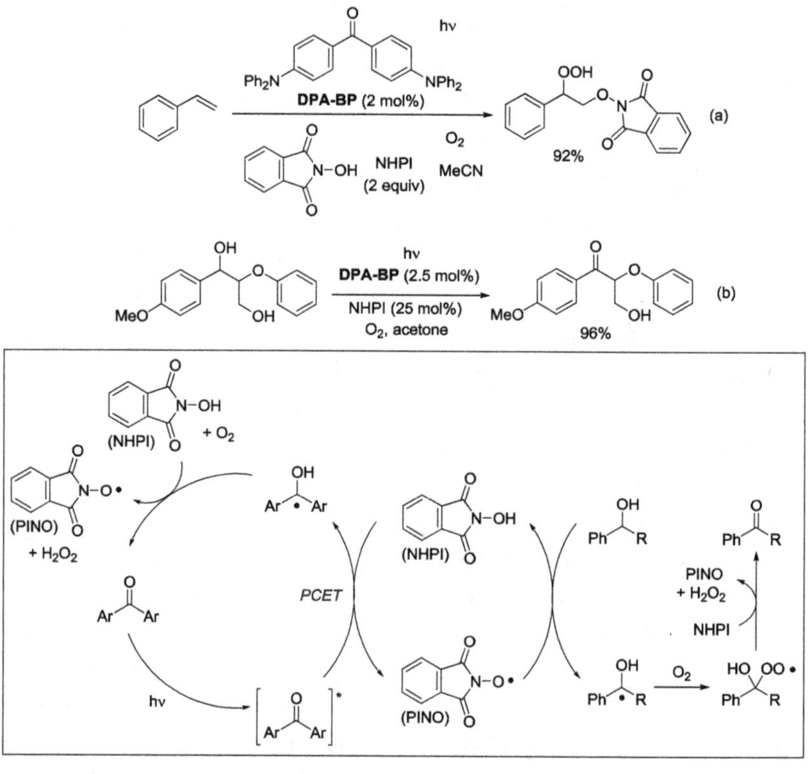

Scheme 1.17. Photoorganocatalyzed oxidation of: (a) an olefin and (b) a lignin model containing a benzyl alcohol group.

to exert a high catalytic ability. When a MeCN solution of styrene (1 equiv) and NHPI (1.2 equiv) under an oxygen atmosphere was irradiated in the presence of **DPA-BP** (2 mol%), the corresponding dioxygenated styrene was obtained in a high yield (Scheme 1.17(a)). This system was further applied to the photoinduced oxidation of lignin models (Scheme 1.17(b)). Notably, the benzyl alcohol moiety in the substrate was chemoselectively oxidized without touching the primary alcohol.

Both transformations were proposed to proceed through a hydrogen-atom transfer between photoexcited ketone (**DPA-BP**) and NHPI *via* PCET to generate the oxyl radical PINO. In the former case, the addition of PINO to styrene, oxygen trapping of the derived benzylic radical, and reduction furnished the dioxygenated product. By contrast, hydrogen abstraction from the benzylic alcohol by PINO to generate a benzylic radical, oxygen trapping, and reduction led to the generation of aryl ketone along with the expulsion of hydrogen peroxide in the latter case.

1.2.7 [2+2] *Cyclization*

Bach and co workers have been intensively investigating enantioselective [2+2] photocycloadditions by elaborating chiral hybrid catalysts. The first example is the enantioselective intramolecular [2+2] cyclization of quinolones tethered to an olefinic part at 4-position (Scheme 1.18) [30]. To attain a high enantioselectivity, a chiral hybrid catalyst containing a 1,5,7-trimethyl-3-azabicyclo[3.3.1]nonane-2-one skeleton and a thioxanthone moiety was designed. Irradiation of the starting quinolone in $PhCF_3$ at −25°C in the presence of the chiral ligand (10 mol%) furnished the expected cyclobutane in 88% yield with

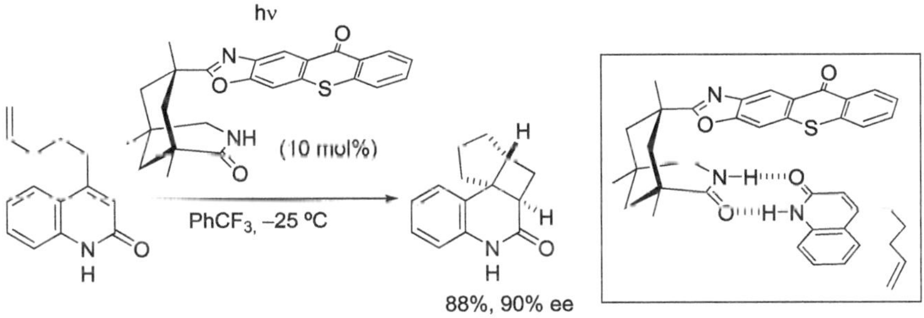

Scheme 1.18. Enantioselective intramolecular [2+2] cycloaddition between a quinolone and an olefin moiety.

90% ee. In the proposed mechanistic model, the pre-coordination between the quinolone substrate and the chiral catalyst through hydrogen bonding was explained to differentiate the enantiofaces, and an EnT then took place from the photoexcited thioxanthone to the quinolone. The same group successfully applied the chiral thioxanthone catalyst to the enantioselective intermolecular [2+2] photocycloaddition between a quinolone and acrylates (Scheme 1.19) [31]. As an example, treatment of the quinolone with excess amounts of methyl acrylate (50 equiv) in $PhCF_3$ at −25°C in the presence of the chiral catalyst (10 mol%) led to the regioselective formation of the desired cyclobutane in 76% yield with 81% ee. This reaction could be conducted with solar irradiation. A similar type of chiral catalyst having a xanthone moiety was designed for the enantioselective intermolecular [2+2] photocycloaddition between pyridones and acetylenedicarboxylates (Scheme 1.20) [32].

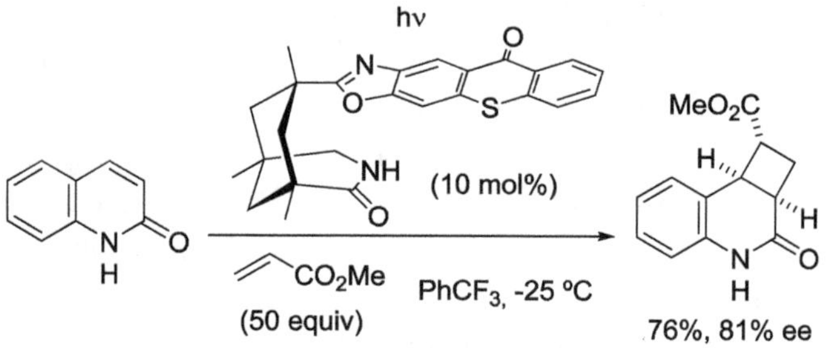

Scheme 1.19. Enantioselective intermolecular [2+2] cycloaddition between a quinolone and an olefin moiety.

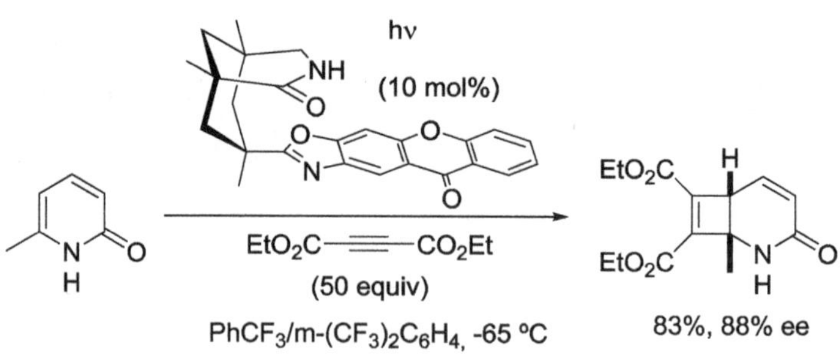

Scheme 1.20. Enantioselective intermolecular [2+2] cycloaddition between a quinolone and an alkyne moiety.

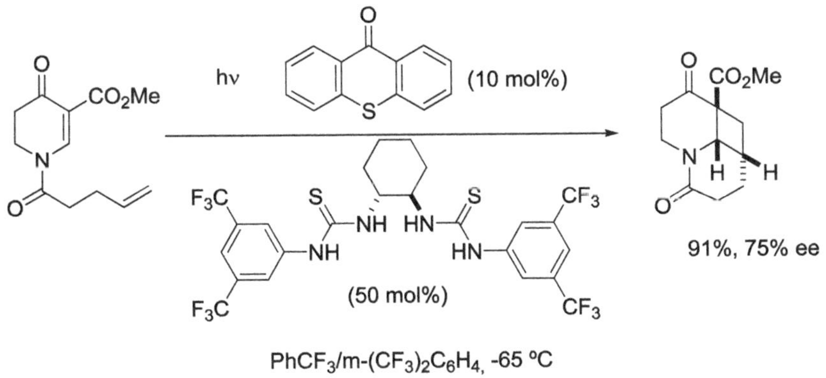

Scheme 1.21. Enantioselective intramolecular [2+2] cycloaddition between a dihydropyridone and an olefin moiety.

Bach and co-workers further developed a dual catalytic system, composed of a chiral bisthiourea and a thioxanthone catalyst for enantioselective intramolecular [2+2] photocycloadditions (Scheme 1.21) [33]. Treatment of the dihydropyridone substrate, the chiral bisthiourea (50 mol%), and thioxanthone (10 mol%) in a $PhCF_3$/hexafluoro-*meta*-xylene (HFX) mixture under irradiation at –65°C led to the formation of a tricyclic product in 91% yield with 75% ee. It was proposed that the chiral bisthiourea coordinated to the starting dicarbonyl compound and shielded the *Re*-face, whereas thioxanthone acted as a photosensitizer.

1.3 Functionalizations of C(sp^3)–H Bonds by Substitution Reactions

1.3.1 *Alkynylation*

Inoue and co-workers developed the introduction of a two carbon alkynyl functional unit using 1-tosyl-2-(trimethylsilyl)acetylene as an alkynylating agent (Scheme 1.22) [34]. A variety of starting substances, such as amine derivatives, cyclic ethers, alcohols, and alkanes was chemoselectively alkynylated in a single step utilizing photoexcited Ph_2CO. In the case of alkynylation at methine C–H bonds, the reaction could be conducted with the starting material as limiting agent in the presence of a semicatalytic amount of Ph_2CO

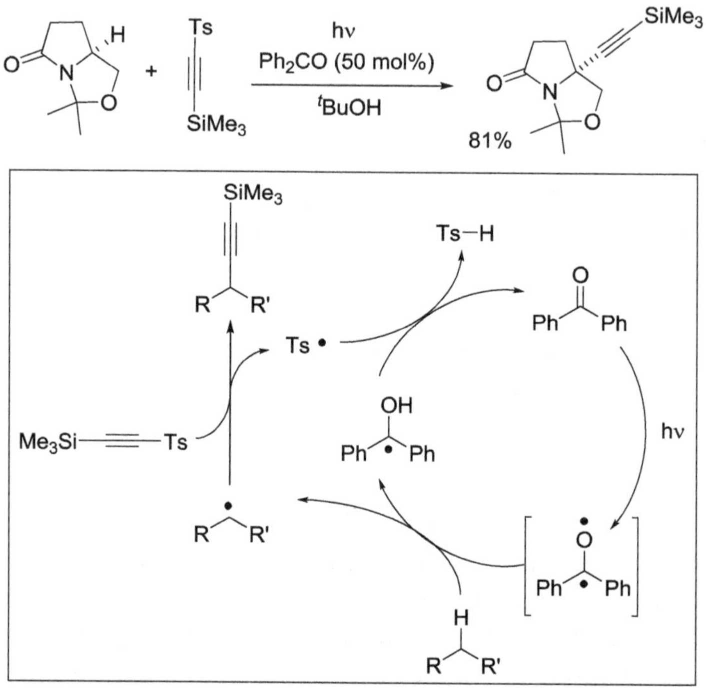

Scheme 1.22. Alkynylation of C(sp^3)–H bonds.

(0.5 equiv). The proposed mechanism involves hydrogen-atom abstraction by the photoexcited Ph_2CO, acting in a manner similar to that of an electron-deficient oxyl radical and leading to the generation of a carbon radical intermediate. The latter was readily trapped with tosyl acetylene to furnish the alkynylated product. The photocatalyst, Ph_2CO, was regenerated with formation of sulfinic acid by hydrogen transfer. The present alkynylation strategy was elegantly applied to the total synthesis of (+)-lactacystin [35], pointing out the power of direct C(sp^3)–H functionalization for the assembly of multiply functionalized structures.

Inoue and co-workers also achieved the enantioselective alkynylation of C–H bonds using a sulfoximine derivative as a traceless chiral auxiliary (Scheme 1.23) [36]. The reaction was conducted under irradiation in the presence of Ph_2CO, and nitrogen-containing substances such as carbamates and amides were converted

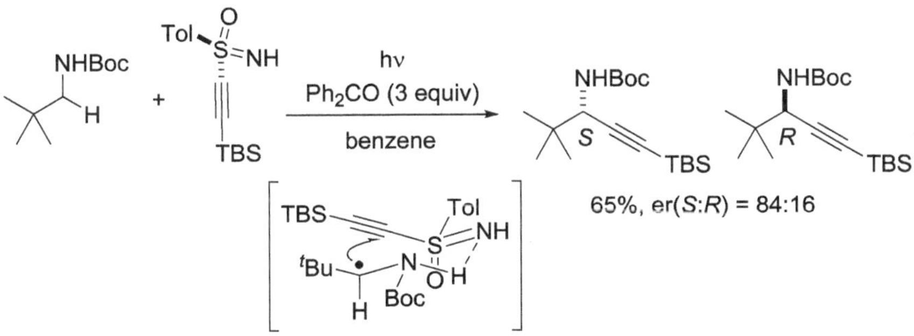

Scheme 1.23. Enantioselective alkynylation of an amine derivative.

successfully to the corresponding propargyl derivatives in a chemo and enantioselective fashion. The strong hydrogen-bond capability between the basic S=NH and protic NHBoc moieties was suggested to be responsible for the enantioselective outcome.

1.3.2 *Vinylation*

Inoue and co-workers then investigated the vinylation of C–H bonds by making use of the C–H bond cleaving ability of photoexcited Ph_2CO (Scheme 1.24) [37]. A wide range of substances, including amine derivatives, ethers, alcohols, and alkanes, were successfully vinylated. A typical reaction employed an excess amount of the starting material (5–10 equiv), with 1,2-bis(phenylsulfonyl)ethylene (1 equiv) as the vinylating agent in the presence of a stoichiometric amount of Ph_2CO (1 equiv). The reaction could, however, be carried out in the presence of a semicatalytic amount of Ph_2CO (0.5 equiv). For some vinylation reactions at methine C–H bonds, the starting material could be the limiting reagent with a slight excess of the vinylating agent (1.2 equiv).

One of the major disadvantages of the photoinduced vinylation and alkynylation discussed in the previous sections is the requirement of a (quasi) stoichiometric amount of Ph_2CO to promote the smooth formation of the corresponding products. Guin and co-workers achieved a catalytic reaction for these C–H functionalizations by using the starting materials as a solvent (Scheme 1.25) [38].

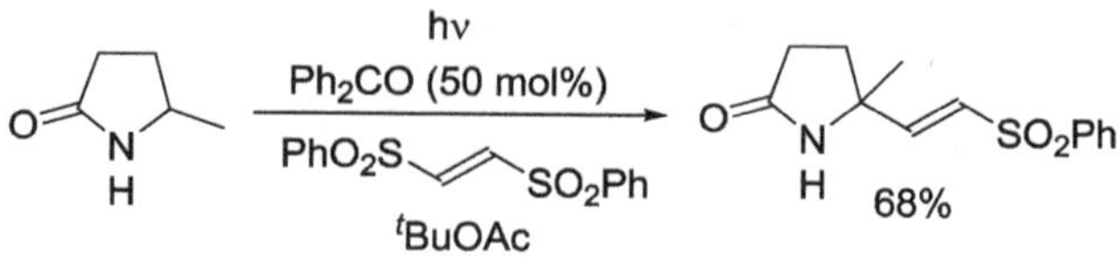

Scheme 1.24. Vinylation of a C–H bond.

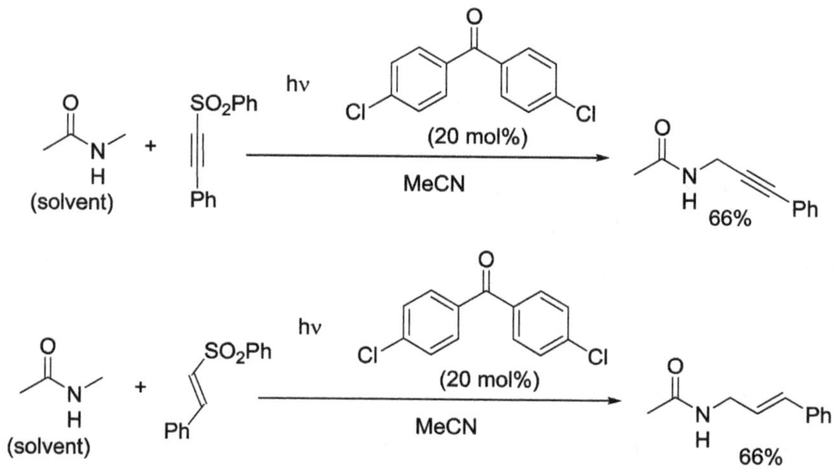

Scheme 1.25. Photoorganocatalyzed alkynylation and vinylation of C–H bonds.

Both transformations were catalyzed by 4,4′-dichlorobenzophenone (20 mol%) instead of Ph_2CO under irradiation with a fluorescent lamp. This study demonstrates that the catalytic functionalizations of C–H bonds can be attained when a large excess amount of starting substances is readily available.

1.3.3 *Allylation*

Kamijo and co-workers developed the catalytic installation of a three-carbon allyl functional unit in the presence of a π-extended aryl ketone as the **POC** under irradiation (Scheme 1.26) [39]. A range of substances, including alkanes, ethers, alcohols, and amine derivatives, were allylated in a chemoselective manner. An

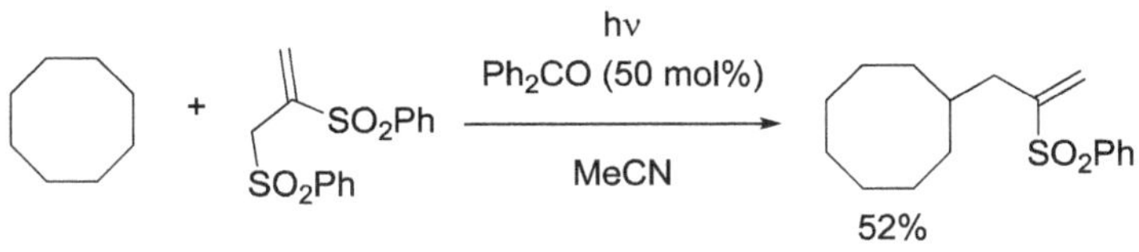

Scheme 1.26. Photoorganocatalyzed allylation of a C–H bond.

allyl sulfone acted as the allylating agent, and the reaction was catalyzed not only by Ph_2CO, but also by 5,7,12,14-pentacenetetrone (**PT**), under UV-light irradiation at 365 nm. Moreover, the π-extended aryl ketone **PT** could catalyze the allylation under visible-light irradiation at 425 nm. A radical mechanism initiated with hydrogen-atom abstraction by photoexcited **PT** was proposed. Indeed, the *in situ*-generated carbon radical intermediate was successfully trapped by 2,2,6,6-tetramethylpiperidine *N*-oxyl radical (TEMPO), which strongly supports the hypothesis.

1.3.4 *Introduction of an aldoxime functional group*

The photoinduced introduction of an aldoxime functionality was reported by Kamijo and co-workers using the sulfonyl oxime as the aldoxime source (Scheme 1.27) [40]. The aldoxime functionality was designed as a masked form of the formyl group, because cleavage of the formyl C(sp^2)–H bond was expected to preferentially occur under the present radical conditions due to the lower bonding energy. Screening of aryl ketones revealed that 4-benzoylpyridine (**4-BzPy**) gave a higher product yield than Ph_2CO. A broad range of starting materials was applicable, including oxygen-, nitrogen-, and sulfur-containing cyclic compounds, as well as alkanes. The reactivity of C–H bonds adjacent to the nitrogen atom could be controlled by a proper choice of the protecting group; indeed, the Boc group allowed the C–H functionalization of the piperazine derivative, while the Ts group inhibited the process. The reaction system was suggested to work also under catalytic conditions, making use of 20 mol% of the aryl ketone.

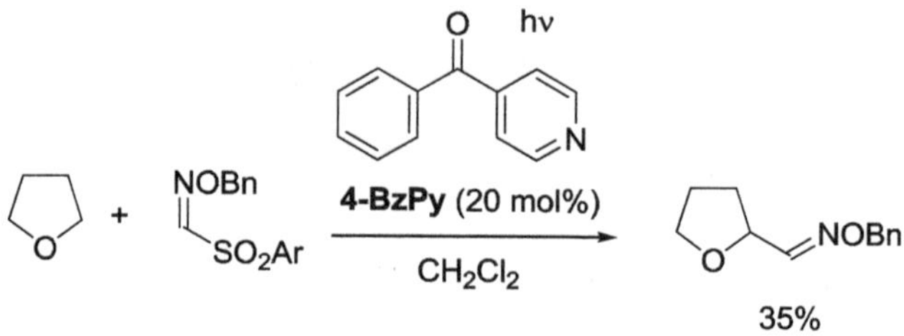

Scheme 1.27. Substitutive introduction of an aldoxime functional group at C–H bonds.

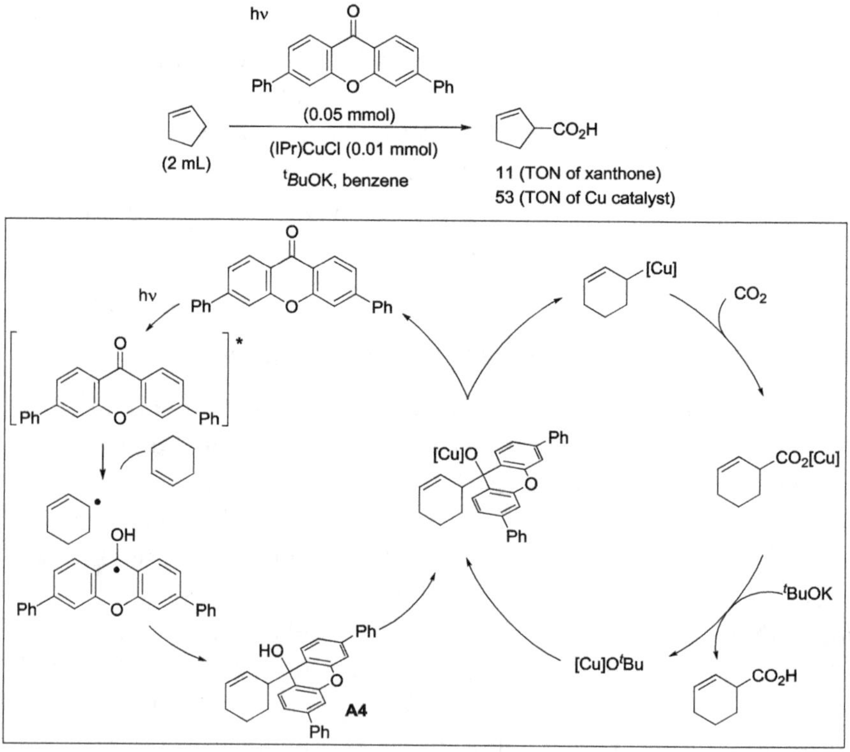

Scheme 1.28. Carboxylation of allylic C–H bonds.

1.3.5 *Allylic carboxylation*

Murakami and co-workers investigated the photoinduced carboxylation of the allylic C–H bonds of simple alkenes with CO_2 in the presence of a ketone and a copper catalyst (Scheme 1.28) [41]. Variously substituted acyclic, as well as cyclic, aliphatic alkenes were converted to the corresponding carboxylic acids in the presence of a catalytic amount of 3,6-diphenylxanthone and the Cu–NHC complex (IPr) CuCl. The photochemical reaction of the olefin and xanthone promoted the formation of homoallyl alcohol **A4**. Moreover, treatment of the derived homoallyl alcohol **A4** with CO_2 in the presence of the copper catalyst furnished the corresponding carboxylic acid, and xanthone was regenerated. Accordingly, the catalytic cycle was proposed to proceed through intervention of the homoallyl alcohol intermediate **A4**.

1.3.6 *Sulfonation*

Kamijo and co-workers investigated the sulfonylation of cyclic ethers in the presence of Ph_2CO under irradiation (Scheme 1.29) [42]. In this transformation, sulfonyl chlorides acted as precursors of the sulfonyl unit. Five- to seven-membered cyclic ethers were successfully sulfonylated, and a chemoselective reaction took place at the ethereal C–H bond, even when ambroxide was used as the starting material. Subsequent treatment of the tosylated substrate with nucleophilic reagents led to the formation of arylated and alkynylated ambroxides. Accordingly, a two-step functionalization protocol of ethers was developed *via* direct ether sulfonylation as the key step.

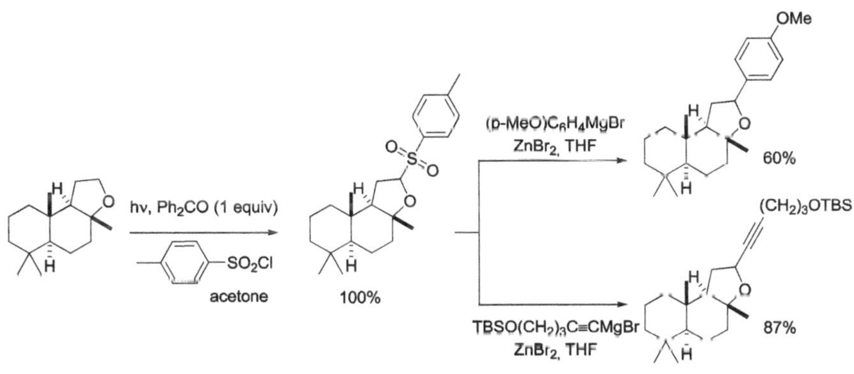

Scheme 1.29. Photoorganocatalyzed sulfonylation of ambroxide.

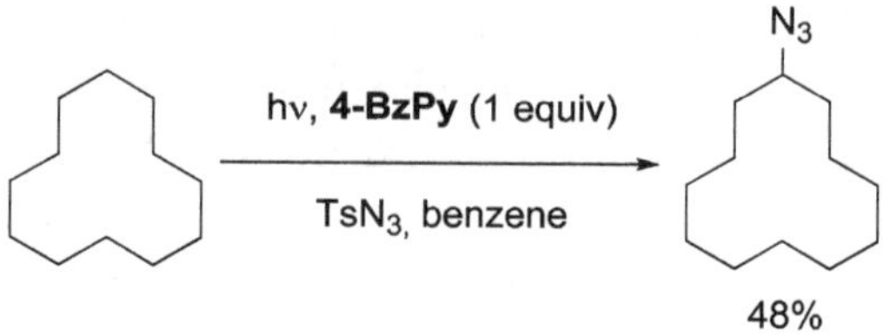

Scheme 1.30. Azidation of C–H bonds.

1.3.7 *Azidation*

The photoinduced synthesis of aliphatic azides was investigated by Kamijo and co-workers, and the single step azidation of cyclic alkanes as well as oxacyclic and azacyclic compounds was achieved (Scheme 1.30) [43]. The reaction took place in the presence of 4-benzoylpyridine (**4-BzPy**) under irradiation by using tosyl azide (TsN_3) as the azide source. Despite the potential photolability of azide compounds, the current transformation provided the corresponding monoazide adducts in a chemoselective manner.

1.3.8 *Chlorination*

With careful and extensive screening of potential chlorinating agents, photochemical chlorination of ethereal C–H bonds was investigated by Kamijo and co-workers (Scheme 1.31) [44]. The derived α-chlorinated ethers were unstable, so that a nucleophilic substitution was combined for the derivatization of ethers in a two-step sequence. A variety of ethers, including five- to seven-membered cyclic ethers, ambroxide, crown ethers, and acyclic ethers, were chlorinated with *N,N*-dimethylsulfamoyl chloride (Me_2NSO_2Cl) in the presence of Ph_2CO under irradiation. Carbon tetrachloride (CCl_4) also acted as a chlorinating agent, although Me_2NSO_2Cl gave the best results. The subsequent treatment of the resulting α-chlorinated ether with a nucleophilic reagent, without isolation, completed the introduction of oxygen and sulfur functionalities as well as an allyl unit at the ethereal carbon center. This one-step ether chlorination allowed rapid access to acetals, *O,S*-acetals, and allylated ethers from simple ethereal compounds.

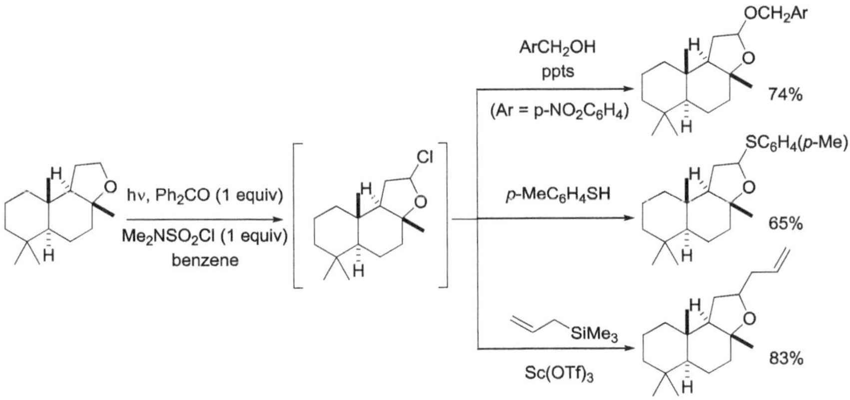

Scheme 1.31. Chlorination of C–H bonds and ensuing derivatization with nucleophiles.

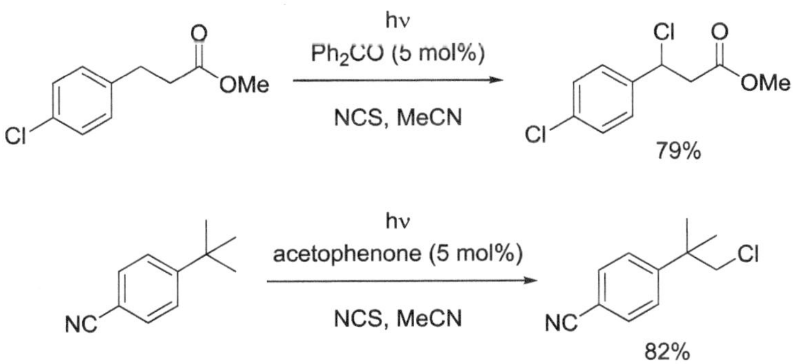

Scheme 1.32. Photoorganocatalyzed chlorination of C–H bonds.

The photocatalytic chlorination of C–H bonds was also investigated by Chen and co-workers employing two types of aryl ketones (5 mol%) and *N*-chlorosuccinimide (NCS) as the chlorinating agent (Scheme 1.32) [45]. Benzophenone (Ph_2CO) was employed as optimal catalyst for the chlorination at benzylic C–H bonds. In contrast, acetophenone was selected as a better catalyst for chlorination at less reactive C–H bonds of aliphatic carbon chains. When NCS was used as a limiting reagent, over-chlorination could be successfully suppressed and the monochlorinated product was selectively obtained. On the other hand, regioselectivity remained to be improved, especially for the chlorination of alkyl chains in acyclic compounds.

1.3.9 *Fluorination*

Chen and co-workers developed a photoinduced catalytic fluorination at benzylic C–H bonds. By choosing a proper aryl ketone as the catalyst, monofluorination and difluorination were successfully controlled [46]. Treatment of the starting alkylbenzene (1 equiv) with Selectfluor (2 equiv) in the presence of a catalytic amount of 9-fluorenone (5 mol%) under irradiation led selectively to the monofluorinated product (Scheme 1.33). On the other hand, treatment of the starting alkylbenzene (1 equiv) with Selectfluor II (3 equiv) in the presence of a catalytic amount of xanthone (5 mol%) was found to provide the difluorinated product in a selective fashion (Scheme 1.34). Kappe and co-workers improved the

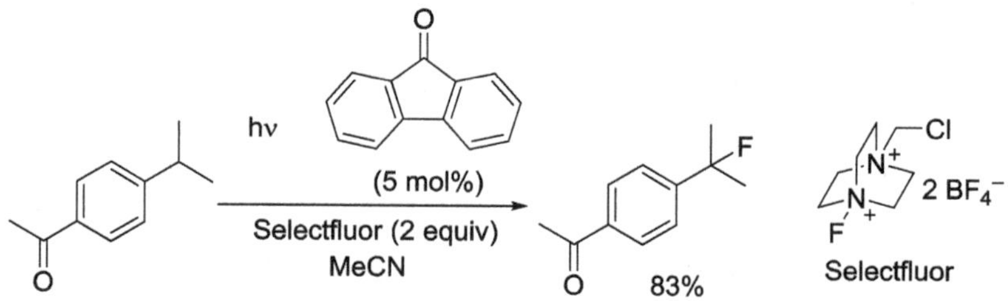

Scheme 1.33. Monofluorination of benzylic C–H bonds.

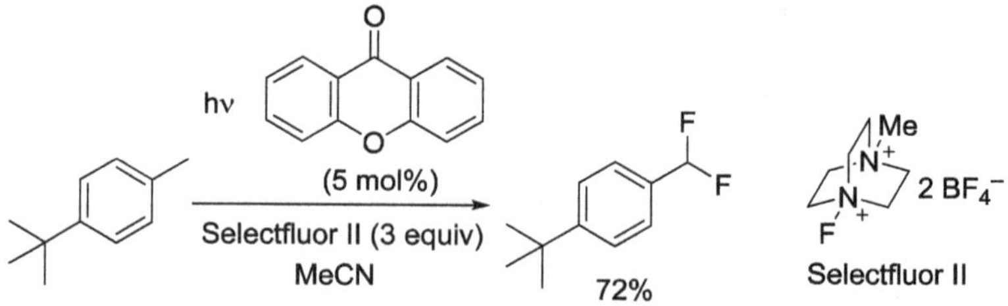

Scheme 1.34. Difluorination of benzylic C–H bonds.

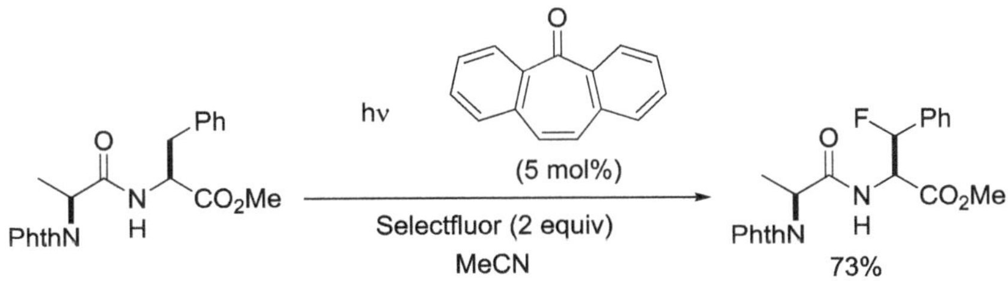

Scheme 1.35. Fluorination of benzylic C–H bonds.

present photoinduced benzylic monofluorination under continuous flow conditions [47]. The reaction using the flow system brought a greater catalytic activity of xanthone and allowed the reduction of the amount of Selectfluor from 2 equiv to 1.2 equiv. Moreover, the monofluorinated products were formed in shorter reaction times and the yields were improved.

Lectka and co-workers applied similar reaction conditions using a photoexcited aryl ketone to the benzylic fluorination of peptides (Scheme 1.35) [48]. After screening of the photocatalytic efficiency of various aryl ketones, 5-dibenzosuberenone arose as the optimal catalyst for the present purpose. The benzylic C–H bond of phenylalanine moieties was chemoselectively functionalized with Selectfluor under irradiation. The present method allowed the fluorination of phenylalanine derivatives as well as phenylalanine-containing dipeptides and tripeptides.

Chen and co-workers achieved the photoinduced catalytic fluorination of C–H bonds of alkanes under slightly modified conditions with respect to their benzylic fluorination reaction (Scheme 1.36) [49]. The reaction was photoorganocatalyzed by acetophenone (5 mol%) and Selectfluor was employed as a fluorinating agent. Not only simple cycloalkanes but also carboxylic acids, ketones, ester, and protected amino acids could be used as suitable starting materials. The formation of regioisomeric mixtures of fluorinated products was observed for substrates having multiple reactive C–H bonds.

Lectka and co-workers developed a different type of photoinduced direct fluorination at benzylic positions that proceeded with the cleavage of unstrained C–C bonds (Scheme 1.37) [50]. The

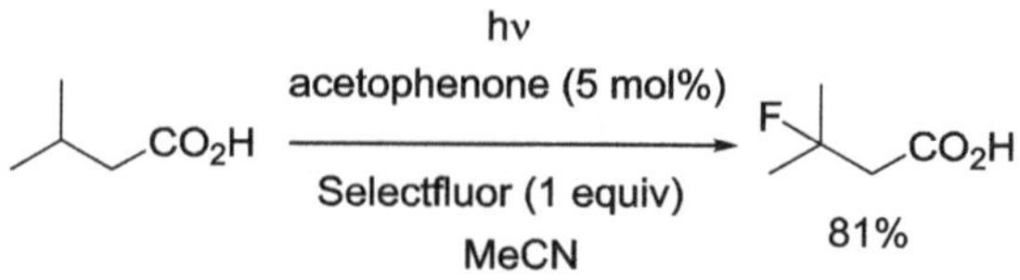

Scheme 1.36. Fluorination of aliphatic C–H bonds.

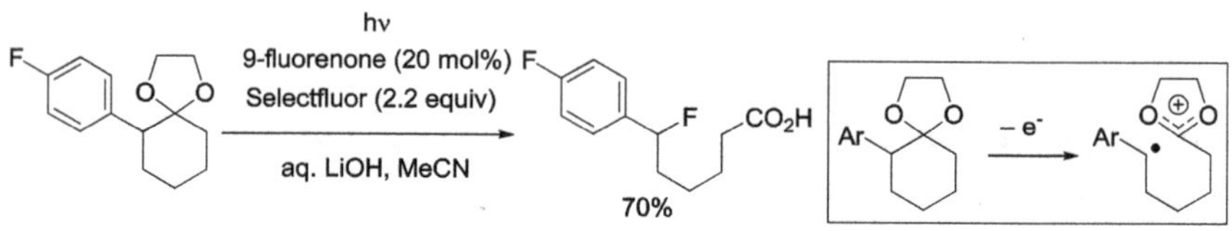

Scheme 1.37. Preparation of a benzyl fluoride through C–C bond cleavage.

irradiation induced the oxidative ring opening of 1-acetal-2-arylated carbocyclic compounds in the presence of a catalytic amount of 9-fluorenone (20 mol%) as a photoxidant, and subsequent hydrolysis furnished the corresponding fluorinated carboxylic acids. Although further investigations were required to elucidate the detailed reaction mechanism, the generation of stable benzyl radicals through the C–C bond cleavage caused by one-electron oxidation was suggested as the key step.

1.3.10 *Oxidation*

Xiao and co-workers achieved the enantioselective oxidation of β-ketoesters under aerobic conditions using bifunctional photocatalysts (Scheme 1.38) [51]. A hybrid structure containing a chiral bisoxazoline and the thioxanthone moiety was designed as a new ligand. This hybrid ligand was complexed with $Ni(acac)_2$ and used as a catalyst (10 mol%). The irradiation of β-ketoesters bearing an adamantyl group in oxygen atmosphere with the chiral Ni catalyst furnished the corresponding hydroxylated products in high yield and high enantioselectivity. In the proposed mechanism, the Lewis acidic Ni/bisoxazoline complex coordinated to the enol-form of the β-ketoester and differentiated its enantioface.

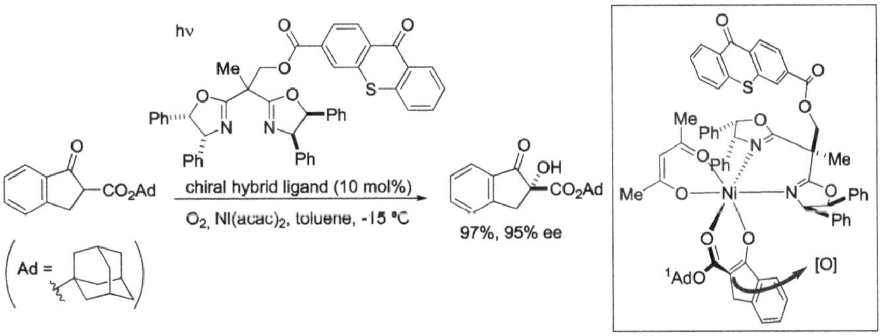

Scheme 1.38. Enantioselective hydroxylation of a ketoester.

Oxygen was activated by photoexcited thioxanthone, and the oxidation took place at the less-hindered *Re*-face of the starting β-ketoester.

1.4 Functionalizations of Aromatic Compounds

1.4.1 *Alkylation*

Opatz and co-workers reported the Ph_2CO-mediated photoinduced coupling reaction of benzazoles, such as benzoxazole, benzimidazole, and benzothiazole, with carbamates, alcohols, and ethers (Scheme 1.39) [52]. When 2-chlorobenzoxazole (1 equiv) and Boc-azepane (2 equiv) were irradiated in aqueous MeCN in the presence of Ph_2CO (1 equiv) and NaOAc (2.5 equiv), the benzoxazole unit was installed at the carbon center adjacent to the nitrogen atom of the starting carbamate. The reaction mechanism was proposed to start with hydrogen-atom abstraction by the photoexcited Ph_2CO from the starting azepane. The resulting carbon radical species added to the chlorinated carbon center of 2-chlorobenzoxazole. Hydrogen-atom transfer between the aminyl radical and ketyl radical followed by the expulsion of HCl completed rearomatization to furnish the coupling product. Coupling could be optimized using a catalytic amount of Ph_2CO (25 mol%) when Boc-azepane, ethanol, or THF were employed as the starting substances.

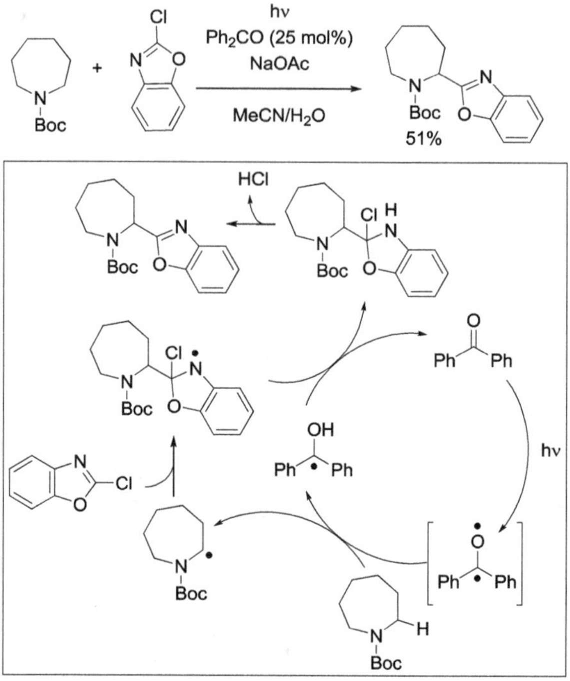

Scheme 1.39. Alkylation of chlorobenzoxazole.

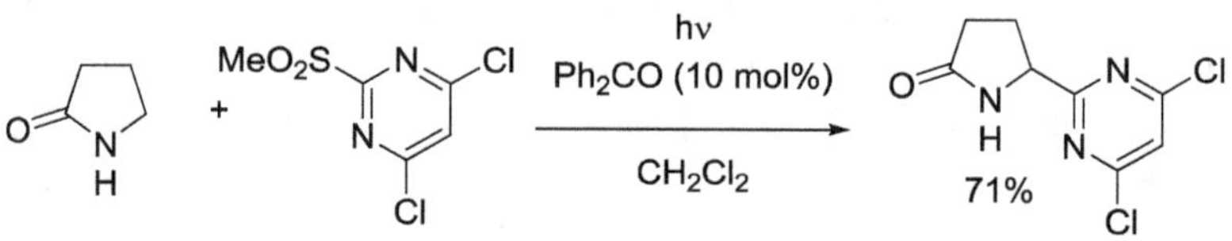

Scheme 1.40. Photoorganocatalyzed alkylation of sulfonylpyrimidines.

Kamijo and co-workers achieved the alkylation of pyrimidines by employing photoexcited Ph_2CO *via* coupling between saturated heterocycles and sulfonylpyrimidines (Scheme 1.40) [53]. The present methodology allowed the introduction of pyrimidine rings at non-reactive C–H bonds proximal to heteroatoms, including oxygen, nitrogen, and sulfur, under neutral reaction conditions at ambient temperature. The catalytic coupling (10 mol%) was realized for highly electron-deficient dichlorinated sulfonylpyrimidines, whereas a stoichiometric amount of Ph_2CO was required when more electron-rich dimethoxy-substituted or non-substituted sulfonylpyrimidines were employed.

1.4.2 *Carbamoylation*

Ji and co-workers achieved the functionalization of heteroaromatic compounds, including benzothiazole, benzimidazole, benzoxazole, thiazole, and pyridine, with the cross-dehydrogenative coupling (CDC) strategy (Scheme 1.41) [54]. The irradiation of benzothiazole in a $HCONH_2$/EtOAc mixture in the presence of $(NH_4)_2S_2O_8$ (APS; 3 equiv) and benzaldehyde (0.5 equiv) afforded benzothiazole-2-carboxamide in excellent yield. The amount of benzaldehyde, the **POC**, could be reduced to 0.1 equiv for the carbamoylation of 4-chloroquinoline. In the proposed mechanism, benzaldehyde promoted the decomposition of APS to generate a sulfate radical prone to abstract a hydrogen from formamide. The

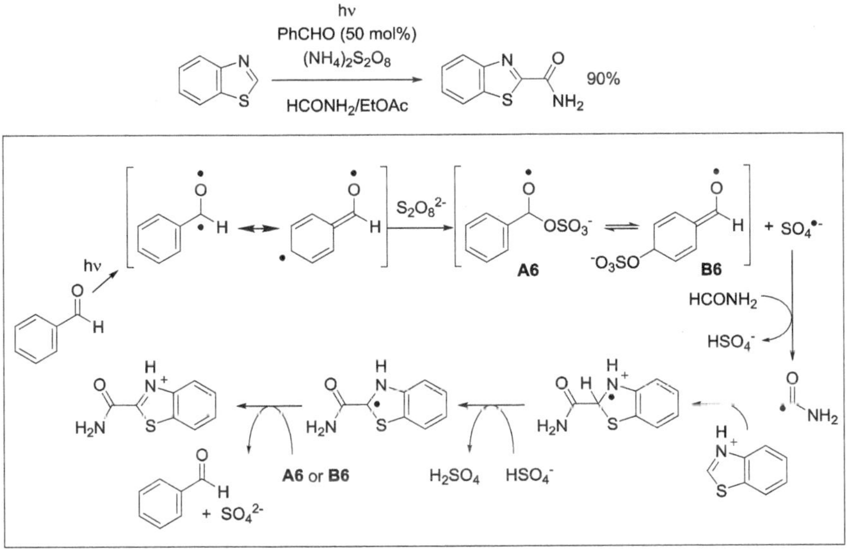

Scheme 1.41. Carbamoylation of benzothiazole.

addition of the carbamoyl radical to protonated benzothiazole took place at the C-2 position, followed by deprotonation and oxidation to provide the coupling product. Intermediates **A6** and **B6** might be involved in the oxidation step, and thus benzaldehyde is regenerated to operate the catalytic cycle. In addition to the carbamoylation employing $HCONH_2$, alkylation was achieved under similar conditions when ethers were employed instead of $HCONH_2$.

1.4.3 *Imidation*

Ooi and co-workers investigated the nitrogen functionalization of aromatic compounds under the catalysis by photoexcited aryl ketones (Scheme 1.42) [55]. With 3,6-dimethoxy-9*H*-thioxanthen-9-one as photocatalyst and 3,5-bis(trifluoromethyl)phenylacyloxyophthalimide as the imidating agent, the imidation of the benzene derivatives naphthalene and a variety of nitrogen containing heteroaromatics, such as pyrrole, indole, pyridine, and caffeine, was achieved. The reaction of caffeine smoothly proceeded by treatment with the imidating agent (1.3 equiv) in the presence of the thioxanthone catalyst (10 mol%) in MeCN under irradiation at room temperature. In the catalytic cycle, thioxanthone acted as an excited-state reductant, and thus an oxidative quenching cycle for this radical aromatic substitution was proposed.

1.4.4 *Biaryl synthesis*

Murakami and co-workers investigated the effect of aryl ketones as photoredox catalysts in the nickel-catalyzed homocoupling of aryl halides (Scheme 1.43) [56]. Such reductive homocoupling generally requires a stoichiometric amount of Ni(0) complex without metal reductants, such as Zn(0) and Mn(0). This study enabled the catalytic formation of biaryl compounds by employing a new cooperative system consisting of light, thioxanthone catalyst, $^{i}Pr_2NEt$, as well as a Ni catalyst. In the proposed mechanism, photoexcited

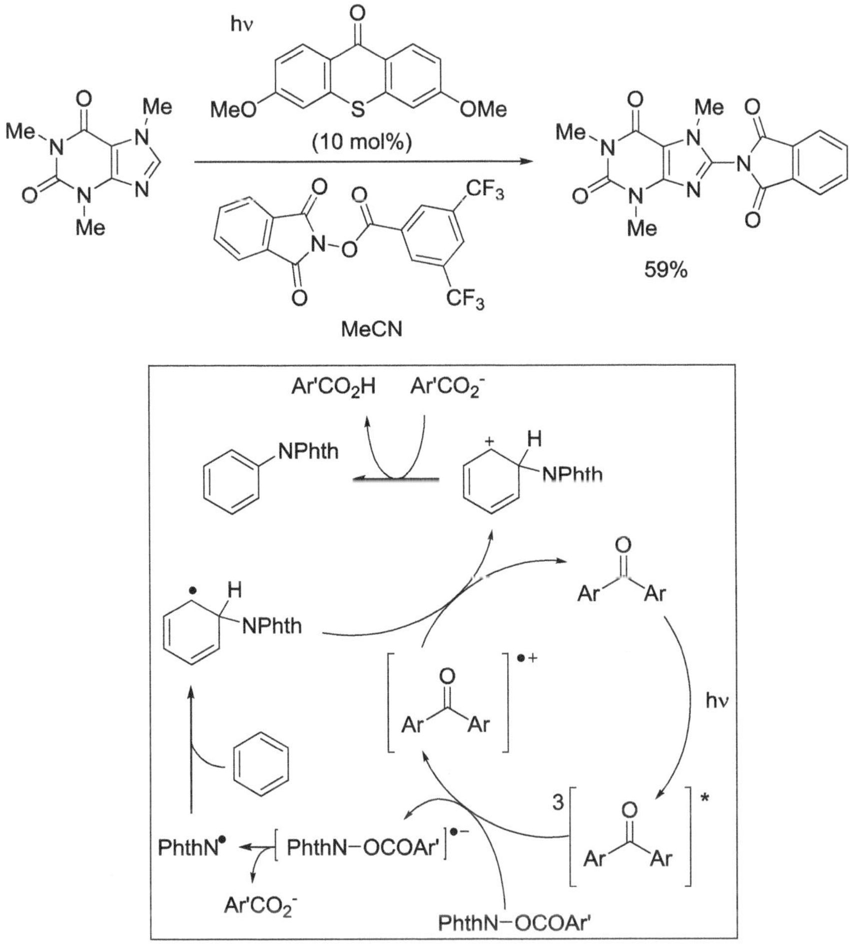

Scheme 1.42. Imidation of caffeine.

thioxanthone was initially reduced by $^{i}Pr_2NEt$ to produce a ketyl-type radical. This species then reduced the $Ni(II)Br_2$ to Ni(0) with the regeneration of thioxanthone. Accordingly, thioxanthone acted as a mediator to deliver an electron from $^{i}Pr_2NEt$ to Ni(II), and this is the key to promote the present catalytic transformation through three consecutive redox cycles under a cooperative strategy.

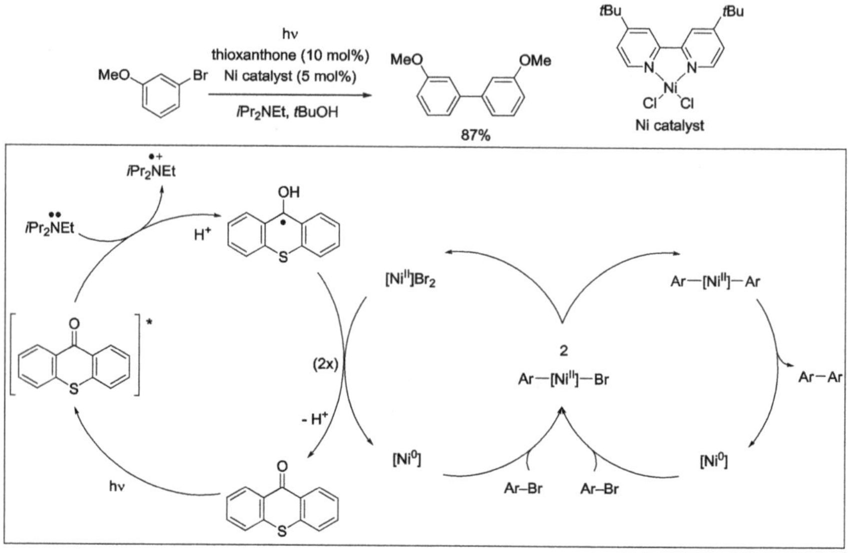

Scheme 1.43. Dimerization of an aryl bromide for biphenyl synthesis.

1.5 Conclusions

As shown in this chapter, ketones and aldehydes were employed in various versatile photochemical transformations to functionalize unsaturated bonds, aromatic compounds, and C(sp^3)–H bonds. Considerable efforts have been devoted to perform these transformations in the presence of a catalytic amount of ketones and aldehydes, although further improvements of their reactivity are still necessary in some transformations requiring a stoichiometric amount of carbonyl compounds. In addition, the design of new catalysts, which can control enantioselectivity and regioselectivity, is highly desirable. Studies along these lines can bring further progress in organic syntheses promoted under photocatalytic conditions.

References

[1] Pérez-Prieto, J.; Galian, R. E.; Miranda, M. A. *Mini-Rev. Org. Chem.* **2006**, *3*, 117–135.

[2] Fagnoni, M.; Dondi, D.; Ravelli, D.; Albini, A. *Chem. Rev.* **2007**, *107*, 2725–2756.

[3] Dormán, G.; Nakamura, H.; Pulsipher, A.; Prestwich, G. D. *Chem. Rev.* **2016**, *116*, 15284–15398.

[4] Romero, N. A.; Nicewicz, D. A. *Chem. Rev.* **2016**, *116*, 10075–10166.

[5] Fraser-Reid, B.; Holder, N. L.; Hicks, D. R.; Walker, D. L. *Can. J. Chem.* **1977**, *55*, 3978–3985.

[6] Benko, Z.; Fraser-Reid, B.; Mariano, P. S.; Beckwith, A. L. J. *J. Org. Chem.* **1988**, *53*, 2066–2072.

[7] Fraser-Reid, B.; Anderson, R. C.; Hicks, D. R.; Walker, D. L. *Can. J. Chem.* **1977**, *55*, 3986–3995.

[8] Graalfs, H.; Fröhlich, B.; Wolff, C.; Mattay, J. *Eur. J. Org. Chem.* **1999**, 1057–1073.

[9] Manfrotto, C.; Mella, M.; Freccero, M.; Fagnoni,M.; Albini, A. *J. Org. Chem.* **1999**, *64*, 5024–5028.

[10] Dondi, D.; Caprioli, I.; Fagnoni, M.; Mella, M.; Albini, A. *Tetrahedron* **2003**, *59*, 947–957.

[11] Dondi, D.; Protti, S.; Albini, A.; Carpio, S. M.; Fagnoni, M.; *Green Chem.* **2009**, *11*, 1653–1659.

[12] Bertrand, S.; Glapski, C.; Hoffmann, N.; Pete, J.-P. *Tetrahedron Lett.* **1999**, *40*, 3169–3172.

[13] Bertrand, S.; Hoffmann, N.; Pete, J.-P. *Tetrahedron Lett.* **1999**, *40*, 3173–3174.

[14] Bertrand, S.; Hoffmann, N.; Humbel, S.; Pete, J.-P. *J. Org. Chem.* **2000**, *65*, 8690–8703.

[15] Marinković, S.; Brulé, C.; Hoffmann, N.; Prost, E.; Nuzillard, J.-M.; Bulach, V. *J. Org. Chem.* **2004**, *69*, 1646–1651.

[16] Harakat, D.; Pesch, J.; Marinković, S.; Hoffmann, N. *Org. Biomol. Chem.* **2006**, *4*, 1202–1205.

[17] Bauer, A.; Westkämper, F.; Grimme, S.; Bach, T. *Nature* **2005**, *436*, 1139–1140.

[18] Doohan, R. A.; Hannan, J. J.; Geraghty, N. W. A. *Org. Biomol. Chem.* **2006**, *4*, 942–952.

[19] Lefebvre, Q.; Hoffmann, N.; Rueping, M. *Chem. Commun.* **2016**, *52*, 2493–2496.

[20] Limnios, D.; Kokotos, C. G. *Adv. Synth. Catal.* **2017**, *359*, 323–328.

[21] Itoh, K.; Kato, R.; Kinugawa, D.; Kamiya, H.; Kudo, R.; Hasegawa, M.; Fujii, H.; Suga, H. *Org. Biomol. Chem.* **2015**, *13*, 8919–8924.

[22] Papadopoulos, G. N.; Limnios, D.; Kokotos, C. G. *Chem. Eur. J.* **2014**, *20*, 13811–13814.

[23] Papadopoulos, G. N.; Kokotos, C. G. *Chem. Eur. J.* **2016**, *22*, 6964–6967.

[24] Papadopoulos, G. N.; Kokotos, C. G. *J. Org. Chem.* **2016**, *81*, 7023–7028.

[25] Arceo, E.; Montroni, E.; Melchiorre, P. *Angew. Chem. Int. Ed.* **2014**, *53*, 12064–12068.

[26] Beniazza, R.; Atkinson, R.; Absalon, C.; Castet, F.; Denisov, S. A.; McClenaghan, N. D.; Lastécouères, D.; Vincent, J.-M. *Adv. Synth. Catal.* **2016**, *358*, 2949–2961.

[27] Beniazza, R.; Douarre, M. D.; Lastécouères, D.; Vincent, J.-M. *Chem. Commun.* **2017**, *53*, 3547–3550.

[28] Li, B.; Fan, D.; Yang, C.; Xia, W. *Org. Biomol. Chem.* **2016**, *14*, 5293–5297.

[29] Luo, J.; Zhang, J. *J. Org. Chem.* **2016**, *81*, 9131–9137.

[30] Alonso, R.; Bach, T. *Angew. Chem. Int. Ed.* **2014**, *53*, 4368–4371.

[31] Tröster, A.; Alonso, R.; Bauer, A.; Bach, T. *J. Am. Chem. Soc.* **2016**, *138*, 7808-7811.

[32] Maturi, M. M.; Bach, T. *Angew. Chem. Int. Ed.* **2014**, *53*, 7661–7664.

[33] Mayr, F.; Brimioulle, R.; Bach, T. *J. Org. Chem.* **2016**, *81*, 6965–6971.

[34] Kamijo, S.; Hoshikawa, T.; Inoue, M. *Org. Biomol. Chem.* **2013**, *11*, 164–169.

[35] Yoshioka, S.; Nagatomo, M.; Inoue, M. *Org. Lett.* **2015**, *17*, 90–93.

[36] Nagatomo, M.; Yoshioka, S.; Inoue, M. *Chem. Asian. J.* **2015**, *10*, 120–123.

[37] Amaoka, Y.; Nagatomo, M.; Watanabe, M.; Tao, K.; Kamijo, S.; Inoue, M. *Chem. Sci.* **2014**, *5*, 4339–4345.

[38] Paul, S.; Guin, J. *Green Chem.* **2017**, *19*, 2530–2534.

[39] Kamijo, S.; Kamijo, K.; Maruoka, K.; Murafuji, T. *Org. Lett.* **2016**, *18*, 6516–6519.

[40] Kamijo, S.; Takao, G.; Kamijo, K.; Hirota, M.; Tao, K.; Murafuji, T. *Angew. Chem. Int. Ed.* **2016**, *55*, 9695–9699.

[41] Ishida, N.; Masuda, Y.; Uemoto, S.; Murakami, M. *Chem. Eur. J.* **2016**, *22*, 6524–6527.

[42] Kamijo, S.; Hirota, M.; Tao, K.; Watanabe, M.; Murafuji, T. *Tetrahedron Lett.* **2016**, *55*, 5551–5554.

[43] Kamijo, S.; Watanabe, M.; Kamijo, K.; Tao, K.; Murafuji, T. *Synlett* **2016**, *48*, 115–121.

[44] Kamijo, S.; Tao, K.; Takao, G.; Murooka, H.; Murafuji, T. *Tetrahedron Lett.* **2015**, *56*, 1904–1907.

[45] Han, L.; Xia, J.-B.; You, L.; Chen, C. *Tetrahedron* **2017**, *73*, 3696–3701.

[46] Xia, J.-B.; Zhu, C.; Chen, C. *J. Am. Chem. Soc.* **2013**, *135*, 17494–17500.

[47] Cantillo, D.; de Frutos, O.; Rincón, J. A.; Mateos, C.; Kappe, C. *J. Org. Chem.* **2014**, *79*, 8486–8490.

[48] Bume, D. D.; Pitts, C. R.; Jokhai, R. T.; Lectka, T. *Tetrahedron* **2016**, *77*, 6031–6036.

[49] Xia, J.-B.; Zhu, C.; Chen, C. *Chem. Commun.* **2014**, *50*, 11701–11704.

[50] Pitts, C. R.; Bloom, M. S.; Bume, D. D.; Zhang, Q. A.; Lectka, T. *Chem. Sci.* **2015**, *6*, 5225–5229.

[51] Ding, W.; Lu, L.-Q.; Zhou, Q.-Q.; Wei, Y.; Chen, J.-R.; Xiao, W.-J. *J. Am. Chem. Soc.* **2017**, *139*, 63–66.

[52] Lipp, A.; Lahm, G.; Opatz, T. *J. Org. Chem.* **2016**, *81*, 4890–4897.

[53] Kamijo, S.; Kamijo, K.; Murafuji, T. *J. Org. Chem.* **2017**, *82*, 2664–2671.

[54] Zhang, Y.; Teuscher, K. B.; Ji, H. *Chem. Sci.* **2016**, *7*, 2111–2118.

[55] Tripathi, C. B.; Ohtain, T.; Corbett, M. T.; Ooi, T. *Chem. Sci.* **2017**, *8*, 5622–5627.

[56] Matsuda, Y.; Ishida, N.; Murakami, M. *Eur. J. Org. Chem.* **2016**, 5822–5825.

Chapter 2

Quinones

Akichika Itoh

Gifu Pharmaceutical University, Gifu, Japan
itoha@gifu-pu.ac.jp

2.1 Photophysical Characteristics of Quinones

Quinones are ubiquitous in biological systems, and the photochemistry of quinones has been investigated for a long time. Scheme 2.1 shows the typical reaction steps of photoreduction of quinones in the presence of donors. Especially, anthraquinone (**AQ**) undergoes fast intersystem crossing to the triplet state (2), which efficiently accomplishes H atom abstraction to form the semiquinone radical of the type •QH (3).

Table 2.1 shows a selection of the photophysical parameters of common quinones. Unfortunately, a systematic report on the photophysical characteristics of quinones is not available yet, and it is sometimes difficult to compare each derivative because several data are missing. However, in general, the excited singlet state of most quinones is extremely short lived.

Since quinones are, in general, cheap and commercially available, several reactions based on their use have been investigated under either thermal or photoirradiation conditions. In particular,

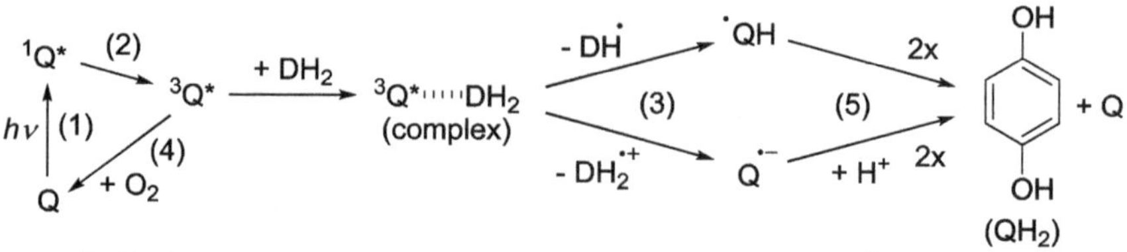

(1) Excitation of the ground state to the excited singlet state ($^1Q^*$)
(2) Decay of $^1Q^*$ to the lowest triplet state ($^3Q^*$)
(3) Generation of long-lived semiquinone radicals
(4) Triplet quenching by oxygen
(5) Generation of dihydroquinone (QH_2)

Scheme 2.1. Photoreduction of quinones in the presence of electron/hydrogen donors.

Table 2.1. Photophysical parameters of quinones (λ_f: maximum of fluorescence emission band; τ_f: lifetime of fluorescence emission; Φ_f: fluorescence quantum yield; τ_p: lifetime of phosphorescence emission; Φ_p: phosphorescence quantum yield; Φ_{isc}: intersystem crossing quantum yield).

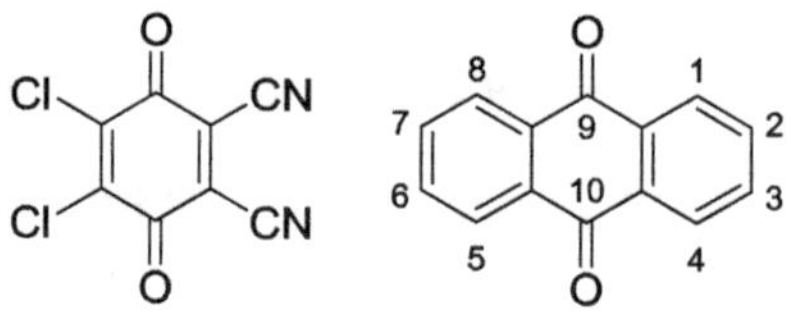

DDQ **AQ**

Quinones	λ_f (nm)	τ_f (ns)	Φ_f	τ_p (μs)	Φ_p	Φ_{isc}
DDQ						1.0[a]
AQ				3200[b]	1[b]	1.04[c]
1-Chloro-**AQ**				160 ± 10[b]	1.58×10^{-2}[b]	
1,5-Dichloro-**AQ**				260 ± 10[b]	1.72×10^{-2}[b]	
1,2-Dihydroxy-**AQ**			0.004[d]			0.16[d]
1-Amino-**AQ**[e]	590	0.94	0.022			
1,4-Diamino-**AQ**[e]	617		0.02			

Note: [a]See Ref. [1]. [b]See Ref. [2]. [c]See Ref. [3]. [d]See Ref. [4]. [e]See Refs. [5–7].

Table 2.2. Electrochemical parameters of the ground and excited states and absorption maxima of quinones (E_{RED} (**POC/POC$^{•-}$**): ground-state reduction potential; E^*_{RED} (**POC*/POC$^{•-}$**): excited-state (T_1) reduction potential).[a]

Quinones	E_{RED} (POC/POC$^{•-}$)	E^*_{RED} (POC*/POC$^{•-}$)	Absorption Maxima (nm)
BQ	−0.52		
DDQ	0.49	+3.18	330
AQ	−0.96	+1.77	326
1,5-Dichloro-**AQ**	−0.12	+2.52	345
1,2-Dihydroxy-**AQ**	−0.05	+2.90	430
1-Amino-**AQ**	−0.19	+1.96	465
1,4-Diamino-**AQ**	−0.16	+1.92	545, 580

Note: [a]Redox potentials are reported in V vs SCE. Data were adapted from those reported in Refs. [8–11] by conversion into values on the SCE scale.

quinones exhibit oxidizing activity in both the ground state and the excited state, and attention has been drawn to their oxidizing power. Table 2.2 shows the electrochemical parameters of both ground and excited states and the absorption maxima of quinones. Most quinones absorb strongly around 300 nm, with weaker absorptions in the near-UV and visible regions (~400 nm); however, substituted **AQ**s absorb light at longer wavelengths. The reduction potentials of their ground states (E_{RED}) are in the range from ca. −0.50 to +0.5 V vs the saturated calomel electrode (SCE). On the other hand, the reduction potentials of their triplet excited states (E^*_{RED}) are in the range from +1.92 to +3.18 V vs SCE, which indicates that they are likely to be very strong oxidizing agents. Thus, excited-state quinones possess sufficient oxidizing ability to react with a variety of organic compounds. Typical reactions of the triplet excited state of quinones include hydrogen-atom transfer (HAT) and oxidative photoelectron transfer (PET). Previously, quinones were typically used as stoichiometric oxidants; however, recently, their use in catalytic applications has been actively investigated. Thus, the use of other terminal oxidants, such as oxygen, nitrogen dioxide, etc., allowed quinones to be included in a complete catalytic cycle. In this chapter, I focus on the production of fine chemicals by exploiting quinones as photoorganocatalysts (**POCs**), and the latest reports on these reactions are introduced below.

2.2 Reactions of Quinones

2.2.1 *Benzoquinones*

2.2.1.1 *1,4-Benzoquinone* (***BQ***)

1,4-Hydroquinone (BQH_2) is a reduced form of 1,4-benzoquinone (**BQ**) and can produce **BQ** *in situ* under oxidation conditions in the reaction system. Mitchell and Moody reported the oxidation of alcohols to the corresponding carbonyl compounds using catalytic amounts of 1,4-hydroquinone (BQH_2), a biorenewable compound, and copper nanoparticles as an electron-transfer mediator (ETM) with molecular oxygen as the terminal oxidant under visible-light irradiation (Scheme 2.2) [12]. Under these conditions, several benzylic and aliphatic primary and secondary alcohols were oxidized, affording the corresponding aldehydes or ketones in moderate to high yields.

Scheme 2.3 shows the working hypothesis for this reaction. Because BQH_2 cannot be oxidized directly by molecular oxygen under neutral reaction conditions, an ETM is required to facilitate this process.

This methodology has also been extended to the oxidative degradation of seven model lignin compounds that underwent C–C bond cleavage to give simple aromatic compounds; however, the yields were modest (Scheme 2.4).

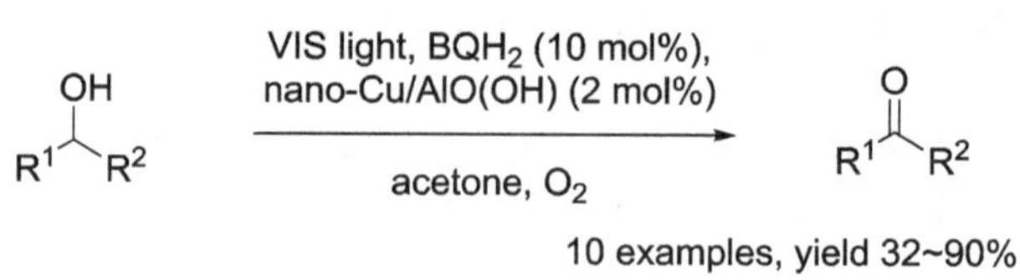

Scheme 2.2. Photochemical oxidation of alcohols with catalytic hydroquinone and copper nanoparticles under O_2.

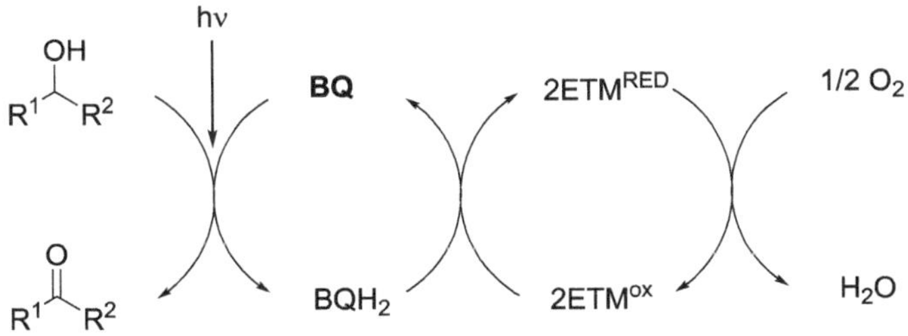

Scheme 2.3. Plausible path of photochemical oxidation of alcohols with catalytic hydroquinone and copper nanoparticles under O_2 atmosphere.

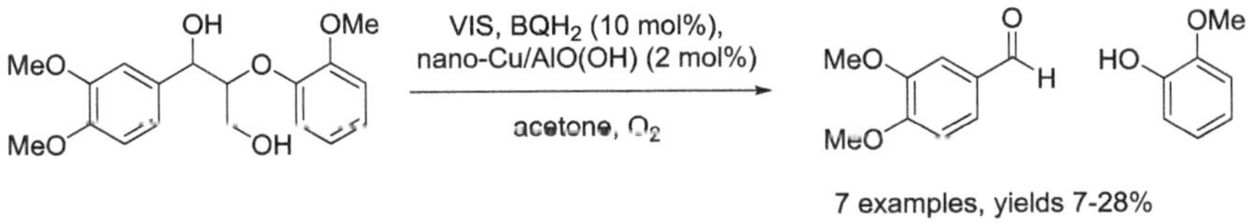

Scheme 2.4. Oxidative degradation of a lignin model.

Scheme 2.5 shows the plausible mechanism for this oxidative degradation reaction. The authors proposed two competing paths. The first one is initiated by the abstraction of a hydrogen atom from an α-C-H bond to afford by-product α-hydroxy ketones (path A). A second path is initiated by the abstraction of a hydrogen atom from an O–H bond by photoexcited triplet quinone (^{3}**BQ**) to give a β-hydroxyalkoxyl radical, which undergoes homolytic fragmentation of the C–C bond to afford an aldehyde and a new α-hydroxyalkyl radical (path B). The cleavage of the C–C bond in β-hydroxyalkoxyl radicals to afford two carbonyl compounds is a known reaction [13, 14].

2.2.1.2 *2,3-Dichloro-5,6-dicyano-1,4-benzoquinone* (*DDQ*)

DDQ is a potent and versatile oxidant owing to the four electron-withdrawing groups on the *p*-benzoquinone ring and forms strong charge transfer complexes with compounds containing electron-rich groups by abstracting electrons and protons. However, drawbacks of the stoichiometric use of this reagent include its considerable toxicity, because hydrogen cyanide is released upon

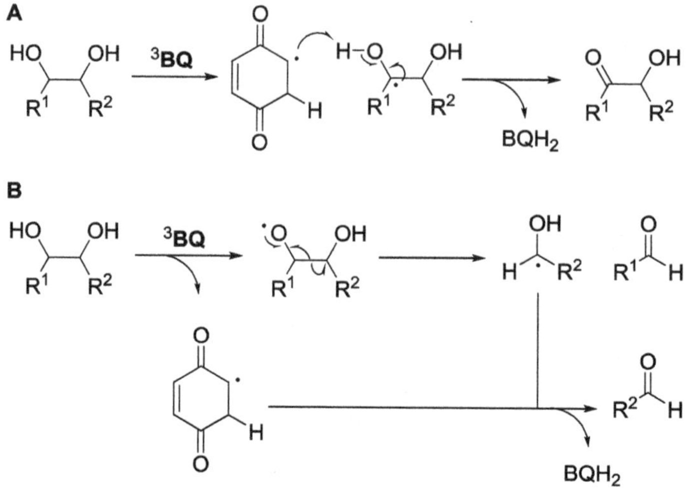

Scheme 2.5. Plausible mechanism of the oxidative degradation of lignin models.

exposure to water, and, occasionally, difficulties in separating the desired product from its reduced form 2,3-dichloro-5,-6-dicyanobenzene-1,4-diol ($DDQH_2$) [8]. These problems have encouraged the development of procedures using a catalytic amount of **DDQ** and a stoichiometric amount of a terminal oxidant. In the case of reactions at the ground state, a variety of terminal oxidants, such as MnO_2, $NaNO_2$, or O_2, are used. For example, the oxidative deprotection of *p*-methoxybenzyl (PMB) ethers was reported by Hu *et al.* [15]. They established an aerobic catalytic system for the selective oxidation of non-sterically hindered benzylic alcohols and electron-rich benzyl methyl ethers. They also employed this system for the oxidative deprotection of PMB ethers **A** to the corresponding alcohols **B** and successfully obtained the desired products with high selectivity (Scheme 2.6).

On the other hand, with excited-state quinones, molecular oxygen has usually been used as the terminal oxidant. The first example of benzene oxygenation to produce phenol using a catalytic amount of **DDQ** under visible-light irradiation of an oxygen-saturated solution of benzene and *tert*-butyl nitrite (TBN) in acetonitrile was reported by Fukuzumi *et al.* (Scheme 2.7) [16].

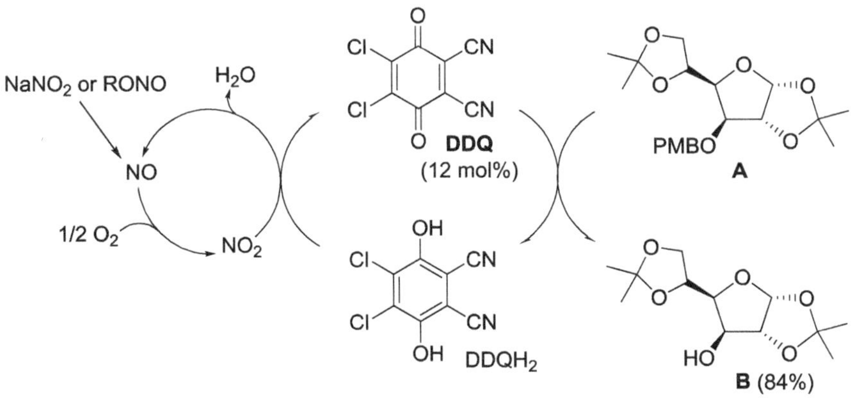

Scheme 2.6. Oxidative deprotection of PMB ethers under thermal conditions.

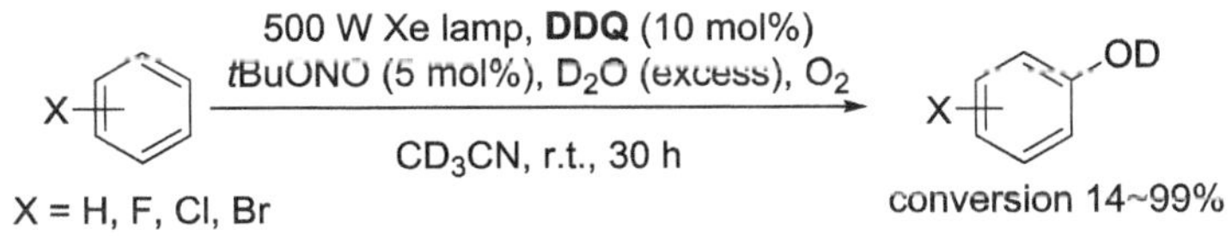

Scheme 2.7. Oxygenation of benzene with a catalytic amount of **DDQ** under visible-light irradiation.

Their typical procedure was as follows: a solution of **DDQ** (3.0 mM) with TBN (1.5 mM) and D_2O (0.5 M) in an NMR tube with a rubber septum was saturated with oxygen by bubbling oxygen through a stainless-steel needle. Subsequently, benzene or halogenated benzene (1.80 μmol; 30 mM) was added to the solution, which was then irradiated with a 500 W xenon lamp through a colored glass filter that transmitted light with $\lambda > 390$ nm at room temperature.

By following this method, arenes were converted to the corresponding phenols in yields ranging from 14% to 99%. In general, *p*-selectivity was predominant when halobenzenes were used as substrates.

Scheme 2.8 shows the plausible mechanism devised by the authors. They investigated the reaction mechanism by examining the results of time-resolved transient absorption experiments. The

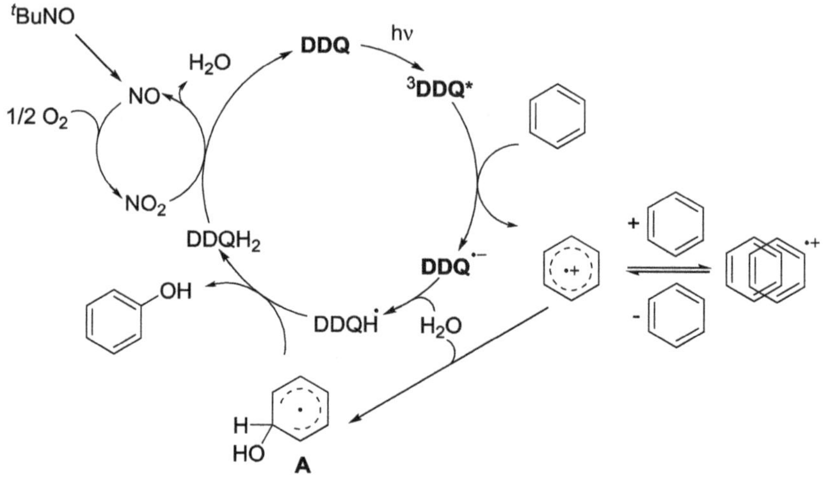

Scheme 2.8. Plausible path of oxygenation of benzene.

transient absorption spectrum indicated the formation of **DDQ**$^{•-}$ and a benzene π-dimer radical cation, which was formed by reaction of the benzene radical cation with neutral benzene. This photooxygenation step was initiated by photoinduced intermolecular electron transfer from benzene to the triplet excited state of **DDQ** (^{3}**DDQ***) driven by the change in free energy for electron transfer. The benzene radical cation reacted with water to produce an OH-adduct radical. On the other hand, **DDQ**$^{•-}$ reacted with water to generate DDQH$^•$, which reacted with the OH-adduct radical **A** to form phenol and $DDQH_2$. $DDQH_2$ is known to be oxidized by NO_2, which was formed from TBN and O_2, to regenerate **DDQ**.

Lei *et al.* reported that the direct activation of the α-C–H bond in thiophene was achieved *via* an oxidation pathway under visible-light irradiation, in which the thiophene radical cation acted as the key intermediate (Scheme 2.9) [17]. A variety of thiophenes and azoles have been converted to the corresponding animated products. Their typical procedure was as follows: a solution of a thiophene derivative (2 equiv), a pyrazole derivative, **DDQ** (15 mol%), and TBN (15 mol%) in dichloroethane was stirred in an air atmosphere under irradiation with 3 W blue LEDs. They reported 27 examples, with yields in the 33%–81% range.

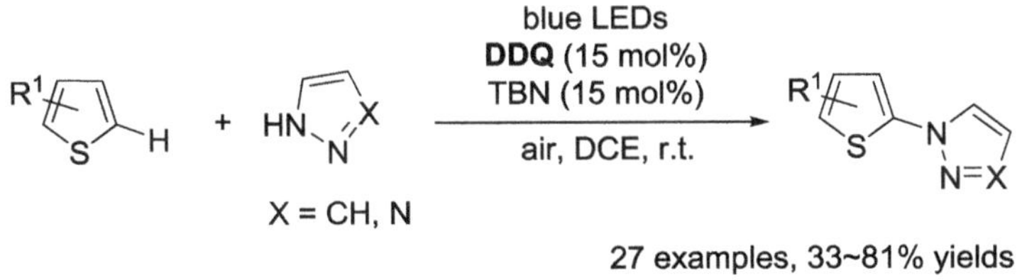

Scheme 2.9. Amination of thiophenes by using DDQ.

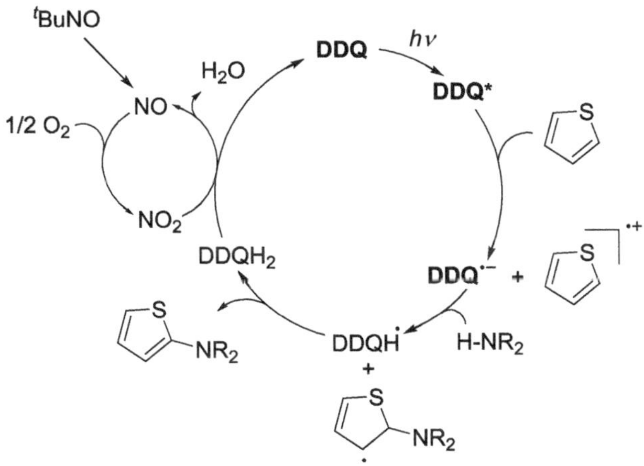

Scheme 2.10. Plausible path involved in the amination of thiophenes.

In this reaction, TBN played the role of ETM and O_2 acted as the terminal oxidant to regenerate **DDQ** and complete the photocatalytic cycle (Scheme 2.10). Intermolecular kinetic isotope effect (KIE) experiments performed to clarify the reaction process indicated that cleavage of the C–H bond in thiophene might not comprise the rate-determining step of the reaction. Radical-inhibiting experiments using 2,2,6,6-tetramethylpiperidine-*N*-oxyl (TEMPO) and butylated hydroxytoluene indicated that the reaction might involve a radical process. On the basis of further control experiments, the authors proposed a reaction mechanism (Scheme 2.10) which is very similar to that devised by Ohkubo *et al.* shown in Scheme 2.8.

Although a stoichiometric amount of **DDQ** was required, Fukuzumi *et al.* also reported the monochlorination of phenol [18].

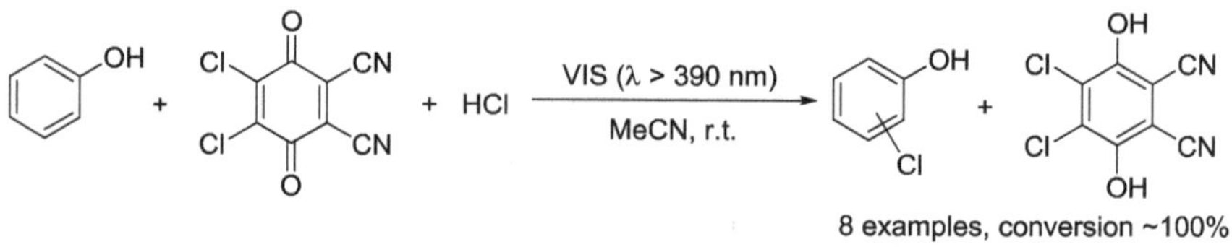

Scheme 2.11. Aromatic monochlorination of phenol with **DDQ**.

Photochlorination of aromatic substrates by hydrogen chloride occurred efficiently to produce the corresponding monochlorinated products selectively in the presence of a stoichiometric amount of **DDQ** under visible-light irradiation (Scheme 2.11). They conducted this reaction in a deaerated acetonitrile solution containing **DDQ** (1.25 equiv), phenol, and HCl (25 equiv) by irradiation with a xenon lamp (500 W, $\lambda > 390$ nm). From phenol, the yields of *p*- and *o*-chlorophenols were 70% and 18%, respectively, with no formation of polychlorinated products. They reacted eight aromatic substrates and obtained the corresponding monochlorinated products in good to high yields, except when biphenyl was used as the substrate.

Scheme 2.12 describes the plausible mechanism of this photochlorination reaction suggested by the authors. They proposed two possible pathways, which are initiated by the transfer of an electron from Cl^- to ^{3}**DDQ*** to produce Cl^- and **DDQ**$^{\bullet-}$, or by the transfer of an electron from phenol to ^{3}**DDQ*** to give the phenol radical cation and **DDQ**$^{\bullet-}$, respectively. They detected the radical intermediate **DDQ**$^{\bullet-}$, which was formed via the transfer of an electron from Cl^- to **DDQ**, by time-resolved transient absorption measurements at 590 nm. In the latter pathway, back electron transfer from **DDQ**$^{\bullet-}$ to the phenol radical cation occurs rapidly. In the former pathway, on the other hand, $Cl^{\bullet}$ reacts with phenol to produce a chlorine-adduct radical. The authors concluded that the selectivity of chlorination of phenol and naphthalene derivatives is controlled by the thermodynamic stability of the chlorinated radical from density functional theory (DFT) calculations. The final products, namely, chlorophenols and $DDQH_2$, are produced by the transfer of a hydrogen atom

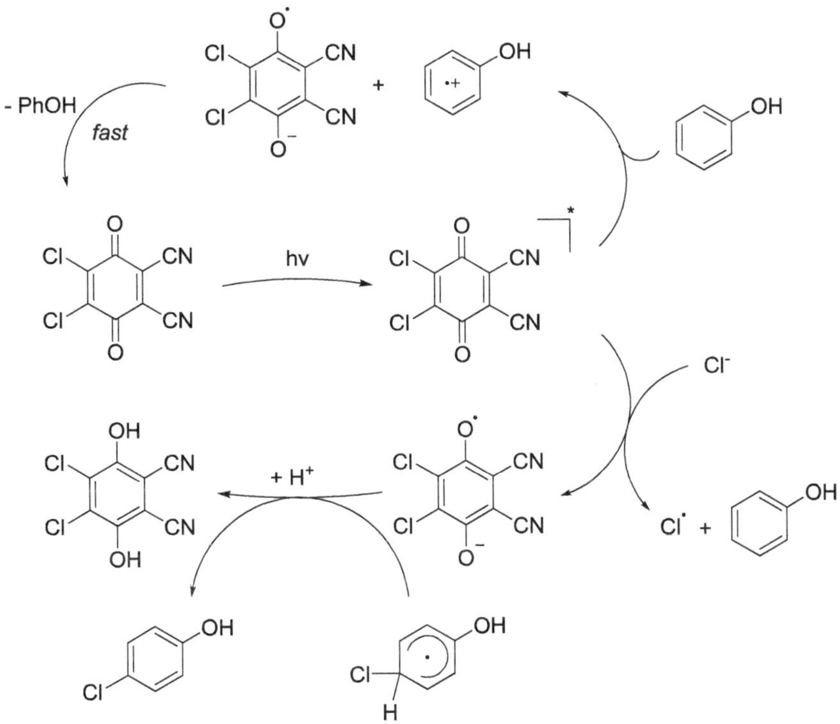

Scheme 2.12. Plausible path of aromatic monochlorination of phenol.

from the chlorine-adduct radical to **DDQ**$^{\bullet-}$ and protonation with HCl. Thus, the quantum yield for the formation of the product is determined by the competition between electron transfer from Cl^- to ^{3}**DDQ*** and electron transfer from phenol to ^{3}**DDQ***.

Brasholz *et al.* reported the visible-light photocatalytic aerobic oxygenation of benzylic C(sp^3)–H bonds using a ^{3}**DDQ**/TBN co-catalytic system [19]. They reacted 10 different substrates, including aromatic hydrocarbons and C3-substituted indoles, by irradiating an acetonitrile solution of **DDQ** (5 mol%) and TBN (50 mol%) with a blue compact fluorescent lamp ($\lambda = 450 \pm 50$ nm) to give the corresponding oxygenated products in moderate to high yields and with high selectivity (Scheme 2.13). This system provided an increase in the reaction rate in comparison with an analogous

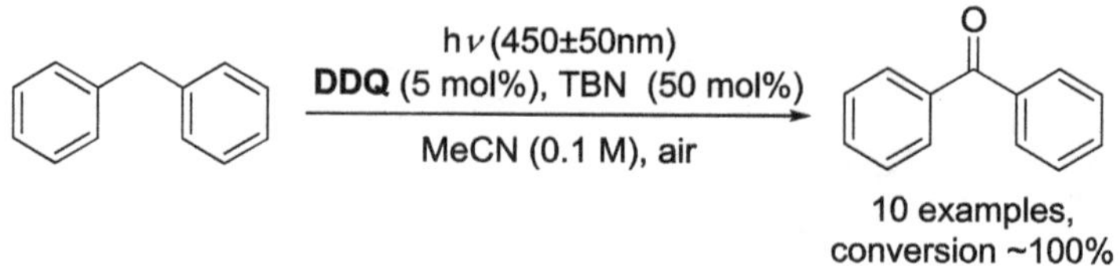

Scheme 2.13. Aerobic benzylic C(sp^3)–H oxygenations.

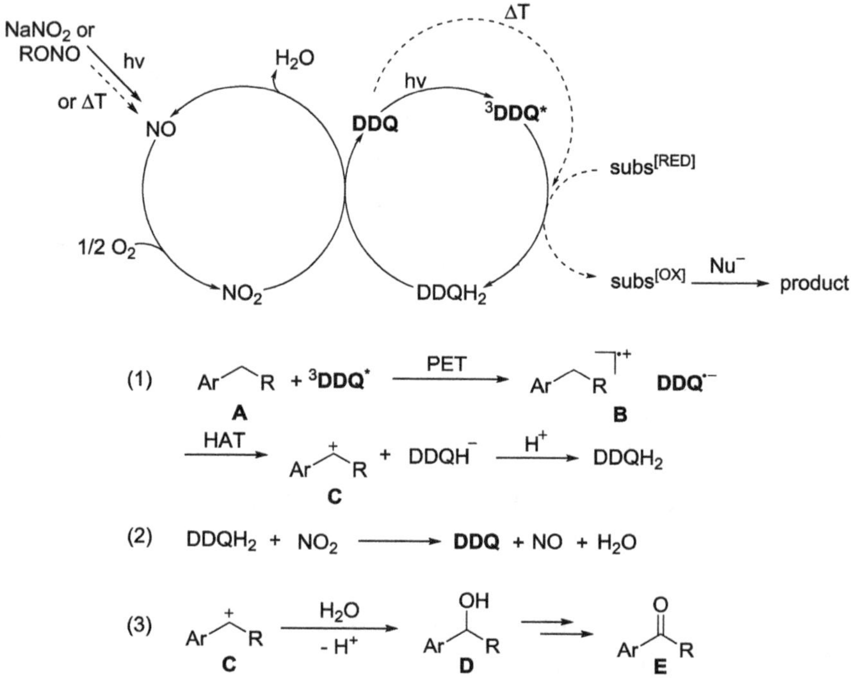

Scheme 2.14. Plausible path of aerobic benzylic C(sp^3)–H oxygenations.

thermal oxidation method carried out at 80°C in the dark. However, a decreased reaction selectivity was observed in some cases, because of the increased oxidizing power of ^{3}**DDQ*** with respect to that of ground-state **DDQ**.

Scheme 2.14 shows the plausible mechanism devised by the authors, which includes the net abstraction of a hydride ion from substrate **A** by **DDQ** [Eq. (1)]. At first, PET between substrate **A** and photoexcited ^{3}**DDQ*** forms the radical ion pair **B**, and subsequent

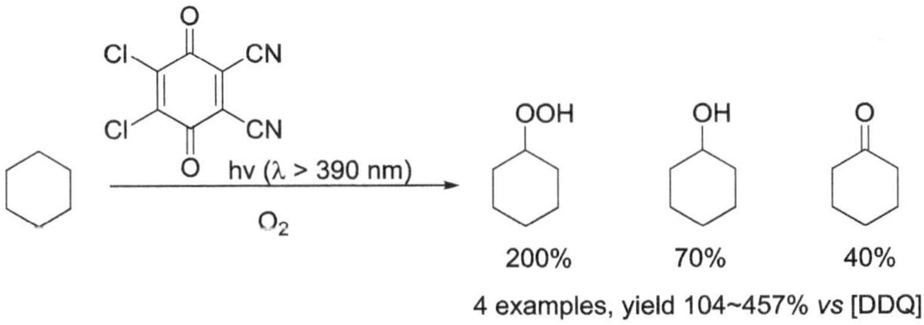

Scheme 2.15. Photooxidation of alkanes.

HAT generates the benzylic cation **C** and $DDQH^-$, which is then protonated to $DDQH_2$. Secondly, $DDQH_2$ reduces NO_2, which gives NO and H_2O [Eq. (2)]. Trapping of the benzyl cation **C** with H_2O produces the alcohol **D**, which is transformed into the final product **E** [Eq. (3)]. The authors performed experiments with $^{18}O_2$ to elucidate this mechanism.

Fukuzumi *et al.* reported the hydroxylation of alkanes, such as cyclohexane, hexane, 3-methylpentane, and pentane, by O_2 with the singlet excited state of **DDQ** (^{1}**DDQ***) under VIS light irradiation (Scheme 2.15) [20].

The typical procedure was as follows: a cyclohexane solution containing **DDQ** with acetonitrile as a co-solvent (2%) in a square quartz cuvette with a silicon septum was saturated with oxygen (the gas was bubbled through a stainless-steel tube for 5 min). The solution was then irradiated with a 500 W xenon lamp through a colored glass filter that transmitted light with $\lambda > 390$ nm at room temperature. They used four substrates, which included linear, branched, and cyclic alkanes. This was the first report of the direct photochemical hydroxylation of alkanes by O_2 *via* photoinduced electron transfer.

The authors examined the mechanism by femtosecond laser-induced time-resolved transient absorption experiments and found that efficient electron transfer occurred from the alkane to ^{1}**DDQ*** to produce the corresponding radical ion pair (alkane$^{\bullet+}$ and **DDQ**$^{\bullet-}$), as shown in Scheme 2.16. The alkane radical cation was

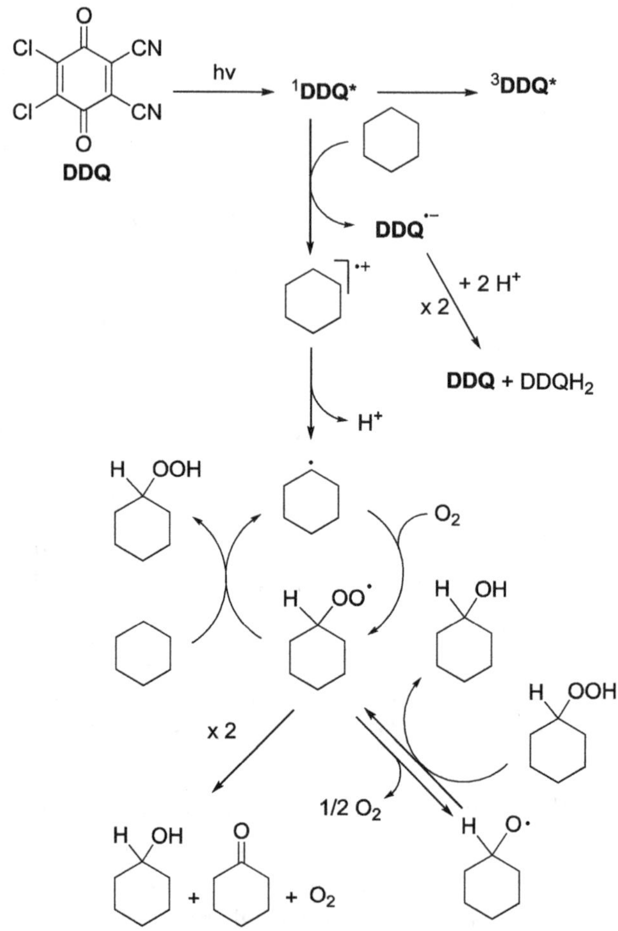

Scheme 2.16. Plausible path of photooxidation of alkanes.

readily deprotonated to produce an alkyl radical, which trapped O_2 to produce an alkylperoxyl radical, which in turn abstracted a hydrogen atom from an additional alkane molecule to afford an alkyl hydroperoxide and regenerate the alkyl radical. The regeneration of the alkyl radical constitutes a radical chain reaction: the so-called autoxidation reaction, which is terminated by the bimolecular reaction of two alkylperoxyl radicals to afford the corresponding alcohol, ketone, and O_2. The **DDQ**$^{\bullet-}$ radical anion produced by

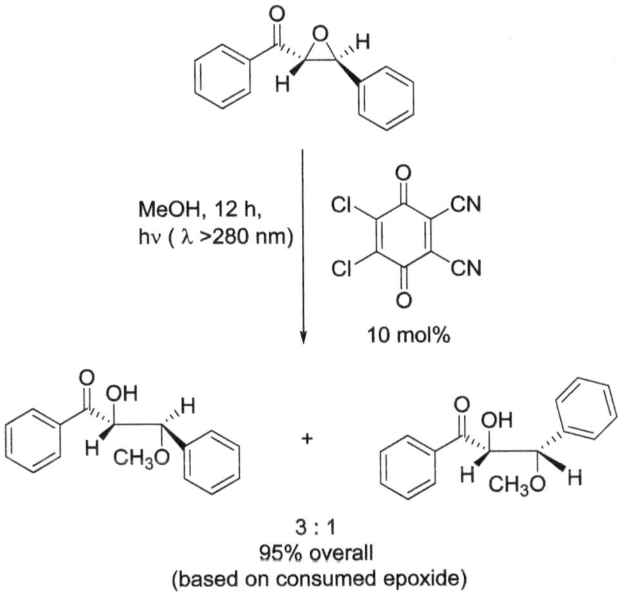

Scheme 2.17. Ring opening reaction of α-epoxyketones in the presence of DDQ.

photoinduced electron transfer from the alkane to [1]**DDQ*** undergoes disproportionation with the addition of protons to yield **DDQ** and $DDQH_2$. Thus, a part of the **POC** is consumed.

The ring opening reaction of α-epoxyketones in the presence of **DDQ** in a methanol solution under irradiation with microwaves or UV light was reported by Memarian and Saffar-Teluri (Scheme 2.17) [21]. The nature and position of additional substituents affected the rate of ring opening. A typical procedure was as follows: a solution of the substrate in methanol with 10 mol% **DDQ** was irradiated ($\lambda >$ 280 nm) at 25°C under running cold water.

The authors proposed an initial electron transfer step between **DDQ*** and the α-epoxyketone. The nucleophilic attack of methanol on the radical cation intermediate formed gave the corresponding α-hydroxy-β-methoxyketone as a mixture of diastereoisomers *via* C_β–O bond cleavage in excellent yields. However, the same process could be carried out under microwave irradiation to give a higher yield after a shorter reaction time.

2.2.1.3 *p-Xyloquinone* (*PXQ*)

Ohkubo *et al.* reported the photooxygenation of alkanes by molecular oxygen using *p*-benzoquinone derivatives such as **PXQ** (Scheme 2.18) [22]. They conducted this photooxidation reaction using **PXQ** with four kinds of alkanes that were used as solvents to produce the corresponding alkyl hydroperoxides, alcohols, and ketones under irradiation with a xenon lamp ($\lambda > 390$ nm) with high quantum yields ($\Phi = 1000\%–1600\%$). However, other *p*-benzoquinones, such as **BQ** and tetramethylbenzoquinone, could produce the desired products, albeit in lower yields than those obtained using **PXQ**.

This photooxygenation reaction starts with the abstraction of a hydrogen atom from the alkane by the triplet excited state of **PXQ** (^{3}**PXQ***; Scheme 2.19), as revealed by laser-induced transient absorption spectroscopy. The authors tried to observe the hydrogen transfer dynamics of ^{3}**PXQ*** by nanosecond laser flash photolysis in cyclohexane and determined that the rate constant of the transfer of hydrogen from cyclohexane to ^{3}**PXQ***, which is the rate-determining step, is $1.3 \cdot 10^7\ M^{-1}\ s^{-1}$. Scheme 2.19 shows the proposed reaction mechanism. From cyclohexane, for example, the photooxygenation reaction is initiated by the abstraction of a hydrogen atom by ^{3}**PXQ*** to form a radical pair consisting of a cyclohexyl radical (Cy$^•$) and a **PXQ** semiquinone radical (**PXQ**H$^•$). Cy$^•$ traps O_2 to produce a cyclohexylperoxyl radical (CyOO$^•$), which abstracts a hydrogen atom from cyclohexane to produce Cy$^•$ and cyclohexyl hydroperoxide (CyOOH). Cyclohexanol (CyOH), cyclohexanone (Cy=O), and O_2

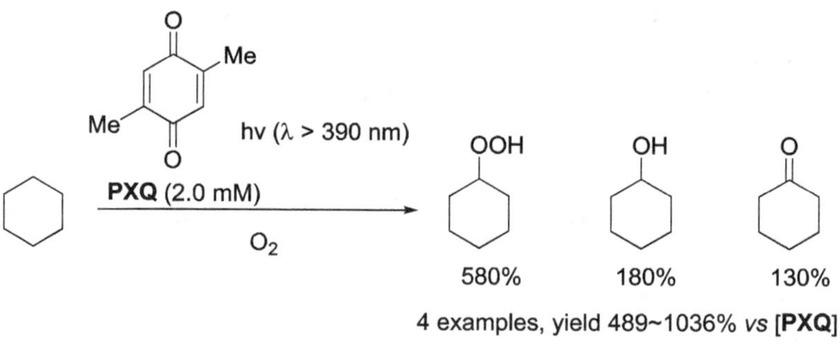

Scheme 2.18. Photooxygenation of alkanes by O_2 with **PXQ**.

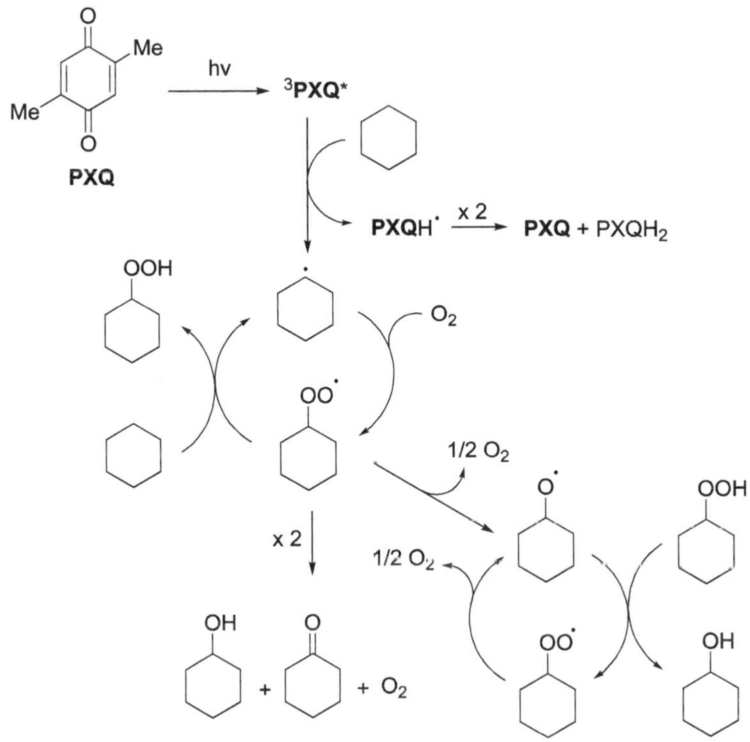

Scheme 2.19. Plausible path of photooxygenation of alkanes with **PXQ**.

are formed by disproportionation of CyOO$^•$, while **PXQ**H$^•$ also undergoes disproportionation to form **PXQ** and PXQH$_2$; however, the yield of CyOH (180%) is higher than that of Cy=O (130%). This additional CyOH may be formed via the decomposition of CyOOH to CyOH *via* radical chain reactions, in which a cyclohexyloxyl radical (CyO$^•$), which is produced by the disproportionation of CyOO$^•$, abstracts a hydrogen atom from CyOOH to produce CyOH.

2.2.2 *Anthraquinones (AQs)*

Tan *et al.* reported the selective fluorination of alkyl C–H bonds with Selectfluor® in the presence of **AQ** (Scheme 2.20) [23, 24]. The

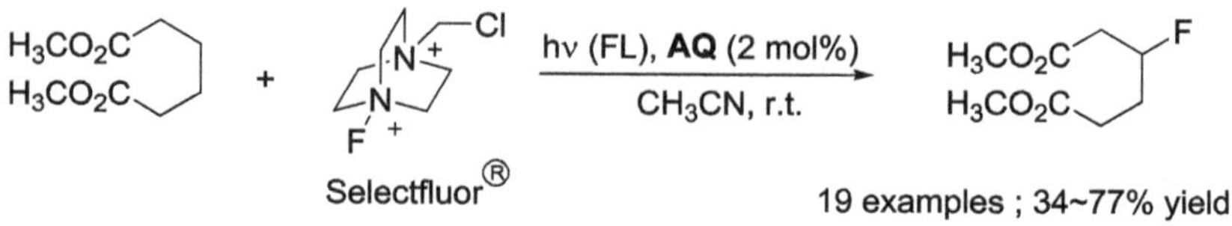

Scheme 2.20. Selective fluorination of alkyl C–H bonds with Selectfluor® in the presence of **AQ**.

typical procedure was as follows: Selectfluor and 2 mol% of the photocatalyst were added to a two-necked Schlenk flask equipped with a magnetic stir bar. The air in the Schlenk flask was replaced with purified N_2, and 1.5 equiv of the substrate was added under N_2 *via* a gas-tight syringe. Acetonitrile was then added under N_2 *via* a gas-tight syringe, and the Schlenk flask was sealed with a glass stopper. The reaction mixture was stirred under irradiation with an 11 W fluorescent lamp. The authors reacted 19 substrates and obtained the corresponding fluorides in yields ranging from 34% to 77%. For linear aliphatic substrates, secondary C–H bonds most distal from an electron-withdrawing group were fluorinated with the highest selectivity. In general, this fluorination reaction is selective for electron-rich sp^3 C–H bonds due to the involvement of cationic *N*-radicals [25], which are generated by the triplet–triplet energy transfer catalyzed by **AQ** during the hydrogen-atom abstraction step [26–28].

The authors elucidated the elementary steps occurring in the photocatalytic fluorination procedure by means of nanosecond transient absorption spectroscopy and DFT calculations [24]. Time-resolved optical spectroscopy suggested that a direct reaction only occurs between photoexcited **AQ*** and Selectfluor, as demonstrated by the spectroscopic evidence of a novel transient **AQ**-Selectfluor species (Scheme 2.21). Furthermore, their DFT calculations suggested that an **AQ**-Selectfluor triplet exciplex formed by photoirradiation is responsible for initiating and sustaining the fluorination reaction.

2.2.2.1 *2-Chloroanthraquinone* (**2-ClAQ**)

Itoh *et al.* developed the direct aerobic photooxidation of 1,3-diketones to the corresponding carboxylic acids in the presence of a catalytic

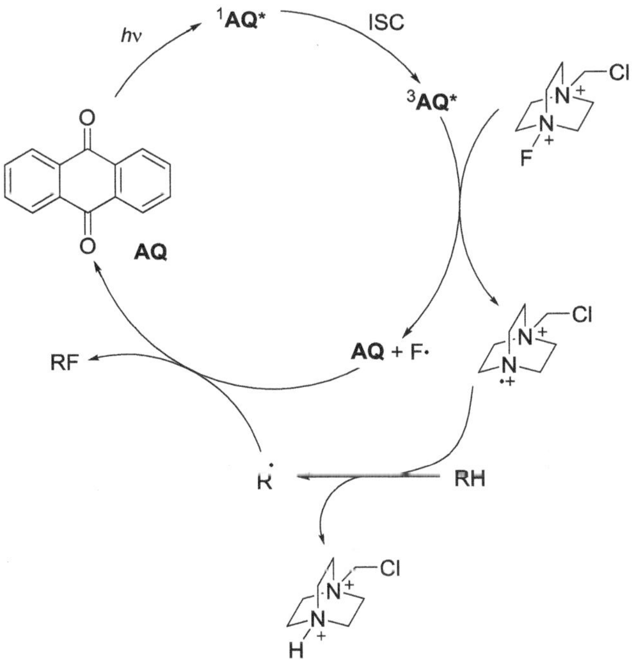

Scheme 2.21. Plausible path of selective fluorination with Selectfluor® in the presence of **AQ**.

amount of **2-ClAQ** under visible-light irradiation (Scheme 2.22) [29]. The typical procedure was as follows: a solution of the substrate, **2-ClAQ** (10 mol%), and Cs_2CO_3 (50 mol%) in dry acetone, purged using an O_2 balloon, was stirred and irradiated externally with four 22 W fluorescent lamps. They mainly used β-ketocarbonyls as substrates and obtained the corresponding carboxylic acids in yields up to 89%.

Scheme 2.23 shows a plausible mechanism for this reaction, which is based on the requirement for continuous irradiation, a catalytic amount of **AQ**, molecular oxygen, and a base, and on the results of a study of the time course of oxidative cleavage. This reaction involves a rearrangement *via* a triketone, which was already reported by Itoh *et al.* [30]. The 1,3 diketone compound **A** is initially oxidized to the triketone **B** by the excited **AQ** in the presence of molecular oxygen. Next, in the presence of a base, a hydrate form of the triketone **B** (**B′**)

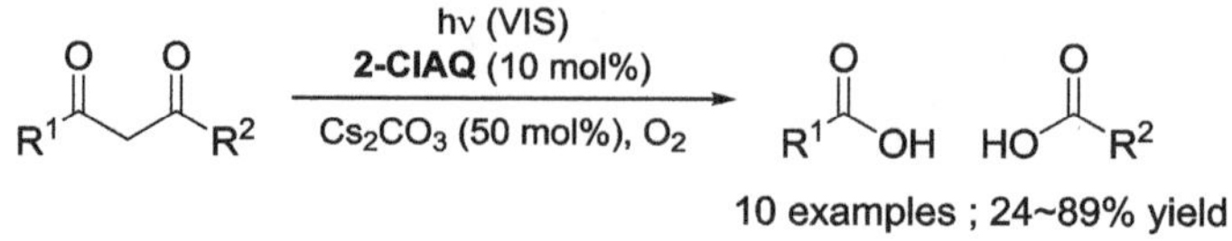

Scheme 2.22. Aerobic photooxidative cleavage of 1,3-diketones.

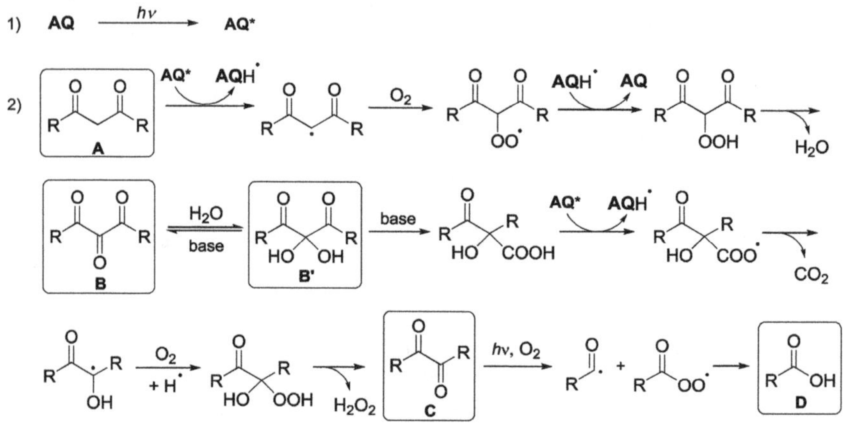

Scheme 2.23. Plausible path of aerobic photooxidative cleavage of 1,3-diketones.

is formed, and **B′** is converted to the 1,2-diketone **C** *via* a base-induced rearrangement, decarboxylation, and further oxidation. Finally, the 1,2-diketone **C** is converted to the carboxylic acid **D** through the generation of an acyl radical and an acylperoxyl radical.

Itoh *et al.* reported the direct activation of a C(sp^3)–H bond to form a C–C bond between tertiary amines and carbon nucleophiles, such as nitroalkanes, ketones, trimethylsilyl cyanide, and indole, under aerobic photooxidizing conditions with **2-ClAQ** as a **POC** (Scheme 2.24) [31]. As an example, a solution of an *N*-aryl-1,2,3,4-tetrahydroisoquinoline, **2-ClAQ** (7 mol%), and $MeNO_2$ (500 mol%) in MeOH in a Pyrex test tube was purged using an O_2 balloon and then externally irradiated with a fluorescent lamp under stirring. A variety of substrates and nucleophiles was employed, allowing the preparation of 18 products in yields ranging from 21% to 80%. This reaction uses the harmless visible light of a fluorescent lamp and

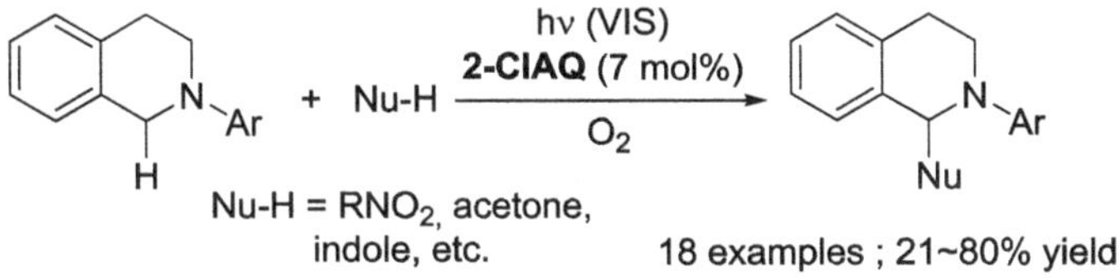

Scheme 2.24. Aerobic photooxidative C–C bond formation between tertiary amines and carbon nucleophiles.

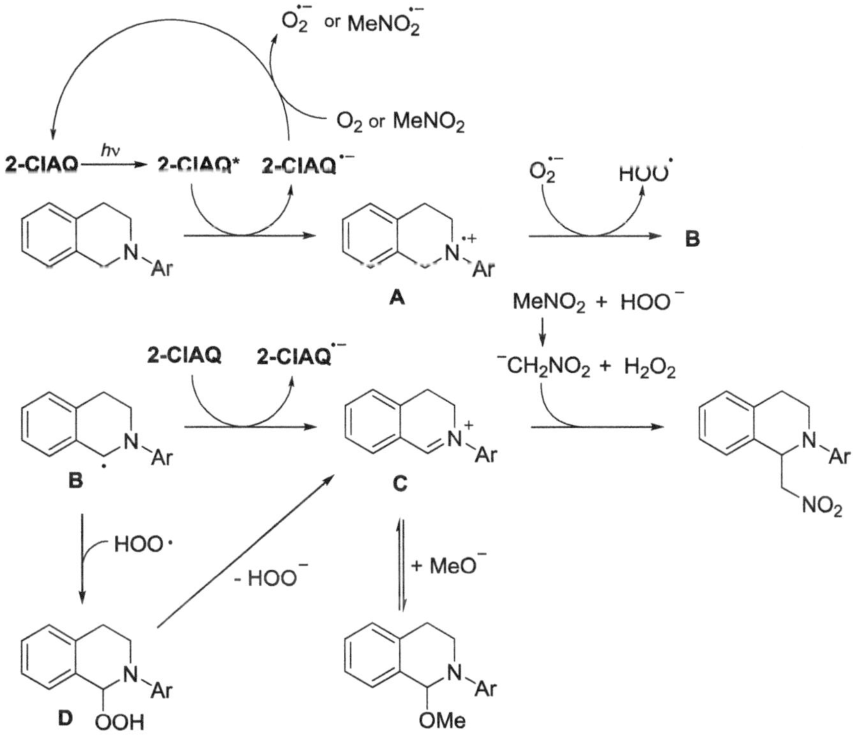

Scheme 2.25. Plausible path of aerobic photooxidative C–C bond formation.

molecular oxygen as the terminal oxidant. The authors also reported the same type of reaction using a combination of iodine and H_2O_2 without irradiation by light.

Scheme 2.25 shows a plausible mechanism for this reaction, which was postulated after considering all the results, including

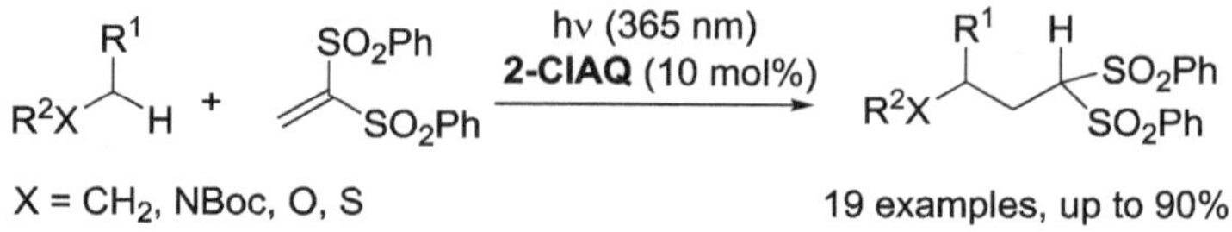

Scheme 2.26. Alkylation of non-acidic C(sp^3)–H bonds by photoinduced catalytic Michael-type radical addition.

those of control experiments, obtained by the authors. At first, excited **2-ClAQ** (**2-ClAQ***) oxidizes the tetrahydroisoquinoline to the radical cationic species **A**. Single-electron transfer (SET) from **B** to **2-ClAQ*** occurs after deprotonation of **A** to give the iminium compound **C**, otherwise **B** can form the hydroperoxyl radical adduct **D**, which is converted to **C** via elimination of the hydroperoxide anion. Iminium compound **C** reacts with $MeNO_2$ to afford the desired product. Electron transfer from **2-ClAQ**$^{\bullet-}$ to O_2 or $MeNO_2$ regenerates **2-ClAQ** and $O_2^{\bullet-}$ or $MeNO_2^{\bullet-}$. The authors detected the generation of the peroxide by iodometry.

Murafuji *et al.* reported the alkylation of non-acidic C(sp^3)–H bonds by a photoinduced catalytic Michael-type radical addition in the presence of a catalytic amount of **2-ClAQ** (Scheme 2.26) [32]. They carried out this reaction using **2-ClAQ** as a catalyst to cleave the C–H bond and 1,1-bis(phenylsulfonyl)ethylene as the olefinic substrate. They irradiated a solution in dichloromethane of the substrate (5 equiv), the olefin, and **2-ClAQ** (10 mol%) with a 365 nm LED lamp under an argon atmosphere. Notably, this method extends carbon chains at non-acidic C–H bonds and installs an active methine site for further transformations.

Scheme 2.27 shows the plausible mechanism proposed by the authors, in which this reaction starts with the abstraction of a hydrogen atom from the substrate by photoexcited **2-ClAQ**, which generates the carbon radical **A** and the semianthraquinone radical **B**. This step is suggested to be the rate-determining step from the observed kinetic isotope effect (KIE = 2.8). The conjugate addition of the radical **A** to 1,1-bis(phenylsulfonyl)ethylene, which is a highly electron-deficient olefin, takes place smoothly to afford the

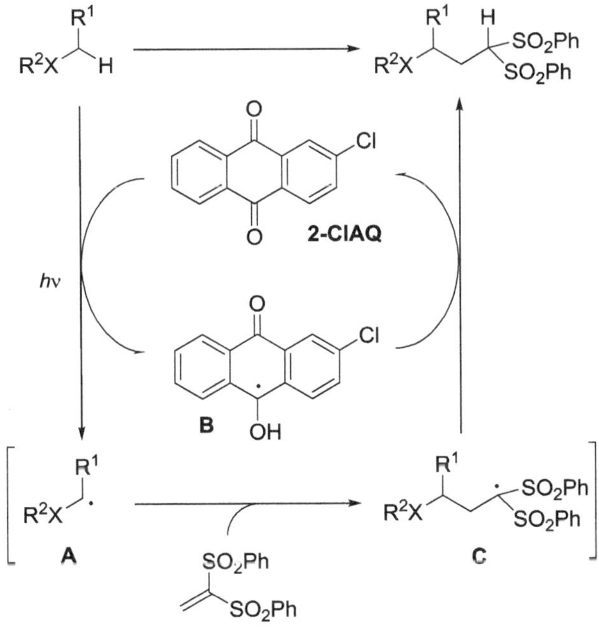

Scheme 2.27. Plausible path for the alkylation of non-acidic C(sp^3)–H bonds by photoinduced catalytic Michael-type radical addition.

radical **C**. Finally, a hydrogen atom should be donated by the semi-anthraquinone radical **B** to **C** to regenerate **2-ClAQ**.

2.2.2.2 *Anthraquinone-2-carboxylic acid* (*AQ-2-CO$_2$H*)

Itoh *et al.* reported the aerobic photooxidative synthesis of peroxybenzoic acids from toluenes using **AQ-2-CO$_2$H** as the **POC** (Scheme 2.28) [33]. They also developed a one-pot epoxidation reaction of alkenes using 4-*tert*-butylperoxybenzoic acid, produced *in situ* by aerobic photooxidative synthesis. The typical procedure was as follows: a solution in dry ethyl acetate and acetonitrile of 4-*tert*-butyltoluene and **AQ-2-CO$_2$H** (3.5 mol%) was placed in a Pyrex tube equipped with an O_2 balloon and then irradiated externally under stirring with a fluorescent lamp, which was placed at a distance of 0.5 cm. Then, an alkene was added to the reaction

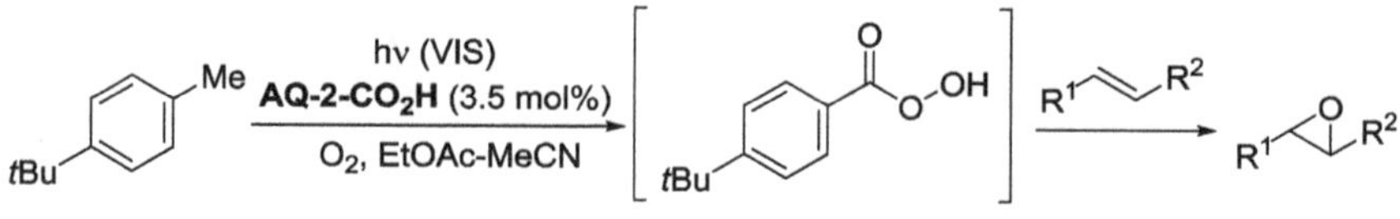

Scheme 2.28. One-pot epoxidation of alkenes using aerobic peroxidation of toluenes.

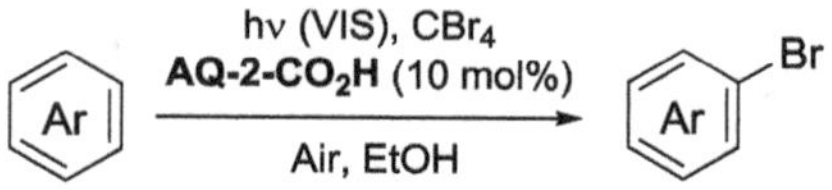

Scheme 2.29. Aerobic photooxidative bromination of aromatic compounds using carbon tetrabromide mediated by **AQ-2-CO$_2$H.**

mixture at room temperature. With their optimized reaction conditions, they obtained the desired epoxides (seven examples) in good to high yields (62%–99%).

Itoh *et al.* developed the aerobic photooxidative bromination of aromatic compounds using CBr_4 in the presence of **AQ-2-CO$_2$H** under visible-light irradiation (Scheme 2.29) [34]. The typical procedure was as follows: a solution of the aromatic compound, CBr_4 (25–100 mol%), and **AQ-2-CO$_2$H** (10 mol%) in dry ethanol was stirred under air atmosphere and irradiated externally with a 21 W fluorescent lamp. The corresponding aryl bromides were obtained in yields ranging from 14% to 92% (14 examples). In general, electron-donating groups, such as alkoxy or amino groups, accelerate this reaction; however, electron-withdrawing groups, such as nitro groups on aromatic rings, inhibit this reaction.

A plausible reaction mechanism was proposed by the authors, based on the necessity for molecular oxygen, continuous irradiation, and **AQ** in this reaction. They postulated that a bromine radical is initially generated *via* the homolysis of a C–Br bond; however, it is not clear what reactive species are involved in the step of carbon–bromine bond formation.

2.2.2.3 *Sodium anthraquinone-2-sulfonate monohydrate* (***AQ-2-SO$_3$Na***)

Itoh *et al.* reported an aerobic photooxidative direct asymmetric aldol reaction using water as the solvent (Scheme 2.30) [35]. In this reaction, primary benzyl alcohols are oxidized to benzaldehydes under an oxygen atmosphere using **AQ-2-SO$_3$Na** as a **POC**. A stereoselective aldol reaction then proceeds using a proline-type organic catalyst. The typical procedure was as follows: a mixture of a benzyl alcohol, **AQ-2-SO$_3$Na** (10 mol%), a ketone (5 equiv), an *L*-proline derivative (10 mol%) in H_2O, and acetic acid (50 mol%) was stirred under an air atmosphere upon irradiation with a 23 W fluorescent lamp. They reacted 21 substrates and obtained the corresponding products in yields ranging from 2% to 73% with very high enantiomeric excess (94%–99% ee). In general, electron-withdrawing groups, such as nitro, ester, or halogens, on the aromatic ring of the benzyl alcohol accelerate this reaction to afford the product in good yields; however, electron-donating groups on the aromatic ring, such as the methoxy group, retard this reaction.

Scheme 2.31 shows the plausible mechanism devised by the authors on the basis of the results obtained from their experiments. Excited **AQ-2-SO$_3^-$** (**A**), which is generated by irradiation with the fluorescent lamp, abstracts a hydrogen atom at the benzylic position of the alcohol to produce the benzyl radical **B**, which in turn traps molecular oxygen to generate the aldehyde via the peroxyl radical

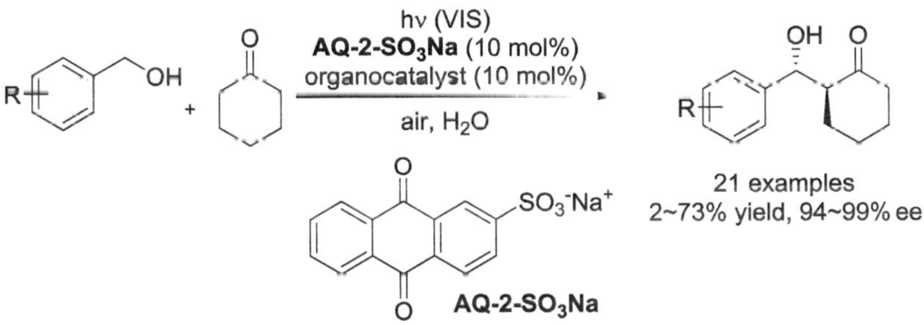

Scheme 2.30. Aerobic photooxidative direct asymmetric aldol reaction.

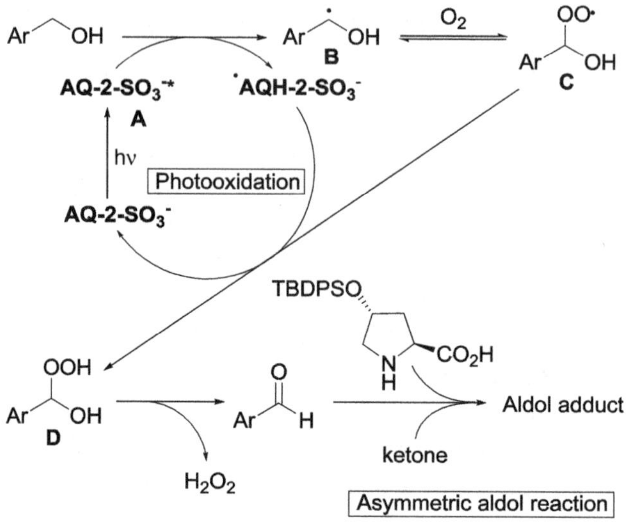

Scheme 2.31. Plausible path of aerobic photooxidative direct asymmetric aldol reaction.

C and the hydroperoxide **D**. Finally, the aldehyde reacts with the ketone to produce the desired aldol adduct in the presence of the proline-type organic catalyst.

2.2.2.4 *2-tert-Butylanthraquinone* (*2-^{t}Bu-AQ*)

Itoh *et al.* reported a novel cross-dehydrogenative amination reaction of C(sp^2)–H bonds in indoles catalyzed by a **POC** (Scheme 2.32) [36]. The reaction is mediated by **2-^{t}Bu-AQ** as the photocatalyst, occurs under harmless visible light, and uses atmospheric oxygen as the sole oxidant, with no transition metal catalyst involved. Thus, a mixture of the aromatic substrate, phthalimide (2 equiv, in the role of nitrogen-source), **2-^{t}Bu-AQ** (10 mol%), K_2CO_3 (60 mol%), and molecular sieves (MS, 4 Å) in dimethylformamide was stirred under air atmosphere and irradiated with a 21 W fluorescent lamp. They reacted 24 substrates and obtained the corresponding products in up to 93% yield.

Scheme 2.33 shows a plausible mechanism for such a reaction, which was postulated after considering all the results of the

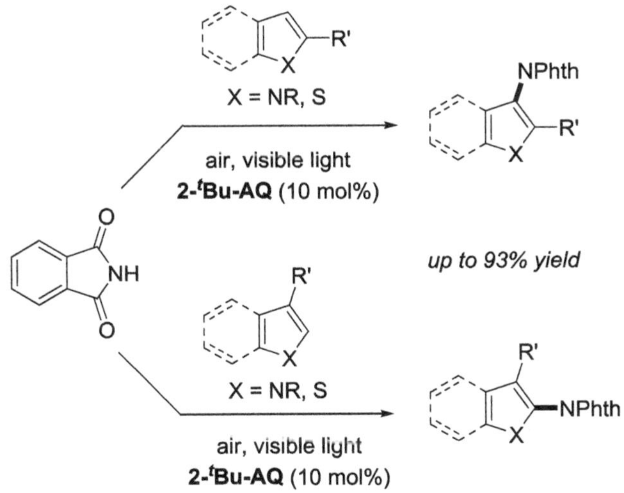

Scheme 2.32. Cross-dehydrogenative C–H amination of indoles under aerobic photooxidative conditions.

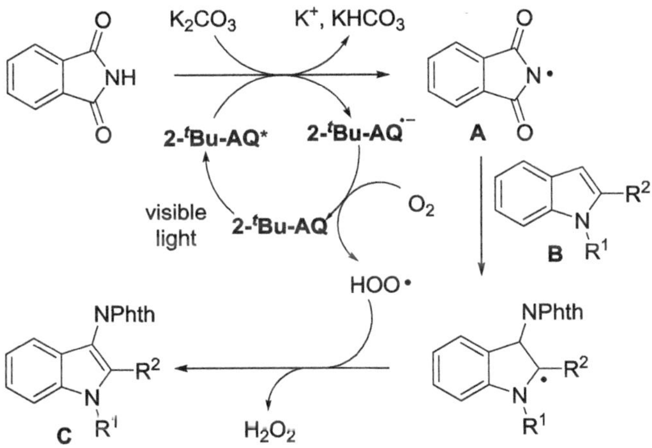

Scheme 2.33. Plausible mechanism for the cross-dehydrogenative C–H amination under aerobic photooxidative conditions.

experiments carried out by the authors. At first, the *N*-centered radical **A** is formed upon deprotonation of phthalimide and SET to **2-^{t}Bu-AQ***, which is excited by irradiation with visible light. The radical species **A** adds to the indole **B** and the final product **C** is then obtained *via* hydrogen-atom abstraction leading to rearomatization of the heterocyclic core. On the other hand, **2-^{t}Bu-AQ$^{•-}$** is reoxidized to the original form by atmospheric molecular oxygen to complete the catalytic cycle; however, the authors did not rule out another pathway, *viz.* the quenching of **2-^{t}Bu-AQ*** by the substrate.

2.2.3 *5,7,12,14-Pentacenetetrone* (*PT*)

Murafuji *et al.* also reported a conjugate radical addition reaction (Scheme 2.34) [37]. They achieved the catalytic introduction of an allyl group at non-acidic C(sp^3)–H bonds under photoorganocatalytic conditions, in which 1,2-bis(phenylsulfonyl)-2-propene acted as a source of allyl groups and 5,7,12,14-pentacenetetrone (**PT**) acted as a catalyst to cleave C–H bonds.

The typical procedure was as follows: a solution in methylene chloride of an alkane (10 equiv), 1,2-bis(phenylsulfonyl)-2-propene, **PT** (5 mol%), and K_2CO_3 (110 mol%) in a Pyrex test tube was purged with argon. The test tube was placed at a distance of approximately 2 cm from an LED lamp (365 or 425 nm) and was irradiated at room temperature. They reacted 19 substrates, including alkanes, carbamates, ethers, sulfides, and alcohols, and obtained the corresponding products, which were chemoselectively allylated, in yields of up to 81% under neutral conditions. Scheme 2.35 shows the plausible mechanism suggested by the authors. The reaction starts with the

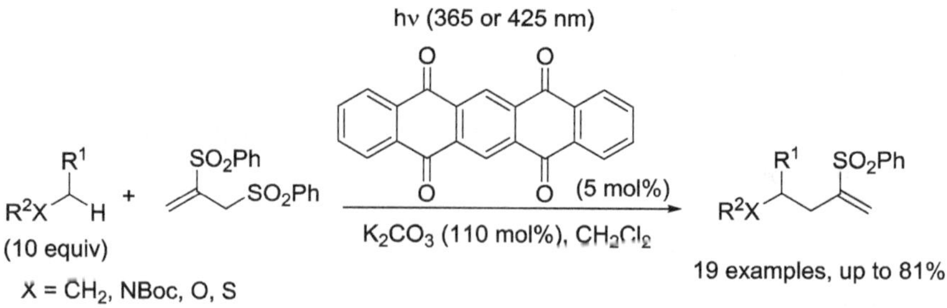

Scheme 2.34. Pentacenetetrone-photoorganocatalyzed radical allylation of C(sp^3–H) bonds.

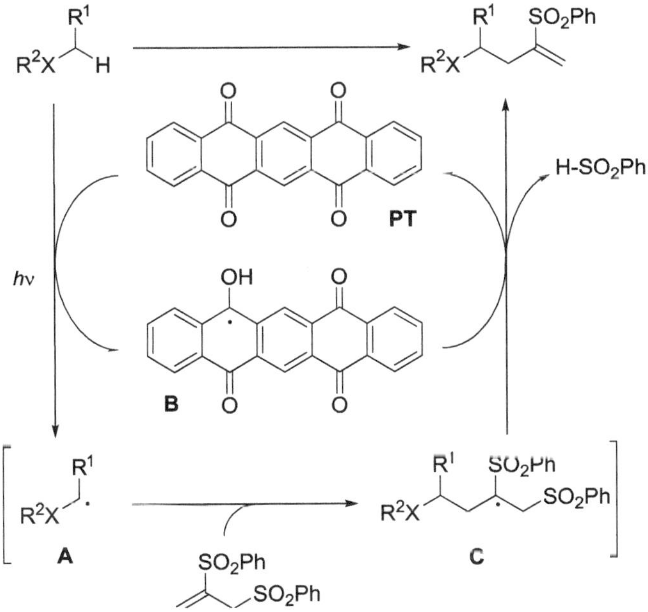

Scheme 2.35. Plausible path of **PT**-photoorganocatalyzed radical allylation of C(sp^3–H) bonds.

abstraction of a hydrogen atom from the substrate by photoexcited **PT**, which generates the carbon radical **A**. Indeed, during the course of the reaction, the formation of the TEMPO adduct of the starting alkane could be detected by radical trapping experiments. This hydrogen abstraction step (C(sp^3)–H bond cleavage) is suggested to be the rate-determining step from the observed kinetic isotope effect (KIE = 3.6). The radical **A** then adds to the electron-deficient vinyl sulfone moiety of the allyl source to give the radical **C**, which is finally transformed into the allylated product via the elimination of a sulfonyl radical. This sulfonyl radical probably accepts a hydrogen atom from **B** to regenerate **PT**, thus completing the catalytic cycle.

2.3 Conclusions

As mentioned above, in recent years, attention has been drawn to the oxidizing power of excited quinones. When in the excited state, they become one of the most powerful classes of oxidants, due to their largely positive reduction potentials. Furthermore, albeit in the past, quinones were typically used as stoichiometric oxidants; recently, a plethora of catalytic applications was reported, where other terminal oxidants were used, including oxygen, nitrogen dioxide, etc. In particular, the reactivity of quinones can be tuned by introducing a variety of substituents in the molecule. Therefore, quinones are a highly promising class of compounds that makes it possible to create an ideal photoorganocatalyzed reaction. Indeed, the introduction of new types of quinones, which will allow to achieve high levels of efficiency and selectivity, is expected in the near future.

Abbreviations

AQ	anthraquinone
ET	electron transfer
DFT	density functional theory
DDQ	2,3-dichloro-5,6-dicyano-1,4-benzoquinone
ETM	electron-transfer mediator
HAT	hydrogen-atom transfer
BQ	1,4-benzoquinone
BQH_2	1,4-hydroquinone
PMB	*p*-methoxybenzyl
MW	microwaves
PET	photoinduced electron transfer
SET	single-electron transfer
TEMPO	2,2,6,6-tetramethylpiperidine-*N*-oxyl

References

[1] Hubig, S. M.; Kochi, J. K. *J. Am. Chem. Soc.* **1999**, *121*, 1688–1694.

[2] Hamanoue, K.; Kajiwara, Y.; Miyake, T.; Nakayama, T.; Hirase, S.; Teranishi, H. *Chem. Phys. Lett.* **1983**, *94*, 276–280.

[3] Serpa, C.; Arnaut, L. G. *J. Phys. Chem. A* **2000**, *104*, 11075–11086.

[4] Gollnick, K.; Held, S.; Martire, D. O.; Braslavsky, S. E. *J. Photochem. Photobiol. A: Chem.* **1992**, *69*, 155–165.

[5] Dahiya, P.; Kumbhakar, M.; Maity, D. K.; Mukherjee, T.; Mittal, J. P.; Tripathi, A. B. R.; Chattopadhyay, N.; Pal, H. *Photochem. Photobiol. Sci.* **2005**, *4*, 100–105.

[6] Dahiya, P.; Kumbhakar, M.; Maity, D. K.; Mukherjee, T.; Tripathi, A. B. R.; Chattopadhyay, N; Pal, H. *J. Photochem. Photobiol. A: Chem.* **2006**, *81*, 338–347.

[7] Dahiya, P.; Kumbhakar, M.; Mukherjee, T.; Pal, H. *J. Mol. Struct.* **2006**, *798*, 40–48.

[8] Lerch, S.; Unkel, L.-N.; Wienefeld, P.; Brasholz, M. *Synlett* **2014**, *25*, 2673–2680.

[9] Rusch, F.; Unkel, L.-N.; Alpers, D.; Hoffmann, F.; Brasholz, M. *Chem. Eur. J.* **2015**, *21*, 8336–8340.

[10] Romero, N. A.; Nicewicz, D. A. *Chem. Rev.* **2016**, *116*, 10075–10166.

[11] Sasaki, K.; Kashimura, T.; Ohura, M.; Ohsaki, Y.; Ohta, N. *J. Electrochem. Soc.* **1990**, *137*, 2437–2443.

[12] Mitchell, L. J.; Moody, C. *J. Org. Chem.* **2014**, *79*, 11091–11100.

[13] Nishinaga, A.; Rindo, K.; Matsuura, T. *Synthesis* **1986**, *6*, 1038–1041.

[14] Gu, X.; Zhang, W.; Salomon, R. G. *J. Org. Chem.* **2012**, *77*, 1554–1559.

[15] Shen, Z.; Dai, J.; Xiong, J.; He, X.; Mo, W.; Hu, B.; Sun, N.; Hu, X. *Adv. Synth. Catal.* **2011**, *353*, 3031–3038.

[16] Ohkubo, K.; Fujimoto, A.; Fukuzumi, S. *J. Am. Chem. Soc.* **2013**, *135*, 5368–5371.

[17] Song, C.; Yi, H.; Dou, B.; Li, Y.; Singh, A. K.; Lei, A. *Chem. Commun.* **2017**, *53*, 3689–3692.

[18] Ohkubo, K.; Fujimoto, A.; Fukuzumi, S. *Chem. Asian. J.* **2016**, *11*, 996–999.

[19] Rush, F.; Schober, J.-C.; Brasholz, M. *ChemCatChem* **2016**, *8*, 2881–2884.

[20] Ohkubo, K.; Hirose, K.; Fukuzumi, S. *Chem. Asian. J.* **2016**, *11*, 2255–2259.

[21] Memarian, H. R.; Saffai-Teluri, A. *J. Mol. Catal. A: Chem.* **2007**, *274*, 224–230.

[22] Ohkubo, K.; Hirose, K.; Fukuzumi, S. *Photochem. Photobiol. Sci.* **2016**, *15*, 731–734.

[23] Kee, C. W.; Chin, K. K.; Wong, M. W.; Tan, C.-H. *Chem. Commun.* **2014**, *50*, 8211–8214.

[24] Kee, C. W.; Shao, H.; Kee, C. W.; Lu, Y.; Soo, H. S.; Tan, C.-H. *Catal. Sci. Technol.* **2017**, *7*, 848–857.

[25] Minisci, F. *Synthesis* **1973**, 1–24.

[26] Lu, Z.; Yoon, T. P. *Angew. Chem. Int. Ed.* **2012**, *51*, 10329–10332.

[27] Zou, Y.-Q.; Duan, S.-W.; Meng, X.-G.; Hu, X.-Q.; Gao, S.; Chen, J.-R.; Xiao, W.-J. *Tetrahedron* **2012**, *68*, 6914–6919.

[28] Farney, E. P.; Yoon, T. P. *Angew. Chem. Int. Ed.* **2014**, *53*, 793–797.

[29] Tachikawa, Y.; Cui, L.; Matsusaki, Y.; Tada, N.; Miura, T.; Itoh, A. *Tetrahedron Lett.* **2013**, *54*, 6218–6221.

[30] Nobuta, T.; Tada, N.; Fujiya, A.; Kariya, A.; Miura, T.; Itoh, A. *Org. Lett.* **2013**, *15*, 574–577.

[31] Yamaguchi, T.; Nobuta, T.; Tada, N.; Miura, T.; Nakayama, T.; Uno, B.; Itoh, A. *Synlett* **2014**, *25*, 1453–1457.

[32] Kamijo, S.; Takao, G.; Kamijo, K.; Tsuno, T.; Ishiguro, K.; Murafuji, T. *Org. Lett.* **2016**, *18*, 4912–4915.

[33] Taguchi, M.; Nagasawa, Y.; Yamaguchi, E.; Tada, N.; Miura, T.; Itoh, A. *Tetrahedron Lett.* **2016**, *57*, 230–232.

[34] Tanaka, M.; Kamito, Y.; Lei, C.; Tada, N.; Itoh, A. *Tetrahedron Lett.* **2015**, *56*, 5886–5888.

[35] Fujiya, A.; Nobuta, T.; Yamaguchi, E.; Tada, N.; Miura, T.; Itoh, A. *RSC Adv.* **2015**, *5*, 39539–39543.

[36] Yamaguchi, T.; Yamaguchi, E.; Itoh, A. *Org. Lett.* **2017**, *19*, 1282–1285.

[37] Kamijo, S.; Kamijo, K.; Maruoka, K.; Murafuji, T. *Org. Lett.* **2016**, *18*, 6516–6519.

Chapter 3

Aromatics and Cyanoaromatics

Stefano Protti* and Davide Ravelli†

PhotoGreen Lab, Department of Chemistry,
University of Pavia, viale Taramelli 12, 27100 Pavia, Italy

**stefano.protti@unipv.it*
†davide.ravelli@unipv.it

3.1 Introduction

The rich redox chemistry accessible upon irradiation of simple aromatics and their electron-rich or electron-poor congeners has been known since many years [1–4]. After a period of underutilization, simple aromatic derivatives are now under the spotlight, thanks to the impressive interest in the area triggered by the development of photoredox catalysis [5]. Indeed, when in the excited state, such compounds are able to promote single-electron oxidations or reductions depending on the substitution pattern of the aromatic ring. The first pathway is typical of electron-poor derivatives, such as cyanoaromatics and aromatic esters, while the second involves compounds bearing strong electron-donating groups, including aromatic amines and ethers. Finally, non-substituted (polycyclic) aromatics can undergo electron transfer steps in both directions. In addition, such photoorganocatalysts (**POCs**) have been sparsely adopted to promote hydrogen-atom transfer (HAT) steps [6,7] as well as energy transfer (ET) processes [8].

This chapter is split into two main sections, focusing on simple aromatic derivatives on one hand, including electron-rich ones, and on (hetero)cyanoaromatics and a few aromatic esters on the other hand, one of the most investigated families of **POCs** in this area. The examined reactions have been classified on the basis of the nature of the chemical bond formed.

3.2 Aromatics

3.2.1 *Photophysics of aromatics and their derivatives*

The excitation of an aromatic derivative typically involves the π system of the chromophore, and the properties of the resulting excited state heavily depend on the size of the system and on its substitution pattern (only unsubstituted and electron-rich derivatives will be considered in the present Section; for cyano-aromatics, see Sec. 3.3).

Emission from the singlet state, as well as intersystem crossing (ISC) followed by emission from the triplet state, plays a significant role in the photophysics of (electron-rich) aromatics, fluorescence being, in most cases, the main event observed (Φ_F up to 0.87 in the case of 9,10-dimethoxyanthracene — **DMA**). A remarkable exception is phenanthrene (**Phen**), where ISC to triplet (Φ_{ISC} = 0.83) competes efficiently with fluorescence (Φ_F = 0.16) [9].

The largest part of the reactions involving aromatics as **POCs** reported in recent years deals with electron-transfer processes. Accordingly, the excited compound can act as either a single-electron oxidant or reductant, depending on the considered photocatalyst and the reaction conditions. Selected redox potentials, both in the ground and excited state, are gathered in Chart 3.1. As an example, **Phen** can undergo either reductive or oxidative quenching leading to the corresponding radical anion or cation, respectively [10]. In another instance, *p*-terphenyl has been converted through the

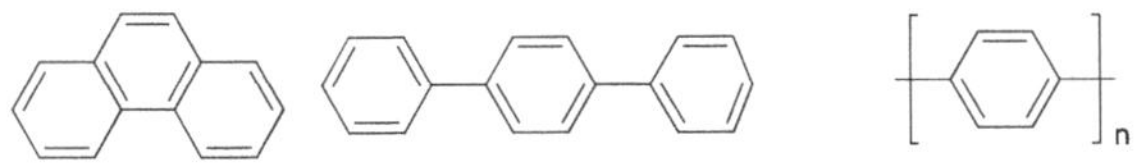

Name	Phenanthrene (**Phen**)[a]	p-Terphenyl[b]	poly-*p*-phenylene (**PPP**)[c]
Absoprtion Edge	300 nm	283 nm (λ_{MAX})	460 nm
E_{OX} (**POC$^{\bullet+}$/POC**)	+ 1.50		
E^*_{OX} (**POC$^{\bullet+}$/POC***)	- 2.1		- 1.7
E_{RED} (**POC/POC$^{\bullet-}$**)	- 2.44	- 2.63	
E^*_{RED} (**POC*/POC$^{\bullet-}$**)	+ 1.14		

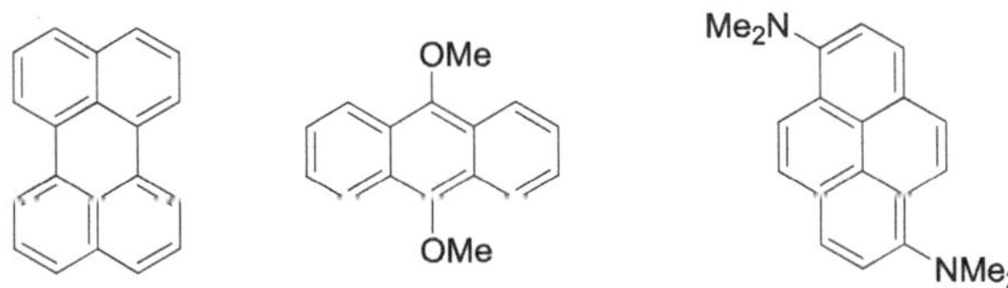

Name	Perylene[d]	9,10-dimethoxyanthracene (**DMA**)[e]	1,6-bis-*N,N*-(dimethylamino)pyrene (**BDMAP**)[e]
Absoprtion Edge	450 nm	430 nm	450 nm
E_{OX} (**POC$^{\bullet+}$/POC**)	+ 0.99	+ 0.98	+ 0.43
E^*_{OX} (**POC$^{\bullet+}$/POC***)	- 1.74	- 2.0	- 2.4
E_{RED} (**POC/POC$^{\bullet-}$**)			
E^*_{RED} (**POC*/POC$^{\bullet-}$**)			

Chart 3.1. Photophysical and electrochemical parameters for selected (electron-rich) aromatics considered in the present section. All the potentials are expressed in V vs SCE (**POC** = Photoorganocatalyst).

Notes: [a] See Ref. [10]. [b] See Ref. [11]. [c] See Ref. [14]. [d] See Ref. [15]. [e] See Ref. [13].

reductive quenching of its excited state to the strongly reducing radical anion, showing a reduction potential E_{RED}(**POC/POC$^{\bullet-}$**) of −2.63 V vs SCE [11–12]. When the aromatic bears electron-donating groups, an oxidative quenching of the excited state typically occurs, with oxidation potentials in the excited state E^*_{OX}(**POC$^{\bullet+}$/POC***) ranging from −2.0 to −2.4 V vs SCE [13].

Furthermore, aromatic derivatives bearing acidic hydrogens (e.g. phenols and related derivatives) can undergo an enhancement of their acidity upon excitation. This feature has been exploited to perform acid-catalyzed reactions [16]. Finally, carbon nanomaterials will be treated as well in this section. Upon excitation, these catalysts can promote different reactivities, including ET or HAT. As an example, C_{70} has been exploited to generate singlet oxygen *via* an ET mechanism involving $^3C_{70}$ and O_2 [17]. Furthermore, the excitation of graphite flakes has been reported to promote the activation of the formyl C–H bond in aldehydes through a direct HAT step [18].

3.2.2 *Carbon–carbon bond formation*

Formation of a C_{sp3}–C_{sp3} bond. Upon monoelectronic reduction, selenides are known to release a radical intermediate *via* mesolytic Se–*X* (*X* = C, Si) cleavage at the radical anion stage [19]. This strategy has been implemented through a photocatalytic approach by exploiting **DMA** (ca. 30 mol%) as the photoexcited reductant, and it has been applied to the alkylation of tertiary amines (Scheme 3.1) [20].

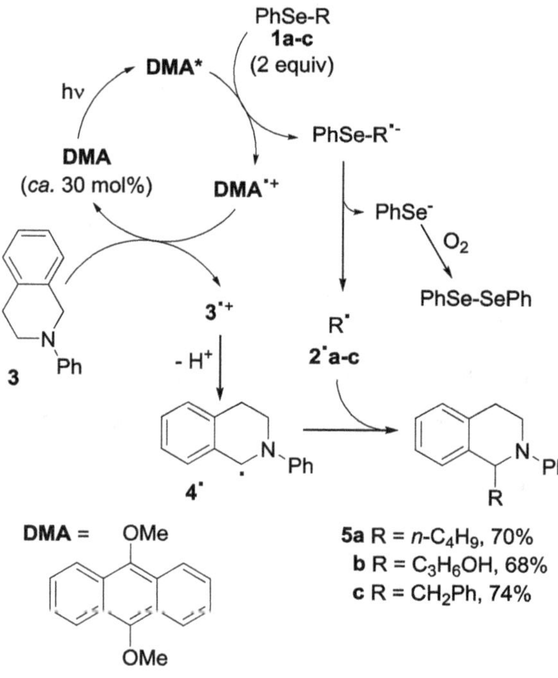

Scheme 3.1. Photocatalytic alkylation of tertiary amines with selenides.

Thus, the excitation of **DMA** allowed for the reduction of alkylphenylselenides **1a-c** and promoted the generation of the corresponding alkyl radicals **2•a-c**, while the phenylselenide anion was converted to dimeric diphenyldiselenide *via* oxidation by O_2. The so-generated **DMA•+** species in turn oxidized monoelectronically *N*-phenyl tetrahydroisoquinoline **3** to **3•+**, leading to α-amino radical **4•** upon deprotonation. The final products **5a-c** resulted from a radical–radical cross-coupling between the two photogenerated intermediates due to the persistent radical effect. Interestingly, an intramolecular version of this protocol was developed as well and applied to the synthesis of Crispine A, a naturally occurring alkaloid possessing the hexahydropyrroloisoquinoline structural skeleton [20].

The possibility of exploiting **Phen** to promote the generation of radical intermediates upon oxidative decarboxylation of carboxylic acids has been widely investigated [21]. A dual catalytic system composed by **Phen** and a cyanoaromatic (both 5 mol%) was exploited for the addition of the photogenerated radical to alkenes in a Michael-type process [22]. As depicted in Scheme 3.2, excitation of **Phen** in the presence of 1,4-dicyanonaphthalene (**DCN**) led to the formation of the corresponding radical ions upon electron transfer. **Phen•+** oxidized monoelectronically the carboxylate anion **6⁻** present in solution, leading to radical **7•** upon decarboxylation. This in turn added to acrylonitrile leading to radical adduct **8•** and then to product **9** by reaction with **DCN•⁻** and protonation. Interestingly, the reaction worked also in the absence of an added base, albeit a longer irradiation time was required. Furthermore, it was demonstrated that chrysene in the presence of 1,4-dicyanobenzene (**DCB**) offered a similar performance in the synthesis of **9** [22].

The oxidative quenching of perylene in the excited state has been recently exploited in the amino-difluoromethylation of aromatic olefins (e.g. **10**) by sulfonium salt **11⁺** (Scheme 3.3). The process worked under irradiation with visible light (425 nm blue LEDs were used) and involved the reduction of **11⁺** to generate the difluoromethyl radical **12•**. This was in turn trapped by **10** to afford benzyl radical **13•**, and then cation **13⁺** upon electron transfer to perylene

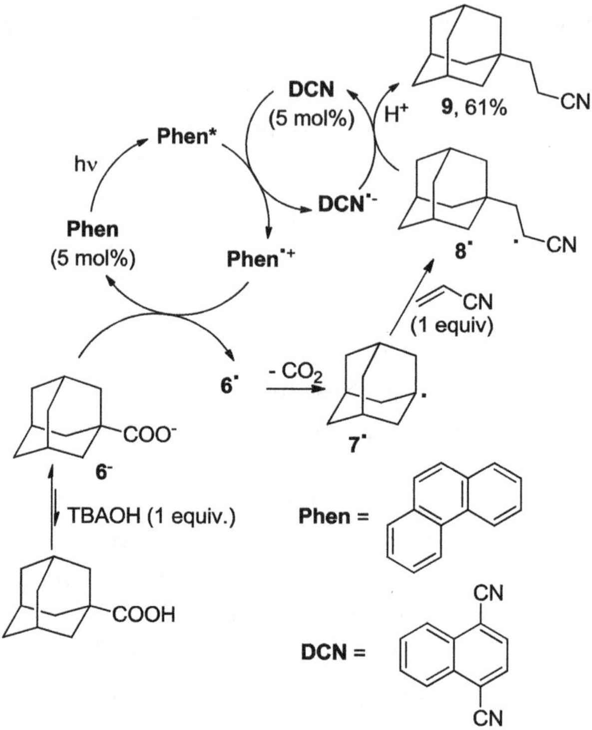

Scheme 3.2. Photocatalytic generation of radicals *via* oxidative decarboxylation of carboxylic acids.

radical cation. In the final step, $\mathbf{13}^{+}$ underwent a Ritter amination with the solvent acetonitrile and afforded the final product **14** in 64% yield on a 1.1 g scale (Scheme 3.3). Interestingly, also 9,10-dimethylanthracene worked to some extent as **POC**, while anthracene and pyrene were not competent in this process, probably due to a lack of absorption at the irradiation wavelength. Furthermore, perylene was successfully applied to promote trifluoromethylation processes as well [15].

The same **POC** was adopted for the reductive coupling of aromatic carbonyls (aldehydes and ketones) [23] and aromatic imines [23, 24] to the corresponding glycols and 1,2-diamines, respectively. The process required the presence of an excess of a tertiary amine in the role of electron donor (e.g. DIPEA). Scheme 3.4 gathers the

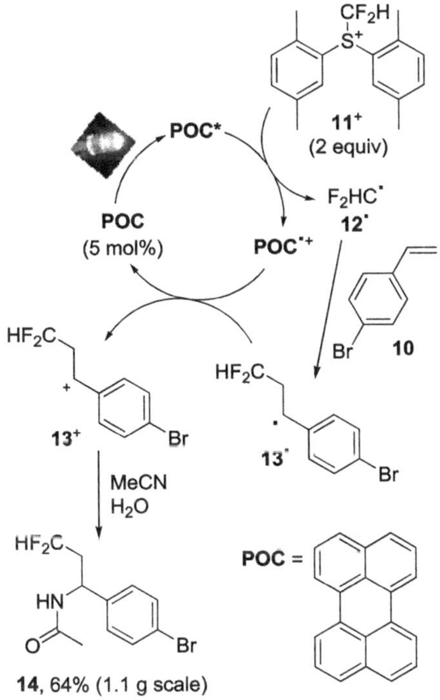

Scheme 3.3. Photocatalytic amino-difluoromethylation of aromatic olefins.

proposed mechanism, involving a different pathway with respect to that reported above in Scheme 3.3. Thus, upon excitation, perylene (12 mol% [23]; in the case of imines, the catalyst loading could be lowered to 1 mol% [24]) promoted the single-electron oxidation of DIPEA to the corresponding radical cation. The photogenerated perylene radical anion was then restored through a second electron transfer step involving the reduction of carbonyl derivatives **15a-d** to **15$^{\bullet-}$a-d**. The latter then underwent a radical–radical coupling step followed by protonation to yield the desired glycols **16a-d** as *dl/meso* mixtures. Furthermore, it has been proposed that the amine radical cation functions as the proton source [23, 24].

The enantioselective α-alkylation of aldehydes was achieved by a dual organocatalytic/photoredox catalytic approach. Anthracene and naphthalene were successfully tested as photocatalysts, albeit the best results have been obtained with Rose Bengal [25].

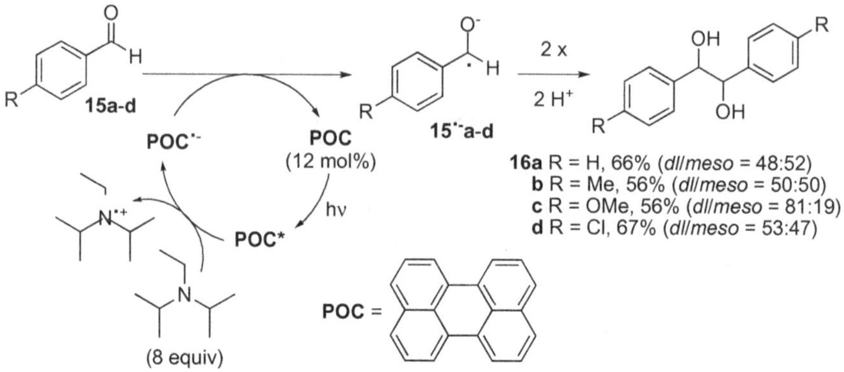

Scheme 3.4. Photoorganocatalyzed reductive coupling of aromatic aldehydes.

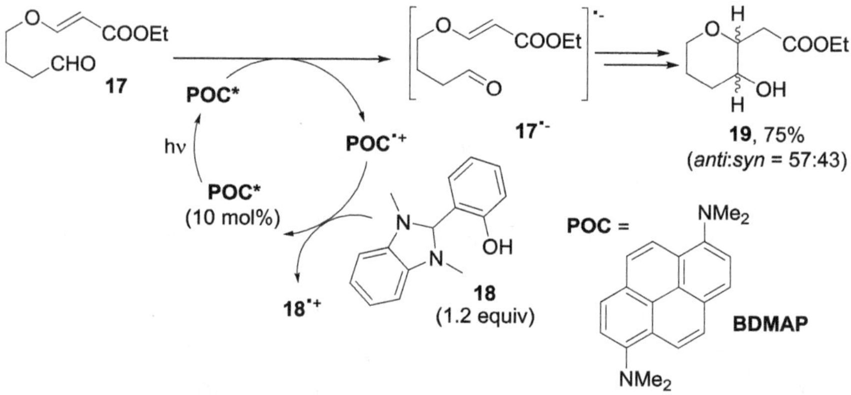

Scheme 3.5. Photoorganocatalyzed reductive cyclization of alkenylketones.

The reductive coupling of carbonyls can be carried out as well by exploiting different systems, including the adoption of: (i) poly-(*p*-phenylene) as the **POC** in the presence of a tertiary amine and a thiol [26] or (ii) electron-rich anthracenes and pyrenes in the presence of a benzimidazoline derivative [13, 27]. As depicted in Scheme 3.5, the latter system could be applied to the reductive cyclization of alkenylketones (e.g. **17**) as well. Thus, the excitation of 1,6-*N*,*N*-bis(dimethylamino)pyrene **BDMAP** promoted the monoelectronic reduction of **17** to **17**$^{\bullet-}$. The photogenerated **BDMAP**$^{\bullet+}$

species was then restored through a second electron transfer step involving benzimidazoline **18**. The final product **19** was obtained in 75% yield as an *anti*:*syn* = 57:43 mixture [28].

Formation of a C_{sp3}–C_{sp2} bond. The photocatalytic approach mediated by aromatic hydrocarbons, in particular **Phen**, has been exploited to promote the coupling of a photogenerated radical intermediate with aromatic nitriles such as **21** (Scheme 3.6). One of the first examples in the area dealt with the decarboxylative photosubstitution of isomeric dicyanobenzenes with aliphatic carboxylate ions. Indeed, it was demonstrated that the reaction also proceeded in the absence of **Phen**, albeit its presence (150 mol%) markedly improved the product yield [29]. More recently, this protocol was optimized, and the scope, in terms of both carboxylic acids (oxalates could be employed as well to form a new C_{sp2}–C_{sp2} bond) and aromatic nitriles, was extended [30, 31]. Despite the catalyst loading was reduced with respect to the original report, it was still employed in an almost stoichiometric amount (75 mol%; 125 mol% in the case of oxalates), as depicted in Scheme 3.6 for the synthesis of compounds **22a,b** from aminoacids **20a,b** [30].

However, it was demonstrated that 90%–96% of the initially employed **Phen** could be recovered and that the reaction worked to some extent also in the presence of 5 mol% only of the **POC** [30].

A variation of this protocol involved the use of tertiary amines in the role of radical precursors and (hetero)aromatic nitriles, resulting in an overall C–C σ-bond metathesis process (Scheme 3.7) [10]. Despite the process works also in the absence of **Phen**, its presence (25 mol%) had beneficial effects in terms of product

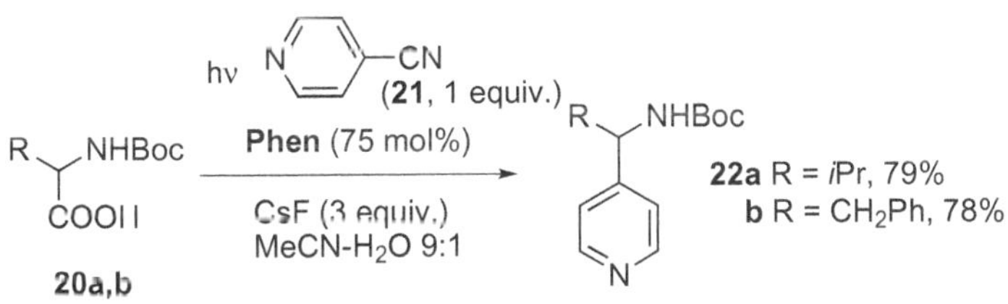

Scheme 3.6. Decarboxylative coupling of carboxylic acids with aromatic nitriles.

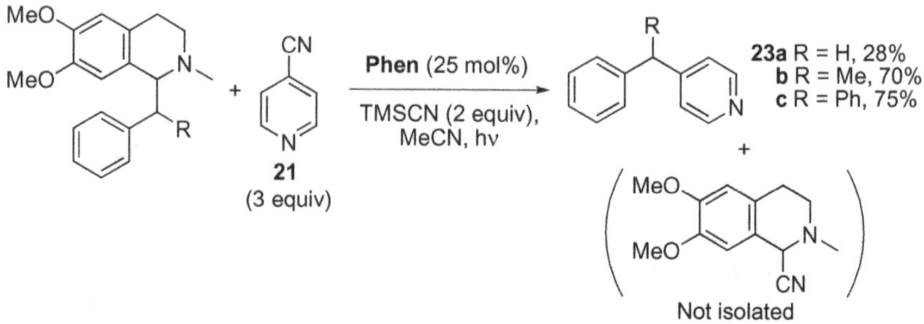

Scheme 3.7. Photoorganocatalyzed C–C σ-bond metathesis process.

yield. A detailed mechanistic investigation revealed that the excited state of **Phen** is quenched both by the tertiary amine and the aromatic nitrile. Accordingly, it was proposed that the process occurred through either an oxidative or a reductive quenching cycle. As a result, the starting nitrile (e.g. **21**) was alkylated through an ipso-substitution of the cyano group to give products **23a-c** in variable yields (Scheme 3.7) [10].

The fixation of CO_2 is an important topic in several fields, particularly in the area of energy storage and depollution. Despite early reports on the use of oligo(*p*-phenylenes) as photoredox catalysts for carbon dioxide monoelectronic reduction dating back to the 1990s [32], only recently this strategy has found application in synthesis. Thus, α-aminoacids (±)**26a-d** were synthesized from tertiary amines **24a-d** and CO_2 (0.34 MPa, corresponding to 3.6 equiv) in DMF in the presence of *p*-terphenyl (20 mol%) as photoredox catalyst and 3 equiv of CF_3COOK. The excited **POC** was reductively quenched by the tertiary amine substrate to give the strongly reducing *p*-terphenyl radical anion, while CO_2 was involved in its restoration *via* a second single-electron-transfer event. The generated amine radical cations **24$^{\bullet+}$a-d** then underwent deprotonation to give α-aminoradicals **25$^{\bullet}$a-d**, in turn prone to undergo a radical–radical anion coupling with $CO_2^{\bullet-}$, finally leading to the desired products **26a-d** upon protonation (Scheme 3.8). Tertiary benzylic amines were elective substrates, albeit the reaction worked as well with purely aliphatic derivatives.

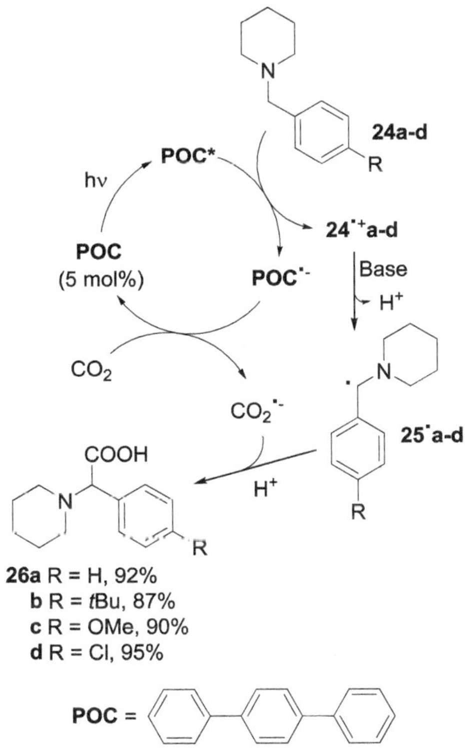

Scheme 3.8. Synthesis of aminoacids *via* CO_2 fixation photocatalyzed by *p*-terphenyl.

Interestingly, the process proceeded upon irradiation with UV light ($\lambda > 280$ nm) for 10 min only in a flow reactor and could be adapted to the carboxylation of biologically relevant molecules, including the adenosine diphosphate (ADP) receptor inhibitor ticlopidine [11]. More recently, the same strategy was exploited for the hydrocarboxylation of styrenes under atmospheric CO_2 pressure to give the corresponding hydrocynnamic acids as products [12].

3.2.3 *Carbon–heteroatom bond formation*

Carbon–Oxygen bond formation. The formation of a C–O bond through the nucleophilic addition of methanol to styrene derivatives in a Markovnikov orientation has been reported to occur in the presence of a variety of aromatic **POCs**, and 1-(*N,N*-dimethylamino)pyrene was demonstrated to afford the best yield in the series, albeit all these derivatives were employed in a stoichiometric amount with respect to the olefinic substrates [33, 34]. More recently, the pyrene derivative was conjugated to a short peptide chain. In the case of arginine-containing peptides, it was possible to avoid the addition of triethylamine, otherwise required in the role of electron shuttle [35].

Carbon–Nitrogen bond formation. A single example in the literature describes the formation of a C–N bond photoorganocatalyzed by an aromatic derivative. This involves the hydroacylation of a dialkyl azodicarboxylate by an aldehyde promoted by a carbon material, such as single and multiwalled carbon nanotubes, fullerenes, graphene oxide, and graphite flakes [18]. As depicted in Scheme 3.9, graphite flakes (10 mg) gave a quantitative yield of product **27** resulting from the addition of heptaldehyde onto diisopropyl azodicarboxylate. The mechanism was proposed to involve the generation of an acyl radical *via* a HAT step promoted by the carbocatalyst, followed by C–N bond formation through addition of the photogenerated radical onto the N=N double bond. The final product **27** was obtained through H-abstraction from another molecule of aldehyde in a radical chain process. Importantly, analysis of

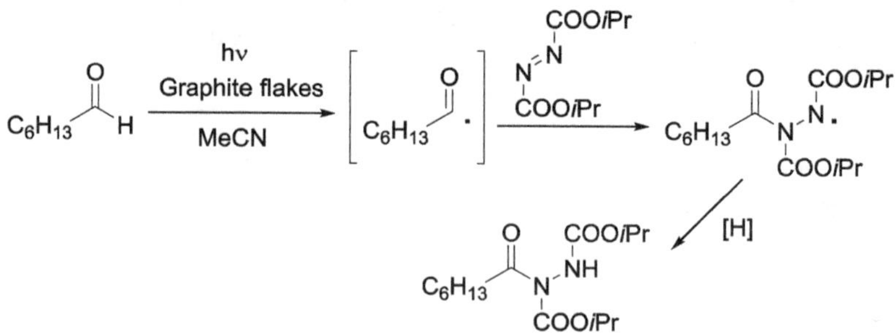

27, > 99%

Scheme 3.9. Graphite flakes-photocatalyzed C–N bond formation.

the catalyst after the reaction revealed a chemical modification of the graphite flakes, explaining the poor performance of the material upon recycling [18].

3.2.4 *Oxidation processes*

Carbon nanomaterials are interesting **POCs** for the realization of oxidation reactions *via* the generation of reactive oxygen species (ROS), such as singlet oxygen ($^1O_2$) or the superoxide anion ($O_2^{\bullet-}$). A recent example deals with the oxidation of secondary benzyl amines to the corresponding imines promoted by C_{70} fullerene. This material proved to be superior to traditional C_{60} and to other commonly employed photosensitizers, such as tetraphenylporphyrin (TPP). Indeed, C_{70} provided fast reaction rates and could operate at very low catalyst loadings (down to 0.01 mol%; however, an increased reaction time had to be used).

As depicted in Scheme 3.10, excitation of C_{70} with visible light followed by ISC and ET to molecular oxygen generated singlet oxygen, prone to trigger the desired amine to imine conversion. The mechanism was demonstrated through dedicated chemical quenching experiments and confirmed the involvement of $^1O_2$, along with a minor contribution by $O_2^{\bullet-}$. Interestingly, the resulting products could be subjected to a subsequent nucleophilic addition in the presence of trimethylsilylcyanide to give the corresponding α-amino nitriles. Given its low solubility, the C_{70} carbocatalyst could be easily recovered by precipitation with methanol and recycled for another

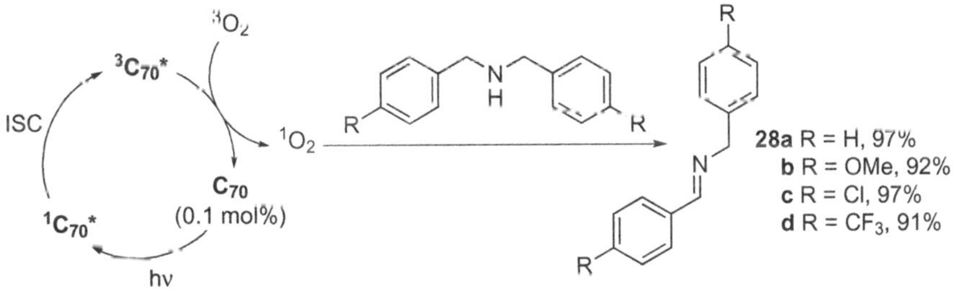

Scheme 3.10. Photosensitized oxidation of secondary benzyl amines.

run, allowing to obtain a similar yield of imine, although a longer irradiation time was required (10 h vs 5 h) [17].

Supported fullerenes (in particular C_{60} adducts) have been likewise prepared, and the application of the resulting materials in oxidation processes reported. These included the oxidation of sulfides to the corresponding sulfoxides [36], different pericyclic processes (ene and Diels–Alder reactions involving $^1O_2$) [37], and the conversion of 1-naphthol to naphthoquinone [37].

3.2.5 *Reduction processes*

A few pyrene derivatives have recently been employed as photocatalysts (1 mol%) to perform the C–F to C–H conversion in fluoroarenes (e.g. **29**) in the presence of DIPEA (1.1 equiv). As described in Scheme 3.11, a detailed mechanistic investigation revealed that the process took place through an initial complexation between **29** and the photocatalyst.

Upon irradiation, this complex underwent local excitation at the pyrene moiety and was then reductively quenched by the amine.

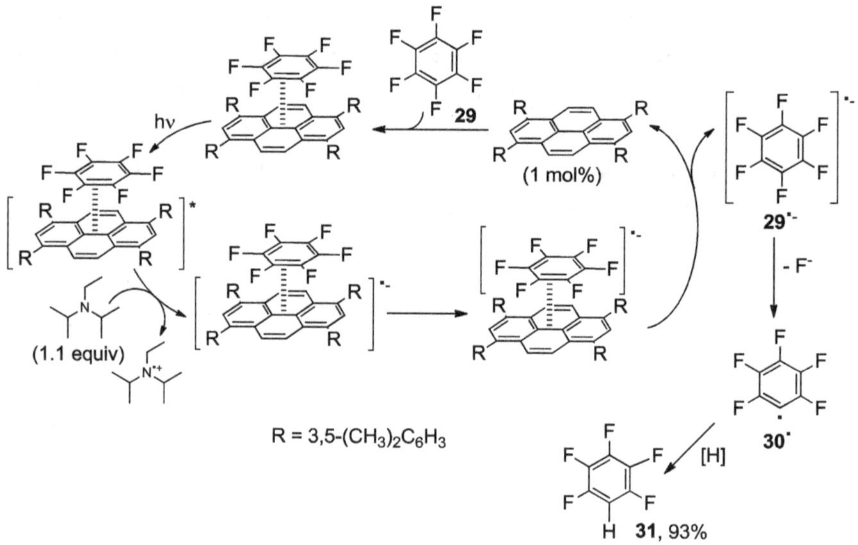

Scheme 3.11. Photoorganocatalyzed hydrodefluorination reactions.

An inner-sphere electron transfer between pyrene radical anion and **29** led to the dissociation of the complex and formation of **29**$^{\bullet-}$. The latter then dissociated leading to the aryl radical **30**$^{\bullet}$ and the fluoride anion. The final product **31** was obtained through a H-abstraction step, probably from the radical cation of DIPEA. Notably, the peculiar inner-sphere electron-transfer mechanism allowed to overcome large underpotential values [38]. Furthermore, it was demonstrated that the photogenerated aryl radical could be exploited in the C–C bond forming reactions in the presence of excess cyclohexene [38].

The reduction of aromatic aldehydes (e.g. **32a-c**; 8 mM) to benzyl alcohols has been performed by exploiting a poly(*p*-phenylene) derivative (10 mg) in the presence of an excess of a tertiary amine (TEA; 1M). Apart from the desired alcohol, the process led to the formation of the corresponding pinacol product (see also Sec. 3.2.2 above). The addition of 2-mercaptoethanol cleanly drove the reaction to the formation of **33a-c**, although only a partial conversion of the starting materials has been observed in some instances (Scheme 3.12) [26]. A similar competition between the formation of a reduced product and the corresponding pinacol was observed when carrying out the reaction in the presence of electron-rich pyrenes or anthracenes (5 mol%) as photocatalysts in the presence of a benzimidazoline derivative as reductant [27, 28]. The latter system was likewise applied to the reductive opening of epoxides in α,β-epoxyketones and led to the formation of the corresponding β-hydroxyketones in good yields [13, 28].

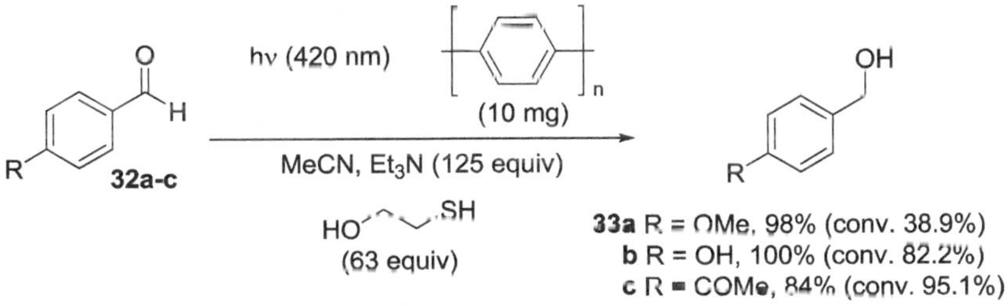

Scheme 3.12. Photoorganocatalyzed reduction of aromatic aldehydes.

3.2.6 *Miscellanea*

Upon excitation, some derivatives undergo deep modifications of their acid–base properties and this feature has been recently exploited for the asymmetric conversion of enol silyl ethers into the corresponding carbonyl derivatives. Indeed, a few enantiopure VANOL (3,3′-diphenyl-2,2′-bi-1-naphthol) dyes endowed with excited state proton transfer (ESPT) properties were exploited to promote the protonation of a series of enol silyl ethers, with formation of the desired products in up to 49% enantioselectivity with 68% yield. The dye, however, had to be used in a stoichiometric amount with respect to the substrate [18].

Conjugated aromatic polymers, including poly-(2-methoxy-5-)2′-ethylhexyloxy-1,4-phenylene vinylene (MEH-PPV), have been used for the degradation of various textile dyes under UV irradiation. Interestingly, the photocatalytic activity of these conjugated polymers was significantly higher than that of Degussa P-25 under similar conditions [39].

3.2.7 *Synthetic processes via photosensitized aromatics*

Recently, a variant of the photocatalytic cycles described above has been reported. As depicted in Scheme 3.13, the widely used $Ru(bpy)_3Cl_2$ (bpy = 2,2′-bipyridine) photoredox catalyst has been adopted to absorb light and then to transfer it to an aromatic hydrocarbon, such as pyrene or anthracene, in turn responsible for promoting the desired redox chemistry [40].

This strategy has been exploited to perform the arylation of (hetero)aryl bromides and chlorides and was based on the generation of the highly reducing radical anion of the employed aromatic in the presence of an electron donor (e.g. DIPEA), in turn also responsible for the activation of the aromatic halide. Overall, this strategy can be considered as a two-center photoredox catalytic approach in which the tasks of light energy collection and electron transfer are assigned to two distinct molecules. In particular, it has been possible to excite "indirectly" pyrene by employing a visible-light source, where the molecule itself has no absorption, *via* ET

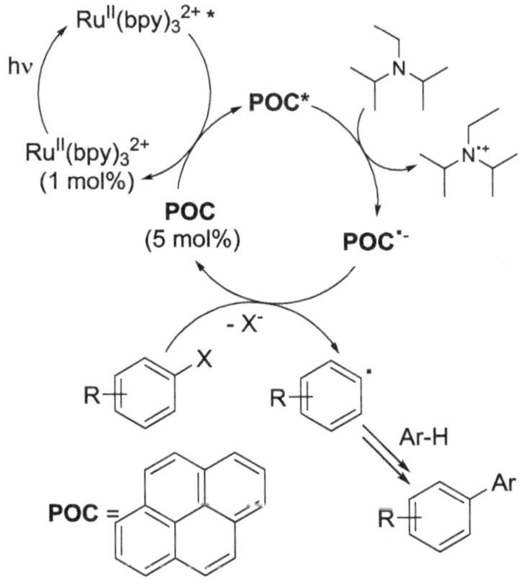

Scheme 3.13. Photoredox catalyzed arylation *via* (hetero)aryl bromides and chlorides.

from the photoexcited Ru-complex [40]. The reported mechanism, however, has been the subject of some controversy and different mechanistic proposals have been later suggested [41, 42].

3.3 Cyanoaromatics

3.3.1 *Photophysics of cyanoaromatics and their derivatives*

Cyanoaromatics (Ar–CN) are electron-poor aromatics which exhibit a rather modest ground-state reduction potential (typically E_{RED}(Ar–CN/Ar–CN$^{•-}$) that ranges from –1.7 V to –0.7 V vs SCE) [5], but possess a highly positive excited state reduction potential value (E^*_{RED}(Ar–CN*/Ar–CN$^{•-}$), often > +2 V vs SCE). Chart 3.2 includes a selection of the most representative photophysical and (photo) electrochemical parameters for such compounds. **DCB**, 1,2,4-tricyanobenzene (1,2,4-**TrCB**), 1,3,5-tricyanobenzene (1,3,5-**TrCB**), and 1,2,4,5-tetracyanobenzene (**TCB**), as well as their derivatives 4,5-dicyanophthalimide (**DCPh**) [43], and methyl mellitate ester (**ME**) [44], are all characterized by a high singlet-excited-state reduction potential (E^*_{RED}(**POC***/**POC**$^{•-}$)) [5]. This allows for the

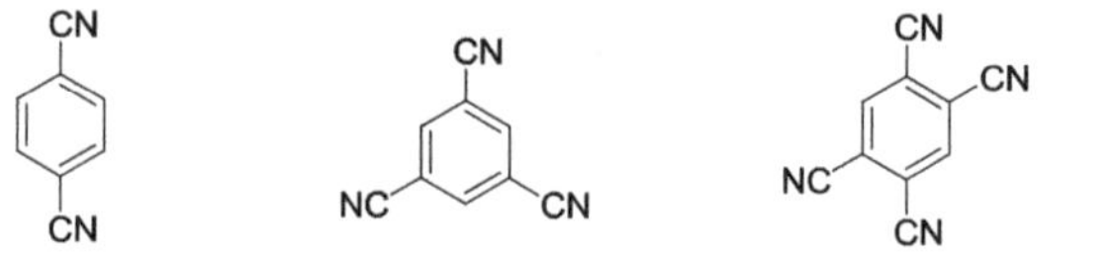

Name	1,4-Dicyanobenzene (**DCB**)	1,3,5-Tricyanobenzene (1,3,5-**TrCB**)	1,2,4,5-Tetracyanobenzene (**TCB**)
Absorption Edge	290 nm (λ_{max})		< 330 nm
E_{RED} (**POC/POC$^{\bullet-}$**)	- 1.46 V[a]	- 1.35 V[b]	- 0.66 V[b]
E^*_{RED} (**POC*/POC$^{\bullet-}$**)	+ 2.55 V[a]	+ 2.68 V[b]	+ 3.44 V[b]

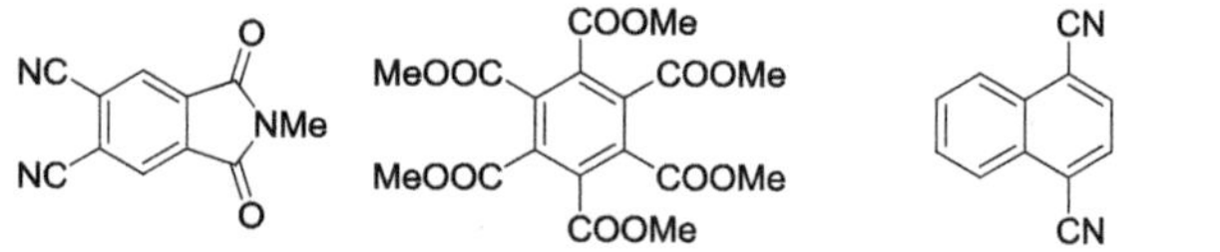

Name	4,5-Dicyanophtalimide (**DCPh**)	Methyl mellitate (**ME**)	1,4-Dicyanonaphthalene (**DCN**)
Absorption Edge			< 350 nm
E_{RED} (**POC/POC$^{\bullet-}$**)	- 0.75 V[b]	- 1.31 V[b]	- 1.27 V[a]
E^*_{RED} (**POC*/POC$^{\bullet-}$**)	+ 3.11 V[b]	+ 1.95 V[b]	+ 2.30 V[a]

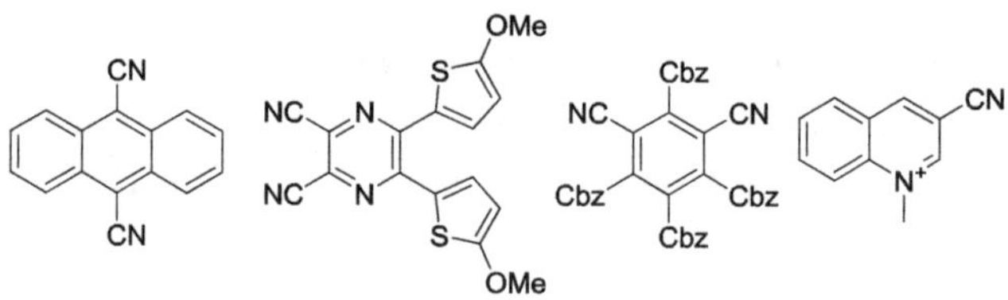

Name	9,10-Dicyanoanthracene (**DCA**)	**DCP**	**4-CzIPN**	**QuCN$^+$**
Absorption Edge	< 450 nm	< 550 nm	< 460 nm	< 500 nm
E_{RED} (**POC/POC$^{\bullet-}$**)	- 0.91 V[a]	- 1.45 V[c]	- 1.21 V[d]	- 0.60 V[e]
E^*_{RED} (**POC*/POC$^{\bullet-}$**)	+ 1.99 V[a]	+ 1.37 V[c]	+ 1.35 V[d]	+ 2.72 V[e]

Chart 3.2. Photophysical and electrochemical parameters of the cyanoaromatics considered in the present section. All the potentials are expressed in V vs SCE.

Note: [a]From Ref. [5e]. [b]From Ref. [45a]. [c]From Ref. [47]. [d]From Ref. [48]. [e]From Ref. [49].

monoelectronic oxidation of redox-stable compounds, including ethers, alkenes, and even alkanes [45].

However, the fact that such compounds absorb exclusively below 350 nm limited their application in modern organic photochemistry. On the other hand, cyanoaromatics with increased benzannulation, such as 9,10-dicyanoanthracene (**DCA**), absorb significantly also in the visible range and have been rather frequently investigated, accordingly. The oxidizing power of the corresponding excited states, however, is lower than that measured for cyanobenzenes [46]. The fluorescence of most cyanoaromatics has a lifetime of about 10 ns [5], with **DCA** exhibiting the highest quantum yield emission value (Φ_F up to 0.9, λ_{em} ca. 460 nm) [9].

By contrast, dicyanopyrazines (**DCPs**) exhibit an absorption spectrum that depends on the nature of the aromatic ring (see, for instance, the pyrazine derivative **DCP** in Chart 3.2) [47]. Analogously, a series of carbazolyl dicyanobenzenes has been synthesized and investigated as donor/acceptor fluorophores. As an example, in the compound labeled as **4-CzIPN**, obtained as a yellow powder, the highest occupied molecular orbital (HOMO) is mainly localized on the donor carbazole units, whereas the lowest unoccupied molecular orbital (LUMO) is centered at the dicyanobenzene core [48]. This photocatalyst was shown to have an excited state reduction potential significantly higher than that of the photoexcited $Ru(bpy)_3^{2+}$ complex [5c]. Finally, the singlet excited state of the 1-methyl-3-cyanoquinolinium ion (**QuCN**$^+$, usually prepared as the perclorate salt) exhibits an impressive oxidizing capability with E^*_{RED}(**QuCN**$^{+*}$/**QuCN**$^•$) = +2.72 V vs SCE [49].

3.3.2 *Beyond photoorganocatalysis: The photochemistry of cyanoaromatics*

Before their use in photoredox catalysis, the peculiar photochemical behavior of cyanoaromatics has been exploited in several ipso-substitutions [50]. As a matter of fact, the reductive quenching of their singlet excited state (Ar–CN* in Scheme 3.14, path a) by a suitable electron donor R-EG results in the formation of a $ArCN^{•-}$/R-$EG^{•+}$ radical ions pair. Thus, different pathways can then ensue, depending on both the nature of the electron donor and the reaction conditions. When the substrate bears a suitable electrofugal group (EG, such as H, COO^- or SiR_3), cleavage of the stable cation EG^+ competes efficiently with back electron transfer (BET, path a′)

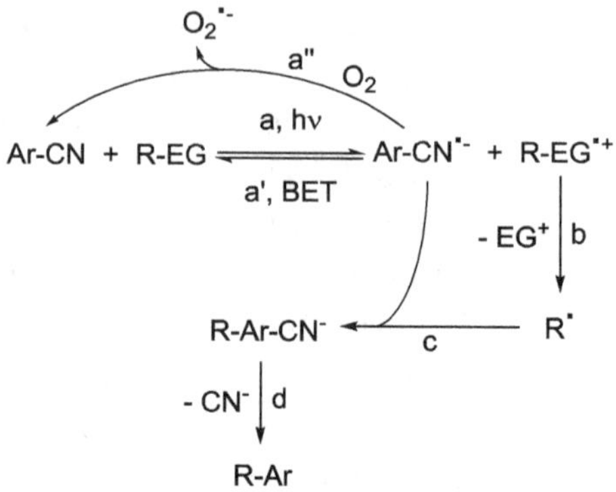

Scheme 3.14. Photochemistry of cyanoaromatics (Ar–CN).

[51] and affords the corresponding carbon-centered radical R• (path b) [45b]. The reaction proceeds *via* coupling of R• with the acceptor ArCN•− (path c), thus forming, after cyanide elimination from the resulting anionic [R–Ar–CN]− intermediate, the ipso-substituted product R–Ar (path d). In some cases, an appropriate co-catalyst (e.g. biphenyl (**BP**), E_{OX}(**BP**•+/**BP**) = +2.53 V vs SCE) [52] is used to suppress BET [5d, 52]. Finally, in aerated solution, the radical anion of cyanoaromatics can also act as reductant, leading to the generation of ROS, e.g. $O_2^{\bullet-}$, path a″, *via* an electron transfer step from ArCN•− to molecular oxygen [52].

A variation of such a process is named photochemical radical–olefin combination aromatic substitution (photoROCAS), occurring in the presence of an electron-poor alkene, which traps R• before coupling with ArCN•−. Accordingly, this allows for the development of a three-component synthetic strategy for the synthesis of complex aromatics [53].

Finally, when the involved radical cation R-EG•+ is not cleavable, another pathway can be observed. A typical example of such photochemical nucleophile–olefin combination aromatic substitution (photoNOCAS) is the irradiation of **DCB** in acetonitrile in the presence of 2-methylpropene **34** as the electron donor (Scheme 3.15) [53]. Thus, the generated intermediate **34**•+ coupled with a molecule of the nucleophilic solvent and the resulting distonic radical

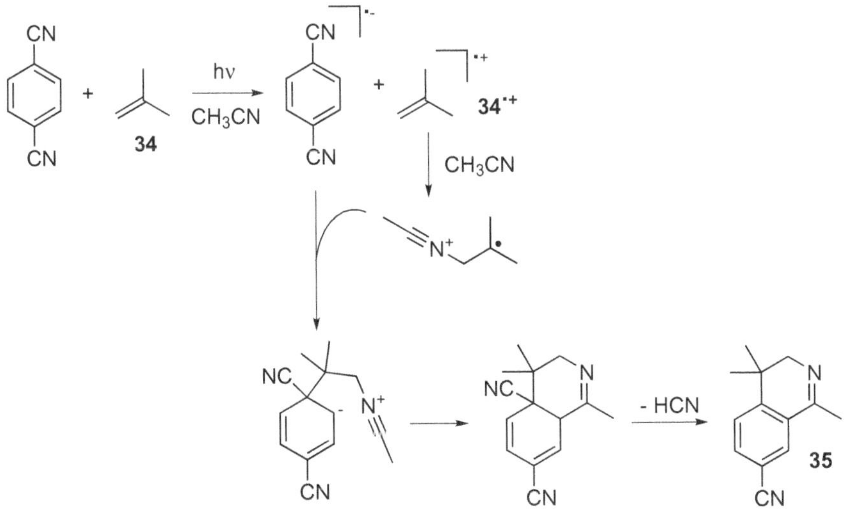

Scheme 3.15. Example of a photoNOCAS.

ion further reacted with Ar–CN$^{\bullet-}$ to form a zwitterion, which smoothly cyclized and finally afforded product **35** upon loss of HCN.

On the basis of the examples previously cited, it sounds surprising that, even though more than 90% of the photocatalytic processes reported in the last decade involved a photoinduced electron transfer as the key-step, and the photoelectrochemical properties of cyanoaromatics have been known for a long time, only little attention has been given to their potentiality as **POCs**. To be fair, cyanoaromatics were often considered during the optimization procedure of photocatalytic processes, but in most cases the results obtained were less satisfactory than those observed with other photocatalysts [54]. Aim of this section is to provide the reader with an overview of the potentialities of cyanoaromatics and their derivatives.

3.3.3 *Carbon–carbon bond formation*

Formation of a C_{sp3}–C_{sp3} bond. One of the earliest examples involving the use of cyanoaromatics as photoredox catalysts deals with a

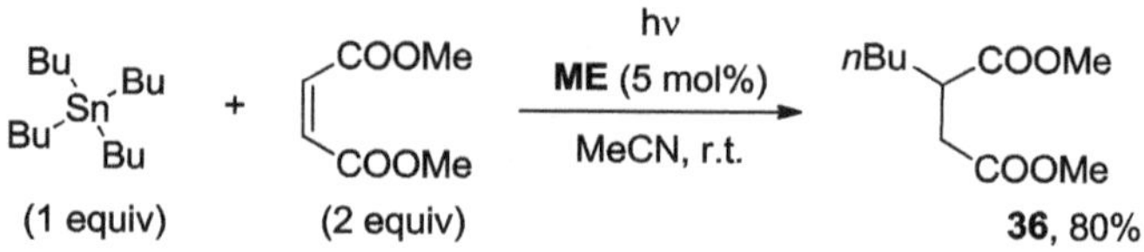

Scheme 3.16. Photocatalyzed alkylation of dimethyl maleate by tetra-alkylstannanes.

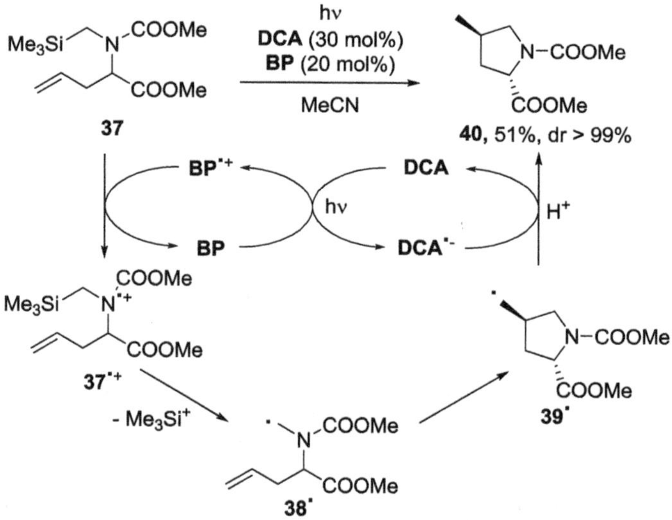

Scheme 3.17. **DCA**-photocatalyzed synthesis of pyrrolidine **40**.

conjugate radical addition process onto electron-poor alkenes. In this case, olefins were alkylated by using a tetraalkylstannane (E_{OX} = +1.6–1.75 V vs SCE [55]) as the radical precursor. Different photocatalysts were tested, including **ME** (an example in Scheme 3.16 for the preparation of **36**) and **TCB** [55]. The same approach has been later applied to the alkylation of α,β-unsaturated ketones [56a] and acetylenedicarboxylate esters [56b].

The radical cyclization of *N*-trimethylsilylmethyl aminoacid derivatives containing an unsaturation (e.g. **37** in Scheme 3.17) was performed by using **DCA** and **BP** as the photocatalyst and the mediator of the electron transfer, respectively. Thus, the photoinduced oxidation of **37** to the corresponding radical cation **37**$^{•+}$

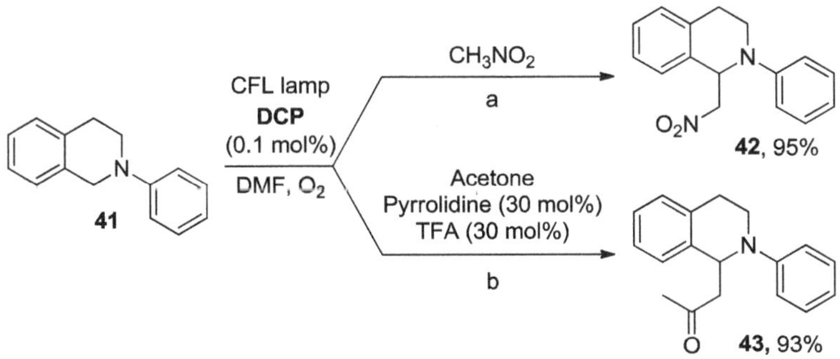

Scheme 3.18. Photocatalyzed functionalization of *N*-phenyl tetra-hydro-isoquinoline.

was followed by cleavage of the Me_3Si^+ cation. The resulting α-aminoradical **38$^\bullet$** smoothly underwent cyclization to **39$^\bullet$**, then reduction by **DCA$^{\bullet-}$**, and ensuing protonation afforded pyrrolidine **40** in 51% yield as a single diastereoisomer [57]. This approach was also exploited to promote structural changes in peptides [57].

The metal-free functionalization of different tertiary/secondary amines with various carbon-based nucleophiles has been achieved by using **DCP** (0.1 mol%) as the organic photocatalyst. Thus, aza-Henry and Mannich condensations were carried out on *N*-phenyl tetrahydroisoquinoline **41** affording the corresponding β-nitro (**42**, Scheme 3.18, path a) and β-keto (**43**, path b) derivatives in satisfactory yields, under either compact fluorescent lamp irradiation or natural sunlight exposition [47].

The same photocatalyst has been employed in the coupling between **41** and *N*-phenyl itaconimide **44**. Interestingly, the reaction conditions can be tuned to form, through four different chemical pathways, a range of valuable *N*-heterocycles, including spyro-derivative **45** (Scheme 3.19, path a), alkylated maleimide **46** (path b), and succinimides **47** and **48** (paths c, d) [58].

The alkylation of imines under redox neutral conditions based on the use of **4CzIPN** as the photocatalyst and ammonium alkyl bis(catecholato)silicates as the radical precursor was reported. By following this approach, a wide range of functionalized amines has

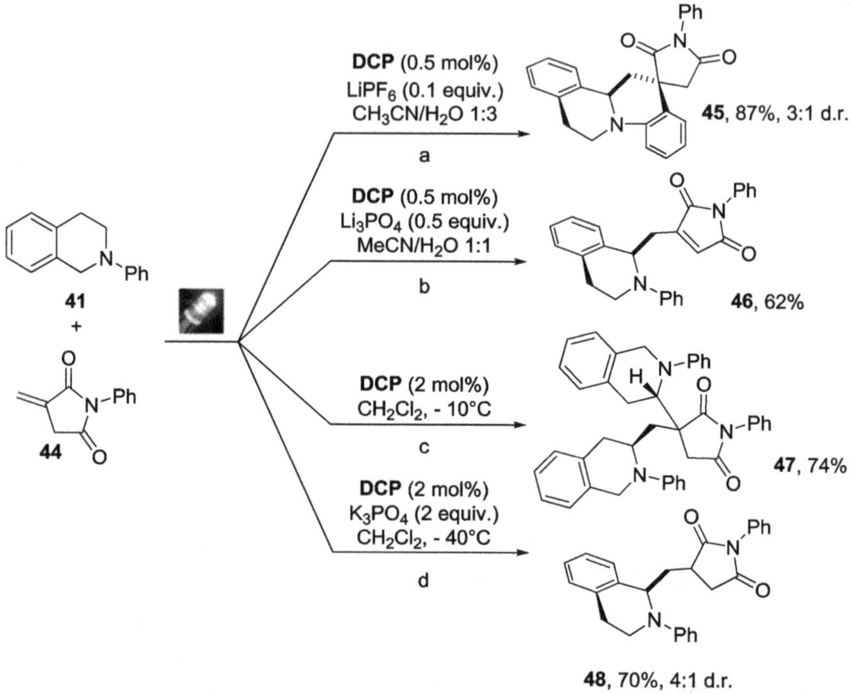

Scheme 3.19. *N*-phenyl itaconimide as the precursor of various functionalized heterocycles.

been obtained in high yields under both metal- and acid-free conditions [59]. Analogously, 1,1-difluoroalkenes (**50**, Scheme 3.20) were prepared from trifluoromethyl-substituted alkenes **49** *via* radical defluorinative alkylation in the presence of alkyl bis(catecholato) silicate or alkyltrifluoroborate salts in the role of radical precursors (an example in Scheme 3.20) [60].

Formation of a C_{sp3}–C_{sp2} bond. The photoredox catalyzed radical hydroformylation of styrenes has been described by using diethoxyacetate salt **51**$^-$ (Scheme 3.21) as the formylating agent. This approach avoided the poor regioselectivity observed in the traditional transition metal-catalyzed processes. The mechanism involved the sequential oxidation–decarboxylation of **51**$^-$ promoted by the photoexcited **4CzIPN*** dye to give **52**$^\bullet$ that was in turn efficiently trapped by styrenes (e.g. **53a-c**).

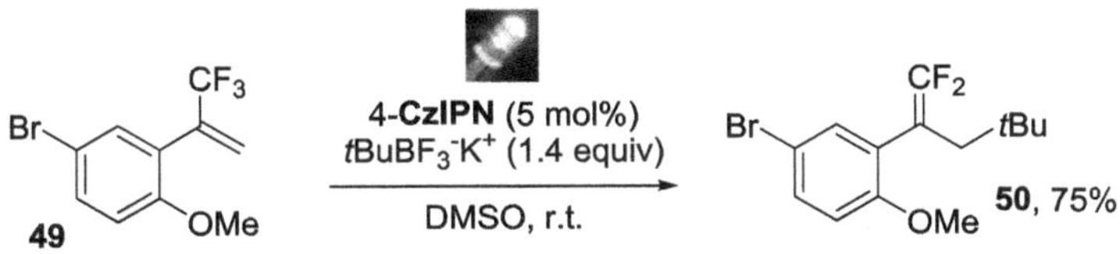

Scheme 3.20. Photocatalyzed synthesis of 1,1-difluoroalkenes.

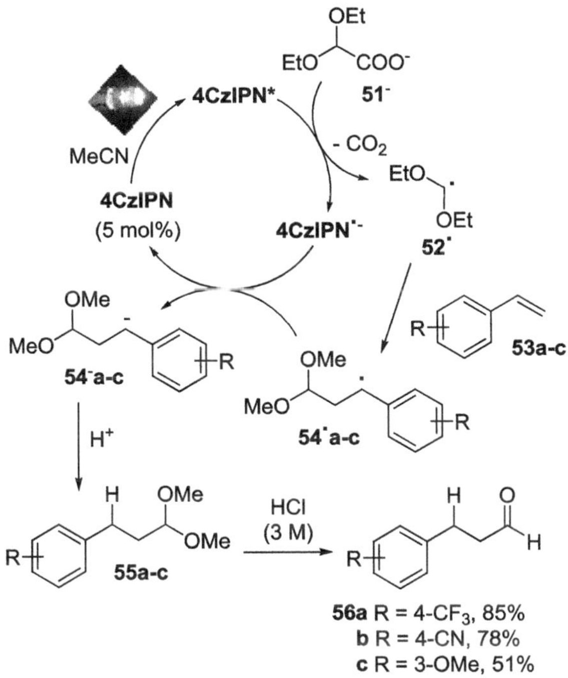

Scheme 3.21. Photoorganocatalyzed radical hydroformylation of styrenes.

Reduction of the thus formed benzyl radicals **54•a-c** (by **4CzIPN•−**), followed by protonation of the obtained anions **54−a-c** and deprotection of acetals **55a-c** gave the desired 3-arylpropionaldehydes **56a-c** in up to 90% yield [61]. Notably, analogs of the sweetener aspartame as well as of non-steroidal antinflammatory indometacine, were obtained. The reaction was also performed under continuous flow conditions and afforded **56a** in 87% yield after 2.5 h irradiation [61].

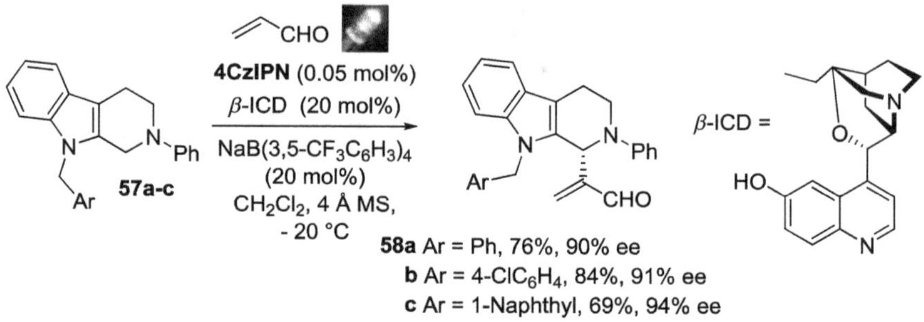

Scheme 3.22. Photocatalyzed enantioselective olefination.

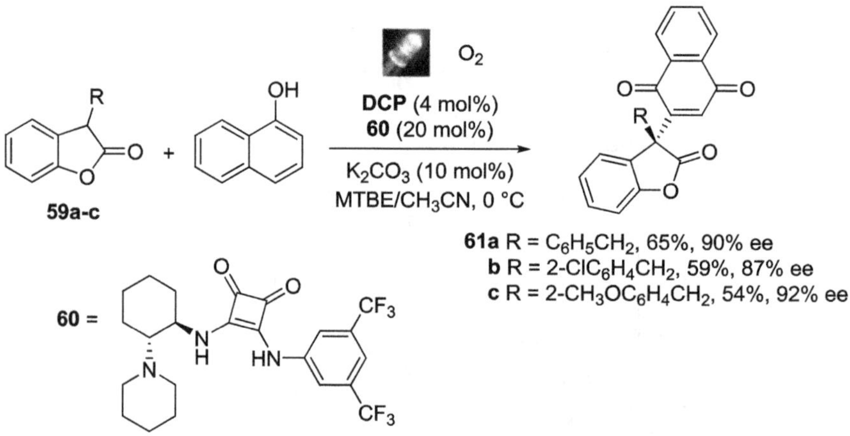

Scheme 3.23. Photocatalyzed functionalization of benzofuran-2(3H)-ones.

4CzIPN was employed as the **POC** in the enantioselective C–H olefination of tetrahydroisoquinolines and tetrahydro-β-carbolines (**57a-c** in Scheme 3.22) under oxidative conditions, by adopting a cooperative photoredox/asymmetric catalytic approach. In this case, quinoline derivative β-ICD was used as the chiral Lewis base catalyst, finally leading to compounds **58a-c** [62].

DCP (4 mol%) was used in the functionalization at the 3-position of benzofuran-2(3H)-ones **59a-c** (Scheme 3.23) by 1-naphthol *via* a cooperative visible-light photoredox/asymmetric hydrogen-bond catalytic approach. The process was carried out in a methyl *tert*-butyl

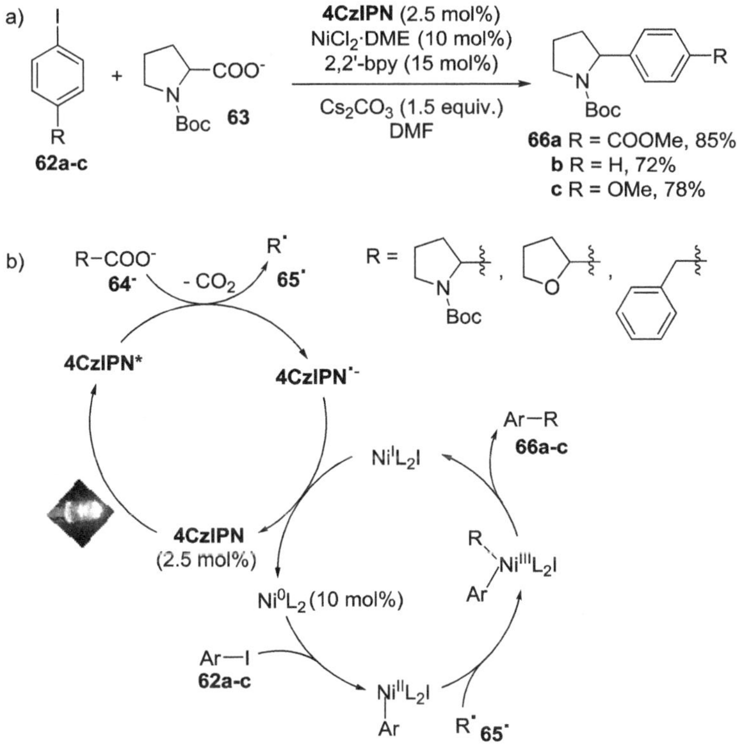

Scheme 3.24. Photocatalyzed alkylation of aromatics.

ether (MTBE)/acetonitrile mixture, with squaramide derivative **60** (20 mol%) in the role of the hydrogen-bond catalyst. The desired adducts **61a-c** were obtained in discrete yields, but with excellent enantioselectivity [63].

The **4CzIPN** photocatalyst was efficiently used in the dual photoredox/Ni catalyzed decarboxylative cross-coupling of aryl halides (mainly iodides, **62a-c**, Scheme 3.24(a)) with α-amino- (see, for instance, **63**), α-alkoxy, and phenylacetic acids. In this case, the highly oxidant photoexcited dye promoted the generation of alkyl radical **65•** *via* oxidative decarboxylation of anion **64**$^-$ (Scheme 3.24(b)). In the meantime, a Ni^0 complex (formed *in situ*) underwent oxidative addition onto the Ar–I bond of **62a-c** and the resulting Aryl–Ni^{II} species reacted with **65•** to form the corresponding Ni^{III} complexes. The latter underwent reductive elimination by releasing the desired coupling products **66a-c**, while a SET between the obtained Ni^{I} species and **4CzIPN**$^{•-}$ restored the starting conditions [48]. Benzyltrifluoroborates can be likewise used as the radical precursors [64]. The scope of the protocol was further extended, since alkyl bis(catecholato)silicates can be exploited as a source of non-stabilized alkyl radicals, in turn allowing for the efficient synthesis of alkylarenes [65].

An analogous Ni/photoredox dual catalytic cross-coupling has been applied to the arylation (in moderate to good yields) of an impressive range of (hetero)aryl bromides in the presence of 1,4-dihydropyridine derivatives as the precursors of the alkyl radicals involved [66]. The potentiality of the process was largely demonstrated through the preparation of the saccharide-containing product **69** (Scheme 3.25), that was isolated in good yield and excellent diastereoselectivity [66].

The mechanism of the intramolecular α-arylation of ketones by enol silyl ether moieties under **DCN** photoorganocatalyzed conditions was investigated in details in the 1990s [67]. The same approach was applied, a decade later, to the multi-step synthesis of the alkaloid 7-deoxypancratistatin **72**, in particular, in the preparation of the key intermediate **71**, that was obtained in 68% yield from the corresponding enol silyl ether **70** (Scheme 3.26) [68].

A similar approach has been exploited in the arylation of chromanones and applied for the synthesis of differently substituted unnatural flavanones [69].

Formation of a C_{sp3}–C_{sp} bond. The decarboxylative alkynylation of carboxylic acids took place under **DCA** photoredox catalyzed

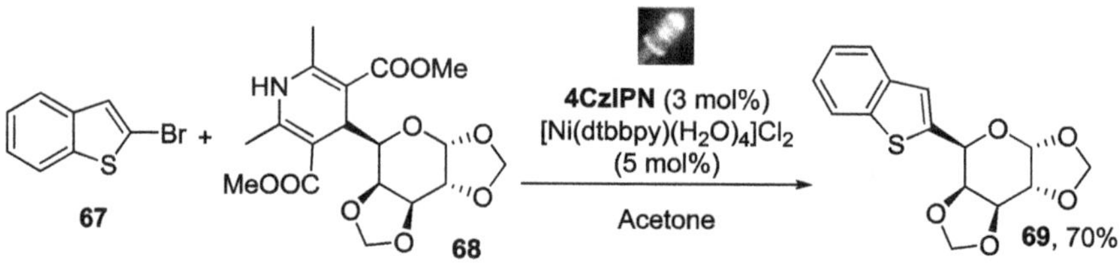

Scheme 3.25. Ni/photoredox dual catalytic cross-coupling.

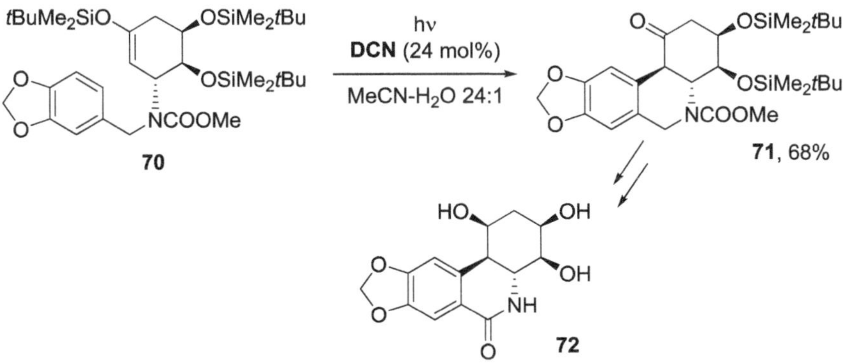

Scheme 3.26. Multi-step synthesis of the alkaloid 7-deoxypancratistatin **72.**

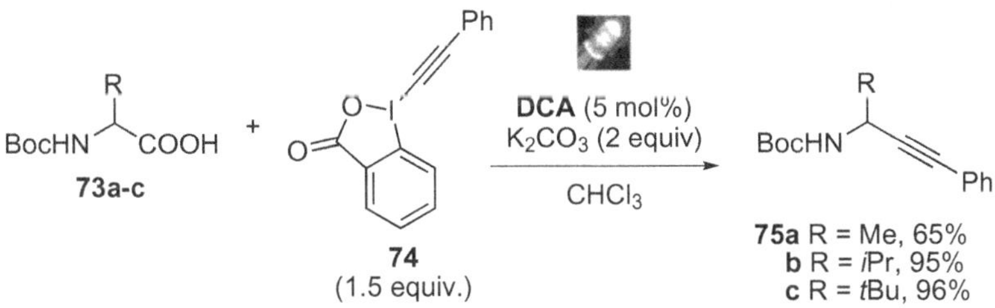

Scheme 3.27. Photocatalyzed decarboxylative alkynylation of carboxylic acids.

conditions. The optimized reaction exhibited a broad scope, since the alkynylation of α-amino acids (see, for instance, **73a-c** in Scheme 3.27) was feasible upon visible-light irradiation in the presence of phenyl ethynylbenziodoxolone **74** as the alkynylating agent, to produce compounds **75a-c** in good yields. α-Keto acids can be also employed, which can afford the corresponding ynones with formation of a new C_{sp2}–C_{sp} bond. Either a blue LED or natural sunlight could be used to promote the process. Interestingly, the use of **DCA**, instead of the rather expensive Ir- and Ru-based complexes, strongly reduced the economic impact of the reaction [70].

The **DCP**-catalyzed oxidative Strecker reaction carried out on *N*-phenyl tetrahydroisoquinoline in the presence of trimethylsilyl cyanide afforded the corresponding α-cyano derivative in 85% yield under compact fluorescent lamp irradiation [71].

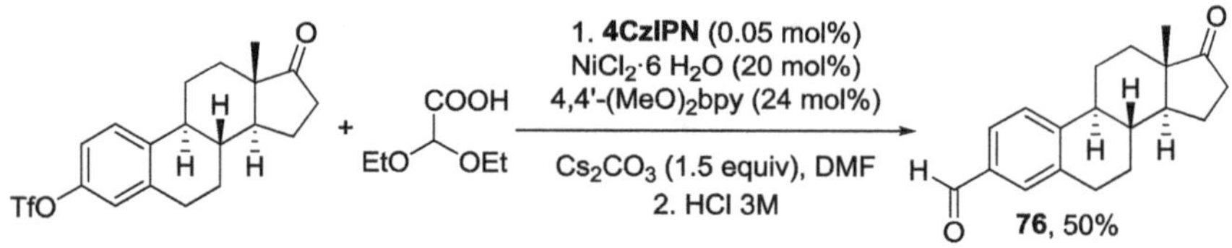

Scheme 3.28. Photoorganocatalyzed formylation of aromatics.

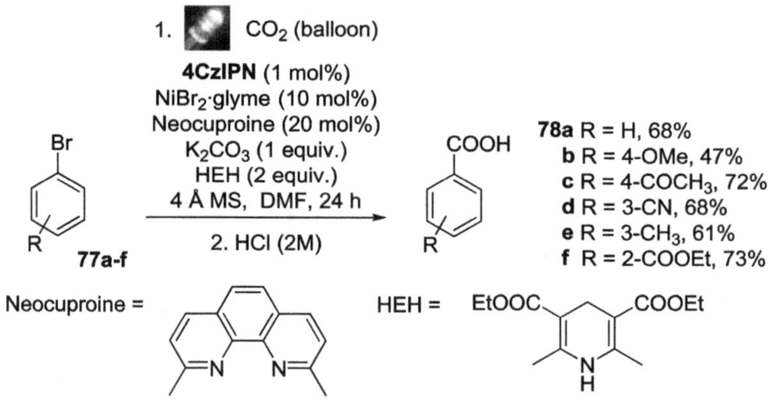

Scheme 3.29. Photoorganocatalytic carboxylation of aryl bromides.

Formation of a C_{sp2}–C_{sp2} bond. The formylation of aryl halides, aryl triflates, and even vinyl bromides took place under synergistic nickel- and **4CzIPN**-mediated photoredox catalysis [72]. As in the procedure described in Scheme 3.21, the diethoxymethyl radical acted as the formyl radical equivalent (see Scheme 3.28). Notably, due to the mild conditions employed, the process tolerated the presence of a plethora of functional groups, thus enabling the access to a wide range of aromatic aldehydes, including estrone derivative **76** (Scheme 3.28) [72].

Very recently, the visible-light induced carboxylation of aryl bromides **77a-f** and triflates to the corresponding benzoic acids **78a-f** (Scheme 3.29), by exploiting K_2CO_3 as the CO_2 source, has been described. The process was carried out in the presence of the **4CzIPN** photocatalyst (1 mol%) and an *in situ*-formed Ni^{II} catalyst [73].

3.3.4 *Carbon–heteroatom bond formation*

Carbon–Oxygen bond formation. The use of cyanoaromatics (in particular, **DCA**) in the oxidation of simple molecules, including sulfines [74], triarylphosphines [50], as well as toluenes [75], is well known. Analogously, one of the early examples exploiting the properties of **DCA** as an oxidant in photoredox catalysis described the quantitative conversion of 2,6-di(4-methoxyphenyl)-1,6-heptadiene **79** in the cyclic peroxide **80** (Scheme 3.30, upper part). Notably, when the same irradiation was carried out under nitrogen atmosphere, the bicyclic derivative **81** was selectively isolated (lower part) [76].

The visible-light-driven metal-free catalytic conversion of alkynes into 1,2-diketones in the presence of **DCA** and **BP** as the co-catalyst has been described (Scheme 3.31) [77]. The initial step involved the excitation of **DCA** to its singlet excited state **DCA***, that underwent reductive quenching by **BP** to afford **BP**$^{\bullet+}$. The latter, in turn, oxidized alkynes **82a-d** to **82**$^{\bullet+}$**a-d**, while the generated **DCA**$^{\bullet-}$ radical anion reacted with molecular oxygen to give the superoxide radical ion. Coupling of $O_2^{\bullet-}$ with **82**$^{\bullet+}$**a-d** followed by rearrangement of the resulting cyclic adducts **83a-d** furnished the desired diketones **84a-d** [77].

The oxidative conversion of *N*-methyl-2-(phenylamino)acetamide **85** into the corresponding unsymmetric oxalamide **86** smoothly

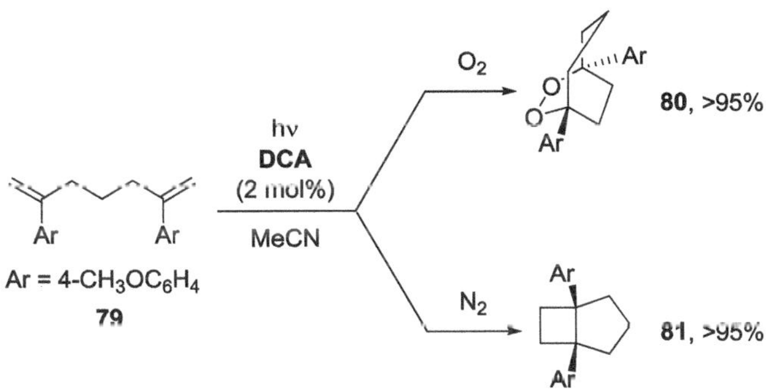

Scheme 3.30. Photocatalyzed cyclization reactions.

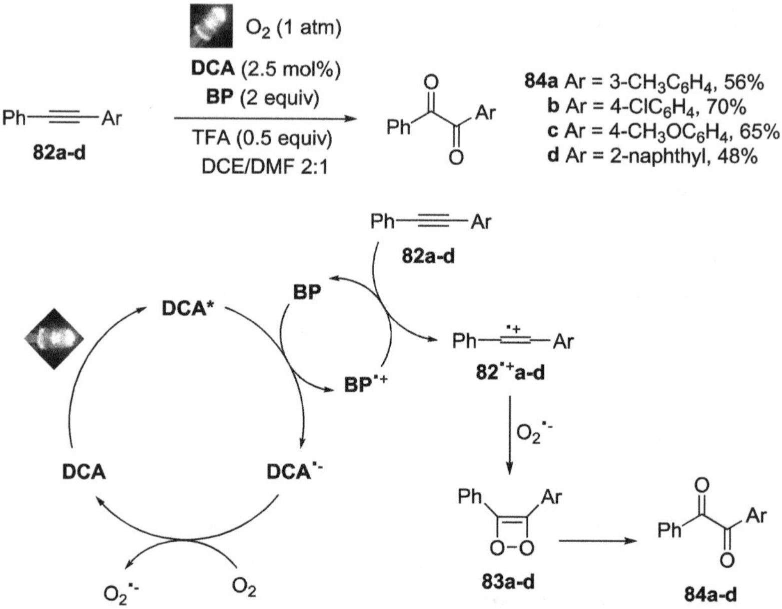

Scheme 3.31. Photocatalyzed conversion of alkynes into 1,2-diketones.

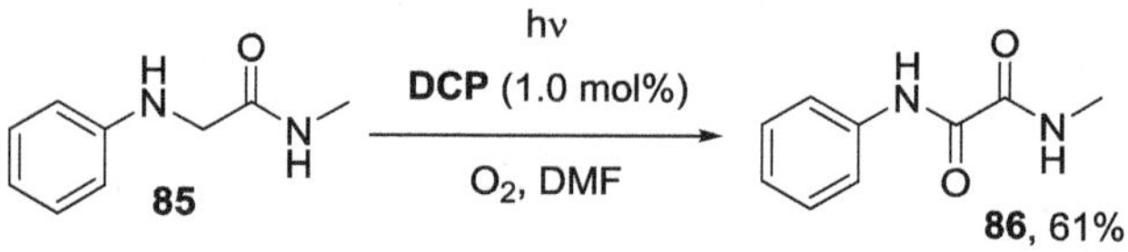

Scheme 3.32. Photoorganocatalyzed synthesis of oxalamides.

occurred under oxygen atmosphere in the presence of **DCP** (1.0 mol%) photocatalyst (Scheme 3.32) [47]. By adopting the same approach, arylboronic acids can be efficiently oxidized to the corresponding phenols in up to 96% yield [47].

Due to the presence of reactive C–H bonds as well as the proximal aromatic ring, the functionalization of the benzyl group is an important goal in organic synthesis [78]. A protocol for the benzylic C–H bond activation *via* an electron–proton–electron (*E–P–E*) loss strategy was recently reported (Scheme 3.33). Thus, irradiation of alkylarenes **87a-c** in acetonitrile in the presence of the photocatalyst

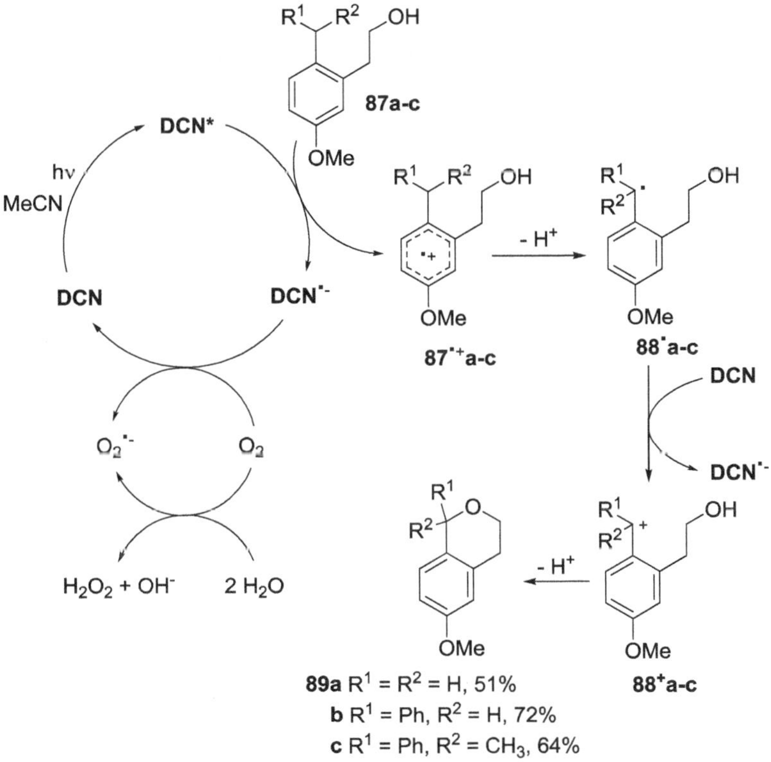

Scheme 3.33. Photocatalyzed benzylic functionalization.

DCN (5 mol%) resulted in the initial formation of the **DCN**$^{\bullet-}$/ **87**$^{\bullet+}$**a-c** radical ions pair. Then, deprotonation from **87**$^{\bullet+}$**a-c** took place to give radicals **88**$^{\bullet}$**a-c**, which were in turn oxidized by a second molecule of **DCN**. The resulting carbocations **88**$^{+}$**a-c** smoothly underwent cyclization to form, after deprotonation, isochromans **89a-c**. The photocatalyst was finally restored by reaction of **DCN**$^{\bullet-}$ with oxygen [71].

The remarkable oxidizing properties of photoexcited 3-cyano-1-methylquinolinium ion have been exploited in the oxygenation of benzene to phenol, that was carried out in aqueous, oxygen-saturated acetonitrile. A 51% yield of phenol (at 73% benzene conversion) was obtained after 5 h irradiation in the presence of

QuCN$^+$ClO$_4^-$ (6.7 mol%) [79]. The same photocatalyst was also used for the alkoxylation of benzene, that took place in rather low yields (<30%) [80].

Finally, the photooxidation of some lignin models in the presence of **DCA** and **DCB** has been investigated through both experimental and computational analyses [81].

Carbon–Nitrogen bond formation. A series of 2′-arylbenzaldehyde oxime ethers (**90a-c**) was employed in the **DCA**-photocatalyzed synthesis of phenanthridines **93a-c**. The best results were obtained in the case of substrates bearing electron-donating substituents.

A detailed investigation of the mechanism carried out by means of Hammet plots pointed out the photooxidation of the oxime ether moiety in **90a-c** to the corresponding radical cations **90$^{\bullet+}$a-c** as the initial step (Scheme 3.34).

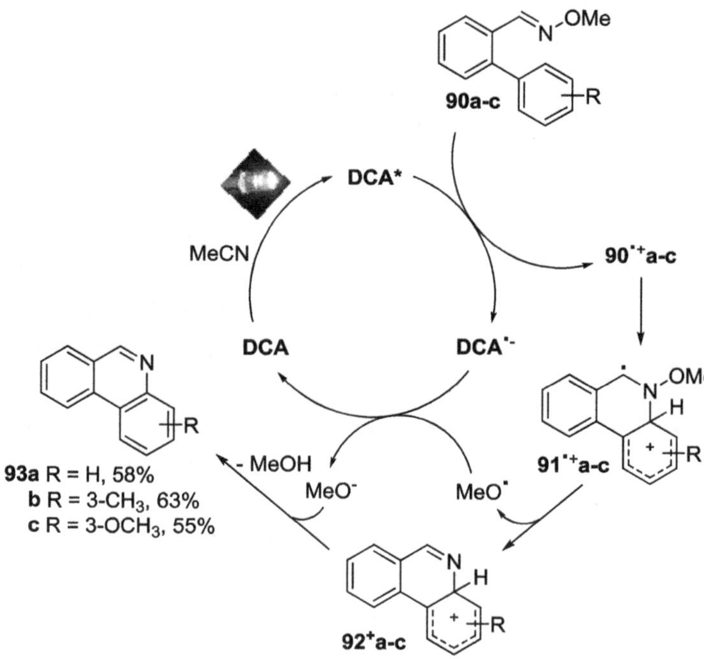

Scheme 3.34. Photoredox-catalyzed synthesis of phenanthridines.

Intramolecular nucleophilic attack of the aromatic ring to the nitrogen atom in **90$^{•+}$a-c** afforded the resonance-stabilized distonic radical cations **91$^{•+}$a-c** that underwent smooth homolysis of the N–O bond. Reduction of the released methoxy radical by **DCA$^{•-}$** restored the starting photocatalyst, while deprotonation of the Wheland intermediates **92^{+}a-c** finally afforded the desired phenanthridine cores **93a-c** [82]. The visible-light-promoted, cross-dehydrogenative benzylic C(sp^3)–H amination reaction was likewise observed to take place under metal- and oxidant-free conditions in the presence of **DCA** as the photoredox catalyst and a Weinreb amide as the nitrogen source [83].

Carbon–Fluorine bond formation. Oxidative benzylic C–H functionalization in the presence of **TCB** as the photocatalyst and Selectfluor as the fluorine source was exploited as a mild, economic, and metal-free route to benzyl fluorides (e.g. **94a-d** in Scheme 3.35). The protocol exhibited a large functional group tolerance, allowing for the quite efficient fluorination of substituted propionaldehydes and propionic acids. Interestingly, the process took place with a satisfactory regioselectivity, steric hindrance being the key factor. The reaction occurred also when electron-poor toluenes were involved, though in lower yields [84]. The photocatalytic fluorination of benzene and haloarenes has been carried out by irradiating the starting substrate in oxygen-saturated acetonitrile (MeCN) in the presence of 2.5 equiv of $Et_4NF·HF$ and **QuCN$^+$ClO$_4^-$** (10 mol%) as the photocatalyst [85].

The tandem photocatalyzed-ring opening/fluorination of cyclopropanols **95a,b** (Scheme 3.36) by **TCB** (10 mol%) as the photocatalyst and Selectfluor as the fluorine source has also been reported

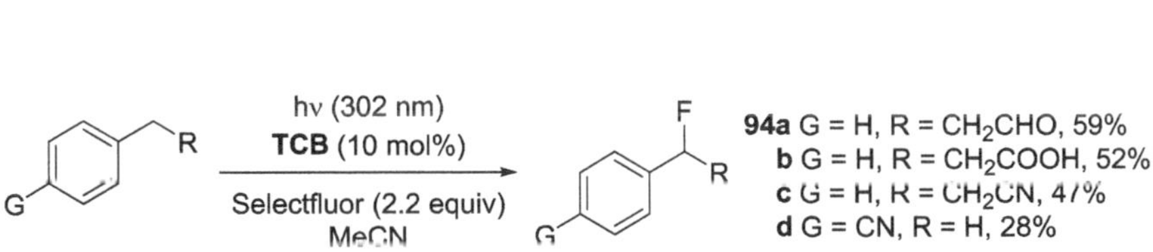

Scheme 3.35. Photocatalyzed synthesis of benzyl fluorides.

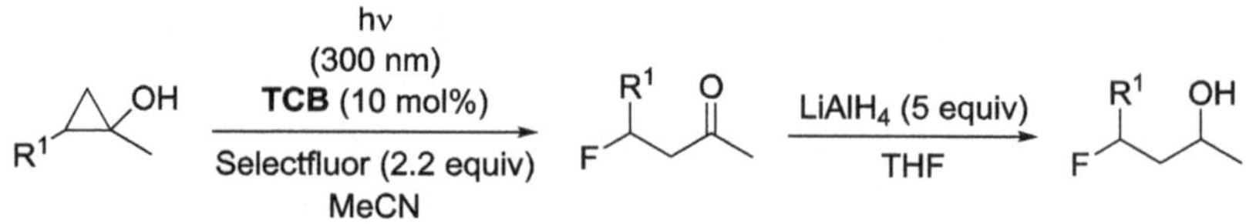

95a,b

96a $R^1 = C_6H_5$, 77%
b $R^1 = C_6H_5CH_2$, 72%

97a, 73%
b, 68%

Scheme 3.36. Photoorganocatalyzed tandem ring opening/fluorination.

as a route for the formation of β-fluorinated ketones (e.g. **96a,b**). Notably, a reductive work-up of the reaction mixture which includes the treatment of the crude product with $LiAlH_4$ furnished the corresponding β-fluorinated alcohols **97a,b** in good to excellent yields [86].

3.4 Conclusions

The photochemistry of aromatic derivatives has been known since a long time, albeit these compounds have been underused as photocatalysts for synthetic applications. However, recent years display a different trend, with many new applications appearing in the literature. These derivatives are particularly appreciated for the redox chemistry accessible through photochemical means. In particular, chemists have started to realize that these compounds offer otherwise inaccessible oxidation and reduction potentials in the excited state, often outclassing those of the corresponding metal-based complexes.

One of the main drawback of this class of photocatalysts is their absorption spectrum, which is often limited to the UV range. The increase of the size of the aromatic system is a possibility to shift the absorption to the visible and, apart from classic examples (e.g. **DCA**), in recent years, many new colored aromatics have been proposed in the role of photoredox catalysts (e.g. **4CzIPN**). Apart from some exceptions, catalyst loadings are quite low, different from other classes of **POCs**.

It is indeed expected that the use of aromatics as photocatalysts will grow at a regular pace in the near future, based on their unique properties in the promotion of single-electron-transfer steps.

References

[1] Fox, M. A.; Chanon, M. (Eds.), *Photoinduced Electron Transfer*, Elsevier, Amsterdam, 1988.

[2] Mella, M.; Fagnoni, M.; Freccero, M.; Fasani, E.; Albini, A. *Chem. Soc. Rev.* **1998**, *27*, 81–89.

[3] Mizuno, K.; Hayamizu, T. Maeda, H. *Pure Appl. Chem.* **2003**, *75*, 1049–1054.

[4] Hoffmann, N. *J. Photochem. Photobiol. C Photochem. Rev.* **2008**, *9*, 43–60.

[5] (a) Narayanam, J. M. R.; Stephenson, C. R. J. *Chem. Soc. Rev.* **2011**, *40*, 102–113; (b) Prier, C. K.; Rankic, D. A.; MacMillan, D. W. C. *Chem. Rev.* **2013**, *113*, 5322–5363; (c) Ravelli, D.; Protti, S.; Fagnoni, M. *Chem. Rev.* **2016**, *116*, 9850–9913; (d) Skubi, K. L.; Blum, T. R.; Yoon, T. P. *Chem. Rev.* **2016**, *116*, 10035–10074; (e) Romero, N. A.; Nicewicz, D. A. *Chem. Rev.* **2016**, *116*, 10075–10166.

[6] Capaldo, L.; Ravelli, D. *Eur. J. Org. Chem.* **2017**, *2017*, 2056–2071.

[7] Protti, S.; Fagnoni, M.; Ravelli, D. *ChemCatChem* **2015**, *7*, 1516–1523.

[8] Dabestani, R.; lvanov, I. N. *Photochem. Photobiol.* **1999**, *70*, 10–34.

[9] Eds. Montalti, M.; Credi, A.; Prodi, L.; Gandolfi, M. T. *Handbook of Photochemistry*, CRC Press, Boca Raton, 2006.

[10] Lipp, B.; Lipp, A.; Detert, H.; Opatz, T. *Org. Lett.* **2017**, *19*, 2054–2057.

[11] Seo, H.; Katcher, M. H.; Jamison, T. F. *Nat. Chem.* **2017**, *9*, 453–456.

[12] Seo, H.; Liu, A.; Jamison, T. F. *J. Am. Chem. Soc.* **2017**, *139*, 13969–13972.

[13] Hasegawa, E.; Takizawa, S.; Seida, T.; Yamaguchi, A.; Yamaguchi, N.; Chiba, N.; Takahashi, T.; Ikeda, H.; Akiyama, K. *Tetrahedron* **2006**, *62*, 6581–6588.

[14] Shibata, T.; Kabumoto, A.; Shiragami, T.; Ishitani, O.; Pac, C.; Yanagida, S. *J. Phys. Chem.* **1990**, *94*, 2068–2076.

[15] Noto, N.; Koike, T.; Akita, M. *Chem. Sci.* **2017**, *8*, 6375–6379.

[16] Das, A.; Ayad, S.; Hanson, K. *Org. Lett.* **2016**, *18*, 5416–5419.

[17] Kumar, R.; Gleißner, E. H.; Tiu, E. G. V.; Yamakoshi, Y. *Org. Lett.* **2016**, *18*, 184–187.

[18] Koutoulogenis, G. S.; Kokotou, M. G.; Voutyritsa, E.; Limnios, D.; Kokotos, C. G. *Org. Lett.* **2017**, *19*, 1760–1763.

[19] Pandey, G.; Rao, K. S. S. P.; Rao, K. V. N. *J. Org. Chem.* **2000**, *65*, 4309–4314.

[20] Pandey, G.; Tiwari, S. K.; Singh, B. *Tetrahedron Lett.* **2016**, *57*, 4480–4483.

[21] Yoshimi, Y. *J. Photochem. Photobiol. A Chem.* **2017**, *342*, 116–130.

[22] Yoshimi, Y.; Washida, S.; Okita, Y.; Nishikawa, K.; Maeda, K.; Hayashi, S.; Morita, T. *Tetrahedron Lett.* **2013**, *54*, 4324–4326.

[23] Okamoto, S.; Kojiyama, K.; Tsujioka, H.; Sudo, A. *Chem. Commun.* **2016**, *52*, 11339–11342.

[24] Okamoto, S.; Ariki, R.; Tsujioka, H.; Sudo, A. *J. Org. Chem.* **2017**, *82*, 9731–9736.

[25] Fidaly, K.; Ceballos, C.; Falguières, A.; Veitia, M. S.-I.; Guy, A.; Ferroud, C. *Green Chem.* **2012**, *14*, 1293–1297.

[26] Zhang, M.; Rouch, W. D.; McCulla, R. D. *Eur. J. Org. Chem.* **2012**, 6187–6196.

[27] Hasegawa, E.; Seida, T.; Chiba, N.; Takahashi, T.; Ikeda, H. *J. Org. Chem.* **2005**, *70*, 9632–9635.

[28] Hasegawa, E.; Chiba, N.; Takahashi, T.; Takizawa, S.; Kitayama, T.; Suzuki, T. *Chem. Lett.* **2004**, *33*, 18–19.

[29] Itou, T.; Yoshimi, Y.; Morita, T.; Tokunaga, Y.; Hatanaka, M. *Tetrahedron* **2009**, *65*, 263–269.

[30] Lipp, B.; Nauth, A. M.; Opatz, T. *J. Org. Chem.* **2016**, *81*, 6875–6882.

[31] Nauth, A. M.; Lipp, A.; Lipp, B.; Opatz, T. *Eur. J. Org. Chem.* **2017**, *2017*, 2099–2103.

[32] Matsuoka, S.; Kohzuki, T.; Pac, C.; Ishida, A.; Takamuku, S.; Kusaba, M.; Nakashima, N.; Yanagida, S. *J. Phys. Chem.* **1992**, *96*, 4437–4442.

[33] Wenner, A.; Bätzner, E.; Wagenknecht, H.-A. *Synlett* **2012**, *23*, 2803–2807.

[34] Weiser, M.; Hermann, S.; Penner, A.; Wagenknecht, H.-A. *Beilstein J. Org. Chem.* **2015**, *11*, 568–575.

[35] Hermann, S.; Wagenknecht, H.-A. *J. Pept. Sci.* **2017**, *23*, 563–566.

[36] Latassa, D.; Enger, O.; Thilgen, C.; Habicher, T.; Offermanns, H.; Diederich, F. *J. Mater. Chem.* **2002**, *12*, 1993–1995.

[37] Jensen, A. W.; Daniels, C. *J. Org. Chem.* **2003**, *68*, 207–210.

[38] Lu, J.; Khetrapal, N. S.; Johnson, J. A.; Zeng, X. C.; Zhang, J. *J. Am. Chem. Soc.* **2016**, *138*, 15805–15808.

[39] Muktha, B.; Madras, G.; Guru Row, T. N.; Scherf, U.; Patil, S. *J. Phys. Chem. B* **2007**, *111*, 7994–7998.

[40] Ghosh, I.; Shaikh, R. S.; König, B. *Angew. Chem. Int. Ed.* **2017**, *56*, 8544–8549.

[41] Marchini, M.; Bergamini, G.; Cozzi, P. G.; Ceroni, P.; Balzani, V. *Angew. Chem. Int. Ed.* **2017**, *56*, 12820–12821.

[42] Ghosh, I.; Bardagi, J. I.; König, B. *Angew. Chem. Int. Ed.* **2017**, *56*, 12822–12824.

[43] Freccero, M.; Fasani, E.; Albini, A. *J. Org. Chem.* **1998**, *58*, 1740–1745.

[44] Mella, M.; Fagnoni, M.; Freccero, M.; Fasani, E.; Albini, A. *Chem. Soc. Rev.* **1998**, *27*, 81–89.

[45] (a) Fagnoni, M.; Protti, S.; Ravelli, D.; Albini, A. *Beilstein J. Org. Chem.* **2013**, *9*, 800–808; (b) Protti, S.; Fagnoni, M.; Monti, S.; Réhault, J.; Poizat, O.; Albini, A. *RSC Adv.* **2012**, *2*, 1897–1904.

[46] Wu, B.; Maroncelli, M.; Castner, Jr, E. W. *J. Am. Chem. Soc.* **2017**, *139*, 14568–14585.

[47] Zhao, Y.; Zhang, C.; Chin, K. F.; Pytela, O.; Wei, G.; Liu, H.; Burês, F.; Jiang, Z. *RSC Adv.* **2014**, *4*, 30062–30067.

[48] Luo, J.; Zhang, J. *ACS Catal.* **2016**, *6*, 873–877.

[49] Ohkubo, K.; Suga, K.; Morikawa, K.; Fukuzumi, S. *J. Am. Chem. Soc.* **2003**, *125*, 12850–12859.

[50] Bonesi, S. M.; Fagnoni, M. *Chem. Eur. J.* **2010**, *16*, 13572–13589.

[51] Bonesi, S. M.; Protti, S.; Albini, A. *J. Org. Chem.* **2016**, *81*, 11678–11685.

[52] Fagnoni, M.; Mella, M.; Albini, A. *J. Am. Chem. Soc.* **1995**, *117*, 7877–7881.

[53] (a) de Lijser, H. J. P.; Arnold, D. R. *J. Org. Chem.* **1997**, *62*, 8432–8438; (b) Tang, J.; Yue, J.-J.; Tao, F.-F.; Grampp, G.; Wang, B. X.; Li, F.; Liang, Z.-Z.; Shen, Y.-M.; Xu, J.-H. *J. Org. Chem.* **2014**, *79*, 7572–7582.

[54] (a) Miyake, Y.; Nakajima, K.; Nishibayashi, Y. *Chem. Eur. J.* **2012**, *18*, 16473–16477; (b) Cui, L.; Matusaki, Y.; Tada, N.; Miura, T.; Uno, B.; Itoh, A. *Adv. Synth. Catal.* **2013**, *355*, 2203–2207.

[55] Fagnoni, M.; Mella, M., Albini, A. *J. Am. Chem. Soc.* **1995**, *117*, 7877–7881.

[56] (a) Fagnoni, M.; Mella, M.; Albini, A. *J. Phys. Org. Chem.* **1997**, *10*, 777–780; (b) Fagnoni, M.; Mella, M.; Albini, A. *J. Org. Chem.* **1998**, *63*, 4026–4033.

[57] Jonas, M.; Blechert, S.; Steckhan, E. *J. Org. Chem.* **2001**, *66*, 6896–6904.

[58] Liu, X.; Ye, X.; Bureš, F.; Liu, H.; Jiang, Z. *Angew. Chem. Int. Ed.* **2015**, *54*, 11443–11447.

[59] Patel, N. R.; Kelly, C. B.; Siegenfeld, A. P.; Molander, G. A. *ACS Catal.* **2017**, *7*, 1766–1770.

[60] Lang, S. B.; Wiles, R. J.; Kelly, C. B.; Molander, G. A. *Angew. Chem. Int. Ed.* **2017**, *56*, 15073–15077.

[61] Huang, H.; Yu, C.; Zhang, Y.; Zhang, Y.; Mariano, P. S.; Wang, W. *J. Am. Chem. Soc.* **2017**, *139*, 9799–9802.

[62] Wei, G.; Zhang, C.; Bureš, F.; Ye, X.; Tan, C.-H.; Jiang, Z. *ACS Catal.* **2016**, *6*, 3708–3712.

[63] Liu, Y.; Li, J.; Ye, X.; Zhao, X.; Jiang, Z. *Chem. Commun.* **2016**, *52*, 13955–13958.

[64] Matsui, J. K.; Molander, G. A. *Org. Lett.* **2017**, *19*, 436–439.

[65] Lévêque, C.; Chenneberg, L.; Corcé, V.; Ollivier, C.; Fensterbank, L. *Chem. Commun.* **2016**, *52*, 9877–9880.

[66] Gutiérrez-Bonet, A.; Tellis, J. C.; Matsui, J. K.; Vara, B. A.; Molander, G. A. *ACS Catal.* **2016**, *6*, 8004–8008.

[67] Pandey, G.; Karthikeyan, M.; Murugan, A. *J. Org. Chem.* **1998**, *63*, 2867–2872.

[68] Pandey, G.; Murugan, A.; Balakrishnan, M. *Chem. Commun.* **2002**, 624–625.

[69] Matsui, J. K.; Molander, G. A. *Org. Lett.* **2017**, *19*, 436–439.

[70] Yang, C.; Yang, J.-D.; Li, Y.-H.; Li, X.; Cheng, J.-P. *J. Org. Chem.* **2016**, *81*, 12357–12363.

[71] Pandey, G.; Pal, S.; Laha, R. *Angew. Chem. Int. Ed.* **2013**, *52*, 5146–5149.

[72] Huang, H.; Li, X.; Yu, C.; Zhang, Y.; Mariano, P. S.; Wang, W. *Angew. Chem. Int. Ed.* **2017**, *56*, 1500–1505.

[73] Meng, Q.-Y.; Wang, S.; König, B. *Angew. Chem. Int. Ed.* **2017**, *56*, 13426–13430.

[74] (a) Lacombe, S.; Cardy, H.; Simon, M.; Khoukh, A.; Soumillion, J. P.; Ayadim, M. *Photochem. Photobiol. Sci.* **2002**, *1*, 347–354; (b) Bonesi, S. M.; Fagnoni, M.; Albini, A. *Eur. J. Org. Chem.* **2008**, 2612–2620.
[75] Ohkubo, K.; Suga, K.; Morikawa, K.; Fukuzumi, S. *J. Am. Chem. Soc.* **2003**, *125*, 12850–12859.
[76] (a) Griesbeck, A. G.; Sadlek, O.; Polborn, K. *Liebigs Ann.* **1996**, 545–549; (b) Sadlek, O.; Gollnick, K.; Polborn, K.; Griesbeck, A. G. *Angew. Chem. Int. Ed. Engl.* **1994**, *33*, 2300–2301.
[77] Qin, H.-T.; Xu, X.; Liu, F. *ChemCatChem* **2017**, *9*, 1409–1412.
[78] Qrareya, H.; Ravelli, D.; Fagnoni, M.; Albini, A. *Adv. Synth. Catal.* **2013**, *355*, 2891–2899.
[79] Ohkubo, K.; Kobayashi, T.; Fukuzumi, S. *Angew. Chem. Int. Ed.* **2011**, *50*, 8652–8655.
[80] Ohkubo, K.; Kobayashi, K.; Fukuzumi, S. *Opt. Express*, **2012**, *20*, A360–A365.
[81] Cho, D. W.; Parthasarathi, R.; Pimentel, A. S.; Maestas, G. D.; Park, H. J.; Yoon, U. C.; Dunaway-Mariano, D.; Gnanakaran, S.; Langan, P.; Mariano, P. S. *J. Org. Chem.* **2010**, *75*, 6549–6562.
[82] Hofstra, J. L.; Grassbaugh, B. R.; Tran, Q. M.; Armada, N. R.; de Lijser, P. H. J. *J. Org. Chem.* **2015**, *80*, 256–265.
[83] Pandey, G.; Laha, R. *Angew. Chem. Int. Ed.* **2015**, *54*, 14875–14879.
[84] Bloom, S.; McCann, M.; Lectka, T. *Org. Lett.* **2014**, *16*, 6338–6341.
[85] Ohkubo, K.; Fujimoto, A. Fukuzumi, S. *J. Phys. Chem. A* **2013**, *117*, 10719–10725.
[86] Bloom, S.; Bume, D. D.; Pitts, C. R.; Lectka, T. *Chem. Eur. J.* **2015**, *21*, 8060–8063.

Chapter 4

Sulfur Heterocycles

Katarzyna Goliszewska, Katarzyna Orłowska, and Dorota Gryko*

Institute of Organic Chemistry, Polish Academy of Sciences, Kasprzaka 44/52, 01-224 Warsaw, Poland

**dorota.gryko@icho.edu.pl*

4.1 Introduction

One of the most important aspects, when designing reactions induced by light, is the choice of the photocatalyst. Sulfur heterocycles as organic dyes are attractive alternatives to metal complexes commonly used in this area of research. Undoubtedly, methylene blue (**MB**$^+$) is the most widely applied sulfur heterocycle due to its ability to catalyze reactions — not only by energy transfer (*EnT*), but also by single electron transfer (*SET*) (Chart 4.1). Other thiazine dyes, such as thiapyrylium salts, are considered to act only through the latter mechanism.

Structural diversity and a plethora of modification possibilities, which can significantly affect their spectroelectrochemical properties, offer a lot of opportunities for sulfur dyes to be used as photocatalysts in various reactions. They have been proved to be efficient in

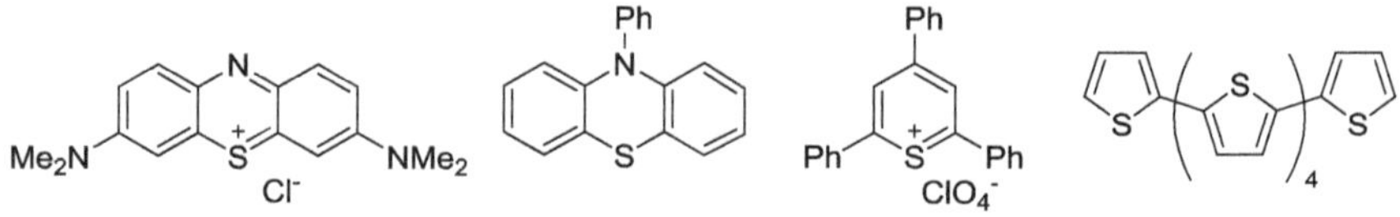

MB⁺ **PTH** **TPTP⁺** **6-TP**

Chart 4.1. Structures of most commonly used sulfur photocatalysts — methylene blue chloride (**MB⁺**), *N*-phenylphenothiazine (**PTH**), 2,4,6-triphenylthiapyrylium perchlorate (**TPTP⁺**), and α-sexithiophene (**6-TP**).

detoxification processes and in the synthesis of biologically active compounds, highlighting the importance of expanding our knowledge of these organic dyes.

In the following chapter, we describe sulfur heterocycles possessing photocatalytic activity (Chart 4.1). Their main modes of action under either photosensitization or photoinduced electron transfer are discussed, along with a number of reactions catalyzed by these dyes.

4.2 MB⁺

MB⁺ is the most widely used phenothiazine dye and, consequently, the most often applied sulfur heterocycle in photochemistry. It was first synthesized by Heinrich Caro in 1876 and was reported as the first fully synthetic drug used in medicine [1]. Initially, applications of **MB⁺** in biology and chemistry were related to its ability to generate singlet oxygen. Recently, emphasis has been put on the utilization of its redox properties under light irradiation.

4.2.1 *Photophysical characteristics*

MB⁺ as the hydrochloride salt is a dark green powder, giving blue aqueous solutions. It absorbs light in the visible region with λ_{max} at ~650 nm (depending on the solvent, protonation, or aggregation processes, Figure 4.1), with a high extinction coefficient value ($\varepsilon = 90000\ M^{-1}\ cm^{-1}$ at 610 nm) [2].

Photon absorption by **MB⁺** generates the short-lived singlet excited state ^{1}(**MB⁺**)* (τ_0 (S_1) ~ 1 ns) of high energy ($E_{0,0}$ (S_1) = 1.89 eV), which undergoes fast intersystem crossing (ISC) to the triplet state

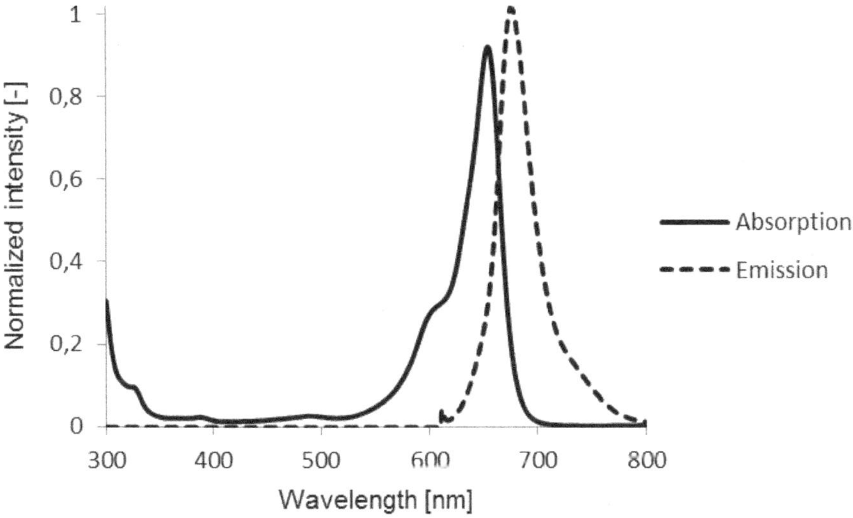

Figure 4.1. Absorption (solid line) and emission (dashed line) spectra of **MB**$^+$ in DCM, $c = 1.1 \times 10^{-8}$ M, $\lambda_{ex} = 610$ nm.

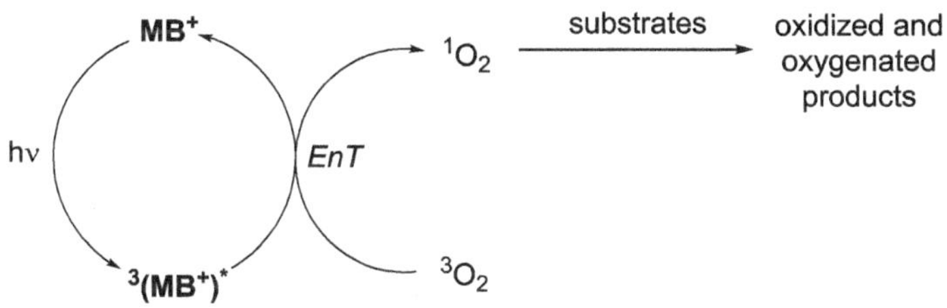

Scheme 4.1. Photosensitization of singlet oxygen.

^{3}(**MB**$^+$)*. The latter is formed with a high quantum yield ($\Phi_{ISC} = 0.52$) and has a significantly longer lifetime (τ_0 (T_1) ~32 μs) as well as lower energy ($E_{0,0}$ (T_1) = 1.5 eV) [2, 3]. As a consequence, the dye is mainly used as a triplet sensitizer.

^{3}(**MB**$^+$)* can be quenched by various compounds *via* two possible pathways — *EnT* or *SET*. Examples of transformations exploiting both quenching patterns concomitantly were also reported.

EnT involves energy transfer between ^{3}(**MB**$^+$)* and the ground-state molecule, often molecular oxygen (Scheme 4.1). Short-lived reactive singlet oxygen ($^1O_2$) generated *via* this pathway readily oxidizes a vast number of organic compounds. The high value of the triplet quenching rate constant ($k_q = 2.46 \times 10^9$ M^{-1} s^{-1}) for $^3O_2$ is beneficial for reactions involving $^1O_2$, however, in *SET* processes a decrease in yield is observed [4].

However, not all oxidations occur with $^1O_2$. Methylene blue radical (**MB$^•$**), formed by *SET* from electron donors to ^{3}(**MB$^+$**)*, can transfer one electron to $^3O_2$ producing superoxide anion $O_2^{•-}$ that also can act as an oxidant [4].

In the presence of electron donors, especially amines, ^{3}(**MB$^+$**)* is reduced to radical **MB$^•$** *via* the *SET* pathway (Scheme 4.2). Transfer of a second electron and protonation, either in a concerted (hydrogen atom transfer (*HAT*)) or stepwise manner, generates colorless *leuco*-methylene blue (**LMBH**). Under aerobic conditions, molecular oxygen oxidizes **LMBH**, regenerating **MB$^+$** in the ground state almost immediately ($k_{obs} = 5.0 \times 10^5$ M^{-1} s^{-1}). The comparison of redox potentials with the most common transition metal photocatalysts shows that in the triplet state ^{3}(**MB$^+$**)* is a stronger oxidant than $Ru(bpy)_3^{2+}$ or $Ir(ppy)_3$, but a weaker reductant (Table 4.1) [3, 5]. As a result, all known photoredox transformations catalyzed by **MB$^+$** proceed through the reductive quenching cycle.

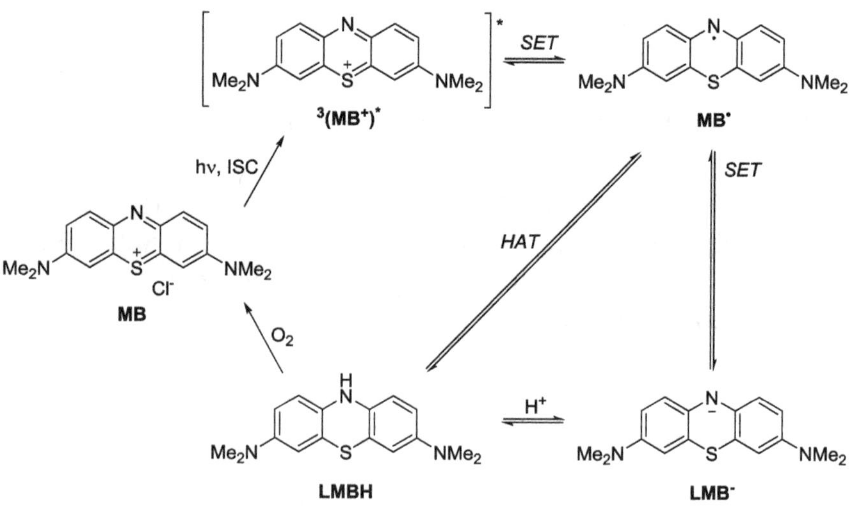

Scheme 4.2. Photoinduced redox cycle of **MB$^+$**.

Table 4.1. Comparison of triplet excited state redox potentials for **MB**$^+$, Ru(bpy)$_3^{2+}$, and Ir(ppy)$_3$.

	E^* of ^{3}**PC** [V vs SCE]	
Catalyst	E^*_{RED} **PC*/PC**$^-$	E^*_{OX} **PC**$^+$**/PC***
MB$^+$	**+1.60**[a]	**–0.68**[a]
Ru(bpy)$_3^{2+}$	+0.77[b]	–0.81[b]
Ir(ppy)$_3$	+0.31[b]	–1.73[b]

Notes: [a]Measured in methanol. [b]Measured in acetonitrile.

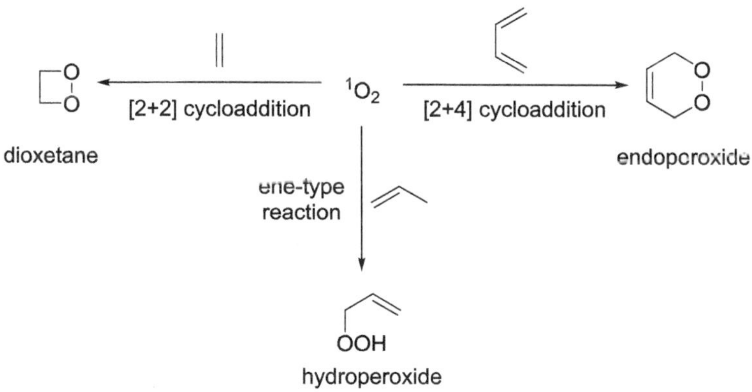

Scheme 4.3. Reactions of singlet oxygen with substrates having C–C double bonds.

4.2.2 *MB*$^+$ *as a photosensitizer for singlet oxygen generation*

Singlet oxygen is a highly reactive, energetically rich form of molecular oxygen. The two main pathways for the generation of $^1O_2$ include chemical transformations and photosensitization by irradiation of oxygen in the presence of organic dyes. **MB**$^+$, along with Rose Bengal, tetraphenylporphyrin, and eosin Y are among the most commonly used sensitizers in this process [6].

In **MB**$^+$-catalyzed reactions, $^1O_2$ reacts with substrates with double bonds following one of three main mechanisms — ene-type reactions or [2+2] and [2+4] cycloadditions (Scheme 4.3). In the following paragraphs, we discuss known photocatalytic transformations proceeding *via* these pathways.

4.2.2.1 *Ene-type reactions*

Alkenes possessing at least one allylic hydrogen atom in the presence of $^1O_2$ form allylic hydroperoxides in which the double bond is shifted to the adjacent position. Due to the high impact of steric and electronic factors on the hydroperoxide formation, the mechanism and regioselectivity of the ene-type reaction of substituted alkenes with $^1O_2$ have been extensively studied [7]. For trisubstituted alkenes, hydrogen-atom abstraction occurs preferentially from the allylic position on the more sterically-hindered side of the double bond (*cis*-effect, Figure 4.2(a)) [8], while for unsymmetrical *cis*- and *trans*-olefins it is on the larger alkyl group of the double bond (Figure 4.2(b)) [9]. Moreover, alkyl- or aryl-substituted cyclic alkenes **1** undergo regioselective hydrogen abstraction from the methylene group next to the substituent, and the ratio of products **2a-f**:**3a-f** is significantly higher for phenyl- and *tert*-butyl-substituted cycloalkenes than for neopentyl derivatives (Table 4.2) [10].

Regioselectivity of the hydrogen abstraction is also influenced by the photocatalyst form, since for reactions occurring in the interior of **MB**$^+$-doped zeolites (e.g. NaMBY), the *cis*-effect decreases (Table 4.3) [11]. In solution, the sum of *cis*-effect products (**a** and **b**) is always higher than 80%, while in zeolites it falls to 44%. On the other hand, in CD_3CN, *gem/trans* selectivity (**a**+**c**/**b**) is low (the ratio of secondary to tertiary peroxides varies from 50:50 to 37:63 in the best case), whereas zeolites enhance the selectivity up to 94:6. It is assumed that these differences arise from the formation of complexes in the interior of the zeolite between Na^+ and the substrate or the pendant oxygen of the perepoxide intermediate

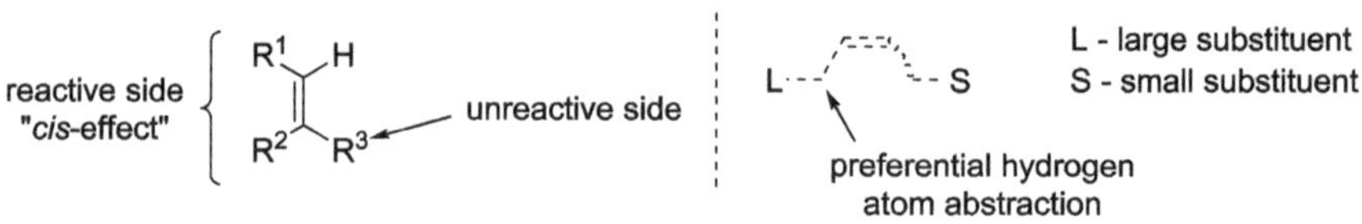

(a) (b)

Figure 4.2. Regioselective aspects of the ene-type reaction of trisubstituted alkenes (a) and *cis*- and *trans*-olefins (b).

Table 4.2. **MB**$^+$-catalyzed regioselective formation of allylic hydroperoxides.

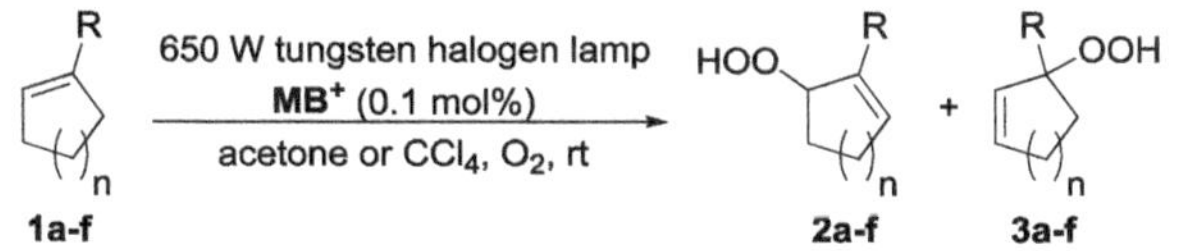

	Substrate	Yield [%] 2	3		Substrate	Yield [%] 2	3
1a	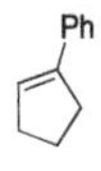	100	—	**1d**	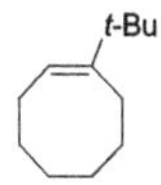	94	6
1b	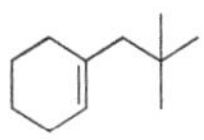	71	29	**1e**	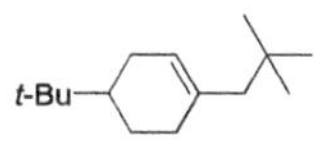	62	38
1c	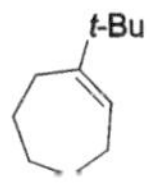	97	3	**1f**	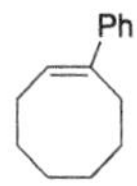	97	3

Table 4.3. Photooxidation of trisubstituted alkenes **4–7** in solution and with zeolites Y.

	Yield [%]				Yield [%]		
	cis-effect		*anti cis*-effect,		*cis*-effect		*anti cis*-effect,
Substrate	*gem*	*trans*	*gem*	Substrate	*gem*	*trans*	*gem*
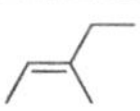	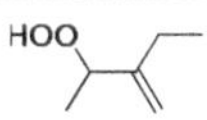	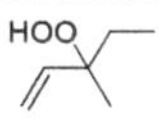	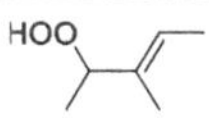	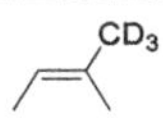	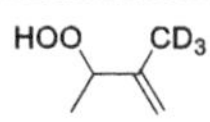	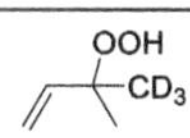	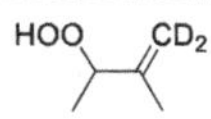
4	**8a**	**8b**	**8c**	**6**	**10a**	**10b**	**10c**
CD_3CN	49	50	1	CD_3CN	38	55	7
NaMBY	57	32	11	NaMBY	36	21	43
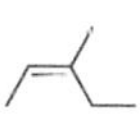	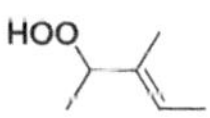	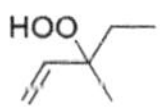	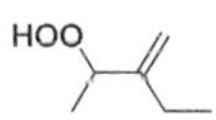	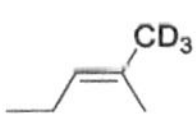	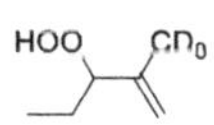	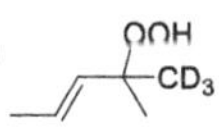	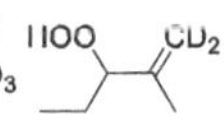
5	**9a**	**9b**	**9c**	**7**	**11a**	**11b**	**11c**
CD_3CN	18	63	19	CD_3CN	48	44	8
NaMBY	9	35	56	NaMBY	61	6	33

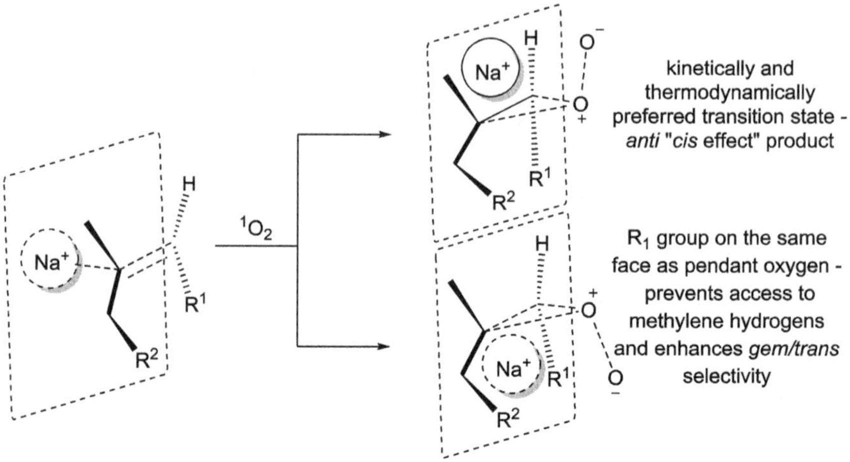

Scheme 4.4. Alkenes photooxidation in the interior of **MB**$^+$-doped zeolites Y.

(Scheme 4.4). The proposed mechanism is also in agreement with intrazeolite photooxidations of allylic alcohols [12].

Although **MB**$^+$-catalyzed photooxidations of simple alkenes and allylic alcohols, both in solution and in zeolites Y, lead to a mixture of products, the regioselective ene-type reaction of α-cyclogeranyl derivatives giving selectively one of the three possible regioisomers and displaying high *cis*-diastereoselectivity has been reported (Scheme 4.5) [13].

Transition states analysis indicates that $^1O_2$ attack occurs preferentially from the upper face of the double bond giving *cis*-product **13a-d** from the more stable **TS1** with the R-substituent in the axial position.

Other substrates for the **MB**$^+$-catalyzed ene-type reaction are 1,3-dicarbonyl compounds **13a-c** that give a mixture of products **15a-c–19a-c** (Scheme 4.6) [14, 15]. In this case, the abstraction of the hydrogen atom from the enol group gives products **15a-c**, while from the 2-methyl group furnishes hydroperoxides **16a-c** that decompose to **17a-c** with the loss of H_2O_2. In the final step, H_2O_2 oxidizes **17a-c** to epoxides **18a-c**. This reaction is strongly solvent-dependent; in alcohols, olefin **17a-c** reacts faster with a solvent affording product **19a-c** instead of epoxide **18a-c** (Table 4.4).

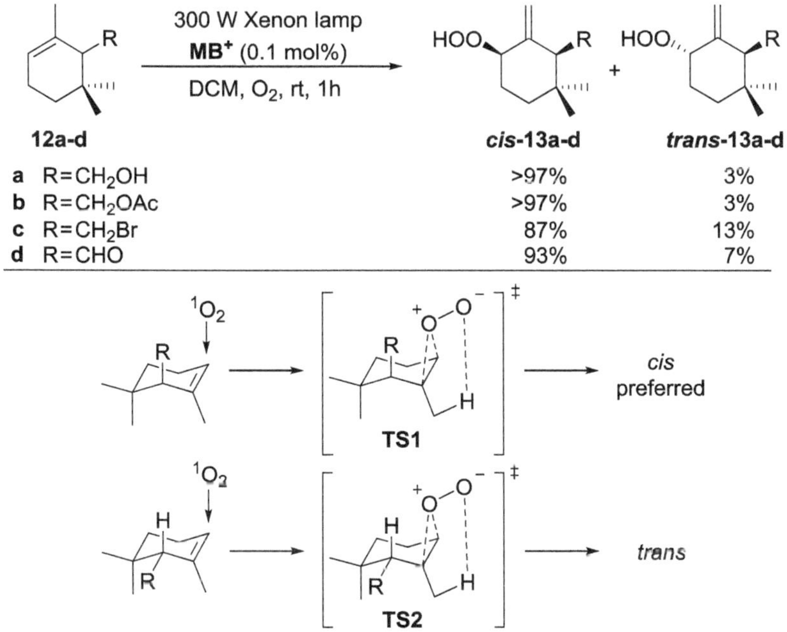

Scheme 4.5. **MB**$^+$-catalyzed regioselective photooxidation of α-cyclogeranyl derivatives.

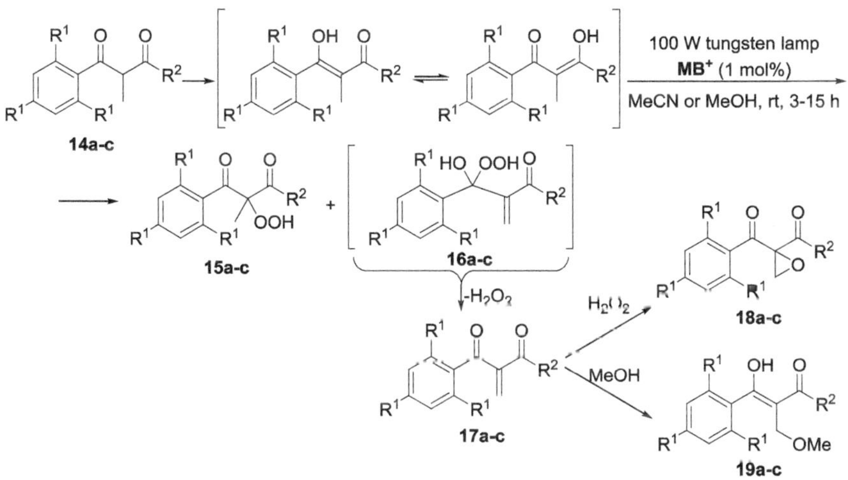

Scheme 4.6. **MB**$^+$-catalyzed reaction of $^1O_2$ with 1,3-dicarbonyl compounds **14a-c**.

Table 4.4. Solvent-dependent photooxidation of 1,3-diketones **14a-c.**

	Substrate		Solvent	Yield [%] **15**	**17**	**18**	**19**
14a	R^1=Me	R^2=Et	MeCN	28	31	10	—
			MeOH	25	—	6	40
14b	R^1=Me	R^2=*i*-Pr	MeCN	32	37	7	—
			MeOH	27	—	7	47
14c	R^1=*i*-Pr	R^2=*i*-Pr	MeCN	36	55	—	—
			MeOH	37	—	—	52

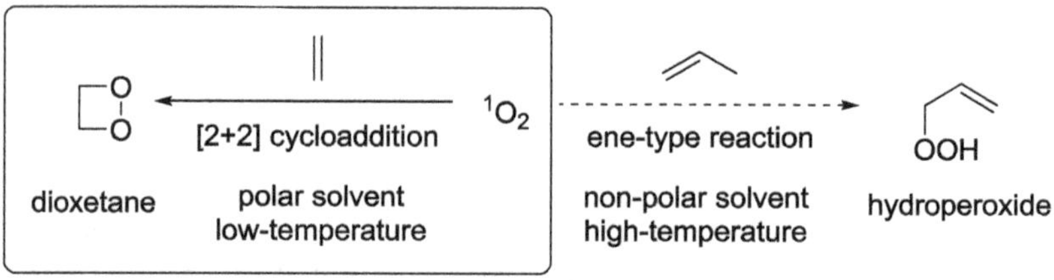

Scheme 4.7. Favorable conditions for dioxetane formation.

4.2.2.2 [2+2] *Cycloaddition*

Next to ene-type reactions, the oxidation with $^1O_2$ photogenerated by $\mathbf{MB^+}$ can follow a [2+2] cycloaddition pathway leading to dioxetane products. For substrates that can also undergo the ene-type reaction, the ratio of products depends on the polarity of a solvent and temperature (Scheme 4.7). Non-polar solvents and higher temperatures favor the ene reaction, while in polar solvents (alcohols), dioxetanes or products obtained from the subsequent nucleophilic addition of the solvent are formed [16]. Notably, dioxetane intermediates readily decompose at an elevated temperature, hence their isolation or even detection is challenging.

For example, the $\mathbf{MB^+}$-sensitized photooxidation of 5,6-dihydro-1,4-dithiins **20a-d** with $^1O_2$ proceeds *via* the [2+2] cycloaddition pathway (Scheme 4.8) [17]. Depending on the electron-withdrawing character of the substituent R, the thermal decomposition of

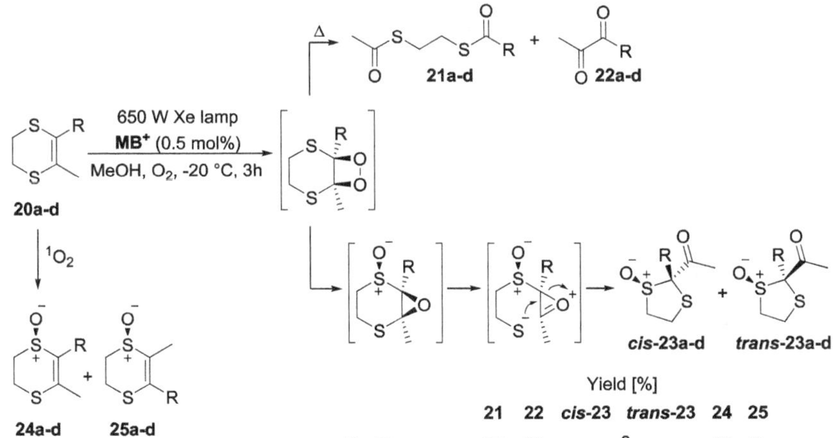

		Yield [%]					
		21	22	*cis*-23	*trans*-23	24	25
a	R=Ph	25	38		8	15	0
b	R=CO_2Me	-	-	67	7	26	-
c	R=CONHPh	20	9	-	3	68	-
d	R=CON(Me)Ph	-	-	44	-	56	-

Scheme 4.8. **MB+**-catalyzed photooxidation of 1,4-dithiins **20a-d**.

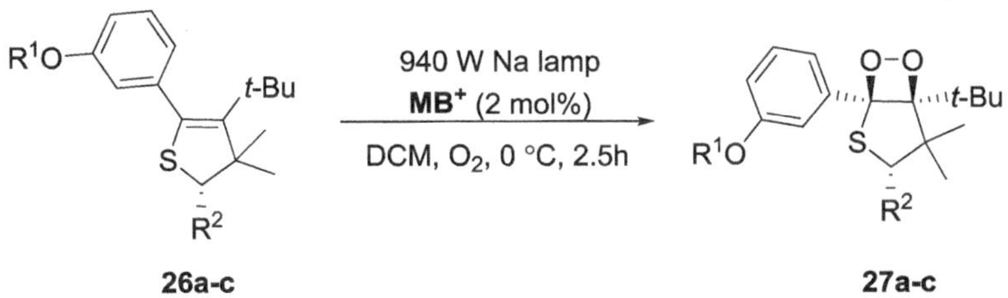

a	R^1=Me	R^2=OAc	88%
b	R^1=Me	R^2=OMe	85%
c	R^1=Ac	R^2=OMe	85%

Scheme 4.9. Synthesis of dioxetanes **27a-c**.

dioxetane to **21a-d** and **22a-d** competes with the formation of keto-sulfoxides **23a-d**. In protic solvents, oxidation of sulfur occurs as a competitive reaction (products **24a-d** and **25a-d**).

Although dioxetanes are usually unstable, the introduction of bulky substituents at the dihydrothiophene ring enables their isolation with no S-oxidation (Scheme 4.9) [18]. $^1O_2$ attacks the double bond exclusively from the less-hindered side, consequently, only *trans*-products are formed.

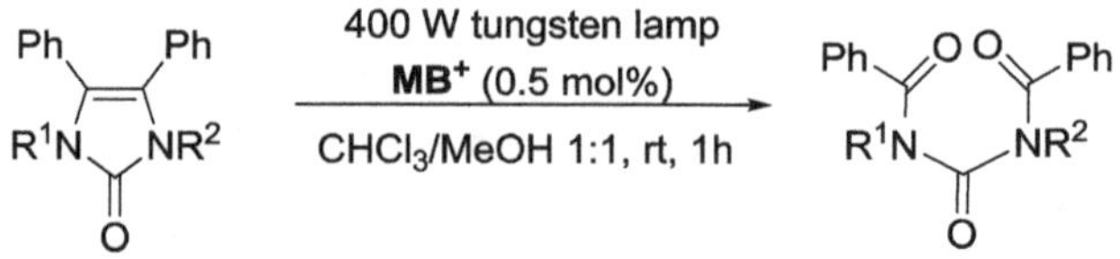

28a-d **29a-d**

a	R^1=H	R^2=H	97%
b	R^1=Ph	R^2=Ph	50%
c	R^1=Me	R^2=Me	90%
d	R^1=*n*-Bu	R^2=*n*-Bu	92%

Scheme 4.10. Synthesis of diacylureas **29a-d** *via* oxidation of imidazolin-2-ones **28a–28d.**

The formation of a dioxetane-type intermediate was also suggested for the **MB**$^+$-sensitized photooxidation of imidazolin-2-ones **28a-d**, though with no clear evidences (Scheme 4.10) [19]. The reaction rate is higher for imidazolin-2-ones **28a-d** with electron-donating substituents as they enhance the nucleophilicity of the double bond, while the oxidation of *N*-phenyl derivatives proceeds slowly due to the nitrogen lone pair being partially conjugated with the phenyl ring.

The [2+2] cycloaddition of $^1O_2$ to pyrrole derivative **30** leads to different products depending on the *N*-substitution pattern [20]. Photooxidation of *N*-methyl pyrrole gives amido diester **31**, while *N*-unsubstituted substrate affords, in the presence of various nucleophiles including alcohols, amines, and *β*-diketones, the corresponding 5-substituted adducts **33a-d** or *α,α′*-coupling products **34** (Scheme 4.11). The more nucleophilic the character of the initially-formed 5-substituted adducts **33a-d**, the higher the yield of *α,α′*-bipyrroles **34a-d**.

4.2.2.3 [2+4] *Cycloaddition*

Among **MB**$^+$-catalyzed oxidations, the most common reactions proceed *via* the [2+4] cycloaddition pathway. They lead to endoperoxides, which, as in the case of dioxetanes, are also highly unstable and undergo subsequent transformations. The course is strongly influenced by the reaction conditions and the structure of the substrate. For example, decomposition of furan-derived endoperoxides follows one of four general pathways leading to products **A-D** (Scheme 4.12).

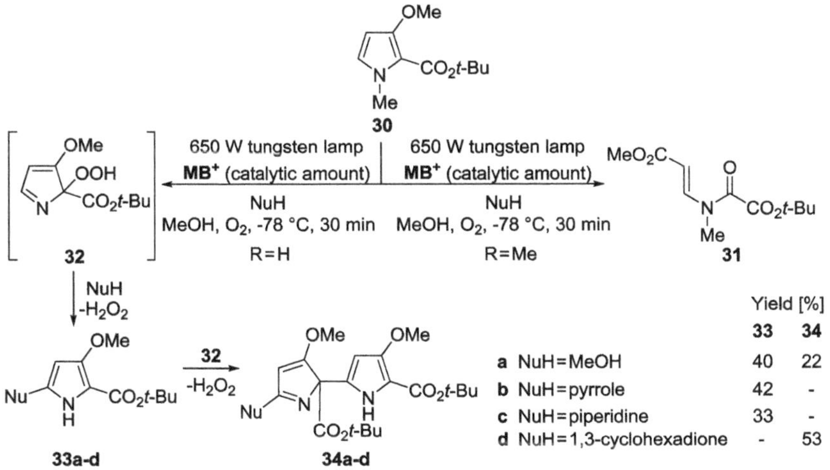

Scheme 4.11. Photooxidation of *N*-substituted and *N*-unsubstituted pyrroles **30**.

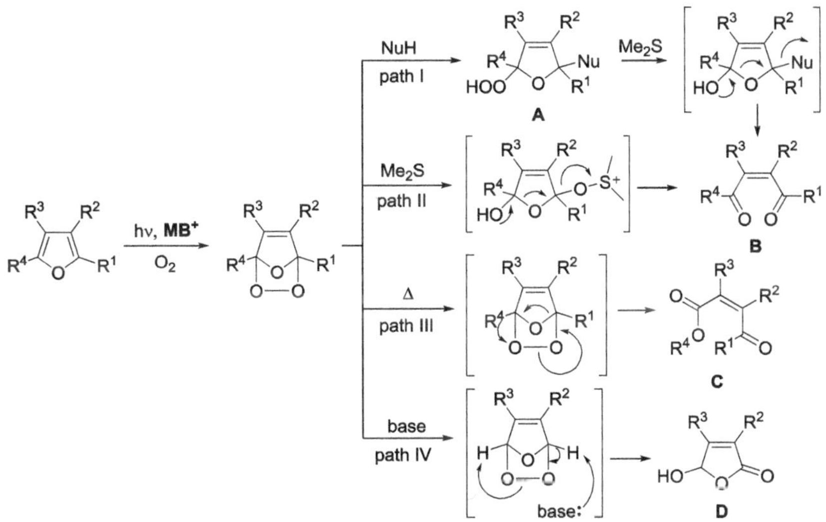

Scheme 4.12. Four pathways for endoperoxide transformation in the [2+4] cycloaddition of $^1O_2$ to furans.

- **Path I**

In alcoholic or aqueous solutions, the oxidation of furan derivatives usually gives hydroperoxydihydrofurans of type **A** (Scheme 4.12,

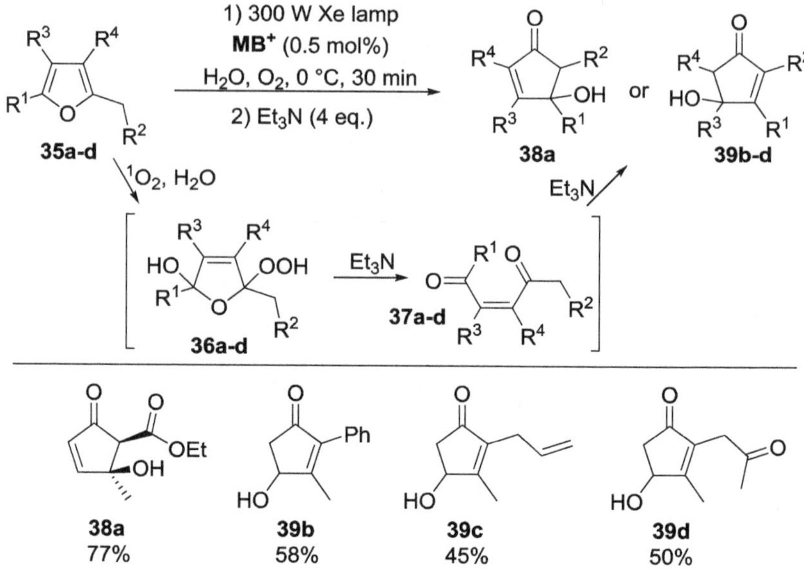

Scheme 4.13. **MB**$^+$-catalyzed one-pot synthesis of 4-hydroxy-substituted-2-cyclopentenones **38a** and **39b-d** from alkyl furans **35a-d**.

path I). This was utilized in the elegant one-pot synthesis of 4-hydroxy-2-cyclopentenones **38a** and **39b-d** (Scheme 4.13) [21]. In this transformation (occurring in the presence of water as the nucleophile), intermediate **36a-d** is formed and then decomposed to enedione **37a-d**. Final cyclization yielding cyclopentanones **38a** and **39b-d** proceeds *via* an aldol reaction catalyzed by the base.

On the other hand, if a nucleophilic group is present in the starting material, an intramolecular reaction preferentially occurs, as in the reaction of 2,5-disubstituted furans **40a-b**, giving bis-spiroketals **43a-b** (Scheme 4.14) [22, 23].

The formation of the first spiro ring results from the intramolecular nucleophilic attack of the hydroxyl group at the endoperoxide ring, whereas the second nucleophilic attack requires prior reduction of hydroperoxide **41a-b** by Me_2S followed by ketalization.

Similarly, bicyclic compounds are formed in the synthesis of pyrrolizidinones **45a-e** (Scheme 4.15) [24]. Notably, $^1O_2$ generated by **MB**$^+$-sensitization reacts exclusively with the furan ring with no amine oxidation.

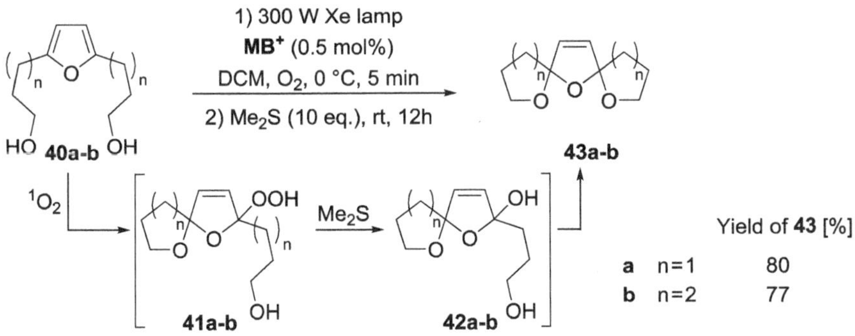

Scheme 4.14. One-pot synthesis of [5,5,5]- and [6,5,6]-bis-spiroketals **43a-b**.

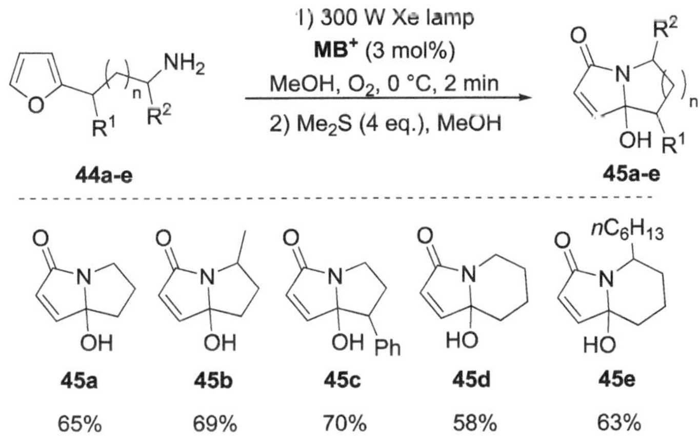

Scheme 4.15. Synthesis of pyrrolizidinone derivatives **45a-e** *via* photoorganocatalyzed oxidation of furylalkylamines **44a-e**.

The synthesis of 5-hydroxy-1*H*-pyrrol-2(5*H*)-ones **50a-c** from alkyl furans **46a-c**, and their dehydrated counterparts, 5-ylidenepyrrol-2(5*H*)-ones **51a-c** proceeding *via* 2-pyrrolidinones **49a-c** is a unique example of transformations utilizing $\mathbf{MB^+}$ as both the photosensitizer and the redox agent in the ground state (Scheme 4.16a) [25]. Lactams **49a-c** generated *via* sequential $^1O_2$ photooxidation, reduction by Me_2S, and then nucleophilic addition of an amine undergo subsequent proton-coupled electron-transfer (*PCET*) catalyzed by $\mathbf{MB^+}$, but this reaction does not require light (Scheme 4.16b). Importantly, such a cascade process is possible only

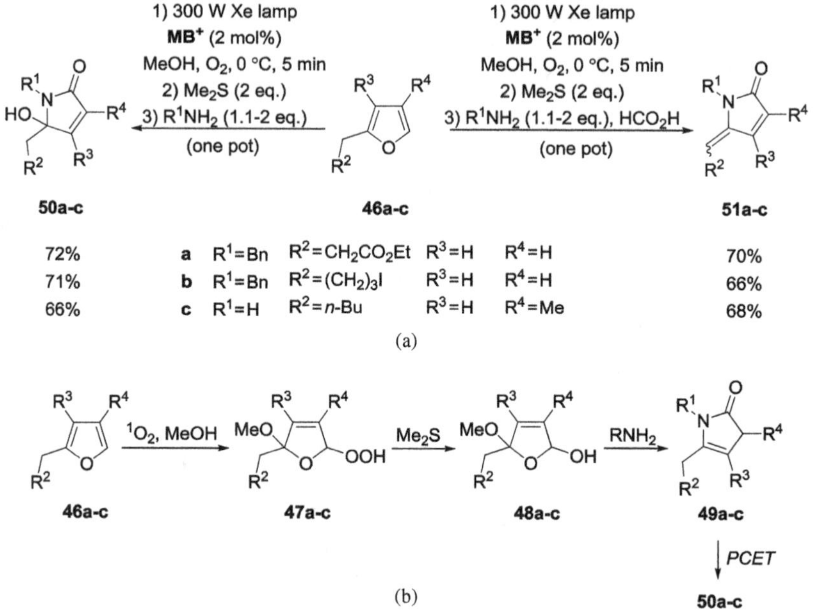

Scheme 4.16. One-pot synthesis of 5-hydroxy-1*H*-pyrrol-2(5*H*)-ones **50a-c** and 5-ylidenepyrrol-2(5*H*)-ones **51a-c** (a) and with a simplified reaction mechanism (b).

in the presence of **MB**$^+$ and does not occur for other sensitizers, such as Rose Bengal.

The path I mechanism was also confirmed for a one-pot stereoselective transformation of 2-methoxyfurans **52a-d** to methyl *cis*-4-oxoalk-2-enoates **53a-d** (Table 4.5) [26]. Interestingly, even furans with substituents sensitive to oxidation, such as the sulfinyl group, are readily transformed to the corresponding enediones [27]. Similarly, 2-thiofurans and 2,5-bis-(thio)furans **54a-d** are oxidized as well, but instead of the expected *S,S*-bis(thio)maleate derivatives **57a-d**, *O,S*-thiomaleates **55a-d** and **56a-d** are formed (Table 4.6) [28]. It is assumed that the fast removal of the phenylthio group enables the rapid decomposition of hydroperoxide A (Scheme 4.12) even in the absence of a reducing agent. *S,S*-Bis(thio)maleates **57a-d** form only in the case of the photooxidation of compound **54c** possessing the methylthio group that does not act as a leaving group.

Table 4.5. Synthesis of methyl *cis*-4-oxoalk-2-enoates **53a-d** *via* **MB**$^+$-sensitized photooxidation.

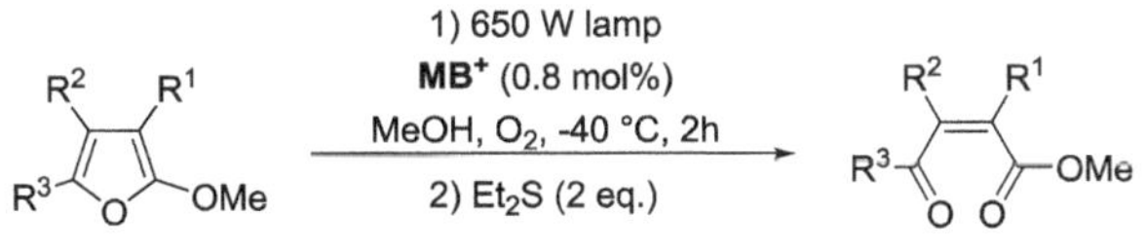

Substrate	R^1	R^2	R^3	Yield of **53** [%]
52a	CO_2Me	Et	Et	77
52b	CO_2Me	Ph	Ph	84
52c	H	CO_2Me	(4-Br)Ph	98
52d	CO_2Me	H	(4-Br)Ph	80

Table 4.6. Synthesis of *O,S*-thiomaleates **55a-d** and **56a-d** *via* **MB**$^+$-sensitized photooxidation.

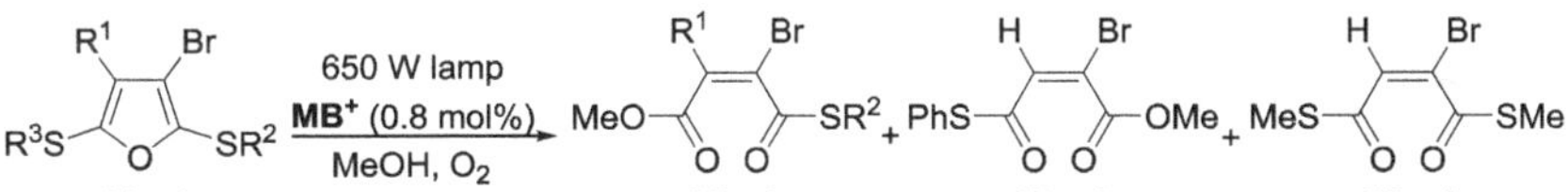

Substrate	R^1	R^2	R^3	Yield [%]		
				55	**56**	**57**
54a	Br	Ph	Ph	96	—	—
54b	H	Ph	Ph	87	6	—
54c	H	Me	Me	59	—	28
54d	H	Me	Ph	84	6	—

- **Path II**

By changing methanol or water to non-nucleophilic solvents, the photooxidation of furans follows path II (Scheme 4.12). In acetonitrile, $^1O_2$-induced rearrangement of furfuryl alcohols **58a-e** leads to enedial intermediate **B** (Scheme 4.12), which in the presence of a strong acid such as TsOH, cyclizes giving predominantly product **59a-e** (Scheme 4.17) [29]. By-products alkylidenefuran-2-ones **60a-e**

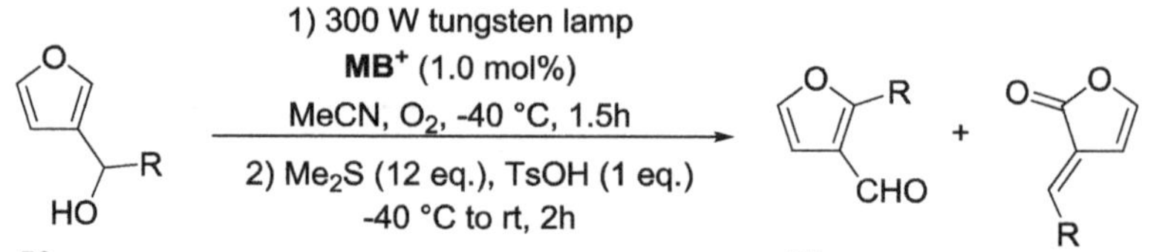

	58a-e	59a-e	60a-e
a	R=Ph	74%	5%
b	R=*t*-Bu	26%	26%
c	R=*n*-Pr	38%	8%
d	R=Bn	73%	<2%
e	R=4-MeOC_6H_4	29%	15%

Scheme 4.17. $^1O_2$-induced rearrangement of furfuryl alcohols **58a-e** to 2-substituted-3-furfurals **59a-e** *via* path II.

result from the reductive cleavage of the endoperoxide and their formation can be enhanced by *in situ* treatment with methanesulfonyl or toluenesulfonyl chlorides.

• **Path II and Path III**

The direction of endoperoxide decomposition can be controlled by the careful selection of the reaction conditions. In the photooxygenation of 2-(*C*-glucosyl)furan **61**, depending on the conditions used in the second step, *cis-C*-glucosyl derivative **62** (*via* path II, Scheme 4.12) or *cis-O*-glucopyranoside **64** (*via* path III, Scheme 4.12) are selectively obtained (Scheme 4.18) [30]. Both products can subsequently isomerize to the corresponding *trans*-derivatives **63** and **65**.

• **Path IV**

In the presence of a hindered base, oxidation of 2- or 2,5-disubstituted furans follows the path IV mechanism. The reaction leads to γ-hydroxybutenolides **D** (Scheme 4.12), which convert easily to acids under basic conditions [31–33]. The **MB**$^+$-sensitized photooxidation of 2-aryl-3,4-dicarbomethoxyfuran **66** to 4-oxoacid **67** also occurs *via* this pathway (Scheme 4.19) [34]. The subsequent acylation in the presence of Tf_2O and anisole furnishes diarylfuranones **68** and **69** in good yield and regioselectivities up to 82:18.

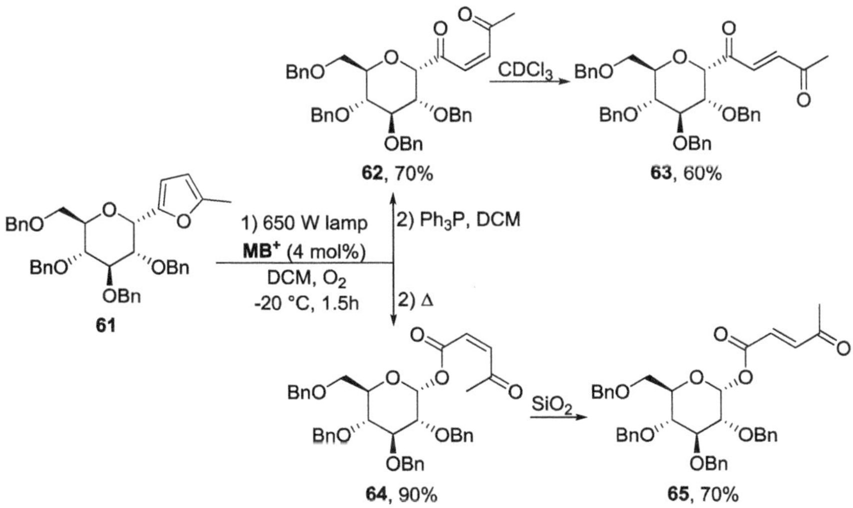

Scheme 4.18. **MB**$^+$-sensitized photooxidation of 2-(*C*-glucosyl)furan **61**.

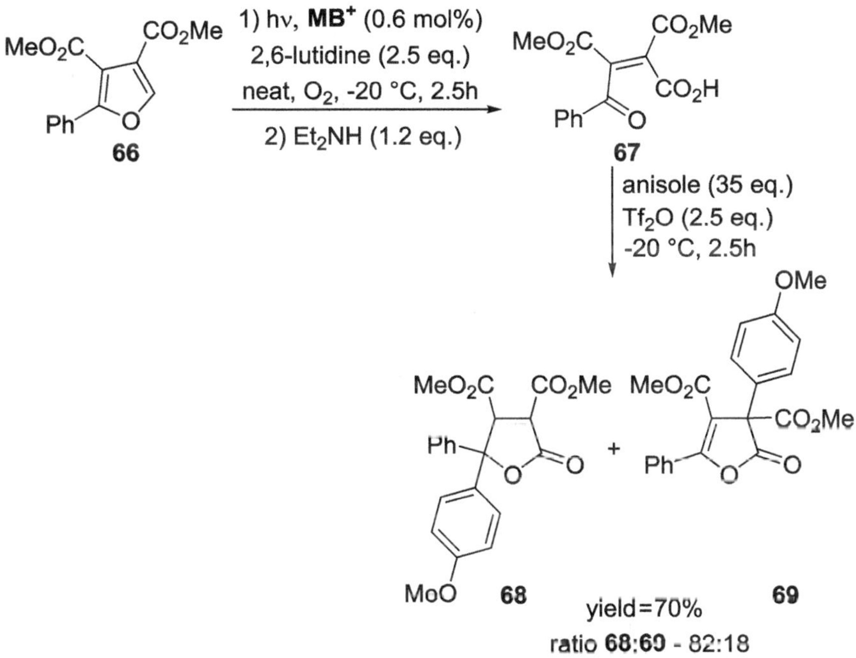

Scheme 4.19. One-pot, three-step synthesis of diarylfuranones **68** and **69**.

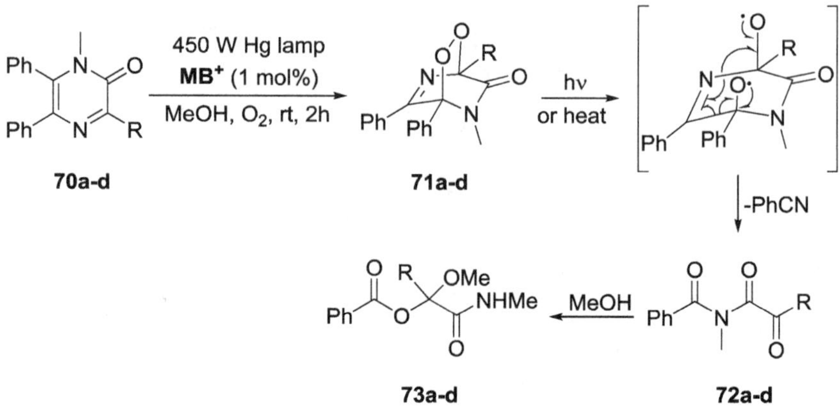

a	R=Me	26%
b	R=Et	43%
c	R=*i*-Pr	18%
d	R=Ph	0%

Scheme 4.20. Photocatalytic oxidation of pyrazine-2-ones **70a-d.**

Other heterocycles also undergo oxidation by **MB**$^+$-generated singlet oxygen. For instance, pyrazine-2-ones **70a-d** undergo [4+2] cycloaddition to $^1O_2$ giving 3,6-epidioxypyrazin-2-ones **71a-d** (Scheme 4.20), reasonably stable in the solid state at room temperature [35]. When heated or irradiated with visible light, these products decompose to α-oxoimides **72a-d** by electrocyclic ring opening with the release of a nitrile. The subsequent nucleophilic attack at the carbonyl group provides rearranged α-methoxyamides **73a-d** in moderate yields.

4.2.2.4 *Oxidation of heteroatoms via* $^1O_2$

Singlet oxygen generated by **MB**$^+$-photosensitization can also oxidize heteroatoms. For example, the visible-light irradiation of 1,3-dithianes **74a-b** in the presence of **MB**$^+$ under an oxygen atmosphere gives desired ketones **75a-b** and sulfur oxidation products, sulfoxides **76a-b** and **77**, and sulfones **78** (Scheme 4.21). The ratio strongly depends on the substitution pattern with preferential formation of the deprotected ketone **75b** for more substituted dithiane **74b** [36].

The problem associated with the selectivity in the oxidation of sulfides is not present when a bentonite-bound **MB**$^+$ is used as a

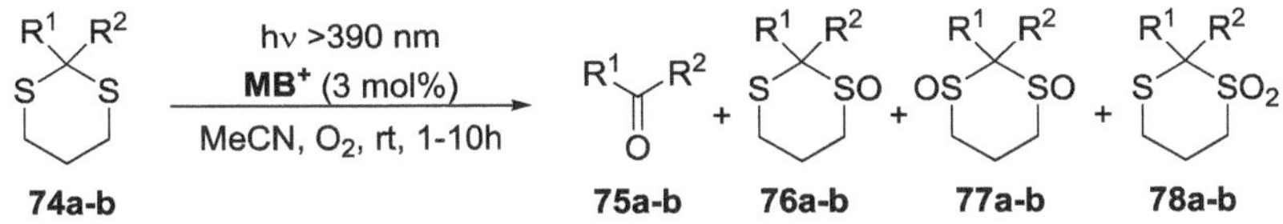

			75a-b	76a-b	77a-b	78a-b
a	R^1=(4-Me)Ph	R^2=H	13%	71%	4%	-
b	R^1=(4-Me)Ph	R^2=(4-Me)Ph	48%	-	-	23%

Scheme 4.21. **MB⁺**-sensitized photooxidation of 1,3-dithianes **74a-b**.

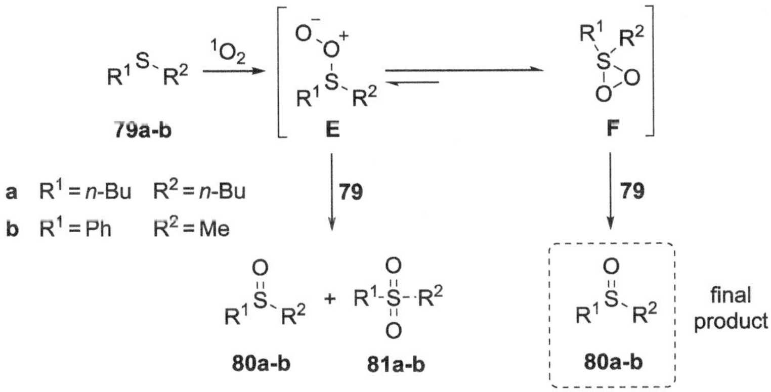

Scheme 4.22. Formation of sulfoxides **80a-b** in the **MB⁺**-catalyzed photooxidation of sulfides **79a-b** in a clay medium.

POC [37]. A persulfoxide intermediate **E** is not well stabilized in a clay media and, as a result, is in equilibrium with the thiadioxirane intermediate **F** (Scheme 4.22). In the presence of an excess of sulfides **79a-b**, the equilibrium is shifted towards the sulfoxides **80a-b**. The reaction is very much concentration-dependent and the ratio of products **80/81** varies from 100:0 for di(*n*-butyl)sulfide (**79a**) in a 0.001 M solution to 86:14 in a 0.01 M solution.

4.2.2.5 *C–X bond formation*

The formation of C–X bonds is not a typical reaction involving singlet oxygen, with only two examples reported. One of them is the tandem transformation of β-oxoesters **82a-d** leading to

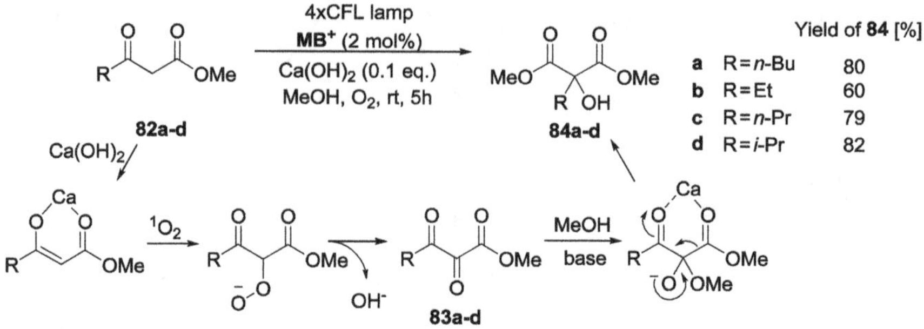

Scheme 4.23. Mechanism of tandem transformation of β-oxoesters **82a-d**.

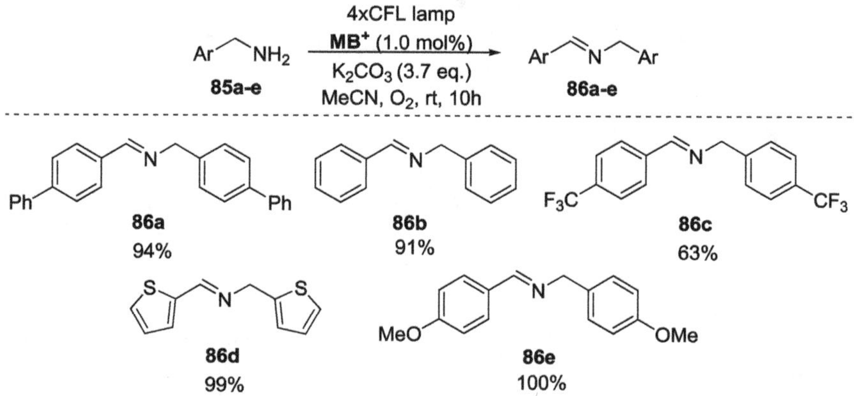

Scheme 4.24. Oxidative coupling of benzylamines **85a-e** catalyzed by **MB⁺**.

2-hydroxymalonic esters **84a-d** *via* oxidation and rearrangement steps (Scheme 4.23) [38]. The reaction involves the nucleophilic addition of the enolate of substrate **82a-d** to $^1O_2$. The final products **84a-d** are formed as a consequence of the base-induced rearrangement of tricarbonyl intermediates **83a-d**. Under similar conditions, β-oxoamides are successfully converted to 2-hydroxymalonic amides [38].

Another example of C–X bond formation in the presence of **MB⁺** and oxygen is the oxidation of benzylamines **85a-e** to imines **86a-e** with various substituents, although amines bearing electron-donating groups are preferred (Scheme 4.24) [39]. The reaction involves hydrogen atom abstraction from the amine leading to phenylmethanimine and subsequent nucleophilic addition of the second molecule of the amine.

4.2.2.6 *Photogenerated* $^1O_2$ *in the synthesis of natural compounds*

Singlet oxygen has been extensively employed in the total synthesis of natural products [6]. Due to space limitations, in this chapter, only a few examples are presented.

Within a multistep total synthesis of biologically active compounds, usually only one stage utilizes $^1O_2$. For example, an ene-type reaction was employed during the total synthesis of the antimalarial agent (+)-yingzhaosu [40], or a [2+4] cycloaddition to bicyclic diene in the construction of acetylcholinesterase inhibitor arugacin A [41]. However, one of the most characteristic features of $^1O_2$ is its ability to promote cascade reactions which allows for the one pot synthesis of complex molecules, significantly reducing the number of synthetic steps. For instance, Vassilikogiannakis *et al.* found an extremely convenient way to synthesize pandamarine (**89**), an alkaloid present in the leaves of the tropical plant *Pandanus amaryllifolius* (Scheme 4.25) [42]. An **MB**$^+$-photocatalyzed transformation is the key step involving the oxidation of *bis*-furan derivative **87** to *bis*-5-hydroxy-1*H*-pyrrol-2(5*H*)-one **88** *via* three stages — [2+4] cycloaddition of $^1O_2$, reduction, and nucleophilic addition of ammonia.

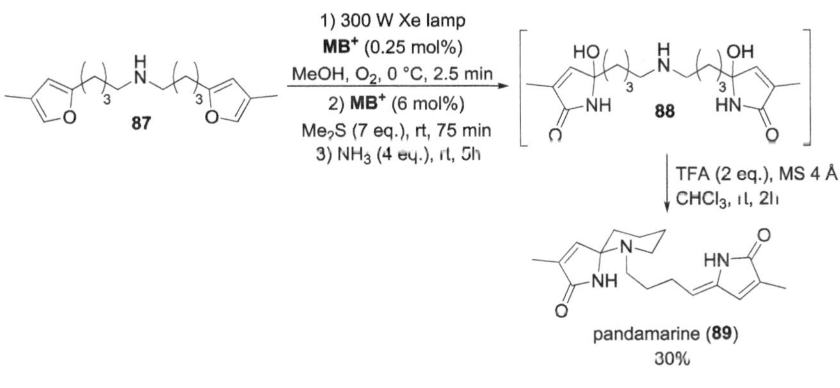

Scheme 4.25. **MB**$^+$-catalyzed step of the total synthesis of pandamarine (**89**).

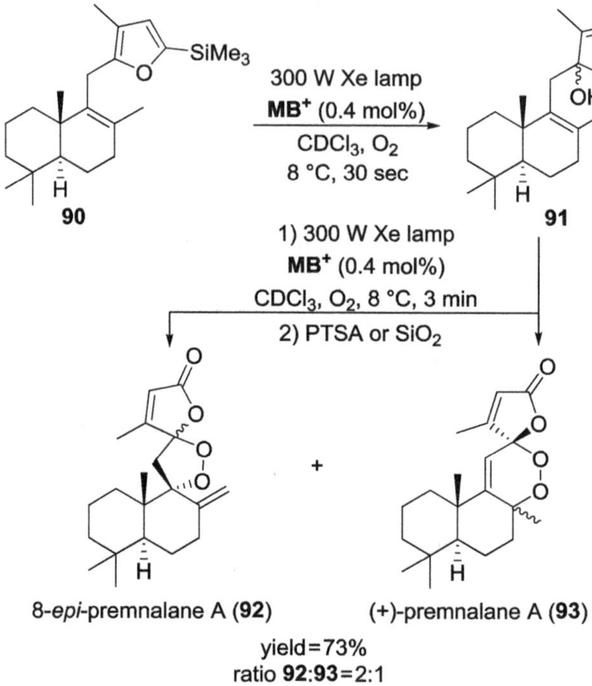

Scheme 4.26. **MB**$^+$-catalyzed reactions in the total synthesis of (+)-premnalane A (**93**).

Subsequent ketalization and dehydration steps under acidic conditions afford pandamarine (**89**) in the 30% overall yield.

A similar "one-pot" approach was employed in the total synthesis of (+)-premnalane A (**93**), an antibacterial natural product isolated from *Premnaoligtricha* (Scheme 4.26) [43]. In this process, **MB**$^+$ promotes the [2+4] cycloaddition of $^1O_2$ to furan followed by an ene-type reaction. The final ketalization step yields a mixture of easily separated products **92** and **93**.

Although in total synthesis, the photooxidation of furan precursors is the most often used transformation catalyzed by **MB**$^+$, other examples have been reported. For example, the oxidative cleavage of indole derivative **94** during the synthesis of two natural products — melohenine B (**95**) and *O*-ethyl-14-*epi*melohenine B involves $^1O_2$ (Scheme 4.27) [44]. It is assumed that these

Scheme 4.27. Photooxidation of indole derivative **94** in the total synthesis of melohenine B (**95**).

transformations proceed *via* a dioxetane intermediate, quantitatively yielding only one diastereoisomer.

4.2.3 *MB*$^+$ *as photoredox catalyst*

Despite its undeniable utility in *EnT*, **MB**$^+$ can also be involved in the photoinduced electron-transfer processes. As a photoredox catalyst, **MB**$^+$ efficiently promotes important chemical transformations such as carbon–carbon or carbon–heteroatom bond formation, removal of protecting groups, and oxidation of various nitrogen compounds. The following paragraphs discuss *SET* processes catalyzed by **MB**$^+$ and reactions involving both electron transfer and *EnT* for $^1O_2$ generation.

4.2.3.1 *C–C bond formation*

The oxidation of *trans*-stilbene to benzaldehyde represents one of the first transformations where singlet oxygen generation in a dye-sensitized reaction was excluded and, instead, *SET* from the olefin to the excited dye was proposed [45]. Fluorescence quenching studies and experiments with $^1O_2$ trapping agents revealed that **MB**$^+$ acts as a photoredox catalyst in this process. This fact was also confirmed by subsequent studies on dye-sensitized oxidation and dimerization of 4-methoxystyrene (**96**, Scheme 4.28) [46]. The reaction proceeds *via* generation of radical cation **96**$^{\bullet+}$ which, under aerobic conditions, can dimerize to cyclobutanes **97** or be oxidized to aldehyde **98** by $O_2^{\bullet-}$. In this transformation, 4-methoxystyrene (**96**) quenches excited state of photocatalysts, with the highest

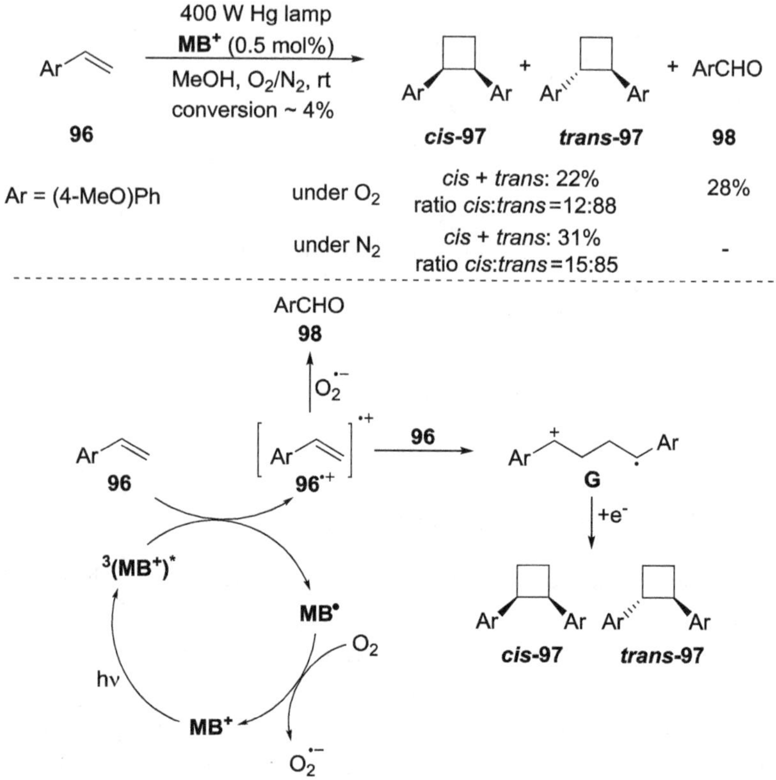

Scheme 4.28. Photoinduced transformation of 4-methoxystyrene (**96**) in the presence of **MB**$^+$ and proposed mechanism.

fluorescence quenching constant value for ^{3}(**MB**$^+$)*. Very low conversion for **MB**$^+$ can be attributed to the most positive value of reduction potential (–0.32 V vs Ag/AgCl) which makes reduction of radical cation **G** less efficient (Table 4.7). These results clearly show that **MB**$^+$-photoorganocatalyzed reactions can represent an interesting alternative to C–C bond formation in the presence of transition metal complexes or other dyes.

The concept of **MB**$^+$ acting through *SET* was not elaborated until 2013 when Pitre *et al.* demonstrated that ^{3}(**MB**$^+$)* can be quenched by aliphatic amines resulting in the formation of two reactive radicals — semireduced **MB**$^•$ and an amine-radical **H** (Scheme 4.29) [4].

Table 4.7. Comparison of photophysical characteristics and conversions for transformation of 4-methoxystyrene (**96**) under light irradiation in the presence of selected dyes.

Dye	Conversion [%]	k_q [10^8 M^{-1} s^{-1}]	τ_f (ns)	E_{RED} [V][a] (**POC/POC**$^-$)
Acriflavine	31	58	4.61	−1.06
Rhodamine 6G	10	18	4.16	−0.83
MB$^+$	**4**	**185**	**0.75**	**−0.32**
Eosin Y	17	4.0	4.44	−1.19
Rose Bengal	18	7.0	2.68	−1.15

Note: [a]Measured vs Ag/AgCl in MeCN.

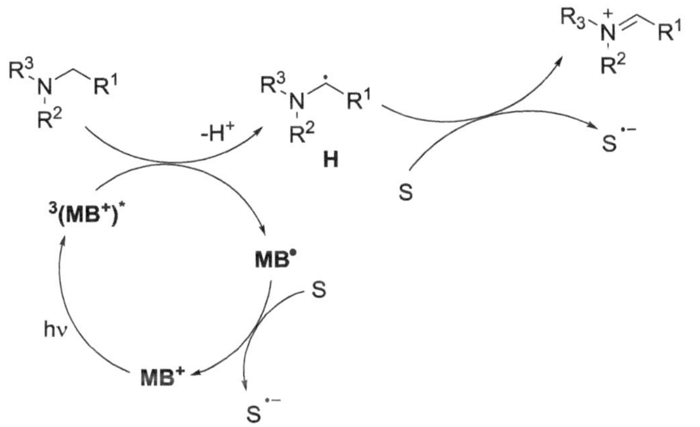

Scheme 4.29. General mechanism for **MB**$^+$-photoorganocatalyzed reactions in the presence of amines.

Subsequently, both species can readily reduce numerous organic substrates initiating synthetically important chemical transformations.

Following this mechanism, radical trifluoromethylation of electron-rich heterocycles **99a-d** and hydrotrifluoromethylation of terminal alkenes **102a-c** or alkynes **103a-c** was accomplished (Scheme 4.30) [47]. The adoption of **MB**$^+$ as **POC**, an amine as a sacrificial reductant, and the electrophilic CF_3 reagent **100** allows for the synthesis of organofluorine compounds **101a-d**, **103a-c**, and **105a-b**,

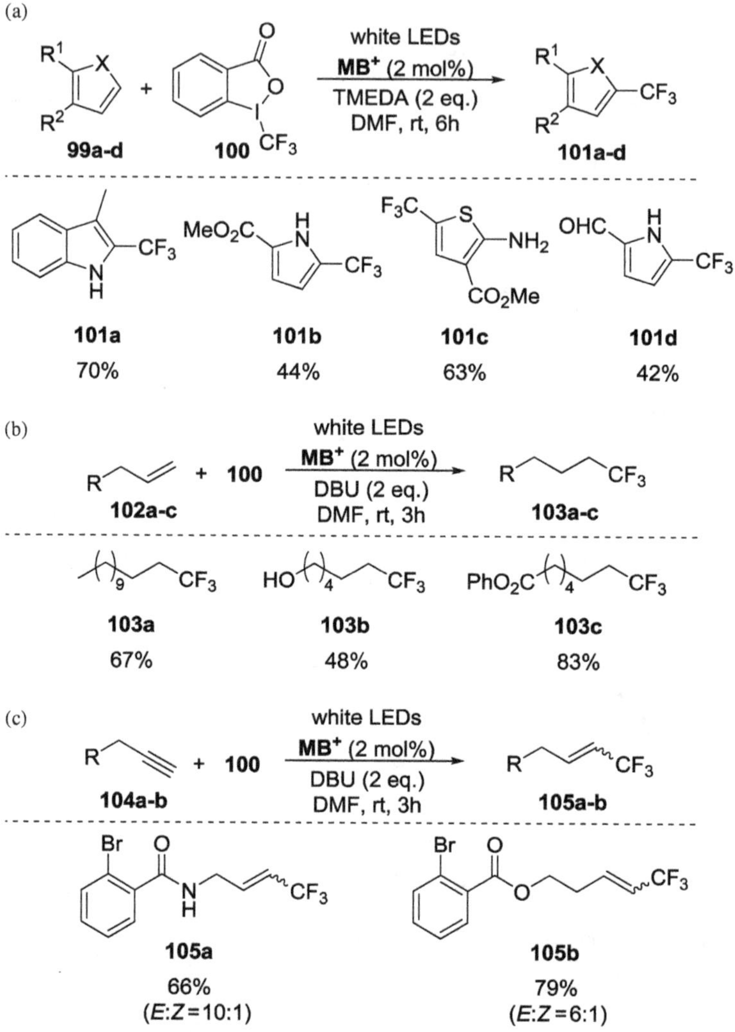

Scheme 4.30. $\mathbf{MB^+}$-catalyzed radical trifluoromethylation of electron-rich heterocycles **99a-d** and hydrotrifluoromethylation of terminal alkenes **102a-c**, or alkynes **104a, 104b**.

respectively, without the use of transition-metal complexes. Kinetic studies and laser flash photolysis analyses reveal that satisfying results are obtained when the amine outcompetes other reagents in $^3(\mathbf{MB^+})^*$ quenching. *N,N,N′,N′*-Tetramethylenediamine (TMEDA), having the highest quenching rate constant out of the examined amines, gives the best yield in the trifluoromethylation of heterocycles **99a-d**. Furthermore, despite having the lowest reduction potential, the Umemoto's reagent is not effective in the production of $CF_3^•$ radicals, as it competes with TMEDA in the quenching of ^{3}(**MB**$^+$)*. In this case, the utilization of Togni's reagent (**100**) with a much lower quenching rate constant gives superior results. Additionally, in the hydrotrifluoromethylation of terminal alkenes **102a-c** and alkynes **104a-b**, the replacement of TMEDA with DBU, having a lower quenching rate constant, leads to a more efficient product formation. This can be attributed to the formation of a charge-transfer complex with **MB**$^+$ facilitating *SET* process. **MB**$^+$ proves also to be suitable for the generation of $CF_3^•$ radicals from CF_3SO_2Na/TFA, however, trifluoromethylation of arenes is low-yielding [48].

Under light irradiation, **MB**$^+$ is able to induce the [3+2] cycloaddition of azomethine ylide **J** (from compound **106**) with dipolarophile **107**, which, after oxidative aromatization, gives pyrrolo [2,1-*a*] isoquinoline derivative **109** (Scheme 4.31) [49]. The photocatalyst

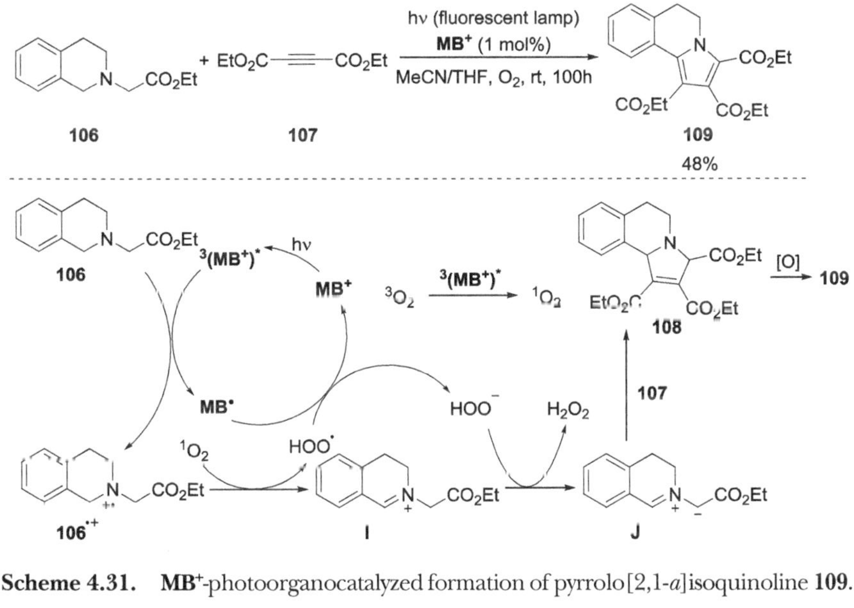

Scheme 4.31. **MB**$^+$-photoorganocatalyzed formation of pyrrolo[2,1-*a*]isoquinoline **109**.

plays a dual role — oxidizes the tetrahydroisoquinoline **106** *via SET* and generates singlet oxygen *via EnT.* $^1O_2$ forms a peroxy anion which deprotonates the imine cation **I** to produce azomethine ylide **J** that undergoes [3+2] cycloaddition with electron-deficient alkenes and alkynes. The subsequent oxidative aromatization occurs without an additional oxidant and may involve either singlet oxygen, hydrogen peroxide generated *in situ,* or a second *SET* process.

4.2.3.2 *C–X bond formation*

Another group of photoredox processes catalyzed by **MB**$^+$ consists of carbon–heteroatom bond forming reactions. These transformations furnish either linear or cyclic compounds, very often of potential pharmacological importance. For example, the photoredox intramolecular cyclization of indole derivative **110** yields 4-benzoxazine **112** (Scheme 4.32) [50]. In this reaction, radical

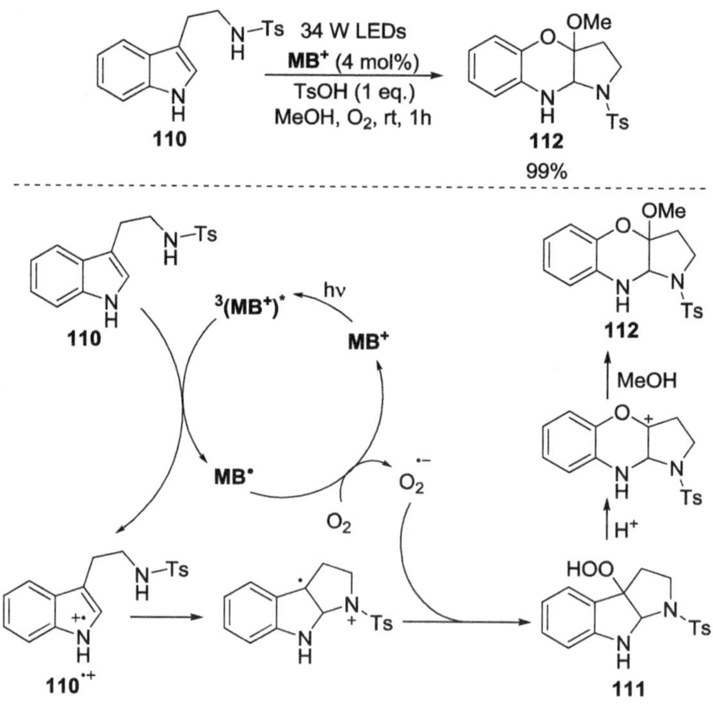

Scheme 4.32. Intramolecular cyclization of indole derivative **110** catalyzed by **MB**$^+$.

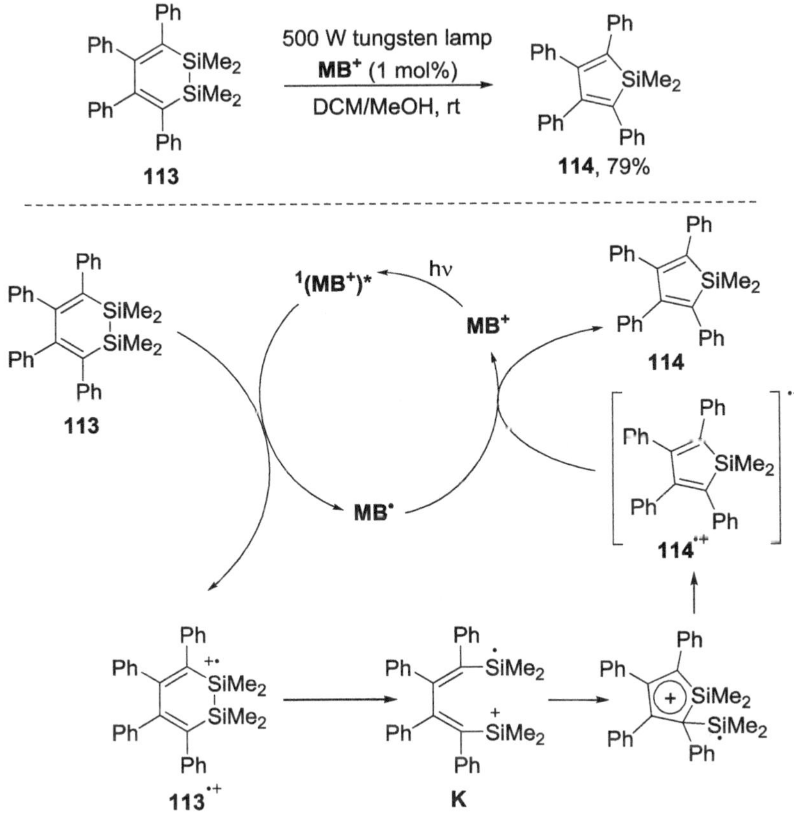

Scheme 4.33. MB^+-catalyzed ring contraction of 1,2-disila-3,5-cyclohexadiene **113**.

cation **110**$^{•+}$ is generated *via SET* and then undergoes an intramolecular attack by a tethered heteroatom and the subsequent addition of the superoxide radical anion $O_2^{•-}$ to afford hydroperoxide **111**. In the last step, an acid-catalyzed rearrangement gives the final product **112**.

Alternatively, **MB**$^+$-catalyzed processes may lead to cyclic products *via* ring contraction. In the presence of the **POC**, 1,2-disila-3, 5-cyclohexadiene **113** undergoes photolysis [51]. As the reaction is not inhibited by ^{3}(**MB**$^+$)* quenchers, ^{1}(**MB**$^+$)* is assumed to participate in *SET* (Scheme 4.33). The possible mechanism starts with a *SET* from the substrate **113** to the singlet excited state ^{1}(**MB**$^+$)*

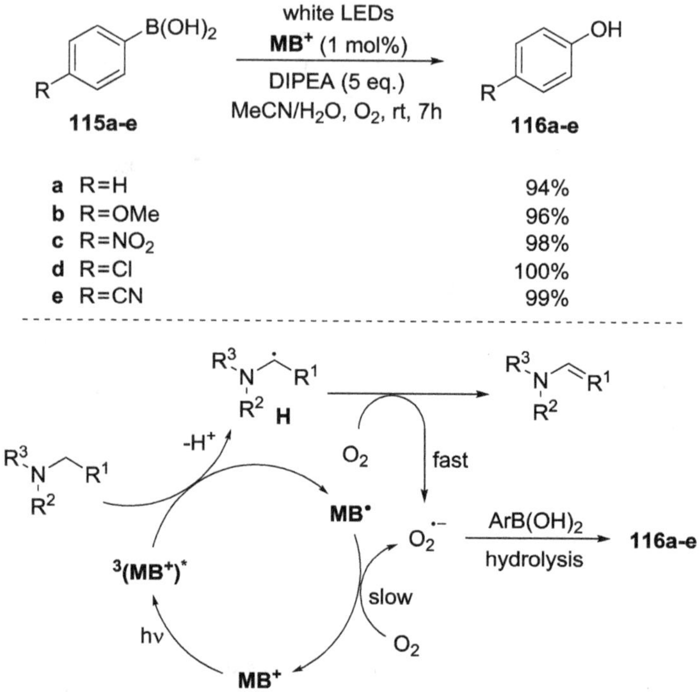

Scheme 4.34. Oxidative hydroxylation of arylboronic acids **115a-e**.

giving a radical cation **113$^{•+}$**. After mesolytic Si–Si bond cleavage, subsequent intramolecular attack of the silyl cation on diene **K**, elimination of the silyl group, and *SET*, gives the final silole **114**.

MB$^+$/amine catalytic systems are efficient not only in carbon–carbon bond forming reactions, but also for oxidations proceeding *via SET* [4]. In this line, visible-light-induced oxidative hydroxylation of arylboronic acids **115a-e** to phenols **116a-e** engages the superoxide anion (Scheme 4.34) [52]. In the catalytic cycle, the *SET* from α-amino radical **H** to oxygen occurs faster than from the semireduced **MB$^•$** to O_2.

MB$^+$ efficiently catalyzes the α-functionalization of aldehydes **117a-d** leading to α-oxyaminated products **119a-d** in good to moderate yields *via* enamine catalysis (Scheme 4.35) [53]. Notably, the reaction works well also for α,β-unsaturated aldehydes and gives the γ-substituted product **119c**.

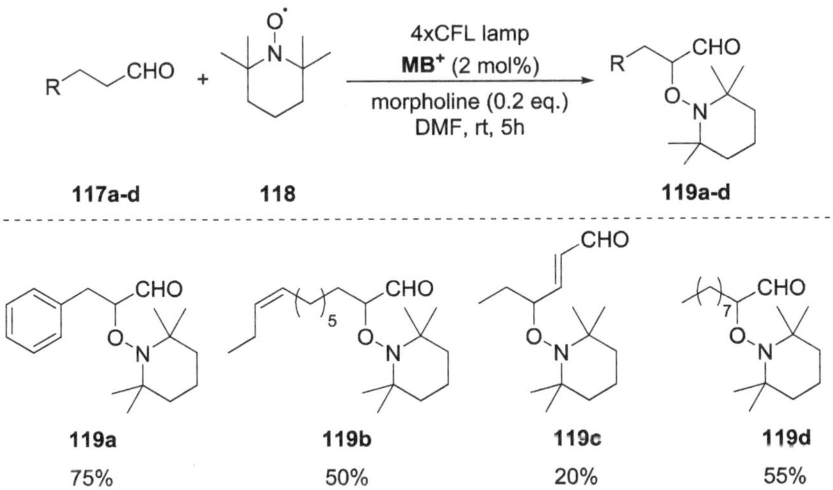

Scheme 4.35. α-Oxyamination of aldehydes **117a-d** catalyzed by **MB**$^+$.

The reaction of alkyl or aryl thiosulfates **121a-d** with diazonium salts is efficiently catalyzed by Ru(bpy)$_3$Cl$_2$ and gives sulfides **122a-d** under basic conditions. However, the catalytic system is not suitable for pyridine derivative **120** [54]. Employing **MB**$^+$ and using Li_2CO_3 as the base significantly improves the yield for this starting material (Scheme 4.36). The differences in the catalytic efficacy between Ru(bpy)$_3$Cl$_2$ and **MB**$^+$ supposedly stems from the Ru coordination to the N-atom of pyridine or lower stability of pyridinium substrate **120** under irradiation with more energetic absorption wavelength required for Ru(bpy)$_3$Cl$_2$.

4.2.3.3 *C–X bond cleavage*

Multistep organic synthesis is not usually possible without a countless number of protection and deprotection steps. The more complex the target molecule, the more selective and milder conditions required for these reactions. In this line, photocatalysis offers great synthetic advantages, often eliminating the need for strong acids or bases. The first reported deprotection reaction catalyzed by a sulfur heterocycle was the cleavage of the dithio protecting

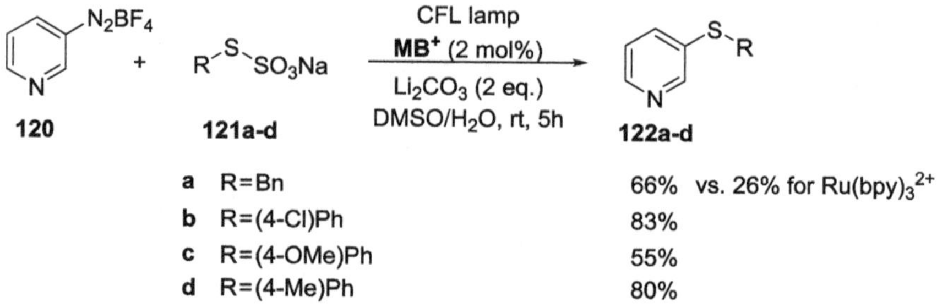

Scheme 4.36. **MB^+**-photoorganocatalyzed reaction of pyridine diazonium salt **120** with alkyl/aryl thiosulfates **121a-d**.

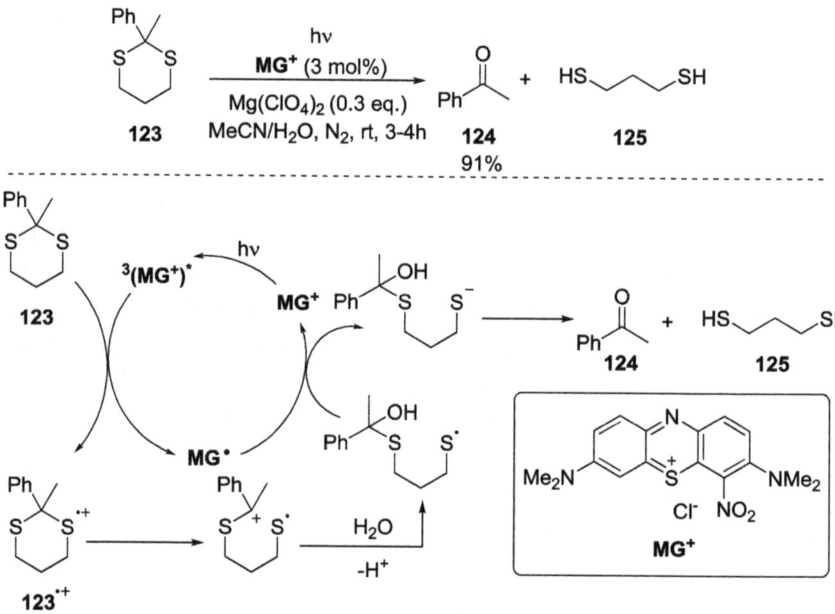

Scheme 4.37. Deprotection of the dithio group catalyzed by **MG^+**.

group in the presence of methylene green (**MG^+**), a derivative of **MB^+** (Scheme 4.37) [55]. Similar to **MB^+**, ^{3}(**MG^+**)* acts in this process as an oxidant and accepts an electron from dithiane **123**. The subsequent C–S bond cleavage and nucleophilic addition of water followed by back-electron transfer leads to acetophenone

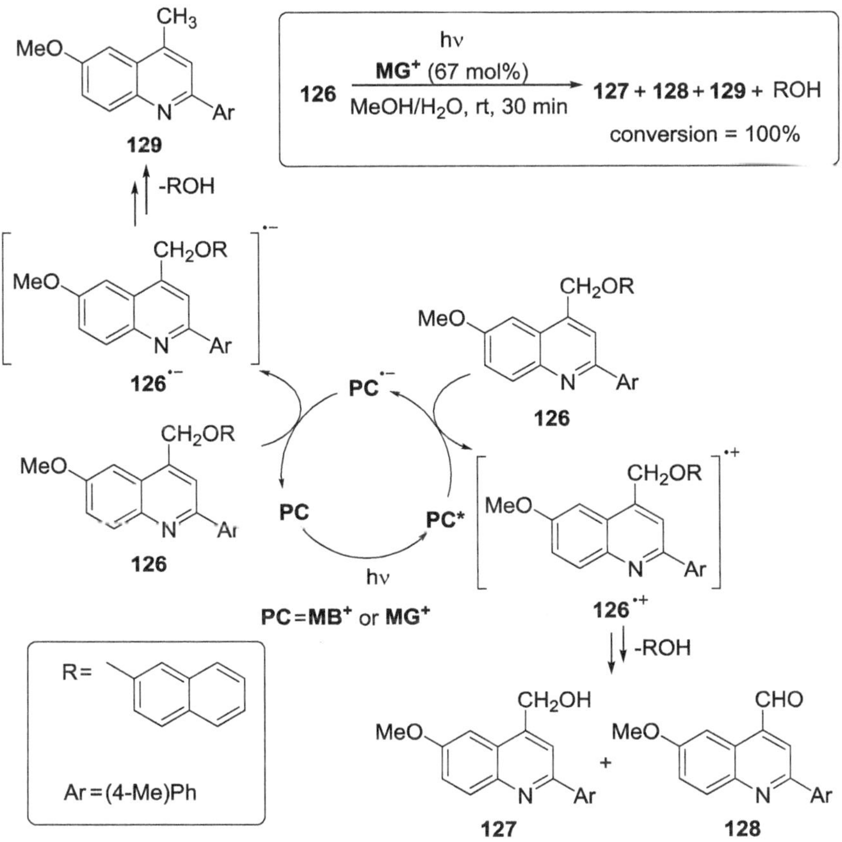

Scheme 4.38. Photocleavage of quinoline methyl ether **126**.

124 and dithiol **125**. The method is efficient for deprotection of both aldehydes and ketones.

Both $\mathbf{MB^+}$ and $\mathbf{MG^+}$ are able to catalyze the photocleavage of methyl ether-protected alcohols [56]. Irradiation of quinoline methyl ether **126** in the presence of a photocatalyst, *D*-sorbitol as a radical scavenger, and Et_3N as the sacrificial reductant, results in a mixture of three products — quinoline alcohol **127**, quinoline aldehyde **128**, and derivative **129** (Scheme 4.38). The initial step for the formation of compounds **127** and **128** involves the generation of radical cation $\mathbf{126^{•+}}$ *via SET* from **126** to the photocatalyst **PC**. The subsequent addition of water to cation $\mathbf{126^{•+}}$ furnishes alcohol **127**, while H• abstraction produces aldehyde **128**. The reduced form **PC**$^{\bullet-}$ goes back to the ground state **PC** by *SET*, along with the formation of radical anion **126**$^{\bullet-}$, which then decomposes to the adequate radical or anion and produces the final product **129** by H• or H^+ abstraction. This mild method for the methyl ether deprotection represents a good alternative for other chemical and electrochemical reactions which often require harsh reaction conditions.

Oxidative cleavage of the C–N bond enables one of the most important metabolic pathways — *N*-demethylation. This biological process can be mimicked by employing visible-light photocatalysis [57]. *N,N*-Dimethylaminophenyl derivative **130**, when irradiated in the presence of a photocatalyst under aerobic conditions, affords secondary amine **131** (Table 4.8, Scheme 4.39). Notably, the use of a photoorganocatalyst proved to be more efficient than $[Ru(bpy)_3]Cl_2$, with the addition of acetic acid accelerating the formation of hydroperoxide **L** *via* the deprotonation and protonation steps. Intermediate **L** decomposes or dehydrates to give, respectively, *N*-hydroxymethyl derivative **132** and *N*-formyl compound **133**. The first decomposes spontaneously to *N*-demethylated product **131**, whereas the latter hydrolyzes to the final product under acidic conditions. The developed deprotection method was applied to the demethylation of various pharmaceutical compounds, giving deprotected amines in ~60% yield even on gram-scale.

Table 4.8. Selecting the photocatalyst for the visible-light-induced *N*-demethylation of 4-(*N,N*-dimethylamino)-acetophenone **130**.

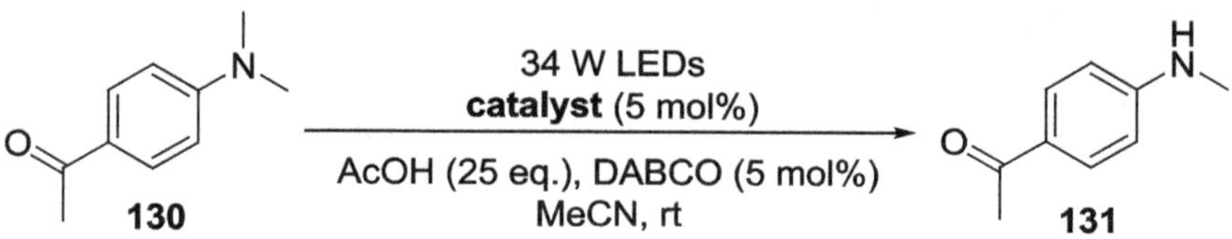

Catalyst	Time [h]	Conversion [%]	Yield of **131** [%]
RB	3	97	91
MB$^+$	**4**	**98**	**86**
Acr–Me$^+$	1.5	>99	79
$[Ru(bpy)_3]Cl_2$	4	<5	—

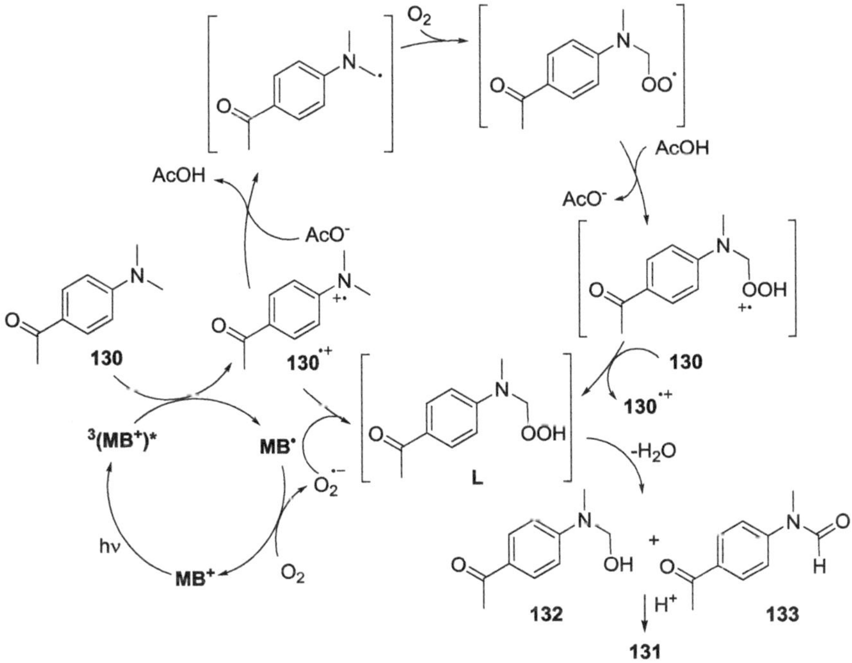

Scheme 4.39. Mechanism of *N*-demethylation of **130**.

4.2.3.4 *Oxidation of nitrogen containing compounds*

Among the reactions catalyzed by **MB**$^+$, the photooxidation of heterocyclic nitrogen compounds and hydrazine derivatives *via SET* is of great importance. The relatively low oxidation potentials of tetraalkylhydrazines, in comparison to tertiary amines, and the unique stability of their radical cations (generated by *SET*), resulting from the equal distribution of the formal charge between the two nitrogen atoms, facilitated photocatalytic processes [58].

The oxidation of *N*-arylamino-l-pyrrolidine or *N*-arylamino-l piperidine **134** in the presence of a catalytic amount of **MB**$^+$, under aerobic conditions, leads to stable radical cations **134**$^{\bullet+}$. The reaction supposedly involves either *SET* from the substrate **134** to ^{3}(**MB**$^+$)* or singlet oxygen generation (Scheme 4.40) [59].

Subsequent processes strongly depend on the nucleophiles present in the reaction. With trimethylsilyl cyanide (TMSCN),

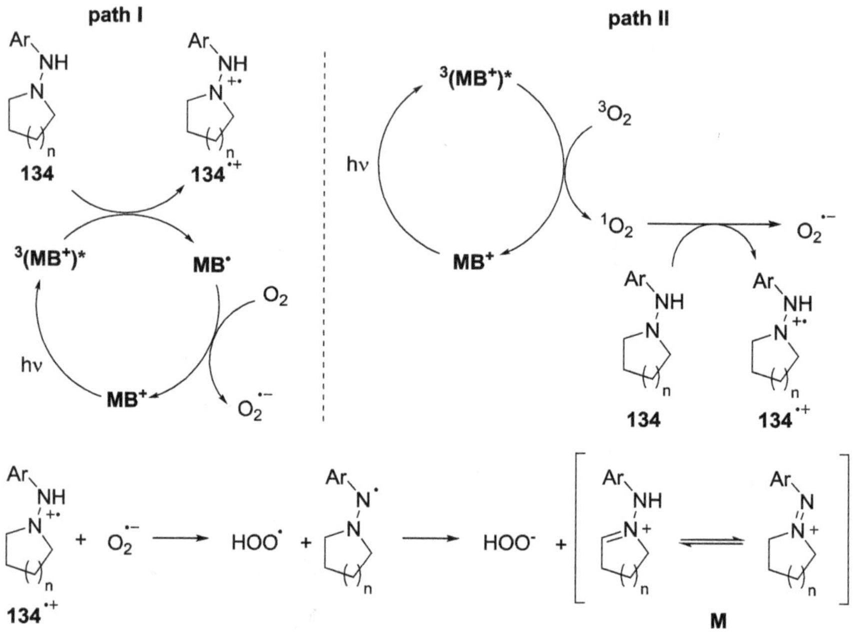

Scheme 4.40. Two possible pathways for the **MB**$^+$-catalyzed generation of hydrazinium radical cation **134**$^{•+}$ followed by the formation of hydrazinium cations **M**.

α-hydrazinonitriles **135a-c** are obtained in a regio and stereoselective manner, and the proton abstraction occurs on the more substituted C–H bond (Scheme 4.41, path I) [60]. When further oxidation is performed under aqueous conditions, lactam **136a-c** is formed [59]. The initial photooxidation of hydrazine derivatives **134a-c** may also lead to a C–N cleavage and ring opening [61]. In the presence of alcohols, the main products are azodialkylacetals **137a-c** (Scheme 4.41, path II), while in water aldehydes **138a-c** are obtained in excellent yields (Scheme 4.41, path III). The synthesis of carbonyls proceeds regioselectively with the preferential cleavage of the less-substituted C–N bond and can be utilized in the synthesis of aminoaldehydes.

The **MB**$^+$-catalyzed photooxidation of arylhydrazines in the presence of alkenes may give the corresponding hydrazones [62] or

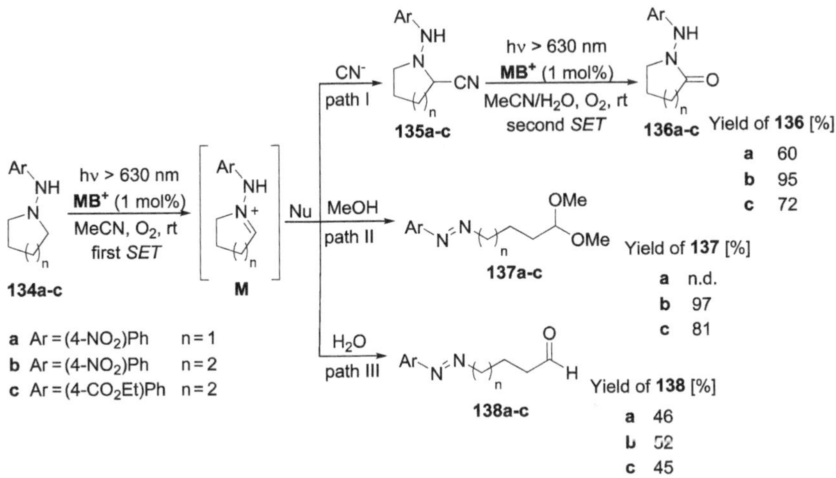

Scheme 4.41. **MB**$^+$-catalyzed oxidation of *N*-arylamino-1-pyrrolidines or *N*-arylamino-1-piperidines **134a-c**.

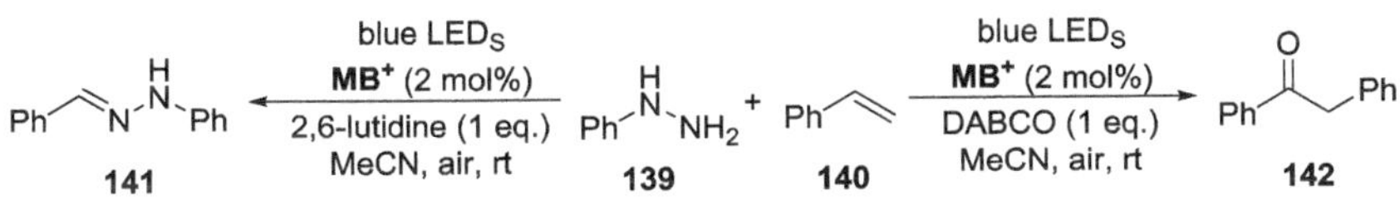

Scheme 4.42. Formation of hydrazones and ketones in the **MB**$^+$-catalyzed photooxidation of arylhydrazines.

ketones (Scheme 4.42) [63]. The reaction outcome mainly depends on the amine required as an additive, with other reaction conditions being the same.

Photooxidation of heterocyclic nitrogen compounds is utilized in bioorthogonal chemistry. In this line, irradiation of **MB**$^+$ with red light allows for the rapid oxidation of dihydrotetrazine **143** to tetrazine **144**, turning on its reactivity toward dienophiles, e.g. compound **145** (Scheme 4.43) [64]. Despite the fact that the reaction can be catalyzed by Rose Bengal, the use of **MB**$^+$ proves to be beneficial due to its low toxicity, high solubility, and suitability for medical applications. The exact reaction mechanism is not known, but a *SET* process was suggested, since the involvement of singlet oxygen was excluded.

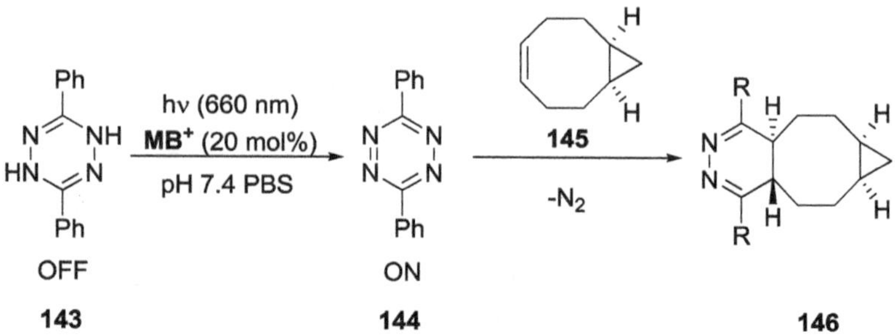

Scheme 4.43. MB^+-catalyzed tetrazine **144** ligation.

4.3 Other Thiazines

MB^+ belongs to the class of thiazine dyes. Their structure is based on a 6-membered ring containing S and N atoms. Phenothiazines possessing two additional aromatic rings are especially worth of attention (Figure 4.3). Functionalizations of these organic dyes include the introduction of alkyl or phenyl groups at the nitrogen atom or modification of aromatic rings at the 3- and 7- positions, for example *via* the Vilsmeier reaction and subsequent condensation with carbonyl compounds [65].

4.3.1 *Photophysical properties*

Photophysical properties of phenothiazine dyes are strongly influenced by the nature of the substituents present in the core structure. Such modifications allow for a fine tuning of their redox properties. The most commonly used **PTH** absorbs in the near-UV (Chart 4.2), while the absorption maxima of 3,7-disubstituted thiazine dyes are usually located between 400–450 nm with the tendency of shifting towards the blue region when strong electron donating groups are introduced [65]. As a result, the oxidation potential is lowered, thus facilitating the oxidation of the photocatalyst (oxidation potentials 0.47–0.53 V vs SCE) and shifting HOMO and LUMO to higher energy levels (−5.26 eV to −5.22 eV and −2.88 eV to −2.60 eV, respectively, Table 4.9) [65]. Besides their oxidizing properties, thiazines can also act as strong reductants (e.g. $E^*_{RED} = -2.1$ V vs

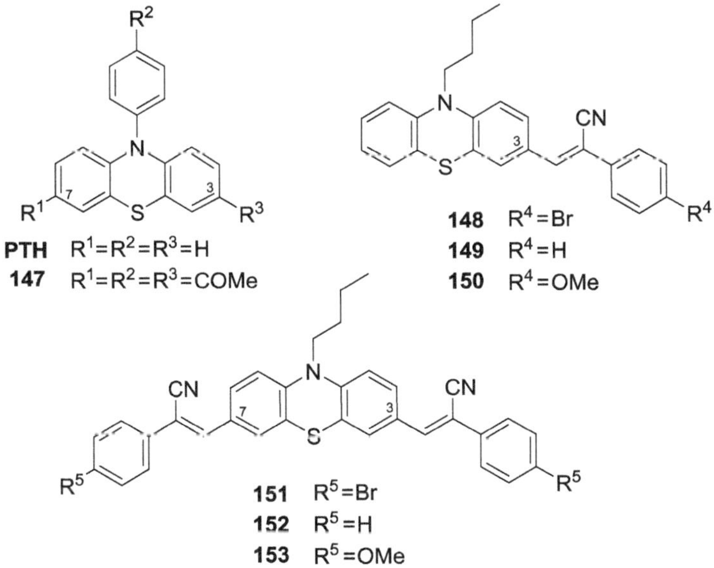

Figure 4.3. Thiazine dyes.

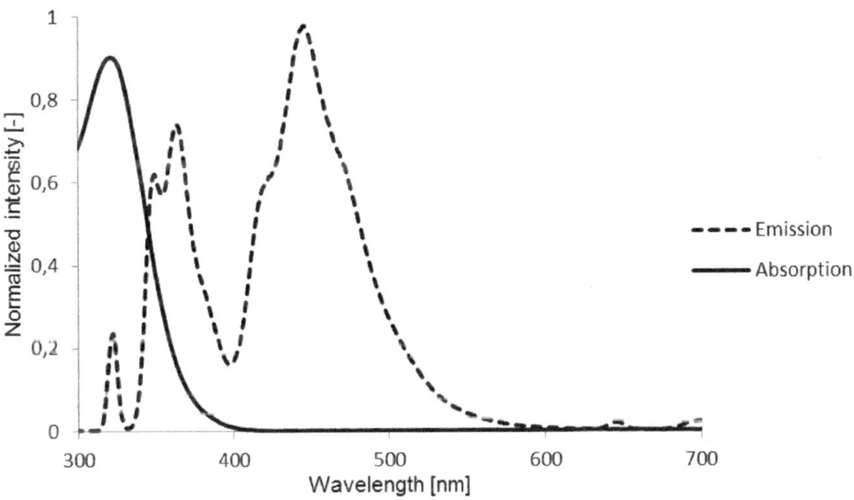

Chart 4.2. Absorption (solid line) and emission (dashed line) spectra of **PTH** in DCM, $c = 2 \times 10^{-7}$ M, $\lambda_{ex} = 321$ nm.

Table 4.9. Influence of aryl substituents on the photophysical properties of dyes **148–153** [65].

Phenothiazine dye	E_{OX} [V][a] **POC⁺/POC**	$E_{0,0}$ [eV]	E^*_{OX} [V][b] **POC⁺/POC***	HOMO [eV]	LUMO [eV]
148	0.48	2.54	−2.06	−5.23	−2.69
149	0.47	2.59	−2.12	−5.23	−2.64
150	0.45	2.62	−2.17	−5.22	−2.60
151	0.53	2.36	−1.83	−5.26	−2.88
152	0.51	2.41	−1.90	−5.25	−2.83
153	0.48	2.43	−1.95	−5.22	−2.78

Notes: [a]Measured vs Ag/AgCl in MeCN. [b]Calculated from E_{OX} and $E_{0,0}$ values.

SCE for **PTH**) [66]. The introduction of electron-withdrawing groups (i.e. acetyl groups in **147**, Figure 4.3) to the aromatic rings of **PTH** leads to an increase of its excited state oxidation potential (E^*_{OX} = −1.5 V vs SCE, in comparison with E^*_{OX} = −2.1 V vs SCE), hence limiting the catalytic ability of compound **147** [67].

4.3.2 *Thiazines as photoredox catalysts*

4.3.2.1 *Thiazines as oxidants*

The oxidative properties of thiazine dyes have not been extensively explored, with only a few examples being reported. The oxidative coupling of primary benzyl amines in the presence of a thiazine photocatalyst gives the desired imines (Scheme 4.44) [65]. 3,7-disubstituted *N*-butyl phenothiazine **153** proves to be the most effective in catalyzing this process, while asymmetric, 3-substituted thiazine dyes **148–150** are less efficient. Additionally, the oxidation of benzylamine (**85b**) proceeds quantitatively when 0.25 mol% of thiazine **153** is used, while application of twice of the amount of **MB⁺** is needed for other amines tested and the resulting yields are slightly lower (Scheme 4.44). Oxidation of amines to imines proceeds under an oxygen atmosphere for a broad range of primary benzyl amines, but not for alkyl amines. Alkyl imines can only be obtained

(a)

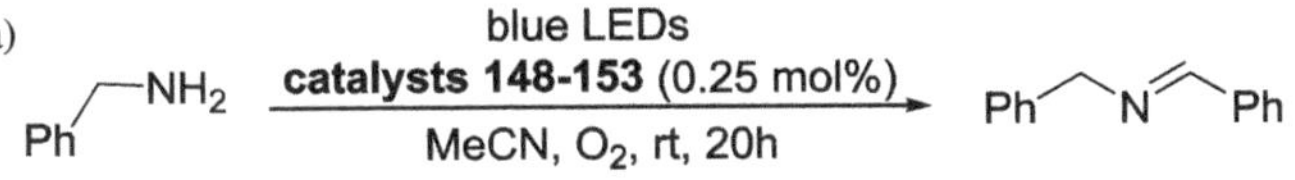

85b **86b**

Catalyst	Yield of **86b** [%]
148	22
149	24
150	46
151	51
152	80
153	83
153 (0.5 mol%)	100

(b)

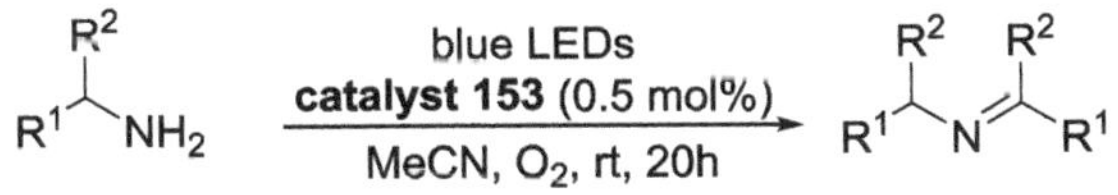

154a-d, 85e **155a-d, 86e**

			Yield [%]	
154a	R^1=(4-Me)Ph	R^2=H	**155a**	94
154b	R^1=(4-F)Ph	R^2=H	**155b**	93
85e	R^1=(4-OMe)Ph	R^2=H	**86e**	88
154c	R^1=2-thiophenyl	R^2=H	**155c**	84
154d	R^1=Ph	R^2=Me	**155d**	30

Scheme 4.44. Thiazine-catalyzed oxidation of primary benzyl amines under light irridation: scope of phenothiazine dyes (a) and amines (b).

by the coupling of an alkyl amine with benzylamine (**85b**). The reaction mechanism involves the reduction of molecular oxygen to the superoxide ion followed by the oxidation of the amine (at the thiazine recovery step).

4.3.2.2 *Thiazines as reductants*

- **C–X bond cleavage**

Phenothiazines are good reductants for aryl halides **156a-f** with a nice correlation with their low oxidation potentials (E^*_{OX} = −2.1 V vs SCE for **PTH**, Scheme 4.45) [66]. With such redox properties, phenothiazine dyes are able to activate challenging C–Br bonds (reduction potentials from −2.57 V to −2.05 V vs SCE) [68, 69].

R-X (**156a-f**) → R-H (**157a-f**)

Conditions: hν (380 nm), **PTH** (5 mol%), HCO_2H (5 eq.), Bu_3N (5 eq.), MeCN, rt

	R	X	Yield
a	R=(4-OMe)Ph	X=I	92%
b	R=(4-CO_2Bn)Ph	X=I	95%
c	R=3-pyridyl	X=Br	94%
d	R=2-thiophenyl	X=Br	92%
e	R=$EtO_2C(CH_2)_3$	X=Br	81%
f	R=(4-CO_2Bn)Ph	X=Cl	83%

Scheme 4.45. Phenothiazine photocatalyzed dehalogenation of aryl halides **156a-f**.

As the reaction is not quenched by oxygen, a popular triplet quencher, **PTH** is likely to act *via* the singlet excited state. The reaction proceeds *via* oxidative quenching of the substrate with the amine additive acting as a source of hydrogen atoms for carbon-centered radical intermediates.

Phenothiazine photocatalysts are suitable for the chemoselective radical dehalogenation of aryl derivatives bearing multiple carbon–halogen bonds [67]. The reduction of C–Cl and C–F bonds requires much stronger reducing agents than the reduction of C–I and C–Br bonds. By modifying the **PTH** structure, hence its redox properties, a high reaction selectivity can be achieved. When different halogen bonds are present in the starting material **158**, analogue **147** catalyzes deiodination selectively, while, under **PTH** catalysis, both deiodination and debromination take place with C–Cl and C–F bonds remaining intact (Scheme 4.46) [67].

- **C–C bond formation**

PTH-catalyzed dehalogenation reactions allow for subsequent C–C bond formation. For example, an excess of a heterocycle favors arylation over hydrogen atom abstraction (Scheme 4.47) [67]. Both mono and diarylation are possible by tuning the properties of the photocatalyst.

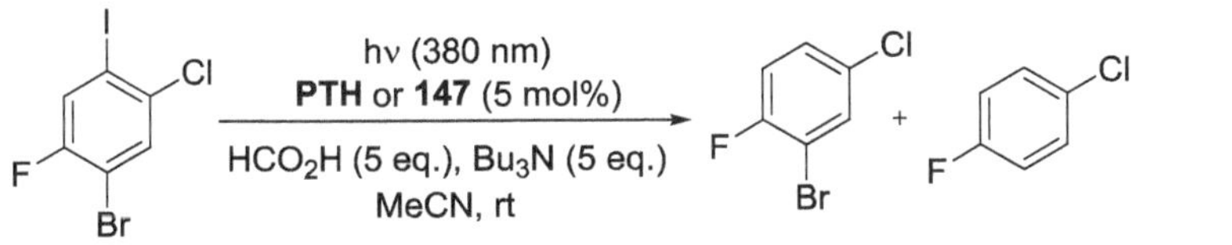

158	catalyst	**159**	**160**	time
	PTH	5%	90%	48h
	147	96%	4%	5h

Scheme 4.46. Chemoselective dehalogenation of arene **158** bearing multiple carbon–halogen bonds.

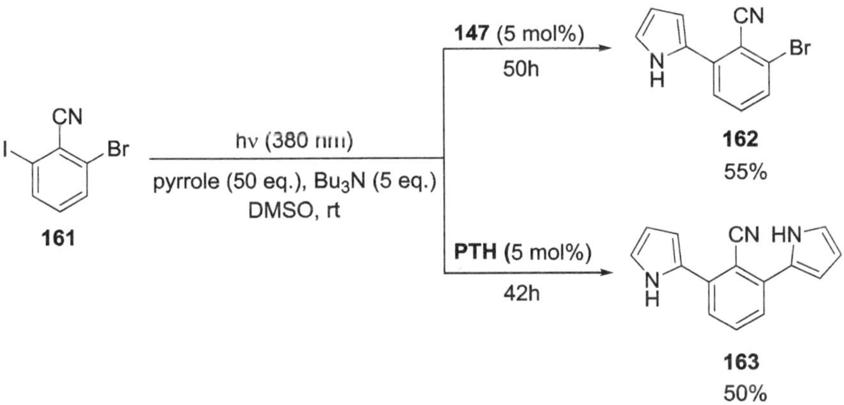

Scheme 4.47. Chemoselective dehalogenation of arenes bearing multiple carbon–halogen bonds leading to C–C bond formation.

4.4. Thiapyrylium Salts

Next to thiazines, thiapyrylium salts represent a group of important **POCs** containing a sulfur heterocycle with the **TPTP**$^+$ being the most commonly used. It is easily obtained from 2,4,6-triphenylpyrylium tetrafluoroborate (**TPP**$^+$) and sodium sulfide on a gram scale (Scheme 4.48) [70].

4.4.1 *Photophysical properties*

TPTP$^+$, a yellow solid, is unstable in neutral and basic media (to prevent degradation of **TPTP**$^+$ in solution, it should be acidified to

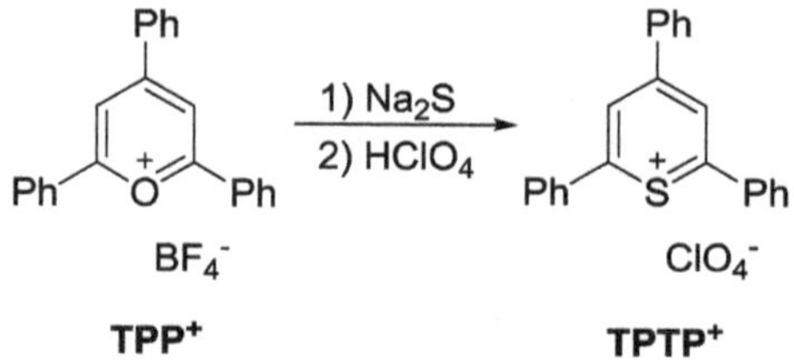

Scheme 4.48. Synthesis of thiapyrylium salt (**TPTP$^+$**) from pyrylium salt (**TPP$^+$**).

Table 4.10. Comparison of **TPTP$^+$** and **TPP$^+$** photophysical properties [3, 72].

Catalyst	$\tau_0(S_1)$ [ns]	$\tau_0(T_1)$ [μs]	E^*_{RED} (S_1) [V vs SCE] **POC*/POC$^-$**	E^*_{RED} (T_1) [V vs SCE] **POC*/POC$^-$**
TPTP$^+$	3.6	6.9	+2.45	+2.09
TPP$^+$	2.7	9.6	+2.55	+2.02

pH ≤ 3.0) [70]. Thiapyrylium salts and their oxygen counterparts, pyrylium salts, have similar photophysical properties, which depend on the nature of the heteroatom, as well as the substituents on the aryl moiety. **TPTP$^+$** has a longer lifetime in the singlet excited state, while the triplet excited state is lower in energy in comparison to **TPP$^+$**. **TPTP$^+$** has lower singlet excited state reduction potential than **TPP$^+$** (Table 4.10), consequently, **TPP$^+$** is more easily reduced than **TPTP$^+$**. Both dyes are not able to reduce molecular oxygen, hence they do not generate the superoxide radical anion $O_2^{\bullet-}$. Since thiapyrylium salt undergoes efficient ISC (Φ_{ISC} = 0.97, calculated $k_{ISC} = 2.7 \times 10^8\ s^{-1}$) [71], it can act as a **POC** in both singlet and triplet excited states (half-wave triplet excited state reduction potential value of 2.09 V vs SCE [3] and ground state reduction potential of −0.19 V vs SCE [3]).

TPTP$^+$ absorbs visible light in the blue region with two absorption maxima at ~370 nm and ~400 nm (depending on the solvent, Chart 4.3).

With 3.6 ns fluorescence lifetime, **TPTP$^+$** encompasses fluorescence with a maximum in ~470 nm ($k_F = 8 \times 10^6\ s^{-1}$), although the

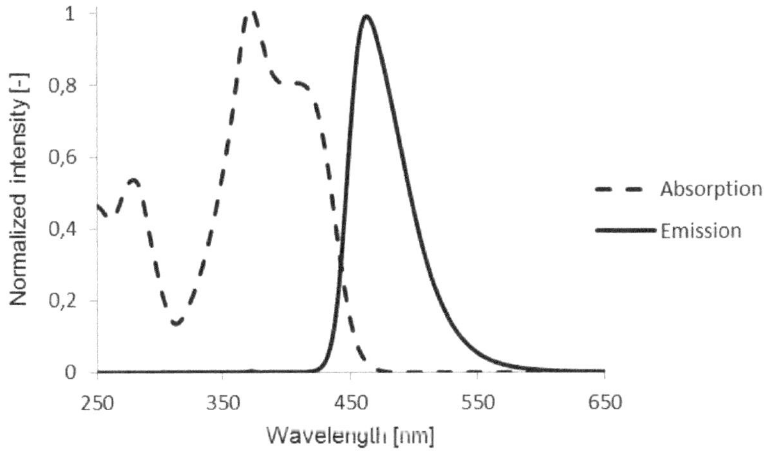

Chart 4.3. Absorption (dashed line) and emission (solid line) spectra of **TPTP**$^+$ in DCM, $c = 7.3 \times 10^{-8}$ M, $\lambda_{ex} = 373$ nm.

fluorescence quantum yield is very low ($\Phi_F = 0.03$) [71]. The T–T absorption spectrum band is located within 460–600 nm with a lifetime of 6.9 μs [72].

4.4.2 *Thiapyrylium salts as photoredox catalysts*

4.4.2.1 *Oxygenations*

Photocatalytic oxygenations are a promising alternative to the application of strong oxidizing reagents [73, 74]. These procedures usually employ singlet oxygen generated photochemically in the presence of light and a photosensitizer [75]. Another mechanism of photooxygenation involves oxygen in the ground state. Thiapyrylium and pyrylium salts absorb light and, in the excited state, oxidize the starting materials through *SET*, generating a radical cation that subsequently reacts with molecular oxygen giving oxidation products (Scheme 4.49) [76].

Following this mechanism, **TPTP**$^+$ efficiently oxidizes heteroatoms like sulfur [77] or phosphorus [78] in organic compounds and promotes detoxification of model pollutants, for example cinnamic acids [79].

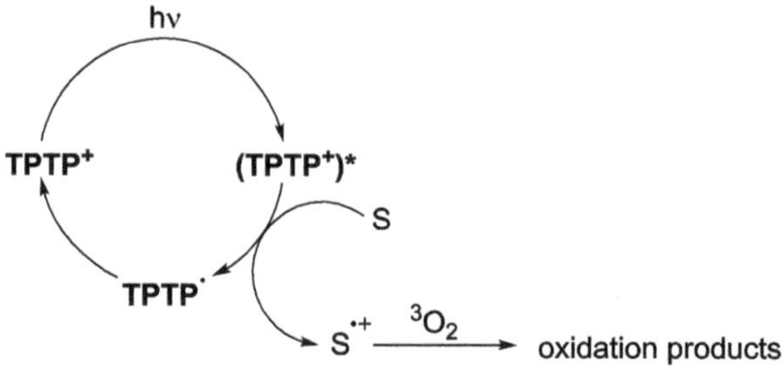

Scheme 4.49. Mechanism of **TPTP**-catalyzed photooxygenation (simplified).

Table 4.11. Photooxygenation of sulfides **164a-c** in the presence of **TPTP**$^+$.

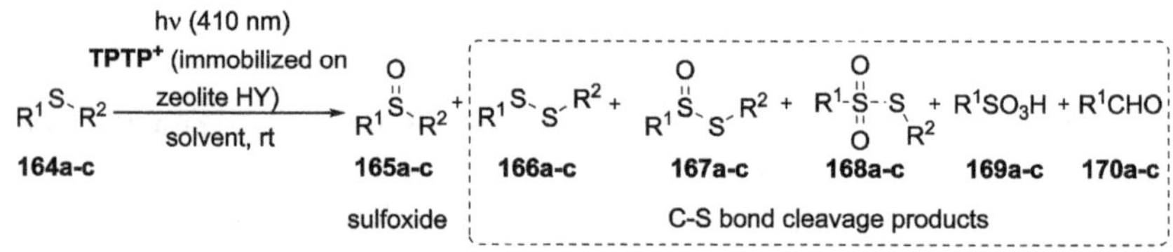

Substrate	R^1	R^2	Solvent	Yield [%]					
				165	**166**	**167**	**168**	**169**	**170**
164a	Et	Et	MeCN	75	—	—	—	24	—
164a	Et	Et	DCE	73	9	3	10	—	—
164b	Bn	Et	MeCN	35	—	—	—	—	18
164c	Ph	*t*-Bu	MeCN	—	35	—	—	—	—

Albini *et al.* reported the photocatalytic oxidation of sulfides **164a-c** in the presence of immobilized thiapyrylium and pyrylium salts [77]. The method is not only suitable for a number of aliphatic, aromatic, and aliphatic–aromatic sulfides leading mainly to sulfoxides **165a-c**, but also for products that are derived from C–S bond which is cleaved, such as compounds **166a-c–169a-c** or aldehydes **170a-c** as by-products (Table 4.11).

In solution, thiapyrylium and pyrylium photocatalysts are susceptible to degradation. Their stability increases by immobilization inside zeolites or adsorption on mesoporous silica, which also enables their easy recovery.

Thiapyrylium salts under visible-light irradiation catalyze the oxidation of triphenylphosphine (**171**) to phosphine oxide (**172**,

Scheme 4.50) [78]. The rate of product **172** formation depends on the solvent used, with acetonitrile being the most efficient. In nucleophilic media, slight photocatalyst degradation (only 3% of the employed amount of **TPTP**$^+$) occurs, but phosphine oxide (**172**) is still obtained in 92% yield. The predominant role of the triplet excited state of **TPTP**$^+$ is assumed based on high *ISC* quantum yield ($\Phi_{ISC} = 0.94$), which is in contrast to the pyrylium analogue (**TPP**$^+$) acting mainly from the singlet excited state [78].

In this reaction, the initial step involves *SET* between the excited thiapyrylium salt ^{3}(**TPTP**$^+$)* and phosphine **171**, leading to the formation of the radical/radical cation pair **N**. Depending on the solvation power of the reaction medium, the oxygenation of the intermediate **N** occurs either at this step (path I, Scheme 4.50), or after solvation (path II, Scheme 4.50). The degradation of the photocatalyst results from oxygen transfer from **TPTP**-OO$^•$ to **171**, and may occur even after formation of pair **O**.

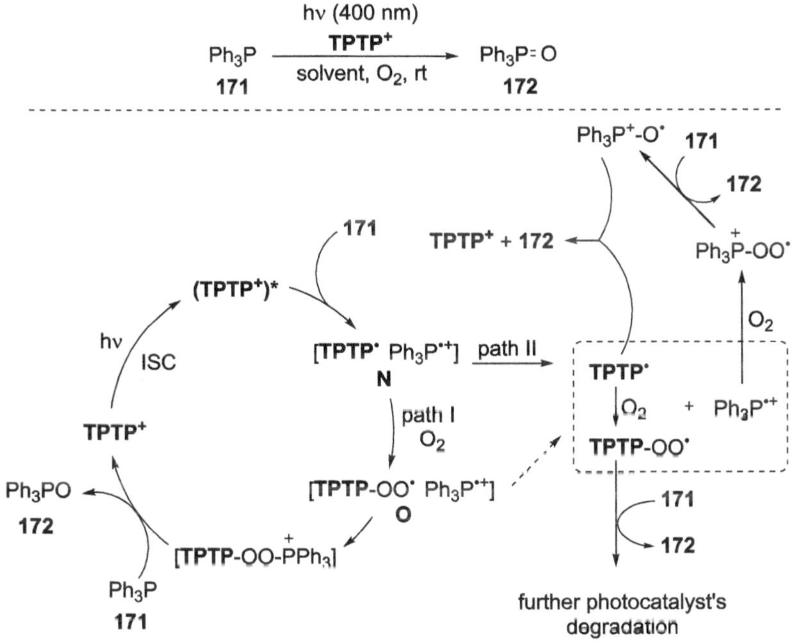

Scheme 4.50. Visible-light-mediated triphenylphosphine (**171**) oxidation in the presence of **TPTP**$^+$.

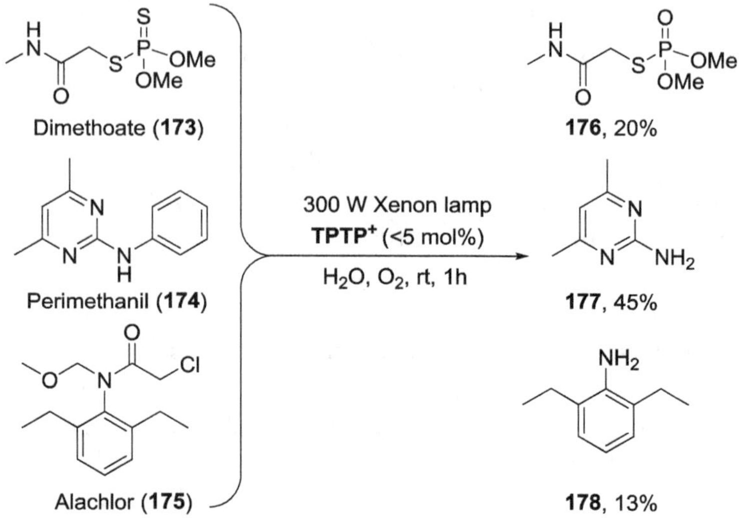

Scheme 4.51. Photodegradation of pollutants **173–175** in the presence of **TPTP**$^+$.

Over the past decade, researchers not only considered thiapyrylium salts as efficient **POCs** in the oxidation processes for synthetic purposes, but also emphasized their potential in the solar-driven photodegradation of pollutants [80]. Mild reaction conditions, with air as the source of oxygen, as well as the use of **POCs** in aqueous media, meet the expectations of green chemistry.

Dimethoate (**173**), perimethanil (**174**), and alachlor (**175**) are popular pesticides that were used as model pollutants to evaluate **TPTP**$^+$ activity in the photodetoxification processes (Scheme 4.51) [81]. These pesticides in aqueous media, in the presence of a catalytic amount of **TPTP**$^+$ (<5 mol%) after a short irradiation time, give degraded products **176**, **177** and **178** in yields of 20%, 45%, and 13%, respectively.

In this process, the singlet excited state of thiapyrylium salt is involved in the electron transfer to the pesticide. Based on Stern–Volmer experiments, the formation of the ground state complex of the photocatalyst and a pesticide within which *SET* occurs upon excitation is suggested.

Subsequently, the detoxification abilities of thiapyrylium salts under air and visible-light irradiation were tested with ferulic acid and the commercial pesticide Senvol bearing the carbaryl active moiety [81]. In the presence of **TPTP**$^+$, 85% of ferulic acid and 80% of Senvol are eliminated, while **TPP**$^+$ is only capable of detoxifying 59% of the ferulic acid.

4.4.2.2 *C–C bond formation*

Apart from oxygenation, thiapyrylium salts in the presence of light, also catalyze C–C bond forming reactions including cycloadditions [71] and dimerizations [72]. The **TPTP**-catalyzed Diels–Alder reaction of indoles **179a-c** with cyclohexadienes **180a-c** affords cycloadducts **181a-c** in yields exceeding 33%, but with selectivity lower than that for **TPP**$^+$ (Scheme 4.52) [71]. Earlier studies revealed that the nitrogen atom of the indole ring should be protected, but the reaction of *N*-acetylindole with cyclohexadiene derivatives gives only traces of the adduct. Hence, it was assumed that the acylation of the product was required. Photophysical experiments (e.g. **TPTP**$^+$ fluorescence quenching by reactants) revealed that both singlet and triplet excited states are involved in the process.

The selectivity of the reaction of aryl-alkenes **182a-c** with thiobenzophenone (**183**) is also unsatisfactory, leading to a mixture of two diastereoisomers *cis*-**184a-c** and *trans*-**184a-c** with only a slight excess of the *trans* isomer being formed (*trans* positions of two aryl substituents, Scheme 4.53) [82]. In this case, the proposed mechanism involves *SET* from thiobenzophenone (**183**) to the triplet-excited

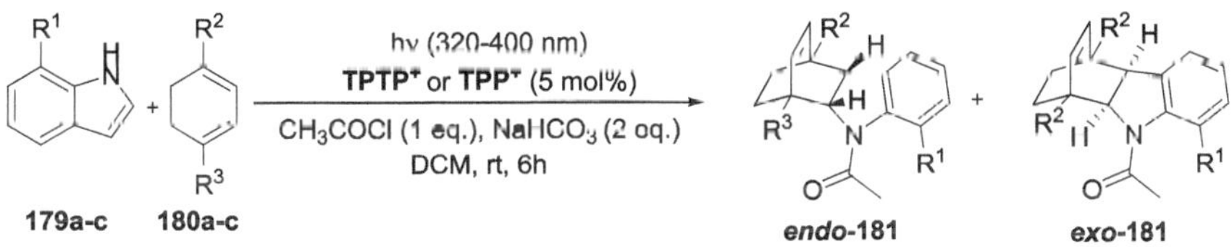

Yield of **181** [%], (*endo*:*exo* ratio)

	R^1	R^2	R^3		TPTP$^+$	TPP$^+$
a	R^1=H	R^2=H	R^3=H	a	62, (2:1)	70, (3:1)
b	R^1=H	R^2=Me	R^3=*i*-Pr	b	66, (2:1)	57, (2:1)
c	R^1=Me	R^2=H	R^3=H	c	33, (2:1)	n.d.

Scheme 4.52. Diels–Alder reaction catalyzed by **TPTP**$^+$.

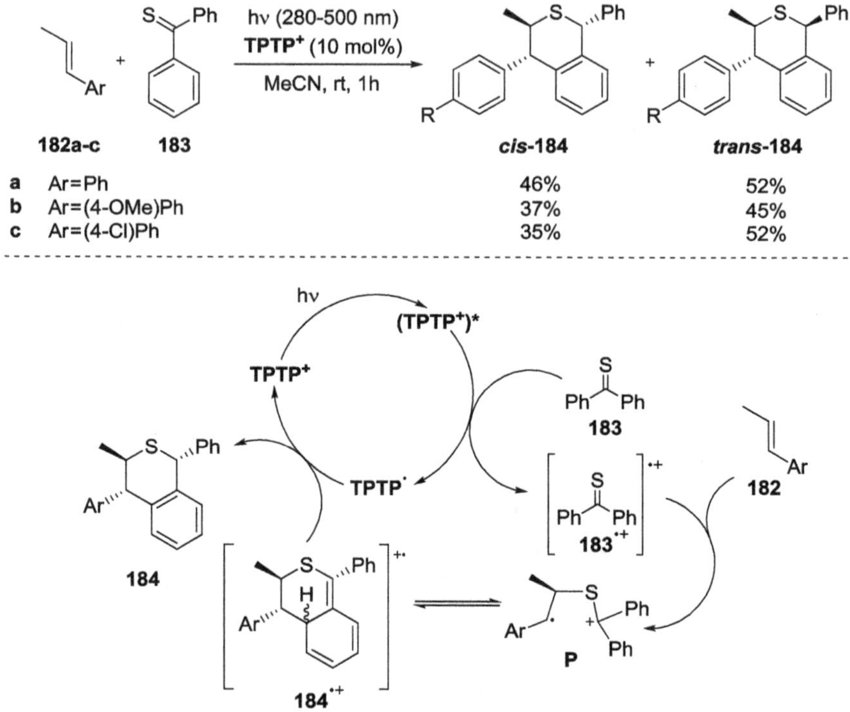

Scheme 4.53. [2+4] cycloaddition of alkenes **182** and thiobenzophenone **183** catalyzed by **TPTP**$^{+}$.

TPTP^{+}. The radical cation **183**$^{\bullet+}$ reacts then with aryl-alkene **182** generating the intermediate **P**. Cyclization and back electron transfer from **TPTP**$^{\bullet}$ to the radical cation closes the catalytic cycle, leading to the recovery of **TPTP**$^{+}$ and cycloadduct **184**.

Dimerization of 1,3-cyclohexadiene (**180a**) may be initiated by **TPTP**$^{+}$, though high solvent dependence is observed (Scheme 4.54) [83]. The photocatalyst-free reaction affords exclusively [2+2] products **186**. On the other hand, the reaction in dichloromethane leads only to [2+4] cycloaddition products **185**, whereas, in chloroform, mainly [2+2] cycloadducts **186** are formed. Both [2+4] and [2+2] cycloadditions lead preferably to the *endo* product.

The use of **TP(4-OMe)TP**$^{+}$ in the reaction of 1,1′-bicyclohexe-1-yl (**187**) with styrene (**140**) efficiently furnishes [4+2] cycloadducts **188**

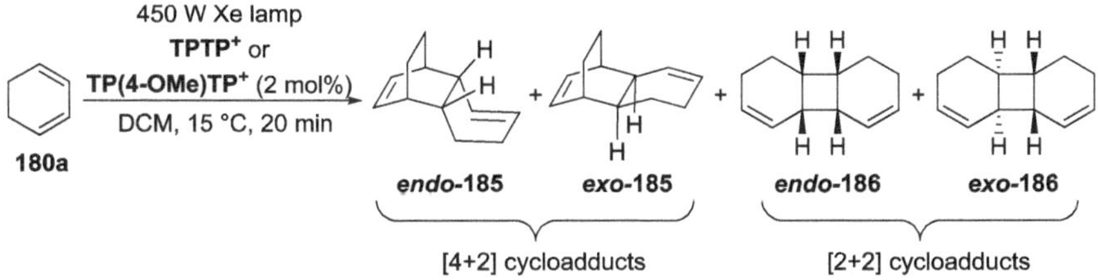

Scheme 4.54. Light-induced dimerization of 1,3-cyclohexadiene (**180a**) catalyzed by **TPTP**$^+$ and **TP(4-OMe)TP**$^+$.

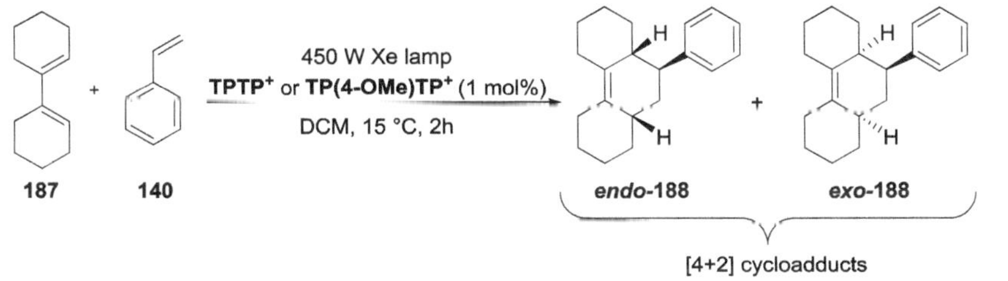

catalyst	Yield of **188**, (*endo*:*exo* ratio)
TPTP$^+$	1%, (1:12)
TP(4-OMe)TP$^+$	68%, (1:11)

Scheme 4.55. Diels–Alder reaction of 1,1′-bicyclohexen-1-yl (**187**) with styrene (**140**) catalyzed by thiapyrylium salts.

with *exo* selectivity in the 68% yield, while unmodified **TPTP**$^+$ results in a significant decrease in the yield (Scheme 4.55) [83]. These results clearly emphasize the structural impact of **POCs** on the photophysical properties and catalytic efficiency.

4.4.2.3 *C–X bond cleavage*

The catalytic properties of thiapyrylium salts enable C–X bond cleavage reactions. In this line, photodeprotection of 1,3-dithianes **189a-d** in the presence of **TPTP**$^+$ leads to the corresponding aldehydes and ketones **190a-d** in moderate to good yields (Scheme 4.56) [84].

The reaction is less efficient for substrates with phenyl rings substituted either with electron-withdrawing or electron-donating groups. The complete inhibition of the reaction by 1,4-benzoquinone,

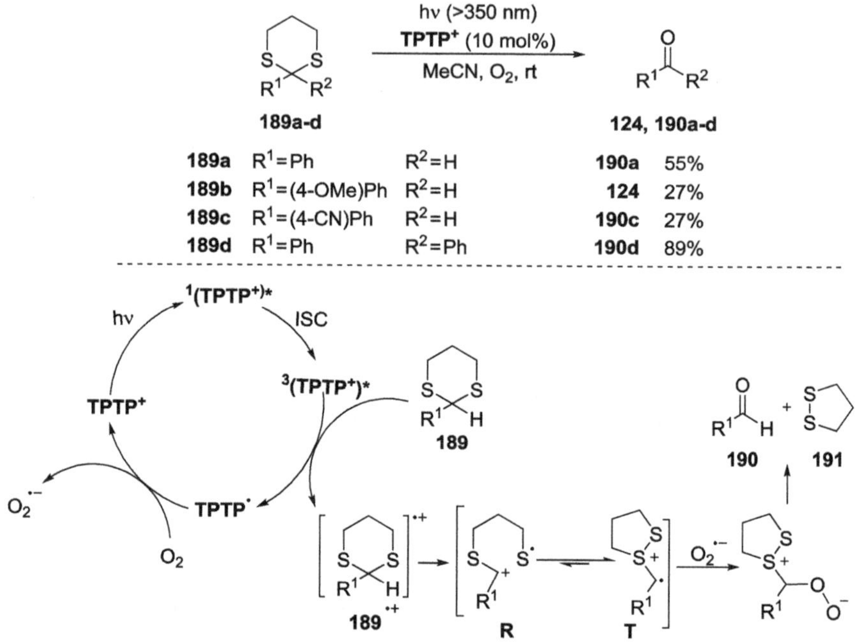

Scheme 4.56. Photodeprotection of 1,3-dithianes **189a–189d** catalyzed by **TPTP⁺**.

a superoxide ion quencher, suggests the importance of oxygen in the deprotection mechanism [85]. Laser flash photolysis experiments revealed the involvement of the triplet excited state of **TPTP⁺** and the formation of radical cation **189**$^{\bullet+}$ *via SET* to the photocatalyst. The radical cation **189**$^{\bullet+}$ degrades to the distonic radical cation **R** which exists in two forms, with the closed ring form **T** being more stable. As such, it immediately reacts with the superoxide ion through nucleophilic attack, which is thermodynamically and kinetically favorable, leading to deprotected carbonyl compounds **190** and by-product **191**.

4.4.2.4 *Cycloreversion*

Cycloreversion reactions, involved in DNA reconstruction after UV irradiation damage [86], are well-known transformations in the

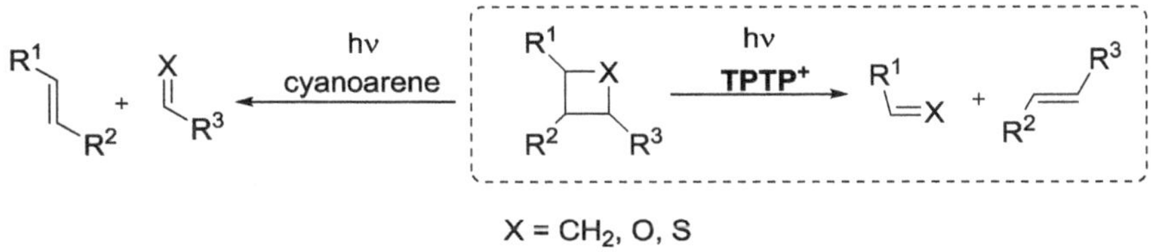

Scheme 4.57. Possible pathways of cycloreversion depending on the photocatalyst used.

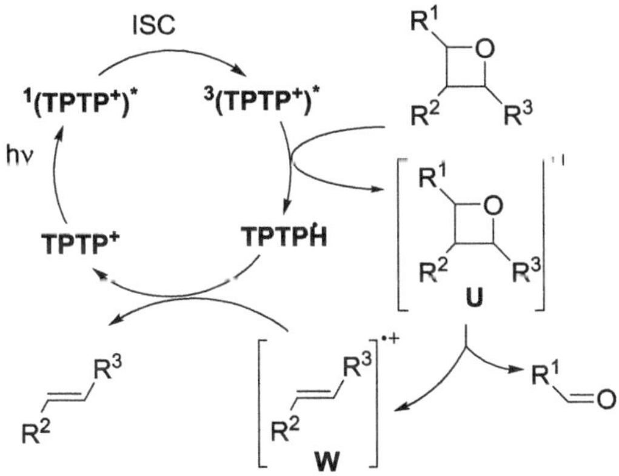

Scheme 4.58. Mechanism of **TPTP**$^{+}$-photoorganocatalyzed cycloreversion of oxetanes.

field of organic chemistry. Miranda *et al.* reported the thiapyrylium photocatalyzed cycloreversion of oxetanes, thietanes, and cyclobutanes [87]. In comparison with cyanoarene photocatalysts, initially used in this type of reaction [88], thiapyrylium salts furnish products derived from different C–X and C–C bond cleavages (Scheme 4.57).

Mechanistic studies suggest *SET* from the oxetane ring to the excited ^{3}(**TPTP**$^{+}$)* (Scheme 4.58) [89]. The resulting radical cation **U** decomposes, releasing an aldehyde and a radical cation **W** which is reduced by **TPTP**$^{\bullet}$ to the final alkene, thus regenerating the photocatalyst. Although thermodynamic calculations imply that both the triplet and singlet excited states of **TPTP**$^{+}$ are likely to participate in *SET*, the high value of Φ_{ISC} indicates the involvement of mainly the triplet excited state of the photocatalyst in the electron transfer [90, 91]. This is in opposition to cyanoarenes which catalyze the cycloreversion reaction from the singlet excited state [88]. An analogous mechanism operates for thietanes and cyclobutanes.

Applying **TPTP**$^+$ in the presence of light for the cycloreversion of oxetane **192** furnishes the acetaldehyde (**193**) and *trans*-stilbene (**194**) quantitatively, while in the case of the pyrylium salt, which is characterized by a lower Φ_{ISC} value, only small amounts of products are formed (Table 4.12) [89].

Nucleophilic trapping experiments were undertaken in order to evaluate which bond, either C–O or C–C, breaks first in the starting material [92]. The formation of a five-membered ring from the substrate **195**, as a consequence of C–C bond rotation, confirms the initial C–O bond cleavage (Scheme 4.59).

Table 4.12. Cycloreversion of oxetane **192** catalyzed by (thia)pyrylium salts.

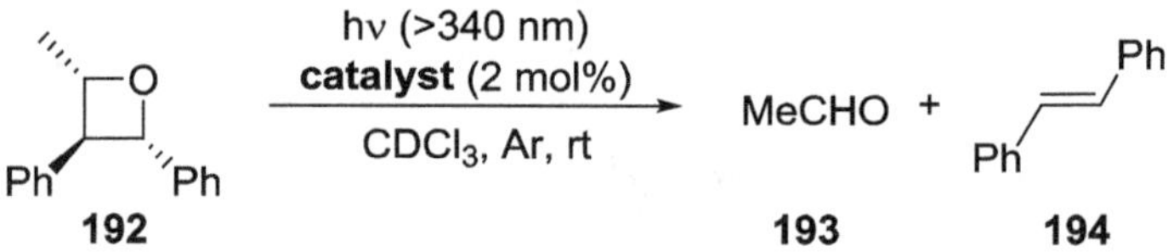

Catalyst	Conversion [%]
TPTP$^+$	100
TPP$^+$	8

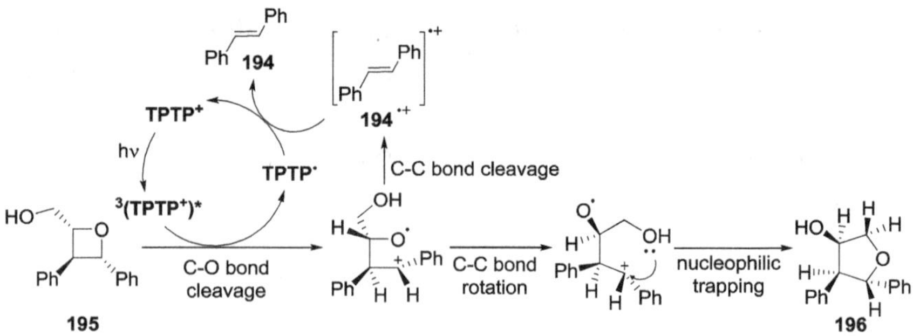

Scheme 4.59. Formation of product **196** *via* a photoorganocatalyzed cycloreversion.

Miranda *et al.* also reported another nucleophilic trapping experiment in which acetonitrile attacks the oxetane radical cation **U** (Scheme 4.58) leading to a six-membered heterocycle, clearly supporting the suggested mechanism [93].

The substitution pattern has a great influence on the bond breaking order and the reaction pathway [94]. Introduction of electron donating groups, such as the methoxy group, to the phenyl substituent at the C_3 position in the oxetane **197c** changes the reaction pathway leading to another bond cleavage, similar to the cyanoarene-catalyzed reaction (Scheme 4.60) [95]. This divergence may be caused by the stabilization of the positive charge in the radical cation **U** by the OMe group. However, substitution of the aryl ring with the electron-withdrawing CN-group makes the oxetane **197b** resistant to cycloreversion, while replacing it with the methyl group leads to products **193** and **198a** resulting from the cleavage of O–C_2 and C_3–C_4 bonds [94].

Unexpectedly, irradiation of thietanes **199a-c** in the presence of **TPTP**$^+$ furnishes two types of products: not only cycloreversion products (the corresponding alkene and thiobenzophenone **183**), but also secondary [2+4] cycloadducts (*cis*-**200** and *trans*-**200**) (Table 4.13) [96].

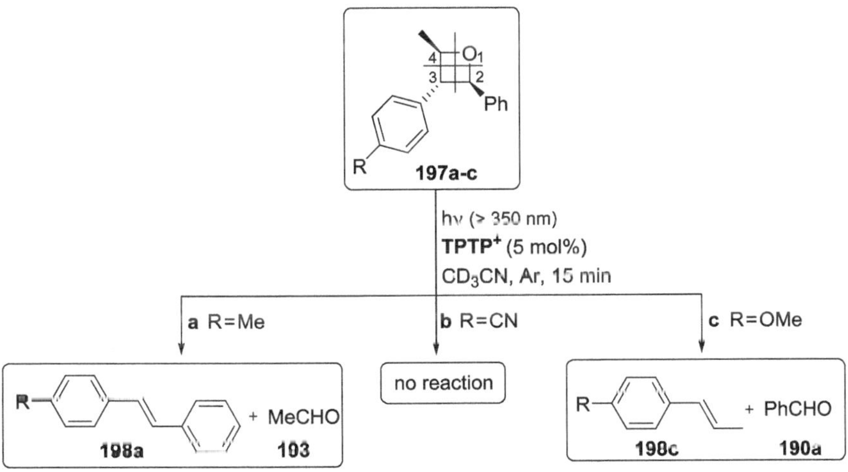

Scheme 4.60. Cycloreversion controlled by the structure of substrates **197a-c**.

Table 4.13. Cycloreversion of thietanes **199a–199c** catalyzed by **TPTP**$^+$.

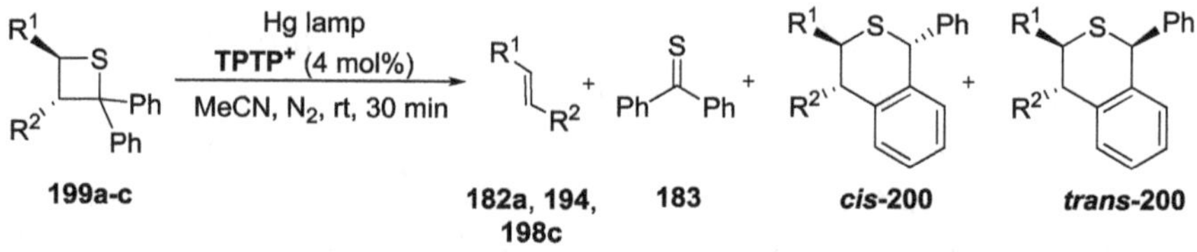

			Yield [%]				
Substrate	R^1	R^2	**alkene**		**183**	***cis*-200**	***trans*-200**
199a	Me	Ph	**182a**	—	—	48	52
199b	Me	(4-OMe)Ph	**198c**	26	28	33	39
199c	Ph	Ph	**194**	100	85	—	—

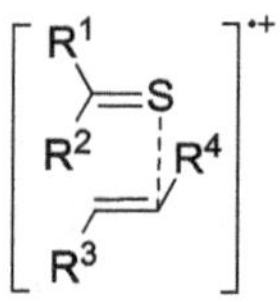

Figure 4.4. Structure of an IMC.

The ratio of the products depends on the structure of the starting material. For thietane **199a**, substituted with one methyl and three phenyl groups in the presence of thiapyrylium, only cycloreversion occurs, while the replacement of the phenyl group at C_4 position with methyl gives exclusively cycloaddition products **200** (Table 4.13). These reactivity differences can be explained based on the ion–molecule complex (IMC, Figure 4.4) formation through the C–C and C–S bonds cleavage. Elimination of the radical cation from the complex leads to cycloreversion, while complex rearrangement results in [2+4] cycloaddition.

In the presence of light, **TPTP**$^+$ catalyzes the cycloreversion of cyclobutanes.1,2,3,4-Tetraphenylcyclobutanes **201a** and **201b** furnish *trans*-stilbene (**194**) as the sole photoproduct (Scheme 4.61) [97]. Laser flash photolysis experiments exclude the formation of a radical cation corresponding to *cis*-stilbene. The mechanism is similar to the one shown in Scheme 4.58, with **TPTP**$^+$ in the triplet excited state being involved in *SET*.

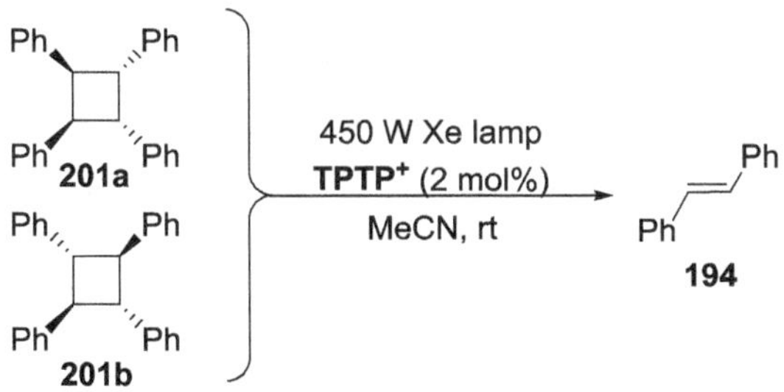

Scheme 4.61. TPTP$^+$-catalyzed cycloreversion of cyclobutanes **201a-b**.

4.5. Sexithiophene and Polymeric Benzothiadiazole Systems

Among sulfur photocatalysts, only a few examples of thiophene applications are reported. For instance, **6-TP** is an efficient photoredox catalyst for dehalogenation of activated vicinal-dibromides **202** (Scheme 4.62) [98]. The reaction proceeds with the heterogenous **6-TP** photocatalyst in the presence of a tertiary amine acting as a sacrificial electron donor.

The **6-TP** lifetimes in the singlet and triplet excited states are respectively, 0.81 ns and 16 μs [2]. Thermodynamic calculations support *SET* from the single excited state of thiophene hexamer to dibromide reagent **202a-g** as the initial reaction step. However, laser flash photolysis reveals that both singlet and triplet excited states of **6-TP** are involved in the photocatalytic cycle. Initially, vicinal dibromides **202a-g** transform to the corresponding alkenes with the release of a bromide anion and a bromide radical. The **6-TP**$^{\bullet+}$ and the bromine radical are reduced by an amine acting as a sacrificial reductant.

Another sulfur photocatalyst is a microporous cross-linked polymeric system based on benzothiadiazole units (**B-BT**) (Figure 4.5). It was obtained in the Sonogashira–Hagihara cross-coupling reaction of 1,3,5-triethynylbenzene with 4,7-dibromo-benzo[*c*][1,2,5]thiadiazole. **B-BT** catalyzes the [2+2] cycloaddition of styrene (**140**) and its derivatives giving products in decent yields with high substrate tolerance and only a small amount of homocycloadducts as by-products (Scheme 4.63) [99]. It also enables the

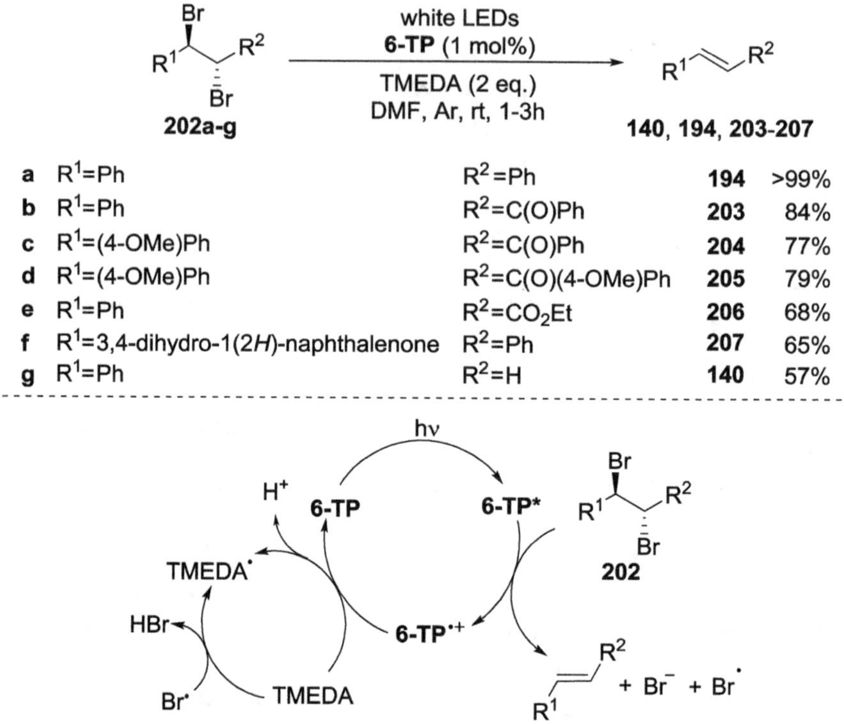

Scheme 4.62. Sexithiophene-catalyzed reductive debromination.

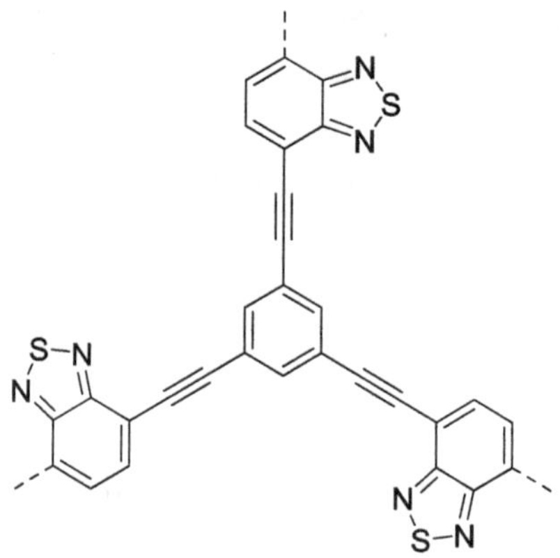

Figure 4.5. Structure of **B-BT**.

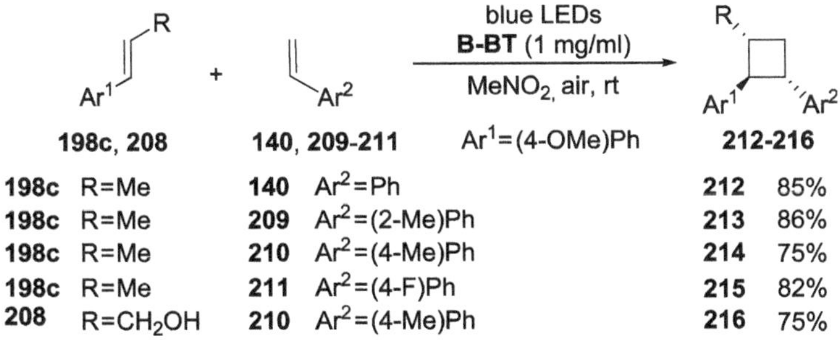

Scheme 4.63. **B-BT**-photocatalyzed [2+2] cycloaddition.

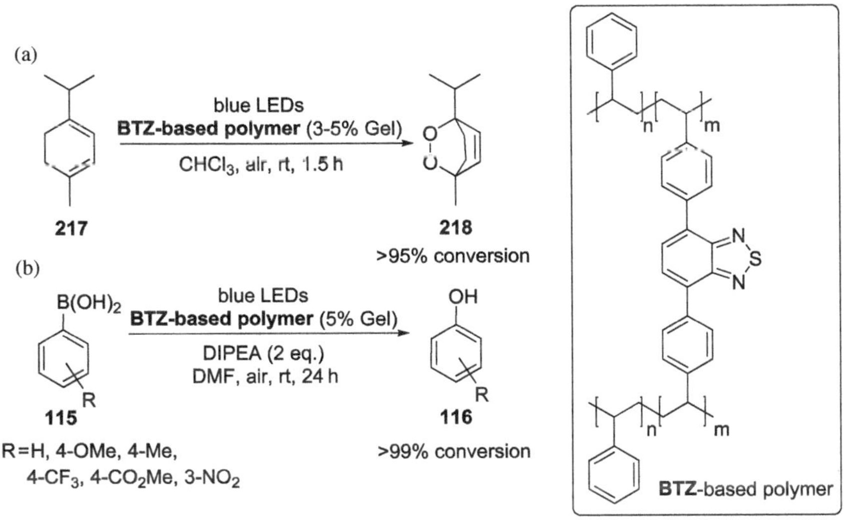

Scheme 4.64. **BTZ**-based polymer as $^1O_2$ photosensitizer in the oxidation of α-terpinene (**217**) (a) and as the photoredox catalyst in the hydroxylation of arylboronic acids **115** (b).

facile synthesis of the biologically active endiandrin A and di-*O*-methylendiandrin A, potent antiinflammatory drugs.

The free radical copolymerization of 2,1,3-benzothiadiazole (**BTZ**)-based vinyl crosslinker with a large excess of styrene (**140**) furnishes the heterogeneous sulfur-containing photocatalyst **BTZ**. **BTZ**-based polymer has been used in a stable form of gel and it enables the photogeneration of $^1O_2$ (Scheme 4.64(a)) as well as the superoxide ion ($O_2^{\bullet-}$) for hydroxylation of arylboronic acids **115** (Scheme 4.64(b)) [100].

Photooxidation of α-terpinene (**217**, Scheme 4.64) to ascaridole (**218**) with $^1O_2$ in the presence of a gel containing 5% of **BTZ** units under light irradiation leads to product **218** with an almost quantitative conversion of the starting substrate. Photocatalytic polymeric systems are stable and resistant towards oxidation by $^1O_2$, which is in contrast to styrene-**BZT** monomer (due to the high reactivity between vinylarenes and $^1O_2$). These reactions can be carried out in a continuous flow manner with sunlight as an alternative to LEDs.

4.6. Conclusions

Sulfur heterocyclic **POCs** consist mainly of thiazines and thiapyrylium salts with **MB**$^+$ being at the forefront. As structural modifications strongly influence the photophysical properties of these dyes, various functionalizations of the sulfur heterocycles provide an impressive set of photocatalysts with distinct characteristics.

Taking into account the absorption maxima of the four most commonly used sulfur catalysts (Table 4.14), one can realize that these compounds can be applied in reactions that proceed with the more energetic near-UV light (**PTH**, **TPTP**$^+$), as well as the near-IR radiation of lower energy (**MB**$^+$). Due to the fast ISC to the long-living triplet state, their photocatalytic activity mainly involves energy or electron-transfer processes from the triplet excited state. As the substitution pattern strongly affects the redox potential values, they span from as low as −2.15 V vs SCE on the cathodic to +2.45 V vs SCE on the anodic site, thus facilitating selective reactions for complex starting materials.

The spectroelectrochemical differences among these sulfur heterocycles are reflected in their catalytic properties. **MB**$^+$ acts as both a photoredox catalyst and a photosensitizer. Ene-type reactions, [2+2] and [2+4] cycloadditions include $^1O_2$ photogeneration through *EnT* from ^{3}(**MB**$^+$)*, while, through electron transfer, **MB**$^+$ catalyzes C–C and C–X bond formation as well as cleavage of C–X bonds. Other phenothiazine dyes include **PTH**, which are the only

Table 4.14. Summary of the photophysical properties of the most commonly used sulfur **POCs** [2, 3].

POC	λ_{max} absorption (nm)	ε (M^{-1} cm^{-1})	λ_{max} emission (nm)	τ_f (ns)	Φ_{ISC}
MB^+	650–660	9400	678	0.60–1.0	0.52
PTH	320	13900	444	0.81–2.3	n.d.
$TPTP^+$	370–400	4500	463	3.1	0.94
α-6T	435	26100	510	n.d.	n.d.

PC	τ_0 (S_1) (ns)	τ_0 (T_1) (μs)	$E_{0,0}$ [eV] $E_{0,0}$ (S_1)	$E_{0,0}$ (T_1)
MB^+	1	32	1.80	1.50
PTH	3	n.d.	2.80	2.40
$TPTP^+$	3.6	6.9	2.64	2.28
α-6T	0.81	16	2.58	1.82

	E of **POC** [V vs SCE]		*E** of ^{1}**POC*** [V vs SCE]		*E** of ^{3}**POC*** [V vs SCE]	
PC	E_{RED} **POC/POC$^-$**	E_{OX} **POC$^+$/POC**	E^*_{RED} **POC*/POC$^-$**	E^*_{OX} **POC$^+$/POC***	E^*_{RED} **POC*/POC$^-$**	E^*_{OX} **POC$^+$/POC***
MB^+	−0.30	+1.13	+1.56	−0.73	+1.60	−0.68
PTH	n.d.	+0.68	n.d.	−2.10	n.d.	−1.70
$TPTP^+$	−0.19	n.d.	+2.45	n.d.	+2.09	n.d.
α-6T	n.d.	+0.42	n.d.	−2.15	n.d.	−1.44

sulfur photocatalysts with the ability to act as both reductants and oxidants *via* a *SET* mechanism. Apart from the participation in oxygenations *via SET*, thiapyrylium salts also catalyze C–C bond formation, C–X bond cleavage, and cycloreversion reactions.

Thanks to their high photocatalytic activity, sulfur heterocycles have found application in total synthesis, drug development, and detoxification processes. To this end, their structural and photochemical diversity enables the rational selection of a suitable photocatalyst for the desired transformation. They were efficiently immobilized in the interior of zeolites or in gel, thus increasing their stability and enabling photocatalyst recovery.

References

[1] Schirmer, R. H.; Adler, H.; Pickhardt, M.; Mandelkow, E. *Neurobiol. Aging* **2011**, *32*, 2325.e7–2325.e16.

[2] Pitre, S. P.; McTiernan, C. D.; Scaiano, J. C. *Acc. Chem. Res.* **2016**, *49*, 1320–1330.

[3] Romero, N. A.; Nicewicz, D. A. *Chem. Rev.* **2016**, *116*, 10075–10166.

[4] Pitre, S. P.; McTiernan, C. D.; Ismaili, H.; Scaiano, J. C. *J. Am. Chem. Soc.* **2013**, *135*, 13286–13289.

[5] Prier, C. K.; Rankic, D. A.; MacMillan, D. W. C. *Chem. Rev.* **2013**, *113*, 5322–5363.

[6] Ghogare, A. A.; Greer, A. *Chem. Rev.* **2016**, *116*, 9994–10034.

[7] Stephenson, L. M. *Acc. Chem. Res.* **1980**, *13*, 419–425.

[8] Orfanopoulos, M.; Grdina, M. B.; Stephenson, L. M. *J. Am. Chem. Soc.* **1979**, *101*, 275–276.

[9] Orfanopoulos, M.; Stratakis, M.; Elemes, Y. *Tetrahedron Lett.* **1989**, *30*, 4875–4878.

[10] Stratakis, M.; Orfanopoulos, M. *Synth. Commun.* **1993**, *23*, 425–430.

[11] Clennan, E. L.; Sram, J. P. *Tetrahedron* **2000**, *56*, 6945–6950.

[12] Clennan, E. L.; Zhang, D.; Singleton, J. *Photochem. Photobiol.* **2006**, *82*, 1226–1232.

[13] Tsangarakis C., Zaravinos, I.-P.; Stratakis, M. *Synlett* **2005**, *12*, 1857–1860.

[14] Yoshioka, M.; Sakuma, Y.; Saito, M. *J. Org. Chem.* **1999**, *64*, 9247–9250.

[15] Yoshioka, M.; Hashimoto, K.; Fukuhara, T.; Hasegawa, T. *J. Chem. Soc. Perkin Trans.* **1998**, *1*, 283–288.

[16] Kwon, B. M.; Foote, C. S. *J. Org. Chem.* **1989**, *54*, 3878–3882.

[17] Cermola, F.; Guaragna, A.; Iesce, M. R.; Palumbo, G.; Purcaro, R.; Rubino, M.; Tuzi, A. *J. Org. Chem.* **2007**, *72*, 10075–10080.

[18] Watanabe, N.; Kikuchi, M.; Maniwa, Y.; Ijuin, H. K.; Matsumoto, M. *J. Org. Chem.* **2010**, *75*, 879–884.

[19] Chawla, H. M.; Pathak, M. *Tetrahedron* **1990**, *46*, 1331–1342.
[20] Wasserman, H. H.; Xia, M.; Wang, J.; Petersen, A. K.; Jorgensen, M.; Power, P.; Parr, J. *Tetrahedron* **2004**, *60*, 7419–7425.
[21] Kalaitzakis, D.; Triantafyllakis, M.; Alexopoulou, I.; Sofiadis, M.; Vassilikogiannakis, G. *Angew. Chem. Int. Ed.* **2014**, *53*, 13201–13205.
[22] Georgiou, T.; Tofi, M.; Montagnon, T.; Vassilikogiannakis, G. *Org. Lett.* **2006**, *8*, 1945–1948.
[23] Tofi, M.; Montagnon, T.; Georgiou, T.; Vassilikogiannakis, G. *Org. Biomol. Chem.* **2007**, *5*, 772–777.
[24] Kalaitzakis, D.; Triantafyllakis, M.; Sofiadis, M.; Noutsias, D.; Vassilikogiannakis, G. *Angew. Chem. Int. Ed.* **2016**, *55*, 4605–4609.
[25] Kalaitzakis, D.; Kouridaki, A.; Noutsias, D.; Montagnon, T.; Vassilikogiannakis, G. *Angew. Chem. Int. Ed.* **2015**, *54*, 6281–6285.
[26] Iesce, M. R.; Cermola, F.; Piazza, A.; Scarpati, R.; Graziano, M. L. *Synthesis* **1995**, *1995*, 439–443.
[27] Arroyo, Y.; Carreño, M. C.; Ruano, J. L. G.; Amo, J. F. R.; Santos, M.; Sanz-Tejedor, A. M. *Tetrahedron Asymm.* **2000**, *11*, 1183–1191.
[28] Arroyo, Y.; Rodríguez, J. F.; Sanz-Tejedor, A.; Santos M. *Tetrahedron Lett.* **2002**, *43*, 9129–9132.
[29] Charbonnet, N.; Riguet, E.; Bochet, C. G. *Synlett* **2011**, *15*, 2231–2233.
[30] Cermola, F.; Iesce, M. R.; Montella, S. *Lett. Org. Chem.* **2004**, *1*, 271–275.
[31] Kernan, M. R.; Faulkner, D. J. *J. Org. Chem.* **1988**, *53*, 2773–2776.
[32] Aquino, M.; Bruno, I.; Riccio, R.; Gomez-Paloma, L. *Org. Lett.* **2006**, *8*, 4831–4834.
[33] Patil, S. N.; Liu, F. *Org. Lett.* **2007**, *9*, 195–198.
[34] DellaGreca, M.; Zuppolini, S.; Zarrelli, A.; Iesce, M. R.; Previtera, L. *Tetrahedron* **2013**, *69*, 4725–4730.
[35] Nishio, T.; Tokunaga N.; Kondo, M.; Omote, Y. *J. Chem. Soc. Perkin Trans.* **1988**, *1*, 2921–2925.
[36] Kamata, M.; Sato, M.; Hasegawa, E. *Tetrahedron Lett.* **1992**, *33*, 5085–5088.
[37] Madhavan, D.; Pitchumani, K. *Tetrahedron* **2001**, *57*, 8391–8394.
[38] Okada, A.; Nagasawa, Y.; Yamaguchi, T.; Yamaguchi, E.; Tada, N.; Miura, T.; Itoh, A. *RSC Adv.* **2016**, *6*, 42596–42599.

[39] Okada, A.; Yuasa, H.; Fujiya, A.; Tada, N.; Miura, T.; Itoh, A. *Synlett* **2015**, *26*, 1705–1709.

[40] Xu, X.-X.; Zhu, J.; Huang, D.-Z.; Zhou, W.-S. *Tetrahedron Lett.* **1991**, *32*, 5785–5788.

[41] Jung, M. E.; Min, S.-J. *Tetrahedron* **2007**, *63*, 3682–3701.

[42] Kalaitzakis, D.; Noutsias, D.; Vassilikogiannakis, G. *Org. Lett.* **2015**, *17*, 3596–3599.

[43] Margaros, I.; Montagnon, T.; Vassilikogiannakis, G. *Org. Lett.* **2007**, *9*, 5585–5588.

[44] Lancefield, C. S.; Zhou, L.; Lébl, T.; Slawin, A. M. Z.; Westwood, N. *J. Org. Lett.* **2012**, *14*, 6166–6169.

[45] Manring, L. E.; Eriksen, J.; Foote, C. S. *J. Am. Chem. Soc.* **1980**, *102*, 4275–4277.

[46] Kojima, M.; Kuriyama, Y.; Sakuragi, H.; Tokumaru, K. *Bull. Chem. Soc. Jpn.* **1991**, *64*, 2724–2729.

[47] Pitre, S. P.; McTiernan, C. D.; Ismaili, H.; Scaiano, J. C. *ACS Catal.* **2014**, *4*, 2530–2535.

[48] Cui, L.; Matusaki, Y.; Tada, N.; Miura, T.; Uno, B.; Itoh, A. *Adv. Synth. Catal.* **2013**, *355*, 2203–2207.

[49] Fujiya, A.; Tanaka, M.; Yamaguchi, E.; Tada, N.; Itoh, A. *J. Org. Chem.* **2016**, *81*, 7262–7270.

[50] Wu, G.; Lv, T.; Mo, W.; Yang, X.; Gao, Y.; Chen, H. *Tetrahedron Lett.* **2017**, *58*, 1395–1398.

[51] Kako, M.; Takada, H. *Tetrahedron Lett.* **1997**, *38*, 3525–3528.

[52] Zou, Y.-Q.; Chen, J.-R.; Liu, X.-P.; Lu, L.-Q.; Davis, R. L.; Jørgensen, K. A.; Xiao, W.-J. *Angew. Chem. Int. Ed.* **2012**, *51*, 784–788.

[53] Ciszewski, Ł. W.; Smoleń, S.; Gryko, D. *Arkivoc* **2017**, *part ii*, 251–259.

[54] Li, Y.; Xie, W.; Jiang, X. *Chem. Eur. J.* **2015**, *21*, 16059–16065.

[55] Epling, G. A.; Wang, Q. *Tetrahedron Lett.* **1992**, *33*, 5909–5912.

[56] Provatas, A.; A. Epling, G.; D. Stuart, J. *Aust. J. Chem.* **2016**, *69*, 763–769.

[57] Wu, G.; Li, Y.; Yu, X.; Gao, Y.; Chen, H. *Adv. Synth. Catal.* **2017**, *359*, 687–692.

[58] Nelsen, S. F.; Peacock, V.; Weisman, G. R. *J. Am. Chem. Soc.* **1976**, *98*, 5269–5277.

[59] Cocquet, G.; Ferroud, C.; Simon, P.; Taberna, P.-L. *J. Chem. Soc. Perkin Trans.* **2000**, *2*, 1147–1153.

[60] Ferroud, C.; Cocquet, G.; Guy, A. *Tetrahedron Lett.* **1999**, *40*, 5005–5008.

[61] Cocquet, G.; Ferroud, C.; Guy, A. *Tetrahedron* **2000**, *56*, 2975–2984.

[62] Ding, Y.; Li, H.; Meng, Y.; Zhang, T.; Li, J.; Chen, Q.-Y.; Zhu, C. *Org. Chem. Front.* **2017**, *4*, 1611–1614.

[63] Ding, Y.; Zhang, W.; Li, H.; Meng, Y.; Zhang, T.; Chen, Q.-Y.; Zhu, C. *Green Chem.* **2017**, *19*, 2941–2944.

[64] Zhang, H.; Trout, W. S.; Liu, S.; Andrade, G. A.; Hudson, D. A.; Scinto, S. L.; Dicker, K. T.; Li, Y.; Lazouski, N.; Rosenthal, J.; Thorpe, C.; Jia, X.; Fox, J. M. *J. Am. Chem. Soc.* **2016**, *138*, 5978–5983.

[65] Park, J. H.; Ko, K. C.; Kim, E.; Park, N.; Ko, J. H.; Ryu, D. H.; Ahn, T. K.; Lee, J. Y.; Son, S. U. *Org. Lett.* **2012**, *14*, 5502–5505.

[66] Discekici, E. H.; Treat, N. J.; Poelma, S. O.; Mattson, K. M.; Hudson, Z. M.; Luo, Y.; Hawker, C. J.; de Alaniz, J. R. *Chem. Commun.* **2015**, *51*, 11705–11708.

[67] Poelma, S. O.; Burnett, G. L.; Discekici, E. H.; Mattson, K. M.; Treat, N. J.; Luo, Y.; Hudson, Z. M.; Shankel, S. L.; Clark, P. G.; Kramer, J. W.; Hawker, C. J.; de Alaniz, J. R. *J. Org. Chem.* **2016**, *81*, 7155–7160.

[68] Fry, A. J.; Krieger, R. L. *J. Org. Chem.* **1976**, *41*, 54–57.

[69] Rondinini, S.; Mussini, P. R.; Muttini, P.; Sello, G. *Electrochim. Acta* **2001**, *46*, 3245–3258.

[70] Wizinger, R.; Ulrich, P. *Helv. Chim. Acta* **1956**, *39*, 207–216.

[71] González-Béjar, M.; Stiriba, S.-E.; Miranda, M. A.; Pérez-Prieto, J. *Arkivoc* **2007**, *part iv*, 344–355.

[72] Akaba, R.; Kamata, M.; Koike, A.; Mogi, K.-I.; Kuriyama, Y.; Sakuragi, H. *J. Phys. Org. Chem.* **1997**, *10*, 861–869.

[73] Muñoz, I.; Rieradevall, J.; Torrades, F.; Peral, J.; Domènech, X. *Sol. Energy* **2005**, *79*, 369–375.

[74] Fukuzumi, S.; Ohkubo, K. *Chem. Sci.* **2013**, *4*, 561–674.

[75] Baptista, M. S.; Cadet, J.; Di Mascio, P.; Ghogare, A. A.; Greer, A.; Hamblin, M. R.; Lorente, C.; Nunez, S. C.; Ribeiro, M. S.; Thomas, A. H.; Vignoni, M.; Yoshimura, T. M. *Photochem. Photobiol.* **2017**, *93*, 912–919.

[76] Miranda, M. A.; Garcia, H. *Chem. Rev.* **1994**, *94*, 1063–1089.

[77] Boncsi, S. M.; Carboncll, E.; Garcia, H.; Fagnoni, M.; Albini, A. *Appl. Catal. B Environ.* **2008**, *79*, 368–375.

[78] Bonesi, S. M.; Protti, S.; Albini, A. *J. Org. Chem.* **2016**, *81*, 11678–11685.

[79] Marin, M. L.; Miguel, A.; Santos-Juanes, L.; Arques, A.; Amat, A. M.; Miranda, M. A. *Photochem. Photobiol. Sci.* **2007**, *6*, 848–852.

[80] Arques, A.; Amat, A. M.; Santos-Juanes, L.; Vercher, R. F.; Marín, M. L.; Miranda, M. A. *Catal. Today* **2007**, *129*, 37–42.

[81] Gomis, J.; Arques, A.; Amat, A. M.; Marin, M. L.; Miranda, M. A. *Appl. Catal. B Environ.* **2012**, *123*, 208–213.

[82] Argüello, J. E.; Pérez-Ruiz, R.; Miranda, M. A. *Org. Lett.* **2007**, *9*, 3587–3590.

[83] Martiny, M.; Steckhan, E.; Esch, T. *Chem. Ber.* **1993**, *126*, 1671–1682.

[84] Oksdath-Mansilla, G.; Hajj, V.; Andrada, D. M.; Argüello, J. E.; Bonin, J.; Robert, M.; Peñéñory, A. B. *J. Org. Chem.* **2015**, *80*, 2733–2739.

[85] Manring, L. E.; Kramer, M. K.; Foote, C. S. *Tetrahedron Lett.* **1984**, *25*, 2523–2526.

[86] Prakash, G.; Falvey, D. E. *J. Am. Chem. Soc.* **1995**, *117*, 11375–11376.

[87] Pérez-Ruiz, R.; Jiménez, M. C.; Miranda, M. A. *Acc. Chem. Res.* **2014**, *47*, 1359–1368.

[88] Nakabayashi, K.; Kojima, J.; Tanabe, K.; Yasuda, M.; Shima, K. *Bull. Chem. Soc. Jpn.* **1989**, *62*, 96–101.

[89] Miranda, M. A.; Izquierdo, M. A.; Galindo, F. *Org. Lett.* **2001**, *3*, 1965–1967.

[90] Miranda, M. A.; Izquierdo, M. A.; Pérez-Ruiz, R. *J. Phys. Chem. A* **2003**, *107*, 2478–2482.

[91] Izquierdo, M. A.; Domingo, L. R.; Miranda, M. A. *J. Phys. Chem. A* **2005**, *109*, 2602–2607.

[92] Miranda, M. A.; Izquierdo, M. A. *J. Am. Chem. Soc.* **2002**, *124*, 6532–6533.

[93] Pérez-Ruiz, R.; Sáez, J. A.; Domingo, L. R.; Jiménez, M. C.; Miranda, M. A. *Org. Lett.* **2012**, *14*, 5700–5703.

[94] Miranda, M. A.; Izquierdo, M. A. *Chem. Commun.* **2003**, 364–365.

[95] Izquierdo, M. A.; Miranda, M. A. *Eur. J. Org. Chem.* **2004**, 1424–1431.

[96] Argüello, J. E.; Pérez-Ruiz, R.; Miranda, M. A. *Org. Lett.* **2010**, *12*, 1884–1887.

[97] Miranda, M. A.; Izquierdo, M. A.; Galindo, F. *J. Org. Chem.* **2002**, *67*, 4138–4142.

[98] McTiernan, C. D.; Pitre, S. P.; Sciano J. C. *ACS Catal.* **2014**, *4*, 4034–4039.

[99] Li, R.; Ma, B. C.; Huang, W.; Wang, L.; Wang, D.; Lu, H.; Landfester, K.; Zhang, K. A. I. *ACS Catal.* **2017**, *7*, 3097–3101.

[100] Tobin, J. M.; McCabe, T. J. D.; Prentice, A. W.; Holzer, S.; Lloyd, G. O.; Paterson, M. J.; Arrighi, V.; Cormack, P. A. G.; Vilela, F. *ACS Catal.* **2017**, *7*, 4602–4612.

Chapter 5

Oxygen Heterocycles: Pyrylium Salts

Sergio M. Bonesi

Departamento de Química Orgánica, CIHIDECAR-CONICET, Facultad de Ciencias Exactas y Naturales, Universidad de Buenos Aires, 3er Piso, Pabellón 2, Ciudad Universitaria, CP1428, Buenos Aires, Argentina
smbonesi@qo.fcen.uba.ar

Al Postigo

Departamento de Química Orgánica, Facultad de Farmacia y Bioquímica, Universidad de Buenos Aires, CONICET, Junín 945, CP1113, Buenos Aires, Argentina
apostigo@ffyb.uba.ar

5.1 Introduction

Single electron redox pathways play a pivotal role in a vast array of processes central to life on the planet. It is no surprise that organic chemists have been intrigued by the possibility that single-electron redox manifolds might be harnessed in the discovery and design of novel organic transformations. In particular, single-electron processes mediated by photoredox catalysis hold significant promise.

Recently, a surge of interest from the synthetic community has brought photoredox manifolds to the forefront of catalysis [1a–1e]. Organic photoredox catalysts have been utilized in this context for several decades, so it is somewhat surprising that only a few recent reports employ this class of catalysts, despite their relatively low cost, wide availability, and properties that can, in some cases, outperform their inorganic and organometallic counterparts [2a–2f]. This chapter will give an account of selected advances in the use of purely organic photooxidants such as pyrylium salts in synthetic transformations.

5.2 Photophysical, Spectroscopic, and Electrochemical Properties of Pyrylium Salts as Photoorganocatalysts

2,4,6-Triarylpyrylium salts are aromatic compounds that are used as excellent electron-transfer (ET) photoorganocatalysts usually employed in organic synthesis. These photocatalysts absorb in the UV–visible (UV–vis) region. When irradiated with visible light ($\lambda > 400$ nm), pyrylium salts are excited to the singlet state and become oxidant species which are able to oxidize a wide variety of organic compounds such as alkenes, alkynes, (hetero)aromatic compounds, thioethers, ethers, and phosphines, among others. This behavior is due to the moderate-to-high reduction potential of the singlet excited states (E^*_{RED}) of the pyrylium salts that depends on the electron-donor or electron-acceptor functional groups attached to the aryl moiety. Table 5.1 collects spectroscopic (λ^M_{abs}: maxima absorption wavelength and *Es:* singlet state energy) and electrochemical ($^1E^*_{RED}$: reduction potential in the singlet excited state) data of several pyrylium salts that are used as photocatalysts [3a–3c].

Other pyrylium salts such as 2-(4-methoxyphenyl)-4,6-diphenylpyrylium and 2,6-diphenyl-4-(4-cyanophenyl)pyrylium tetrafluoroborate have been employed as efficient photocatalysts and their spectroscopic data are shown in Table 5.1. As can be seen from the table, these photocatalysts have similar $^1E^*_{RED}$ values, as those observed for the case of 2,4,6-triarylpyrylium salts. Some pyrylium ions bearing one or two dialkylaminophenyl substituents show very low fluorescence quantum yields ($\Phi_f < 0.010$) that are advantageous for photoredox reactions when used as photocatalysts. In fact, their

Table 5.1. Spectroscopic and electrochemical data of several 2,4,6-triarylpyrylium tetrafluoroborate salts.

Pyrylium salts	X	λ^{M}_{abs}/nm	$^{1}E^{*}_{RED}$ V (vs SCE)[a]	E_s/eV[b]	Acronym
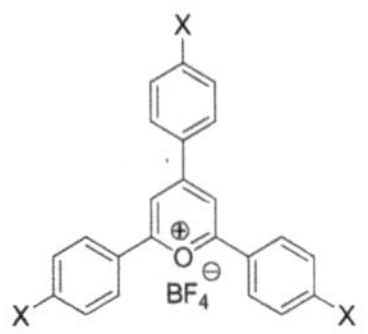	H	368	+2.28	2.66	**Py**
	Me	394	+2.03	2.55	**4-MePy**
	OMe	422	+1.74	2.34	**4-MeOPy**
	Br	397	+2.24	2.52	**4-BrPy**
	Cl	394	+1.84	2.55	**4-ClPy**
	F	372	+2.28	2.63	**4-FPy**
	Ph	443	+1.82	2.25	**4-PhPy**
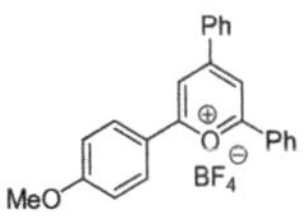		450	+2.03	2.39	**2-(4-MeOPh)Py**
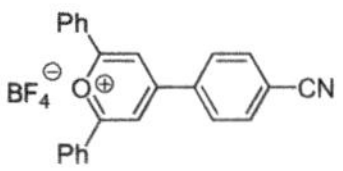		415	+2.12	2.30	**4-(4-CNPh)Py**
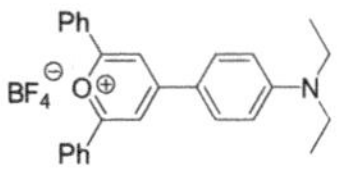		545	+2.01	2.37	**4-(4-Et$_2$NPh)Py**
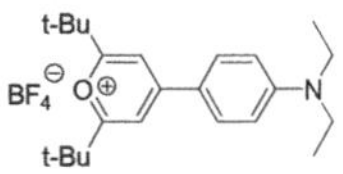		492	+1.86	2.40	**2,6-di*t*-BuPh-4-Et$_2$NPhPy**
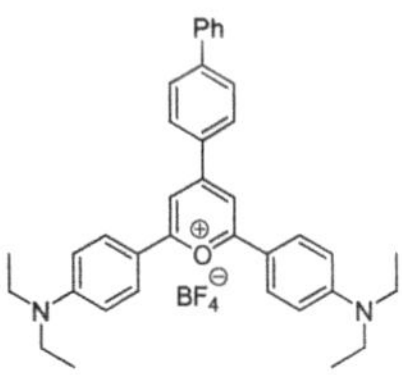		612	+1.81	2.27	**2,6-diEt$_2$NPh-4-PhPy**

Notes: [a]Calculated as $^{1}E^{*}_{RED} = E_{RED} + E_s$. [b]Estimated from the λ^{M}_{fluo}.

$^{1}E^{*}_{RED}$ range between 1.8 V and 2.0 V vs SCE (see Table 5.1) [4]. These photocatalysts can also react from their triplet excited states as in the case of 2,4,6-triphenylpyrylium tetrafluoroborate (**PyBF$_4$**) where the triplet energy (E_T) is 2.30 eV and the reduction potential from the triplet excited state ($^{3}E^{*}_{RED}$) is +1.92 V vs SCE.

Generally, 2,4,6-triarylpyrylium tetrafluoroborates exhibit two characteristic absorption bands located around 360 nm and 450 nm with high molar extinction coefficients. Due to the different charge distributions involved in the transition from the solvated ground state to the excited unsolvated state, increasing the solvent polarity, a hypsochromic shift of the absorption maxima is observed. This behavior can be ascribed to two independent chromophores: the 2,6-diarylpyrylium substructure (longer wavelength) and the 4-arylpyrylium moiety (shorter wavelength) [2a]. Pyrylium salts also have characteristic emission spectra. Both fluorescence and phosphorescence can generally be observed [5, 3a]. Notably, triarylpyrylium salts bearing methoxy groups in the aryl moiety show strong fluorescence emission with fluorescence quantum yields near unity [5a–5d].

These photocatalysts are neither able to photosensitize singlet oxygen through energy transfer nor can they produce superoxide ion as demonstrated by laser flash photolysis (LFP) studies [2a]. This behavior is useful in the case of photooxygenation reactions of a variety of electron-rich substrates such as alkenes, thioethers, and phosphines. In fact, the photooxygenation products come from pure ET processes instead of a singlet oxygen photosensitization.

On the other hand, the lack of superoxide formation during the photocatalysis of pyrylium salts is due to thermodynamic reasons, being $\Delta G = +0.49$ eV (if the Coulombic factor associated to charge separation is neglected). This was confirmed by LFP experiments which showed no influence of oxygen in the transient spectra of pyranyl radical formed after the reduction of the excited 2,4,6-triphenylpyrylium salt (singlet or triplet) by several electron-donors [2a].

Pyrylogens, a series of new dicationic catalysts that are hybrids of pyrylium salts and viologens, have been recently synthesized [3c]. The electrochemical and photophysical properties of these pyrylogens are collected in Table 5.2 in sufficient detail to allow the rationale design of new photoinduced ET reactions. The range of their reduction potentials (from +0.37 V to +0.05 V vs SCE) coupled with the singlet and triplet energies demonstrates that they are potent oxidizing agents in both their singlet and triplet excited states, capable of activating substrates with oxidation potentials as high as 3.1 eV. Such molecules show long wavelength absorptions,

Table 5.2. Spectroscopic and electrochemical data of some **Pr** photoorganocatalysts.

Pr salts	X	λ^{M}_{abs}/nm	$^{1}E^{*}_{RED}$ (S)/V[a]	E_s/ eV	$^{3}E^{*}_{RED}$ (T)/V[b]	E_T/eV	Acronym
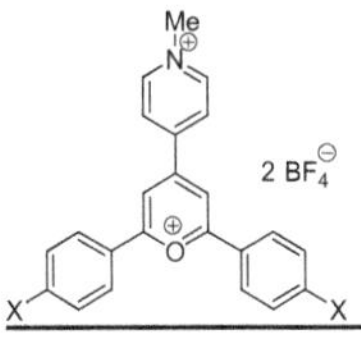	OMe	523	+2.03	2.08	+2.03	2.08	**4-MeOPr**
	H	441	+2.39	2.56	+2.30	2.47	**Pr**
	Cl	454	+2.24	2.43	+2.11	2.30	**4-ClPr**
	CF_3	413	+2.36	2.73	+2.10	2.47	**4-CF_3Pr**

Notes: [a]Reduction potential given vs SCE. Calculated as $^{1}E^{*}_{RED} = E_{RED} + Es$. [b]Reduction potential given vs SCE. Calculated as $^{3}E^{*}_{RED} = E_{RED} + E_T$.

between 413 nm and 523 nm, out of the range where most substrates can competitively absorb. However, these new photocatalysts do react with water, hence requiring the use of dry solvents.

The fluorescence of all the pyrylogens at room temperature in acetonitrile (MeCN) is characterized by a single peak devoid of any fine structure at wavelengths between 494 nm and 659 nm. The fluorescence quantum yields are much smaller (0.18–0.42) than that of 0.60 reported for 2,4,6-triphenylpyrylium salt. It is tempting to suggest that the lower fluorescence quantum yields are due to the absence of the γ-chromophore in pyrylogens. Additionally, these compounds show low triplet quantum yields, i.e. 0.18 for the case of pyrylogen (**Pr**) bearing no substituent (X = H in Table 5.2).

The radical cations formed during the reduction of **Pr** salts do not react with oxygen on the cyclic voltammetry time scale (50–500 mV – s^{-1}), in contrast to the rapid reaction of the radical formed upon the reduction of **$PyBF_4$**. This is most likely a result of the enhanced electrophilic character of **Pr** salts in comparison to pyrylium salt.

As it was pointed out above, these photocatalysts are capable of efficiently oxidizing a variety of donor substrates. Scheme 5.1 shows the reduction potentials of some excited photocatalysts and the oxidation potentials of some representative electron-donors that have been used in photocatalytic reactions.

An excited photocatalyst (Scheme 5.1) oxidizes a substrate when E_{OX} is less positive than $^{1}E^{*}_{RED}$. The variety of substituents attached to the aryl moieties of pyrylium and **Pr** derivatives

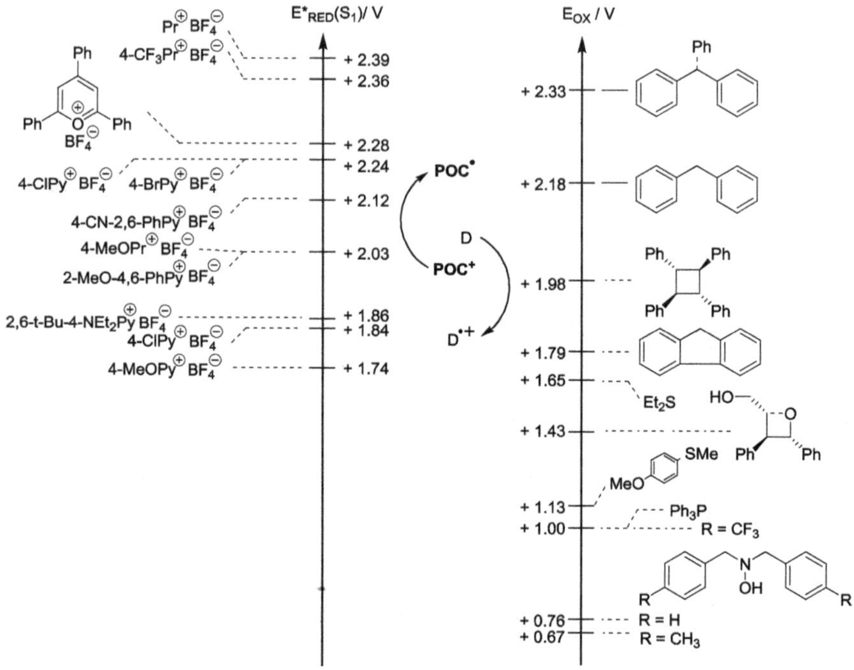

Scheme 5.1. Reduction potentials of some singlet excited photocatalysts ($^1E^*_{RED}$) and the oxidation potentials (E_{OX}) of some electron-donors (*D*). (**Py** = 2,4,6-triphenylpyrylium. **Pr** = pyrylogen).

modulates the span of reduction potentials in the excited state between +1.39 V and +2.39 V allowing the selection of the optimal photocatalyst to perform the photocatalytic reaction. The mode of action of photoexcited molecules, from a mechanistic viewpoint, can be described through a general reaction pathway (Scheme 5.2).

The reaction mechanism shown in Scheme 5.2(a) describes the typical scheme of Diels–Alder, [2+2]-cycloaddition, oxygenation, and endoperoxidation reactions. Once the substrates are oxidized to the corresponding radical cations ($D^{•+}$), another reagent R such as an alkene or molecular oxygen reacts efficiently with $D^{•+}$ to provide, in some cases, the products as radical cation $P^{•+}$. In the last step, these intermediates are reduced efficiently to the desired products by the reduced photocatalyst, which is in turn oxidized closing the photocatalytic cycle. In other cases, the intermediate $P^{•+}$

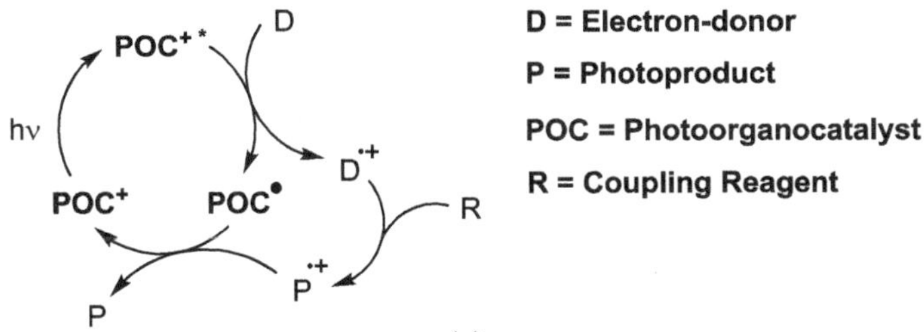

(a)

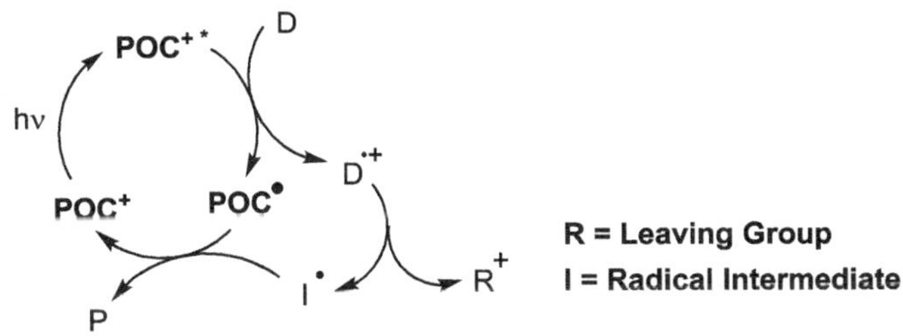

(b)

Scheme 5.2. General reaction mechanisms of the photoorganocatalytic reactions of **Py** or **Pr**.

undergoes a rearrangement, loss of a proton or a leaving group, etc. Scheme 5.2(b) describes a general reaction mechanism which is associated with chemical transformations that involve fragmentation of C–C or C–X bonds. Again, a photocatalytic ET process provides the radical cation $D^{\bullet+}$ that loses a leaving group, i.e. a proton, giving a radical intermediate such as $I^{\bullet}$. Reduction of the radical closes the photocatalytic cycle providing the desired product. Notably, photocatalysts are used in very low amount (*viz.* 5 mol% to 20 mol%), and only in a few cases they are consumed to a very low extent.

The synthetic applications of 2,4,6-triarylpyrylium and pyrylogens as photocatalysts are briefly surveyed in the following sections. These selected examples show that the photocatalytic reactions take place under mild and metal-free conditions to provide interesting products in good to excellent chemical yields, often with noticeable stereoselectivity and in reasonable reaction times. Some of these reactions would occur thermally only under harsh conditions, whereas, under photocatalytic conditions, they are carried out under mild conditions, in accordance with the paradigms of green chemistry.

5.3 Pyrylium Salts Photoorganocatalyzed Reactions

5.3.1 *Diels–Alder cycloadditions*

Steckhan and co-workers have prepared a series of 2,4,6-triarylpyrylium salts and have tested their photocatalytic ability in the dimerization reaction of 1,3-cyclohexadiene (**1**). With dichloromethane (DCM) as the solvent, all the tested photocatalysts have been able to catalyze the dimerization of **1** in almost quantitative yields and the endo/exo ratio of the dimer **2** varied from 6.1:1 to 8.3:1 (see Scheme 5.3) [**6**]. The efficiency of the cycloaddition reaction and the product distribution was strongly affected by the solvent.

Cross Diels–Alder reaction between 1,3-cyclohexadiene and indene in the presence of **Py** in DCM provides cycloadduct **3** in 27.5% yield, exclusively in the endo configuration [**6**]. The yield of **3** can be improved up to 40% by using **4-BrPy**. Similarly, when 1,2-dihydronaphthalene is used instead of indene, the cycloadduct **4** is obtained in poor yield (10% yield) with an endo/exo ratio of 4.5:1 (Scheme 5.4) [6].

The photocatalyst **4-MeOPy** can be employed in the preparation of cross Diels–Alder adducts (**6a-d**) when 1,1′-biscyclohexenyl

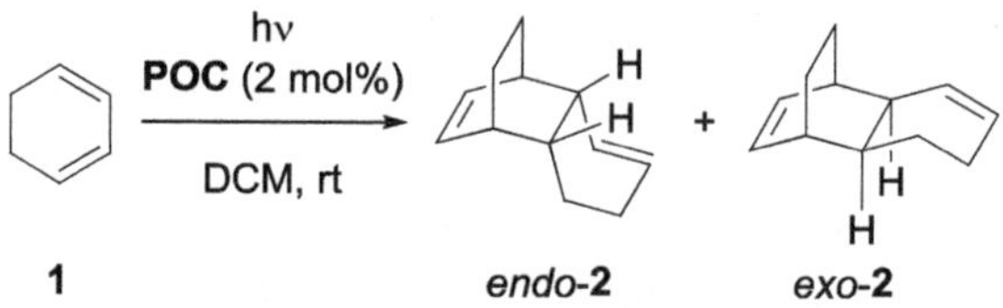

POC	*Yield of* **2** *(%)*	*endo/exo* ratio
Py	100.0	7.1:1
4-MeOPy	99.0	7.5:1
4-MePy	91.7	8.3:1
4-BrPy	100.0	6.1:1
4-ClPy	100.0	6.8:1
4-FPy	100.0	6.0:1
4-PhPy	100.0	6.8:1

Scheme 5.3. Photocatalytic cycloaddition reaction of 1,3-cyclohexadiene.

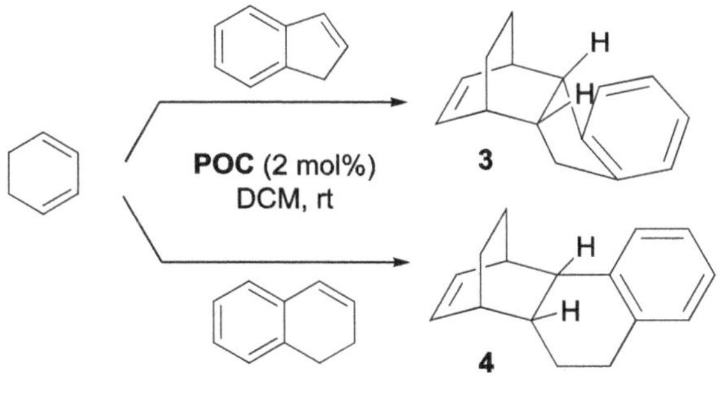

POC	Yield (%)
Py	27.5
4-BrPy	40

POC	Yield (%)
Py	10.2 (*endo/exo* 4.5/1)

Scheme 5.4. Photocatalytic cross cycloaddition reaction of 1,3-cyclohexadiene with indene and 1,2-dihydronaphthalene.

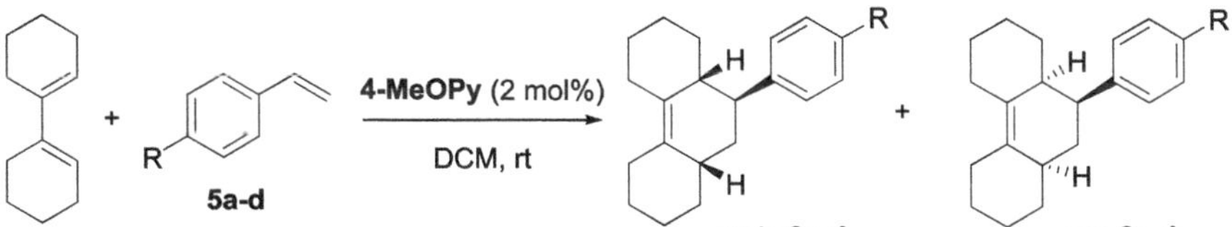

R		Yield (%)	*endo/exo* ratio
H	**6a**	33.6	1:9.8
Cl	**6b**	30.1	1:9.2
Me	**6c**	48.3	1:8.9
OMe	**6d**	52.8	1:5.4

Scheme 5.5. Cross cycloaddition reaction of 1,1′-biscyclohexenyl with *p*-substituted styrenes.

derivatives are reacted with electron-rich styrenes (**5a-d**) in DCM (Scheme 5.5). The cycloadducts **6a-d** are obtained in good yields showing an endo/exo ratio of ca. 1:9 with the exception of styrene **6d** [6].

Steckhan and co-workers [6] have used **4-MeOPy** in the cycloaddition reactions of heterocycles, such as 2-vinylbenzofuran, 2-isopropenylbenzofuran, with different olefins such as 1,3-cyclohexadienes [7], styrene derivatives, and 2,4-hexadiene derivatives. The cycloadducts formed during the irradiation are obtained in modest to good yields with significant diastereoselectivity, at least in some cases (Scheme 5.6).

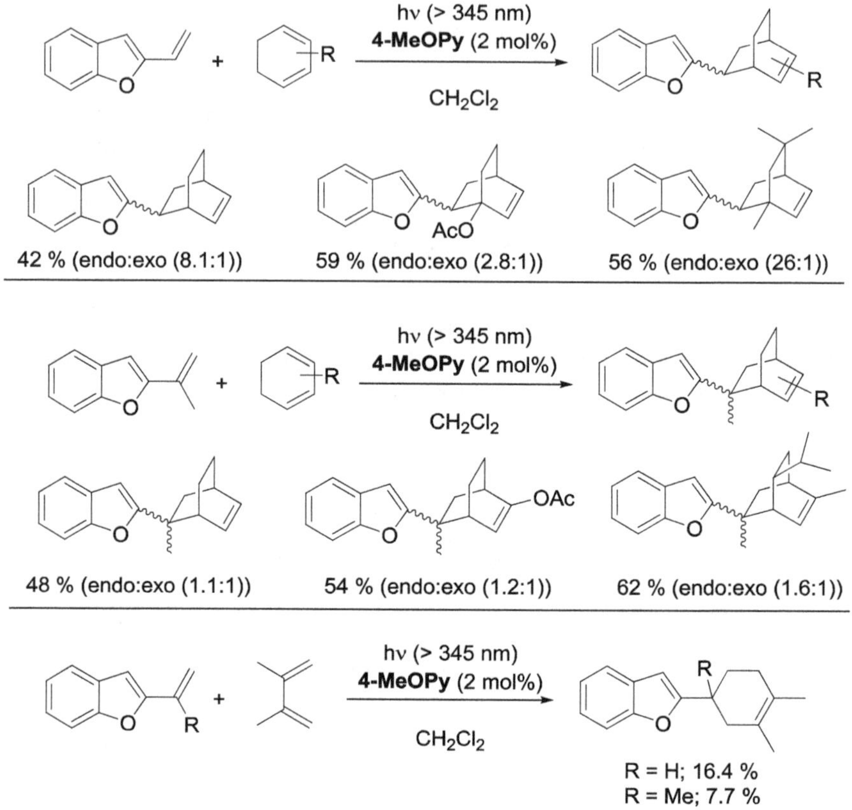

Scheme 5.6. Photocatalytic cycloaddition reaction of 2-vinyl and 2-isoprenylbenzofuranes.

The photoreaction is initiated by an ET process affording the 2-alkenylbenzofuran radical cations which can act as dienes or dienophiles and follow diverse pathways. In the presence of styrenes, a mixture of [2+2] and [4+2] adducts are formed, involving a cyclobutyl radical cation intermediate. When cyclic and acyclic 1,3-hexadienes are used, [4+2] cycloadducts are formed, involving a 1,3-sigmatropic C-shift of the vinylic double bond in the last step.

These observations are contrasting with the examples studied earlier by the same research group [8], where 2-vinyl indoles act as dienes in the pyrylium salts-photocatalyzed reactions in the

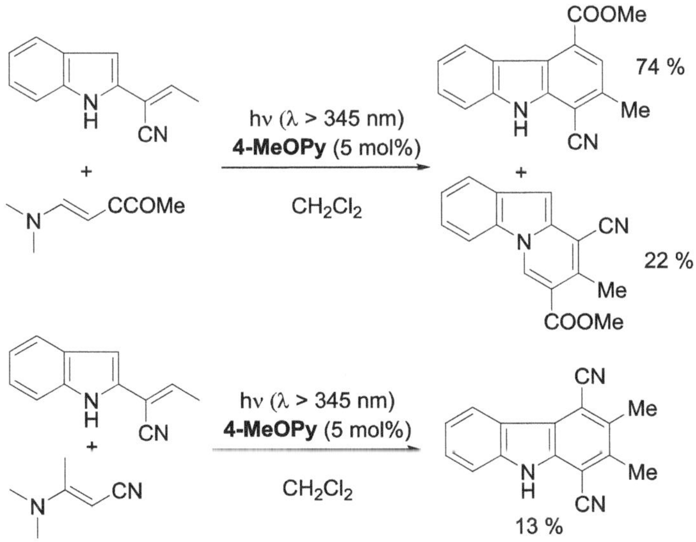

Scheme 5.7. Photocatalytic formation of carbazole and pyridoindole derivatives.

presence of 1,3-cyclohexadiene derivatives [9]. [4+2]-Cycloadditions between 2-vinylindoles acting as dienes and α-acceptor substituted cyclic and acyclic enamines (or β-enamidoesters [10]), are induced by the formation of a 2-vinylindole radical cation either *via* anodic oxidation or photocatalytic ET process using catalytic amounts of triarylpyrylium tetrafluoroborates. In this way, pyrido[1,2-*a*]indoles or indolo[1,2-*a*]hexahydro-1,8-naphthyridines are formed in one step with complete regio and stereochemical control [11]. Scheme 5.7 illustrates the two examples of the photocatalytic formation of carbazole and pyridoindole derivatives.

Steckhan and co-workers have also reported cycloaddition reactions of indoles with exocyclic dienes [12, 13] (and 1,3-cyclohexadienes [14], vide infra) photocatalyzed by **2-MeOPy**, which yields [*b*]-annelated tetrahydrocarbazoles that are not accessible by neutral cycloadditions. With dienes bearing a (hetero)aromatic substituent, the reactions proceed with complete regioselectivity.

Wiest and co-workers [15] have also accomplished the [4+2] cyclization of indole and 1,3-cyclohexadiene, employing **Py** as the

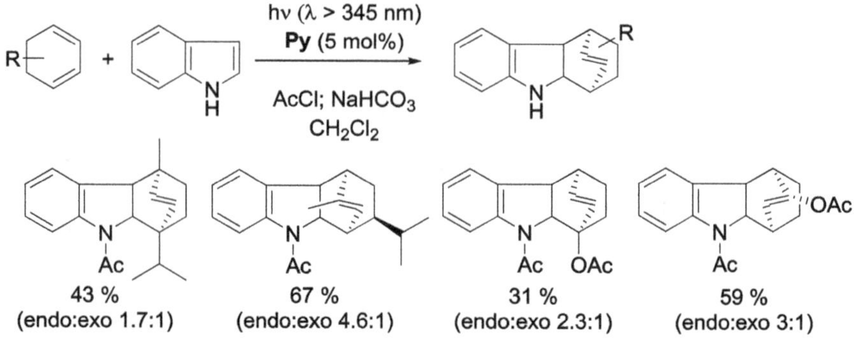

Scheme 5.8. Photocatalytic cycloaddition between indoles with 1,3-cyclohexadienes.

photocatalyst, obtaining a (3.3:1) stereoisomeric mixture of exo, endo, tetrahydro-*1H*-1,4-ethanocarbazole [16]. The same group had previously reported on the photochemically-induced radical-cation Diels–Alder of indoles and electron-rich dienes (Scheme 5.8) [17].

Cycloaddition reactions of poly(vinyl)cinnamates (PVCin) mediated by pyrylium salts in solution have been investigated by Miranda and co-workers [18]. **Py** produces highly selective [2+2] cyclodimerization between cinnamate units of PVC in head-to-head dimers (truxinates), accounting for 99% of the cycloaddition products. Among all the identified dimers, ξ-truxinate seems to be specifically associated with the presence of the pyrylium salt, since not even traces of this compound can be detected under photocatalyst-free conditions. ξ-Truxinate and β-truxinate are the major products, but their yields are inverted in comparison with the unsensitized reaction. 2,4,6-Triphenylthiapyrylium perchlorate has been shown to be three to four times more efficient than pyrylium salt, as evidenced by UV–vis and ^{1}H NMR experiments. The use of various cross-linking additives such as 1,4-butanediol dicinnamate, pentaerithritol tetracinnamate, dimethyl *p*-phenylenediacrylate, diethyl cinnamylidenemalonate, and pentaerithritol tetraacrylate, results in a marked enhancement of the intermolecular cycloaddition [18].

Miranda and co-workers [19] have investigated the [4+2] cycloaddition between thiobenzophenone and arylalkenes in the

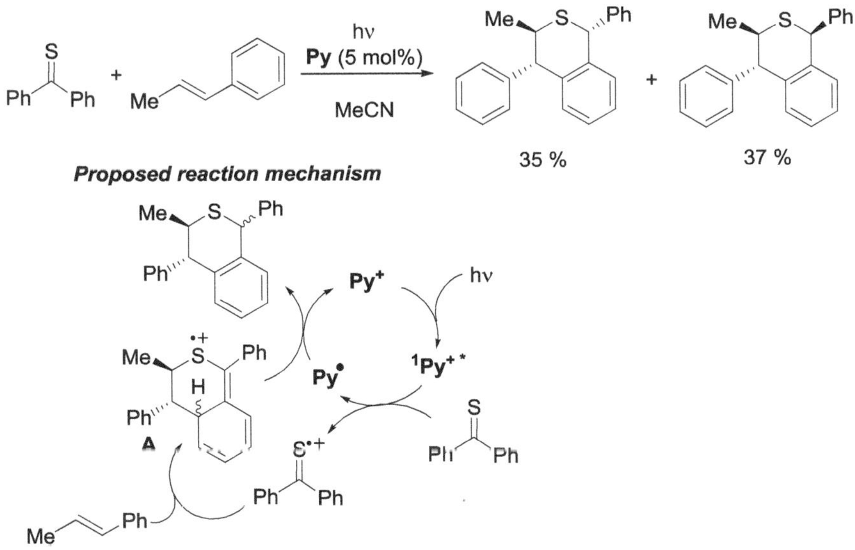

Scheme 5.9. Photocatalytic [4+2] cycloaddition between thiobenzophenone with *trans-β*-methylstyrene and the proposed reaction mechanism.

presence of (thia)pyrylium salts as photocatalysts. The photoreaction with **Py** as the photocatalyst affords a mixture of diastereomers in fairly good yield (see Scheme 5.9) with a consumption of thiobenzophenone up to 82%. LFP has allowed to propose a reaction mechanism for the photosensitized [4+2] cycloaddition of thiobenzophenone with *trans-β*-methylstyrene which is illustrated in Scheme 5.9.

The photoreaction starts with the formation of the radical cation of thiobenzophenone. Addition of this radical cation to the neutral alkene, in a concerted or stepwise fashion gives rise to the radical cation **A**. Back ET to the pyranyl radical with the recovery of the photocatalyst, followed by [1,3]-hydrogen migration, explains the formation of the obtained diastereomeric cycloadducts [19].

The photocatalytic imino-Diels–Alder reaction between *N*-benzylideneaniline and *trans*-anethole in the presence of **Py** has been demonstrated by a combination of product, LFP studies, and DFT theoretical calculations [20]. The photoreaction occurs

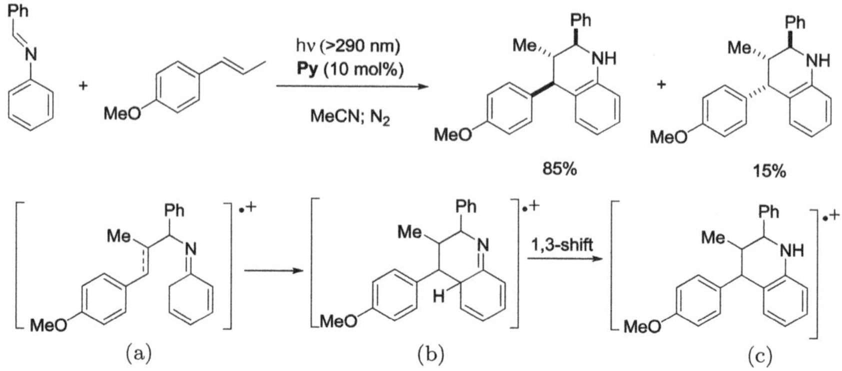

Scheme 5.10. Photoorganocatalyzed imino-Diels–Alder cycloaddition between *N*-benzylideneaniline and *trans*-anethole.

efficiently and a diastereomeric mixture is obtained (Scheme 5.10). However, irradiation of *N*-benzylideneaniline and *trans*-stilbene under the same catalytic conditions affords no photoproducts. From a mechanistic viewpoint, the reaction proceeds through the formation of an ion–molecule complex, which evolves to the intermediate **A**. In the subsequent step, ring closure occurs, with the formation of a new heterocyclic intermediate **B**. Finally, H shift affords the cycloadduct radical cation **C** (see Scheme 5.10).

Arylimines have also been employed in photochemically-promoted Diels–Alder reactions with *N*-vinylpyrrolidinone and *N*-vinylcarbazole, by using **Py** as the photocatalyst to produce the corresponding 2-oxopyrrolidin-1-yl and carbazol-9-yl-tetrahydroquinoline in high yields [21a, 21b]. The scope of the transformation is illustrated in Scheme 5.11.

In 2017, Meng and co-workers [22] have reported a highly regioselective [2+2+2] cyclization including aromatic alkynes and nitriles for the preparation of 2,3,6-trisubstituted pyridines under irradiation with blue LEDs, using **4-ClPy** as the optimal photoredox catalyst. This cycloaddition is achieved through a photooxidative single-ET process occurring at room temperature and under metal-free conditions. A variety of aromatic alkynes and nitriles can be employed to furnish the annulation products in good yields (see Scheme 5.12).

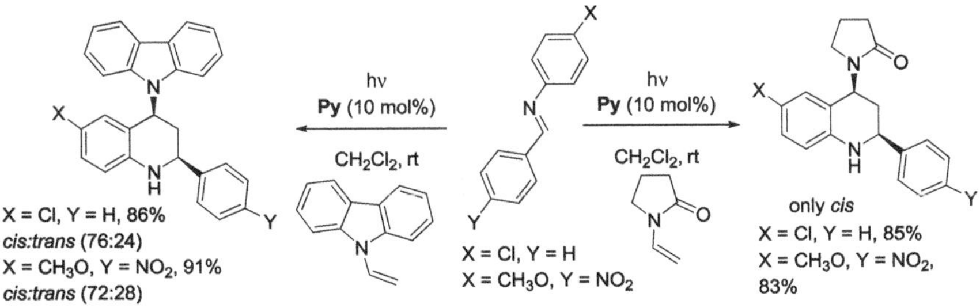

Scheme 5.11. Photocatalytic imino-Diels Alder cycloaddition of *N*-benzylideneaniline with *N*-vinylpyrrolidinone and *N*-vinylcarbazole.

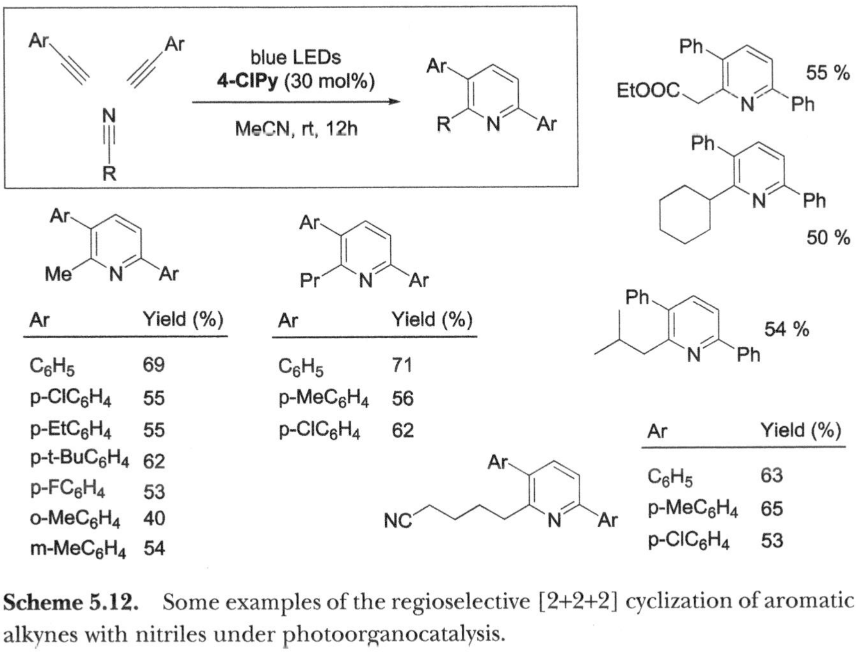

Scheme 5.12. Some examples of the regioselective [2+2+2] cyclization of aromatic alkynes with nitriles under photoorganocatalysis.

The authors have screened different substituted pyrylium salts as **POCs** such as **4-MeOPy**, **4-MePy**, **4-FPy**, and **4-ClPy** salts. It has been found that the electron-acceptor groups favor the photocatalytic cyclization, whereas the electron-donor substituents hinder the reaction at all or give only traces of the photoproducts. The proposed reaction mechanism involves the formation of three consecutive

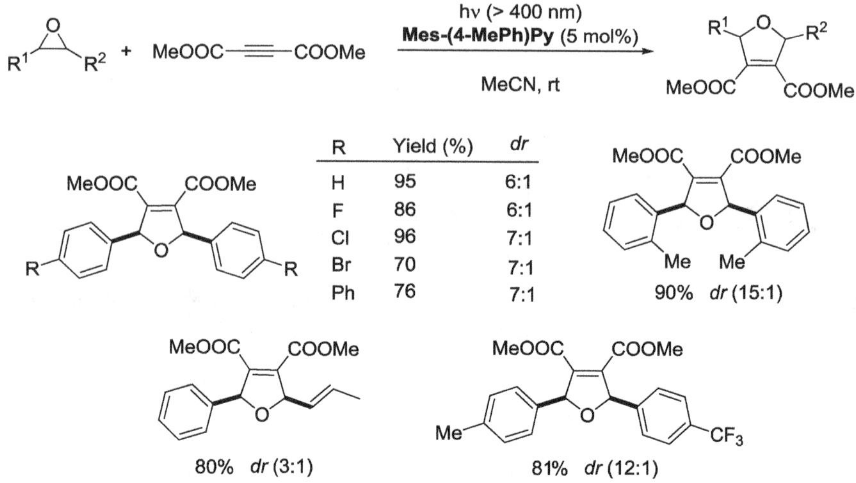

Scheme 5.13. Scope of the [3+2] dipolar cycloaddition.

radical cations that finally regenerate the photocatalyst through an ET process affording the desired photoproducts [22].

Beeler and co-workers have recently reported the [3+2] dipolar cycloaddition between a carbonyl ylide (formed from a diaryl epoxide) in the presence of dimethyl acetylene dicarboxylate by using 4-mesityl-2,6-di-*p*-tolylpyrylium tetrafluoroborate (**Mes-(4-MePh)Py**) [23]. Exploring the scope of this photocatalytic reaction, the authors found that a variety of electron-donor and electron-acceptor groups are well tolerated, providing the cycloadducts in good yields and, in some cases, with high diastereomeric ratios (see Scheme 5.13).

5.3.2 [2+2]-*Cycloaddition reactions*

Nicewicz and co-workers [24] have employed **4-MeOPy** as the photocatalyst for the stereoselective synthesis of C2-symmetric cyclobutane alkene dimers. The strategy has been applied successfully to the total syntheses of the lignans *magnosalin* [25a–25c], *endiandrin A* [26(a)], and the lignan-like cyclobutane *pellucidin A* [26(b)], whose structures have been revised as a result of this study. The improvement of the methodology is due to the presence of

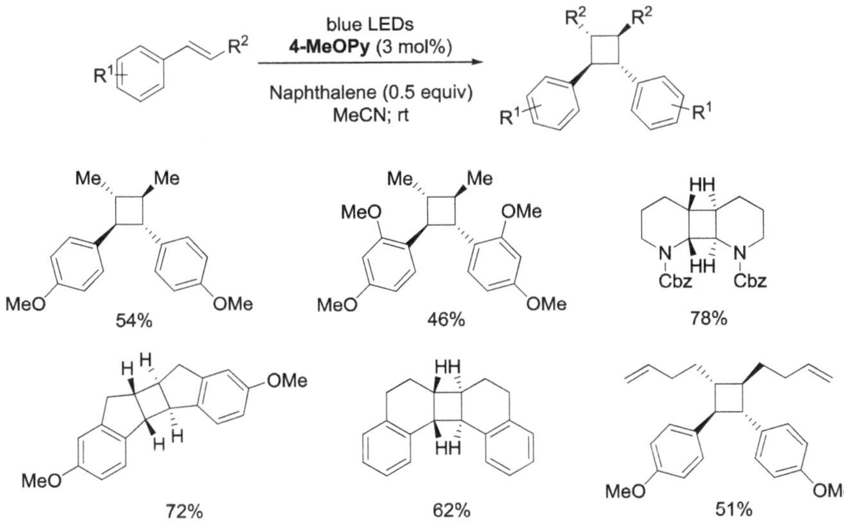

Scheme 5.14. Photocatalytic [2+2]-cycloaddition of styrenes in the presence of an electron relay (ER).

electron relay (ER) additives, such as naphthalene, anthracene, or triphenylamine, that prevent cycloreversion and polymerization of β-substituted styrenes and terminal styrenes, respectively. The photocatalytic reaction proceeds efficiently when naphthalene is used as the optimal ER and the [2+2]-cycloadducts are formed in good to excellent yields (see Scheme 5.14).

The photocatalytic method has also been applied to the head-to-head dimerization of terminal styrenes. The photoreaction involves the use of anthracene and **4-MeOPy** in the role of ER and photocatalyst, respectively (Scheme 5.15(a)).

The proposed mechanism for this photocatalytic transformation is also depicted in Scheme 5.15(b). Excitation of the photocatalyst with blue LEDs followed by oxidation of the ER provides the active oxidant ($ER^{\bullet+}$), which in turn oxidizes the alkene. After cycloaddition with another molecule of alkene, the reduced photocatalyst most likely reduces cyclobutane cation radical **I** (see Scheme 5.15(b)). In the absence of the ER, the cycloaddition process is reversible.

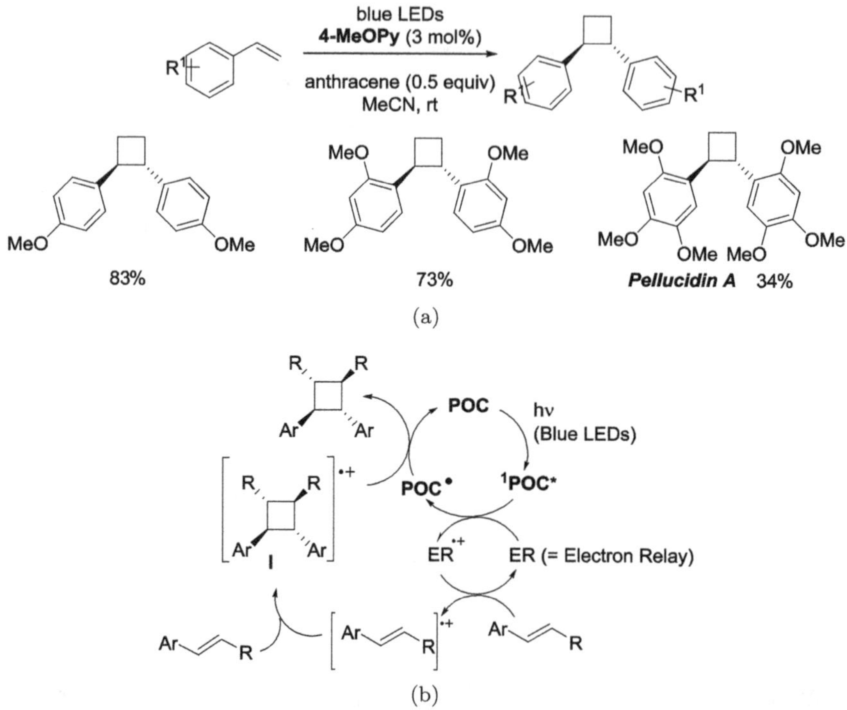

Scheme 5.15. (a) Photoorganocatalytic [2+2]-cycloaddition of terminal styrenes. (b) Proposed reaction mechanism for the ER-photocatalyzed cycloaddition of styrene derivatives.

Mattay and co-workers [27] have used pyrylium salts in the ET-catalyzed dimerization and mixed cyclization of 1,3-cyclohexadienes [28] and phenyl vinyl ether derivatives. In addition, **Py** has been shown to operate as an efficient photocatalyst for the dimerization and photooxygenation of 1,1-diphenylethene.

5.3.3 *Cycloreversion reactions: C–C and C–heteroatom bond cleavage*

Cycloreversion of four-membered rings is one of the most important ET-catalyzed pericyclic reactions. It has attracted much attention in recent years, mainly focused on theoretical calculations about the

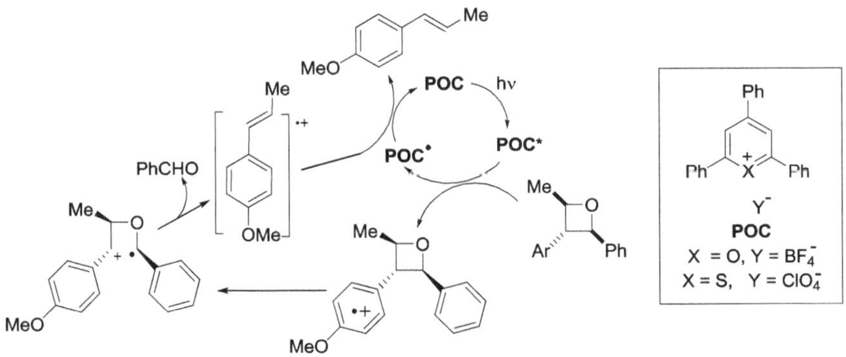

Scheme 5.16. Photoorganocatalyzed ring splitting of *trans,trans*-3-(4-methoxyphenyl)-2-phenyl-4-methyloxetane.

reaction pathways, stereoelectronic effects, and biological implications in DNA repair [29]. Miranda and co-workers have reviewed several photoinduced heterocycloreversion reactions such as the ring opening of oxetanes, azetidines, and thietanes in the presence of different photoorganocatalysts including 2,4,6-triaryl(thia)pyrilium salts [30]. The general reaction mechanism involved in the photocatalytic ET ring splitting of the heterocycles is illustrated for the case of *trans,trans*-3-(4-methoxyphenyl)-2-phenyl-4-methyloxetane in Scheme 5.16.

In principle, both the singlet and the triplet excited states of the photocatalysts can be involved, and indeed they are effectively quenched by the oxetane. Characterization of the alkene radical cation has been achieved by LFP. Initial ionization occurs at the methoxy-substituted aromatic ring, and subsequent C–C cleavage generates a distonic 1,4-radical cation and, in turn, a C–O bond splitting affords the alkene radical cation. ET to the formed radical cation by the reduced form of the photocatalyst provides the desired alkene and closure of the photocatalytic cycle. The molecular mechanism for oxetane cycloreversion has also been studied at the UB3LYP/6-31G* level [31]. Calculations are in agreement with an asynchronous process [32], which allows a favorable rearrangement of the spin electron density from the oxygen atom of the oxetane radical cation [33] to the π system of the alkene radical cation.

trans,trans-2,3-Diphenyl-4-methyloxetane has been subjected to react with catalytic amounts of 2,4,6-triarylpyrylium tetrafluoroborate (2 mol%) in MeCN at room temperature [34]. The photoproducts are obtained in low yields, such as 8% by using **Py** salt, while the yield is 33% with the **4-BrPy** salt. Best results are obtained with 2,4,6-triphenylthiapyrylium perchlorate. Irradiation of 1,2,3,4-tetraphenylcyclobutanes with $\lambda > 340$ nm in the presence of catalytic amounts of **Py** or **4-BrPy** salts in MeCN affords *trans*-stilbene as the only photoproduct in a good yield and with a noticeable cycloreversion quantum yield (up to 0.40) [35].

Irradiation of 1-hydroxymethyl-substituted oxetane **7** in the presence of catalytic amounts of **Py** or **4-BrPy** salts as photocatalysts produces stilbene **8**, 2,5-dihydroxy-1,4-dioxane (the dimer of hydroxyacetaldehyde), and tetrahydro-4,5-diphenylfuran-3-ol **9** [36]. With both photocatalysts, stilbene is formed as the main photoproduct (see Scheme 5.17(a)) and, again, quantitative conversion of oxetane is achieved with 2,4,6-triphenylthiapyrylium perchlorate as the photoorganocatalyst. Stepwise cycloreversion of oxetane radical cation with an initial C–O bond cleavage has been proposed as the reaction mechanism (see Scheme 5.17(b)). This gives an open

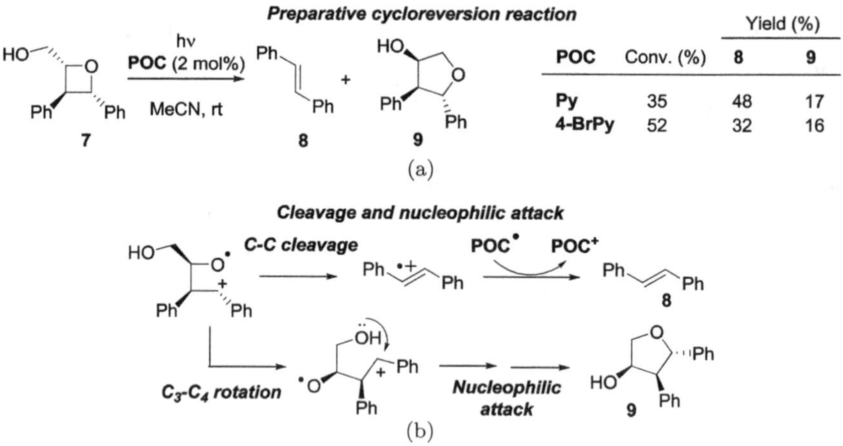

POC	Conv. (%)	Yield (%) 8	Yield (%) 9
Py	35	48	17
4-BrPy	52	32	16

Scheme 5.17. Photocatalytic cycloreversion of a 1-hydroxymethyl substituted oxetane.

1,4-distonic radical cation where the spin and charge are located on the oxygen and carbon (C2) atoms, respectively. Then, two competitive pathways operate and a second C–C fragmentation affords stilbene (**A**), while the C_3–C_4 rotation followed by the nucleophilic attack affords cyclic compound **9**.

Product distribution analysis and transient spectroscopic evidence for C_2–C_3 cleavage in 2,3-diaryloxetane radical cations have been reported by Miranda and co-workers [36]. Methoxy substitution at the 3-aryl group induces the C_2–C_3 fragmentation of 2,3-diaryloxetane radical cation showing a noticeable regioselectivity that can be controlled by modifying the nature of the substituent [37]. DFT calculations correlates well with the experimental results [38–40].

Mukai and co-workers have reported the cycloreversion of a cage molecule and related cyclobutanes using **Py** as photocatalyst with visible light in MeCN, affording the expected photoproducts in quantitative yield in very short reaction times (up to 20 min) (Scheme 5.18) [41].

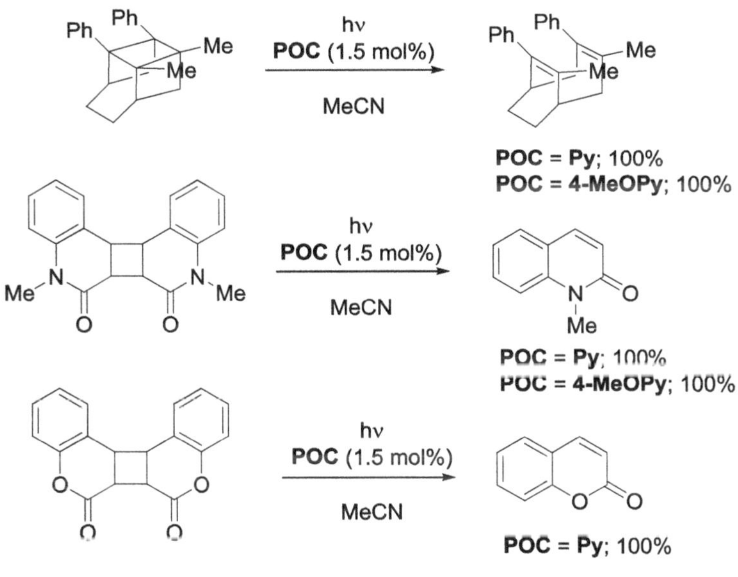

Scheme 5.18. Photoorganocatalytic cycloreversion of a cage compound and related cyclobutanes.

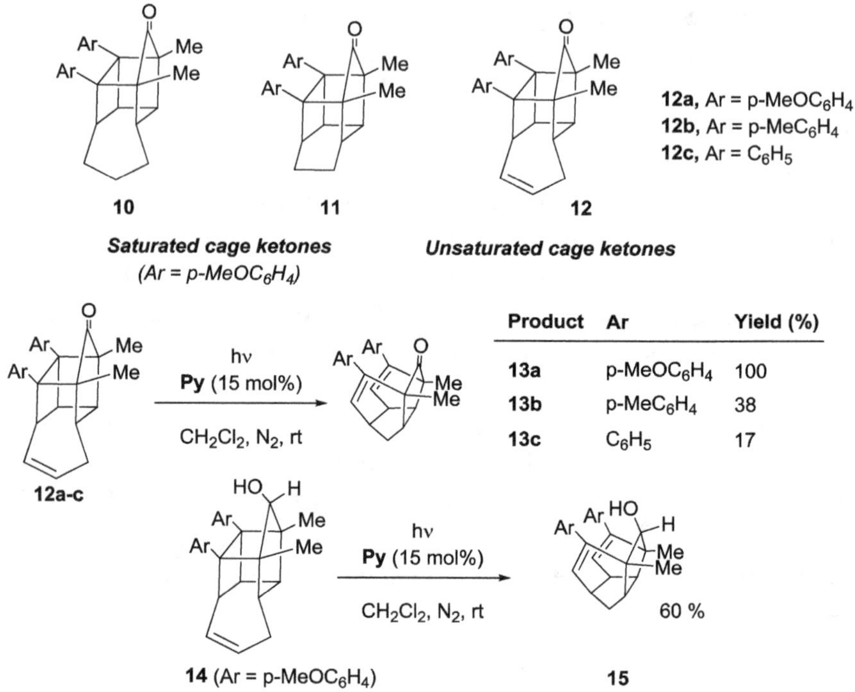

Scheme 5.19. Photoinduced cycloreversion of saturated and unsaturated cage ketones.

Cage ketones having a 1,2-diaryl-substituted bicyclo[2.2.0]-hexane moiety have been investigated by using **Py** (15 mol%) as the photocatalyst in DCM (see Scheme 5.19) [42]. Although saturated cage ketones **10** and **11** are recovered unchanged under the reaction conditions, cage ketones **12a-c**, which contain a double bond in the bridge, undergo novel pericyclic reactions to give photoproducts **13a-c** in moderate to quantitative yields (see Scheme 5.19). The photoreaction is sensitive to the substituents' effect, which can be rationalized by the different oxidation potentials of the ketones; the lower the oxidation potential, the easier the reaction. Furthermore, 1,2,4,5-tetramethoxybenzene works as a quencher of the reaction. Thus, irradiation of a DCM solution of **12a** in the presence of the photocatalyst and the quencher gives **13a** in 24% yield, while irradiation of solutions of compounds **12b** and **12c** does not give any product. On the other hand, addition of co-catalysts such as pyrene, phenanthrene, and biphenyl, to the photoreaction of **12b** and **12c** in the presence of the photocatalyst shows that phenanthrene increases the production of **13b** and **13c**, respectively, giving these products in 59% and 46% yields (compare these data with those shown in Scheme 5.19). Such co-sensitization is considered to take place *via* ET from cage ketones **12b, 12c** to the radical cation of the co-catalyst, which more easily escapes from the solvent cage of the geminate radical ion pair.

The reactions discussed above are the skeletal rearrangements of the 1,2-diaryl-3-vinylbicyclohexane moiety. The ketonic bridge of **12** does not participate directly in the reactions.

In order to investigate the effect of the ketonic bridge, the photocatalytic reaction of alcohol **14** has been investigated. Alcohol **14** has a lower oxidation potential than ketone **12a**. This fact indicates that the ketonic bridge shows some electronic effects on the diaryl-substituted cyclobutane bond. The photocatalytic reaction of **14** gives **15** in 60% yield (see Scheme 5.19). The fact that alcohol **14** undergoes the same rearrangement as ketones **12** indicates that only the 1,2-vinylbicyclohexane moiety is essential for the occurrence of the novel pericyclic reaction.

5.3.4 *C–C and C–X bond formation*

Irradiation of **Py** or **4-MeOPy** in the presence of 4-methoxybenzyl alcohol in MeCN or DCM leads to the formation of bis(4-methoxyphenyl)methane as the main photoproduct (up to 78% yield), where a new Csp^3–Csp^2 bond is formed [8] (Scheme 5.20). Photocatalyzed formation of the *p*-anisyl cation promotes the ipso attack on the 4-methoxybenzyl alcohol with the liberation of formaldehyde. No oxidation products of the easily oxidizable 4-methoxybenzyl alcohol can be observed even in the presence of air or oxygen. However, benzyl and *p*-chlorobenzyl alcohols are oxidized to the corresponding benzaldehydes.

A mild and photocatalytic method for the cyclization of polyenes has been reported with **Py** as the photocatalyst [43]. The

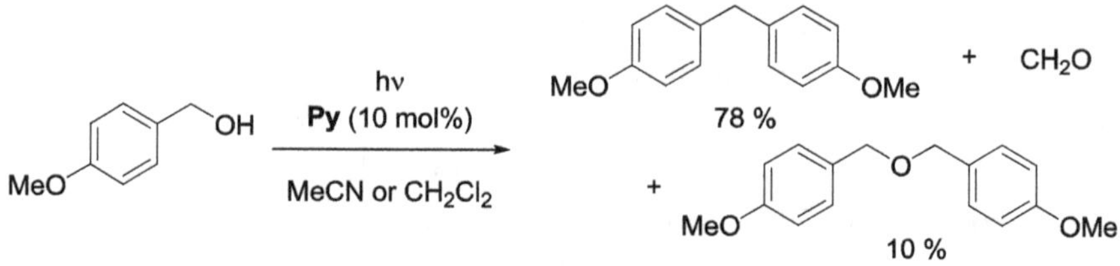

Scheme 5.20. Photocatalytic reaction of *p*-methoxybenzyl alcohol.

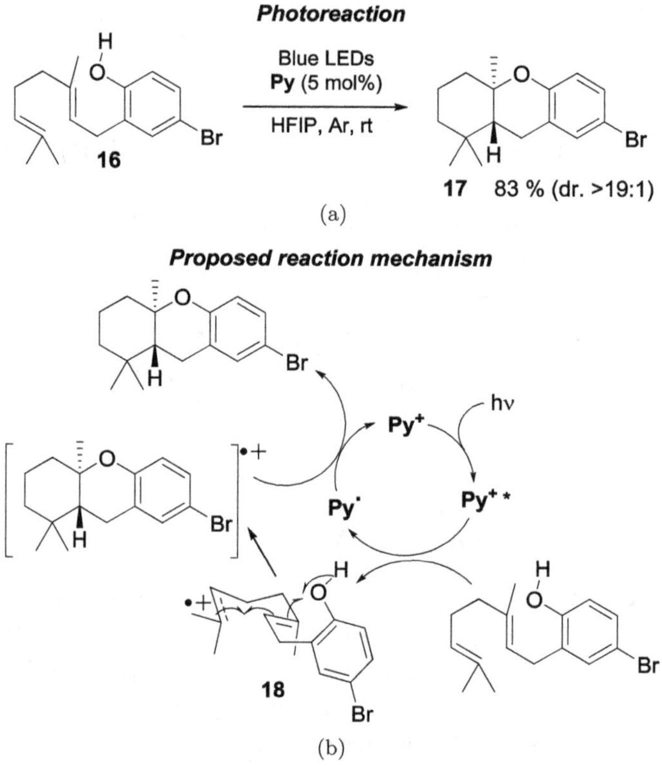

Scheme 5.21. Photoorganocatalytic cyclization of 1,5-dienes.

photoinduced cyclization of 1,5-diene **16** has been carried out in hexafluoro-2-propanol (HFIP) at 450 nm under N_2 atmosphere. Product **17** is obtained in 83% yield with noticeable diastereoselectivity (>19:1 d.r., Scheme 5.21(a)). Similar results are obtained when the photocatalyst is 9-mesityl-10-methylacridinium perchlorate or N-methylquinolinium hexafluorophosphate [44a, 44b]. The proposed reaction mechanism involves the formation of an alkene radical cation that undergoes radical cascade cyclization alongside simultaneous hydrogen shift to give the radical cation intermediate **18**. Finally, intermediate **18** is reduced to the desired product (see Scheme 5.21(b)).

This methodology is an interesting synthetic approach because of the concomitant formation of two new bonds, *viz.* a C–C and a C–O bond, in one-pot three-step reaction and with significant diastereoselectivity. Also, the H-shift is favored by the reaction solvent, *viz.* hexafluoro-2-propanol.

Photoinduced ET to **Py** from terpenoid polyalkenes bearing electron-withdrawing substituents, e.g. 2,6-dimethylhepta-1,5-diene-1,1-dicarbonitrile (**19**) as well as from other donors, such as biphenyl and *trans*-stilbene, has been studied in solution under steady-state irradiation and complemented by LFP spectroscopy [45]. Irradiation of compound **19** at 366 nm in argon-saturated MeOH in the presence of **Py** affords 35% of the corresponding ether as the major photoproduct (see Scheme 5.22). When the irradiation is carried out in argon-saturated water, the corresponding alcohol is formed as the main photoproduct in 30% yield (see Scheme 5.22). The photoreaction involves an ET process leading to

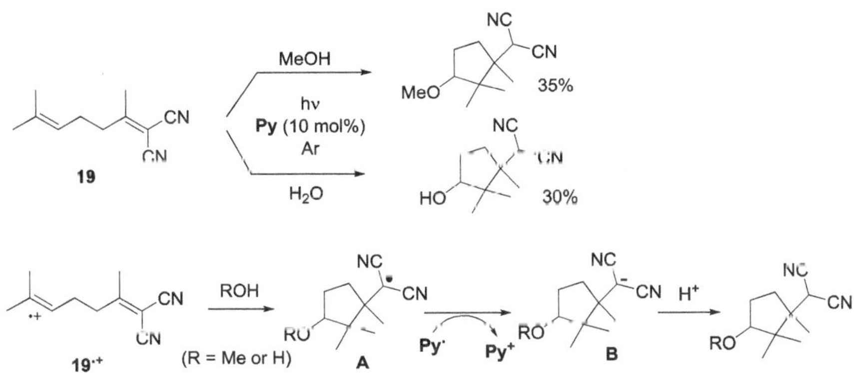

Scheme 5.22. Photocatalyzed cyclization of a polyene.

the formation of the radical cation of polyene **19**. Cyclization of the radical cation followed by the addition of alcohol (or water) or addition of a nucleophile followed by cyclization leads to the formation of radical **A** that in turn is reduced by the **POC** to the carbanion **B**. Proton capture affords the cyclized photoproduct. This photocatalytic reaction allows for the formation of two new bonds, *viz.* a C–C and a C–O bond, in a one-pot procedure.

The photocatalytic oxidation of 7-(p-methoxyphenyl)cycloheptatriene in the presence of **Py** (50 mol%) in a deoxygenated MeCN solution at room temperature gives bitropyl in high yields. The primarily formed radical cation deprotonates and the resulting cycloheptatrienyl radical undergoes a self-reaction, thus forming a new C–C bond [46]. By contrast, other aryl-substituted cycloheptatrienes, such as 7-, 1- and 3-(*p*-N,*N*-dimethylaminophenyl)cycloheptatrienes do not form bitropyl products.

Phenylacetylene reacts in the presence of **Py** and **4-CNPy** in MeCN upon irradiation at 405 nm to give pyridine and dimer adducts in 14:1 and 16:1 ratios for the two **POCs**, respectively [47]. When the same reaction is carried out on a mixture of phenylacetylene and 1-hexyne, 1-phenylnaphthalene, 1-butylnaphthalene, and two isomeric pyridines (1.5:1 ratio) are obtained.

An unprecedented method that makes use of the cooperative interplay between molecular iodine and photoredox catalysis has been developed for dual light-activated intramolecular benzylic C–H amination. Iodine serves as the catalyst for the formation of a new C–N bond by activating a remote Csp^3–H bond (1,5-hydrogen atom transfer process) under visible-light irradiation, while the organic photoredox catalyst **Py** effects the reoxidation of the molecular iodine catalyst [48]. To explain the compatibility of the two involved photochemical steps, the key N–I bond activation was proposed. The new cooperative catalysis has important implications for the combination of non-metallic main group catalysis with photocatalysis. This method constitutes an effective procedure for the synthesis of various 2-arylpyrrolidines in good to excellent yields (see Scheme 5.23 for some selected examples).

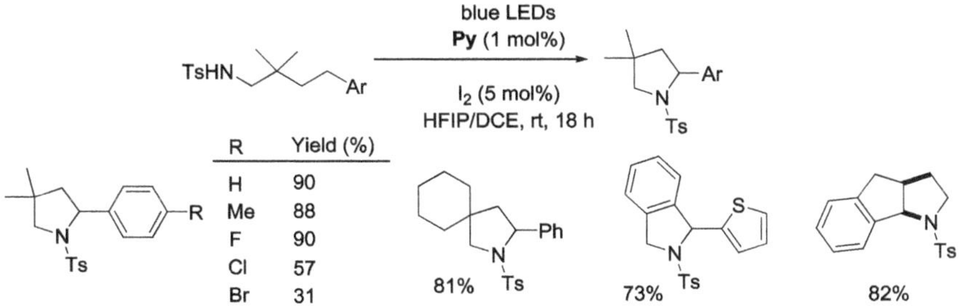

Scheme 5.23. Photoorganocatalyzed Csp^3–N bond formation.

The syntheses of substituted oxazepanes, benzomorpholines, thiomorpholines, and thiazepanes, involving a photocatalytic coupling of aldehydes and silicon amine protocol (SLAP) in the presence of **Py** (5–10 mol%) as the photocatalyst and trimethylsilyl triflate (1.30 equiv) as the Lewis acid [49], have been recently reported. The photocatalytic reaction proceeds efficiently in MeCN/HFIP (10:1) under blue light, providing the heterocyclic compounds in good to excellent yields (up to 94%) and, in some cases, with noticeable diastreroselectivity.

The aerobic dehydrogenative lactonization of alkenoic acids facilitated by a cooperative non-metallic catalysts' pair has been reported [50]. The procedure relies on the adjusted interplay of a photoredox catalyst such as **4-MeOPy** (5 mol%), and a selenium-π-acid catalyst such as $(PhSe)_2$ (5 mol%) in MeCN at room temperature, which allows for the regiocontrolled construction of five- and six-membered lactone rings in yields up to 96%. Notable features of this method are pronounced efficiency and practicality, good functional group tolerance, and high sustainability, since ambient air and visible light ($\lambda = 465$ nm) are adequate for the clean conversion of alkenoic acids into the corresponding lactones.

5.3.5 *Oxidation and oxygenation reaction*

5.3.5.1 *Miscellaneous oxidation reaction*

The photodegradation of coumaric acid has been studied in water in the presence of **Py** hydrogen sulfate (5 mol%) under oxygen atmosphere [51]. After 8 h of irradiation, *p*-hydroxy benzaldehyde and oxalic acid are formed in low yields. Also, *cis/trans* photoisomerization of coumaric acid takes place.

The photocatalyst **Py** encapsulated inside the supercages of Zeolite Y has been used as an efficient **POC**, to perform the photodegradation of chlorophenoxy acetic acid and methyl parathion in aqueous solution using visible or solar light [52a–52c]. Photodegradation of 2-isopropoxyphenyl *N*-ethylcarbamate (commonly known as Propoxur and commercialized with the registered name of Baygon®) in the presence of the above-mentioned photocatalyst has been reported by Garcia and co-workers [53]. Two main photoproducts, *viz.* o-isopropoxyphenol and o-dihydroxybenzene, are formed in 48% and 47% yields, respectively, after irradiation of an aqueous solution of Propoxur for 5 h.

Py photocatalyzes the oxygenation of cycloalkenes to allylic hydroperoxides in oxygen-saturated MeCN [54]. The hydroperoxides are obtained in 70%–80% yield. Treatment of the reaction mixture with PPh_3 affords the corresponding allyl alcohols in 65%–70% yield. The photoreaction involves an ET process giving the cycloalkenyl radical cation that reacts with molecular oxygen, finally providing the allylic hydroperoxides (Scheme 5.24).

Recently, Breder and co-workers have reported a metal-free catalytic protocol for the oxidative coupling of non-activated alkenes with carboxylic acids [55]. This method is based on the cooperative interaction of a diselenide and a **POC**, *viz.* **4-MeOPy**. The terminal oxidant involved in the photoreaction is air or O_2. The allylic esterification coupling provides the desired products with good yields (up to 89%) and excellent regioselectivity, as well as good functional group tolerance (Scheme 5.25).

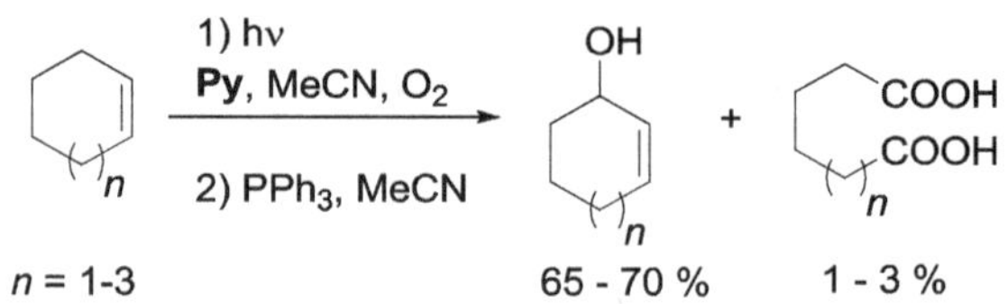

Scheme 5.24. Photooxygenation of cycloalkenes.

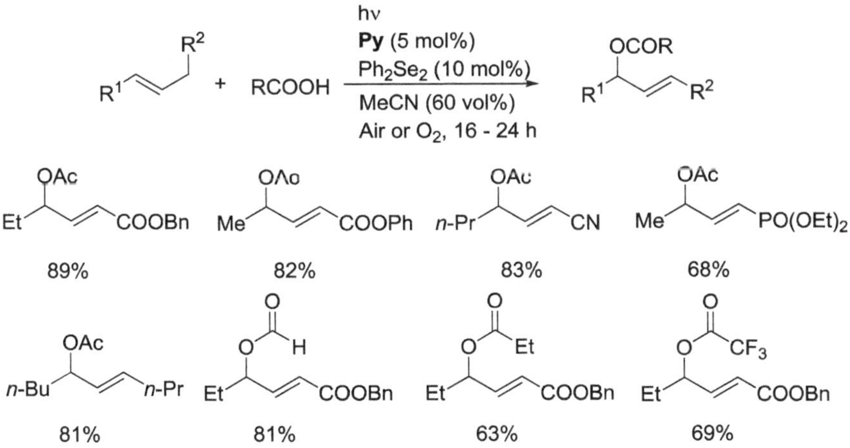

Scheme 5.25. Photooxidative allylic esterification.

Gopidas and co-workers have studied the photocatalytic ET oxygenation of furan derivatives by using **Py** perchlorate in the presence of oxygen in DCM at room temperature [56]. The photoreaction takes place by irradiation of the solution at 350 nm for 10 min. The reaction gives a mixture of *trans-* and *cis-*1,4-enediones, the latter being the main product (up to *ca.* 85%). Because a thermal *cis/trans* isomerization occurs, the main photoproduct is, in any case, the *trans*-isomer. Dibenzofuran does not react under photooxygenation conditions. LFP demonstrates that the photoreaction involves the formation of the radical cation of the substrate as the primary intermediate which reacts with molecular oxygen to finally give the 1,4-enedione.

Liang and co-workers described the oxidative dimerization of arylakynes in the presence of molecular oxygen and **4-(4-CNPh)Py**, which is an efficient photocatalyst [57]. The photoreaction provides the synthesis of *Z*-1,4-enediones with high stereoselectivity and with yields up to 75%, showing good functional group tolerance (Scheme 5.26). However, when aliphatic terminal alkynes are used, the photoreaction is unsuccessful.

Generation of a radical cation and its deprotonation produce carbon-centered radicals that react efficiently with molecular

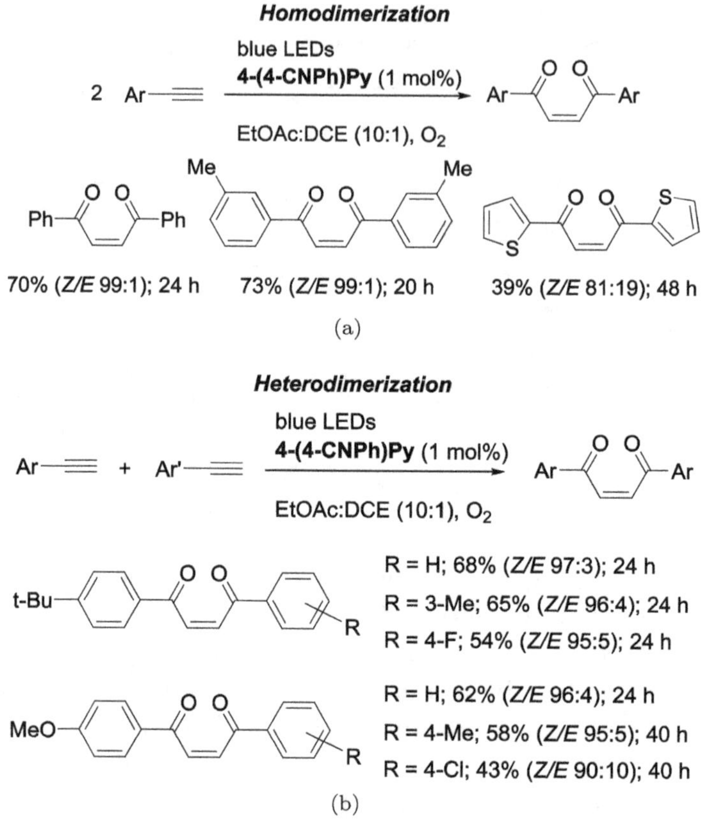

Scheme 5.26. Oxidative dimerization of terminal arylakynes.

oxygen to provide carbonyl derivatives. Akaba and co-workers reported the photocatalytic ET oxygenation of several arylalkanes in DCM using **Py** as the **POC** [58]. Diphenylmethane and triphenylmethane afford, under photocatalytic conditions, benzophenone in 94% and 78% yields, respectively, in only 2.5 h, with a quantitative consumption of the starting material. The photoreaction involves an ET process between the excited photocatalyst and arylalkane providing the arylalkane radical cation that in turn deprotonates to give the corresponding arylmethyl radical. LFP corroborates this proposed reaction mechanism. Finally, molecular oxygen reacts with the arylmethyl radicals to afford the corresponding carbonyl derivatives.

Also, alkene radical cations react with molecular oxygen to afford the corresponding ketones in good yields. Akaba and co-workers have also studied the photocatalyzed oxygenation reaction of adamantanylideneadamantane in DCM in the presence of **Py** for 30 min, which affords adamantanone in 75% yield [59a, 59b]. However, short reaction times favor the formation of the endoperoxide as the exclusive photoproduct in 78% yield. Thermal decomposition of the endoperoxide is proposed for the formation of adamantanone.

The photooxygenation of aryl olefins upon photoredox-catalyzed conditions have been intensively studied [60] and, in a number of cases, the formation of oxygenation products has been rationalized through the reaction of the olefin radical cations with molecular oxygen. This appears to be the preferred photooxygenation pathway when **Py** is used as the photocatalyst, since the excited photocatalyst generates neither singlet oxygen nor superoxide radical anion [2a]. Akaba and co-workers studied the triplet-mediated ET oxygenation of *trans*-stilbene derivatives with **Py** [61]. The corresponding carbonyl compounds are obtained in good yields with almost quantitative conversion of the *trans*-stilbenes (see Scheme 5.27).

Miranda and co-workers have reinvestigated the photooxygenation of *cis*- and *trans*-stilbenes as well as the corresponding stereoisomeric epoxides in the presence of **Py** in DCM at room temperature. Irradiation of *trans*-stilbene gives benzaldehyde, 1,1-diphenylacetaldehyde, and *cis*-stilbene in quite similar yields (ca. 33% each) [62]. Likewise, irradiation of *cis*-stilbene affords *trans*-stilbene in 80% yield together with low yields of benzaldehyde

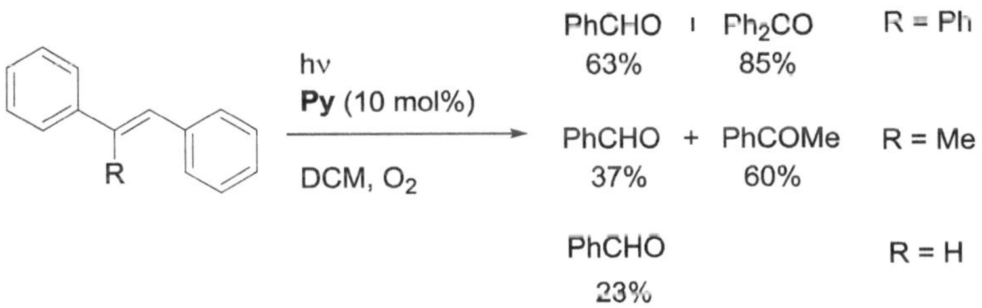

Scheme 5.27. Photooxygenation of trans-stilbene derivatives.

and 1,1-diphenylacetaldehyde. On the other hand, irradiation of *cis*- or *trans*-1,2-diphenyloxirane in the presence of the same photocatalyst affords 1,1-diphenylacetaldehyde as the main photoproduct with both oxiranes in 55% and 75% yields, respectively.

The same authors have also reported the photooxygenation of 5-methylene-10,11-dihydro-*5H*-dibenzo[*a,d*]cycloheptene in the presence of **Py** in DCM at room temperature, which gives a mixture of dibenzosuberane **20** (31%) and dibenzosuberone **21** (29%) as the main photoproducts. The photoreaction and the possible intermediates proposed for the formation of the photoproducts, *viz.* dioxetane **22** and epoxide **23**, are shown in Scheme 5.28.

5.3.5.2 *Oxidation of benzyl alcohol, benzyl ethers, and hydroxylamines*

The photooxidation of various benzyl methyl ethers (XPhCH$_2$OMe, with X = H, 4-MeO, 4-Me, 3-Cl, and 4-CF$_3$) photocatalyzed by **Py** in MeCN, in the presence of oxygen has been investigated by Del Giacco and co-workers [63]. The reaction is suggested to proceed through a benzylic radical cation intermediate, formed *via* ET from the ether to the excited photocatalyst. Depending on the nature of the substituents, two photoproducts are formed, *viz.* benzaldehydes and methyl benzoates, in fairly good yields (Scheme 5.29). The formation of the products can be rationalized as follows: the radical

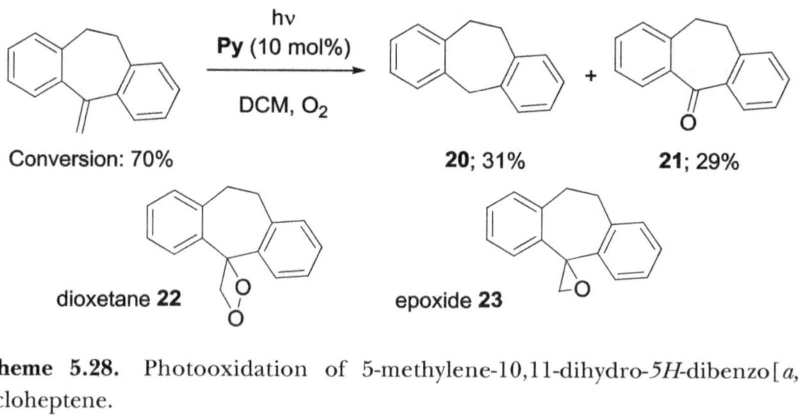

Scheme 5.28. Photooxidation of 5-methylene-10,11-dihydro-*5H*-dibenzo[*a,d*] cycloheptene.

Product distribution

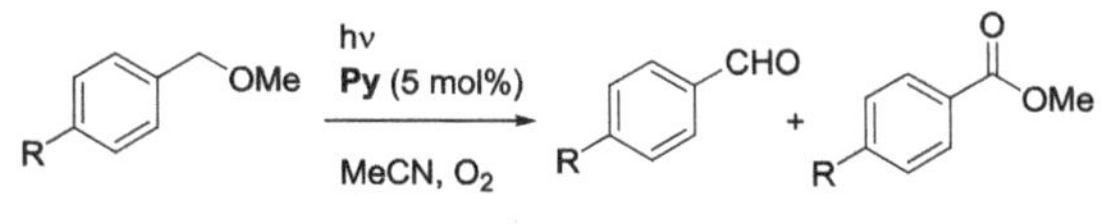

R	Yield (%)	
	Aldehyde	Ester
4-MeO	29	14
4-Me	22	33
H	16	33
3-Cl	12	35
4-CF_3	1.1	5.4

(a)

Benzyl methyl ether radical-cation reactivity

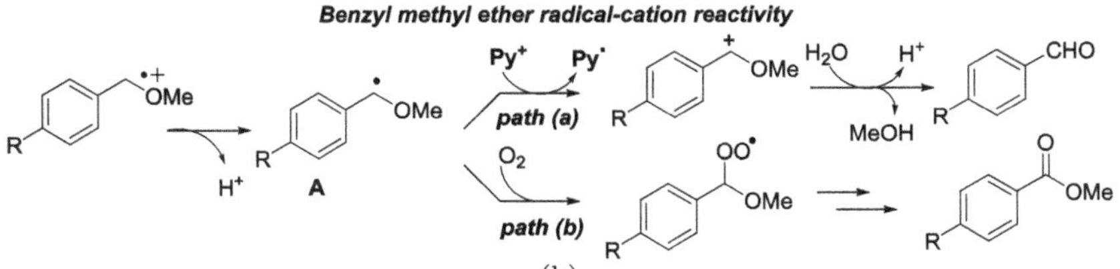

(b)

Scheme 5.29. Photooxygenation of substituted benzyl methyl ethers and reactivity of the substituted α-methoxybenzyl radical **A**.

cation of the ether, produced by ET to the photocatalyst, deprotonates to give the substituted α-methoxybenzyl radical (**A**), which is the primary intermediate. The radical intermediate may undergo two competitive paths: (i) oxidation to the corresponding carbocation (path a, Scheme 5.29), which leads to benzaldehyde and methanol through a hemiacetal intermediate; (ii) capture of the α-methoxybenzyl radical by oxygen to form the benzylic peroxide radical (path b, Scheme 5.29), which, by further oxidation to the carbonyl group, gives substituted methyl benzoates. These competitive pathways are also depicted in Scheme 5.29.

Bietti and co-workers have also reported the photooxidation of several benzyl alcohols with a variety of methoxy groups attached to the aromatic ring in the presence of **Py** (10 mol%) in DCM under O_2 atmosphere, from preparative and mechanistic viewpoints [64]. Irradiation of *p*-methoxybenzyl alcohol under the above experimental conditions affords *p*-methoxybenzaldehyde in 96% yield after 30 min of reaction. However, other substituted benzyl alcohols were not oxidized to the corresponding aldehydes. Notably, irradiation of bis(4-methoxybenzyl) ether in the presence of the photocatalyst gives *p*-methoxybenzaldehyde in 71% yield when irradiated under air atmosphere for 30 min. Likewise, irradiation of *p*-methoxytoluene provides smoothly only *p*-methoxybenzaldehyde in 19% yield (irradiation time: 30 min; air atmosphere). In all the cases studied, the photoreaction involves an ET process giving the radical cation as the key intermediate. The irradiation of **Py** in MeCN under molecular oxygen atmosphere provides the photooxidation of *N,N*-dibenzylhydroxylamine to the corresponding *N*-benzylidenebenzylamine *N*-oxide derivative in 64% yield [65]. However, the authors also state that *N,N*-di(*p*-methylbenzyl)- and *N,N*-di(*p*-trifluoromethylbenzyl) hydroxylamine are photooxidized in the presence of pyrylium salt to the corresponding *N*-oxides, but no chemical yields have been reported.

5.3.5.3 *Oxidation of thioethers and phosphines*

Dialkyl, diaryl, and arylalkyl sulfides (or thioethers) are an interesting class of compounds that have been thoroughly investigated during the past years, and can be converted to the corresponding sulfoxides and sulfones under both energy transfer (type II mechanism) and ET (type I mechanism) photosensitization. Both the proposed reaction mechanisms are depicted in Scheme 5.30.

In type II mechanism, conventional sensitizers such as Rose Bengal, Methylene Blue, or tetraphenylporphyrin have been extensively used [66–69]. On the other hand, in type I mechanism, most results have been obtained with photocatalysts such as cyanoaromatics [70–74] or 4-benzoyl benzoic acid [75], a water-soluble derivative of benzophenone, already known to react with sulfides [76–80]. The formation of a persulfoxide intermediate between the sulfide radical cation and the superoxide radical anion (Scheme 5.30) has often been put forward to account for products' formation, mainly sulfoxides and sulfones [66, 70, 71]. The same intermediate is also suggested in type II mechanism. It follows from these assumptions that sulfoxide and sulfone can arise from both types of mechanism, although this point has been argued by Baciocchi and co-workers [80]. Albini and co-workers have demonstrated that an alternative pathway involved in the photosensitized oxidation of sulfides (and

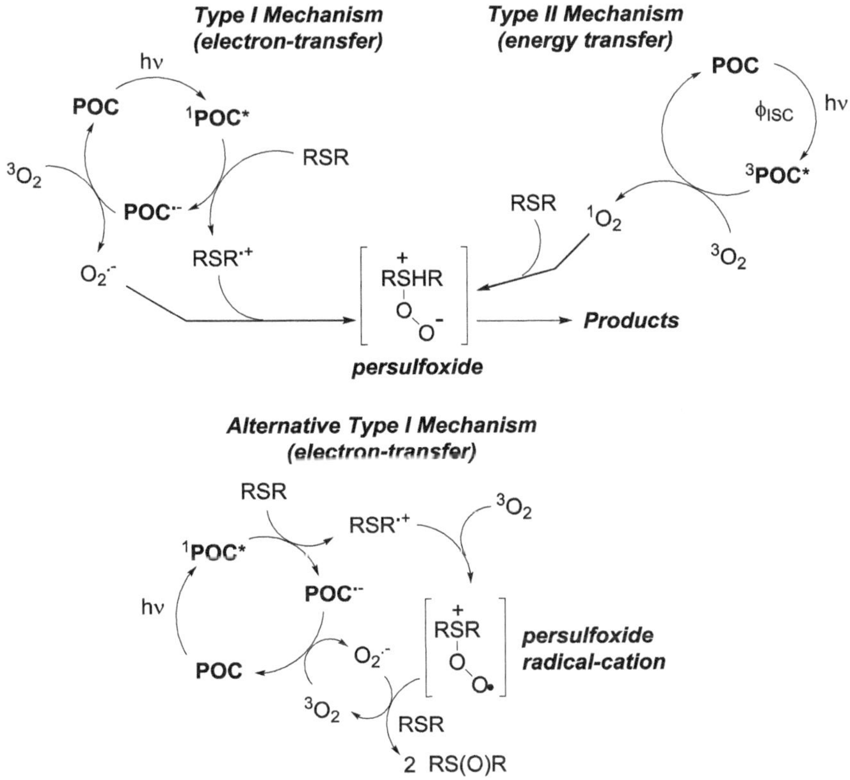

Scheme 5.30. Proposed photoorganocatalyzed oxidation of sulfides by energy and ET reactions.

also of aryl phosphines) is the reaction between the sulfide radical cation and molecular oxygen (Scheme 5.30) to give the persulfoxide radical cation intermediate, while the superoxide ion can be considered as an electron carrier [81, 82].

Lacombe and co-workers have reported the photosensitized oxidation of di-*n*-butylsulfide in oxygenated MeCN in the presence of **Py** (0.4 mol%) to afford only 5% yield of the sulfoxide after 150 min, while the photocatalyst is totally consumed [79]. However, a higher photocatalyst loading leads to higher sulfide consumption, mainly producing sulfoxide in 55% yield and significant amounts of acidic products, such as *n*-butyl sulfonic acid and sulfuric acid in a 20%

total yield, with complete bleaching of the photocatalyst. The photoreaction is initiated by an exergonic ET process from sulfide to the excited state of the **POC** ($\Delta G = -0.74$ eV) leading to an RSR$^{\bullet+}$/ **Py**$^{\bullet}$ pair.

Zhao and co-workers have reported the photooxidation of dibenzothiophene and 4,6-dimethyldibenzothiophene with visible light in O_2-saturated MeCN in the presence of **2-(4-MeOPh)Py** [83]. After 6 h of irradiation, the corresponding sulfoxides are obtained in good yields together with the sulfones (Scheme 5.31). In the case of 4,6-dimethyldibenzothiophene, a formyl derivative is also formed in 7.3% yield (Scheme 5.31). Electron spin resonance (ESR) and LPF experiments confirmed an ET process. The reaction mechanism proposed is also depicted in Scheme 5.31.

Albini and co-workers have studied the photooxidation of diphenyl and diethyl sulfides in the presence of **Py** and O_2 in polar and non-polar solvents with $\lambda_{exc} = 410$ nm [81]. Both sulfides afford the corresponding sulfoxides as the only photoproducts, and the photoreaction proceeds efficiently in all of the solvents studied, *viz.*

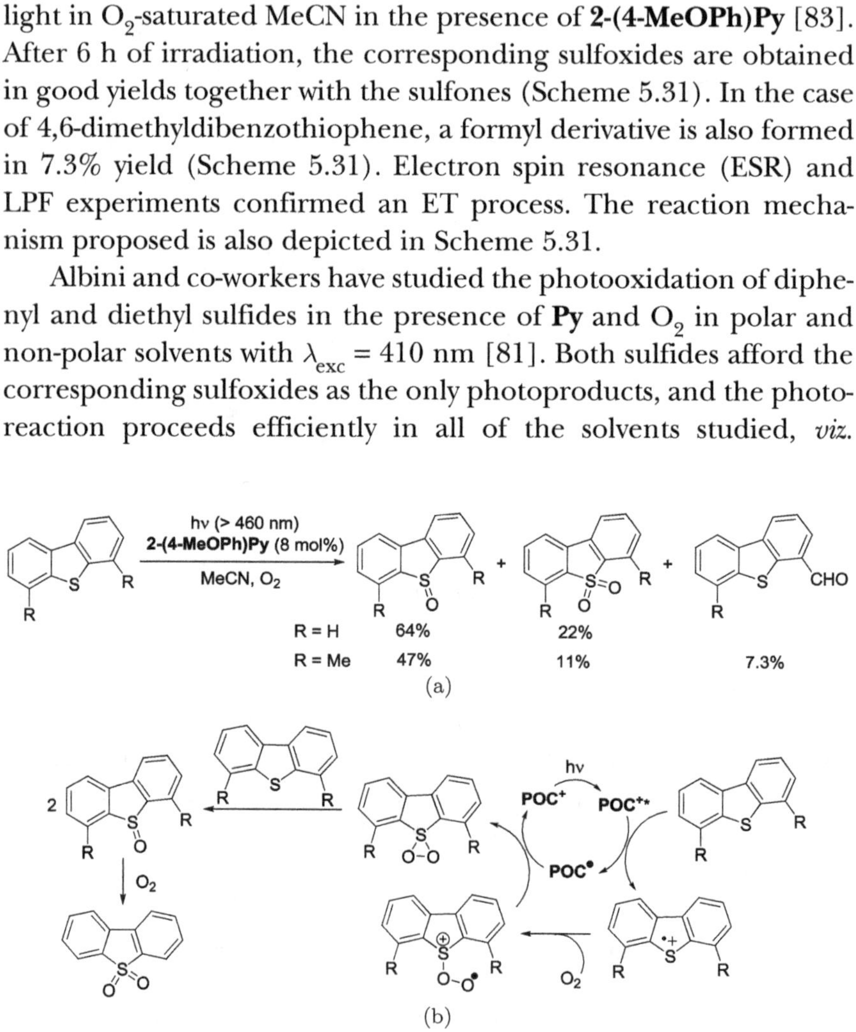

Scheme 5.31. (a) Photocatalytic oxidation of some dibenzothiophenes. (b) Proposed reaction mechanism.

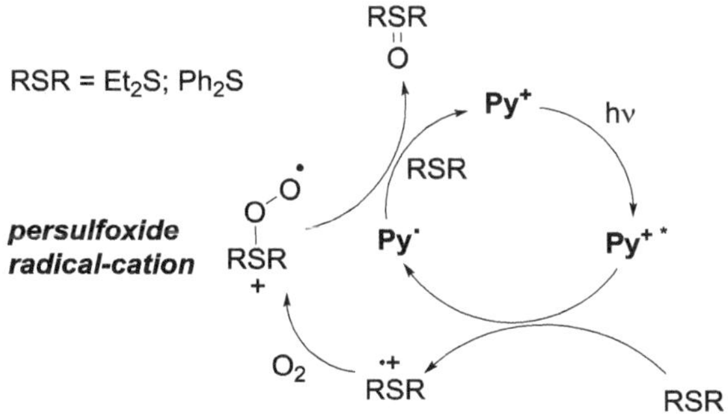

Scheme 5.32. Photocatalytic oxidation of diethyl and diphenyl sulfides.

methanol, MeCN, and 1,2-dichloroethane. LFP and chemical quenching experiments show that the radical cations of the sulfides formed during the irradiation react with molecular oxygen to give the persulfoxide radical cations that further react with another molecule of the sulfide to afford the sulfoxide. No consumption of the photocatalyst was observed during the irradiation. The photocatalytic oxidation reaction mechanism is depicted in Scheme 5.32.

Photocatalytic ET reactions of aryl benzyl sulfides using **Py** result in the oxidation of these sulfides to the sulfoxides and, in most cases, in the competitive C–S bond cleavage, thus resulting in the formation of aromatic aldehydes [84]. The sulfide radical cation is the key intermediate which undergoes competitively either oxidation to the sulfoxide or C–S bond cleavage to aromatic aldehydes. On the other hand, aromatic aldehydes are formed in good yields (up to 46%). Irradiation of dibenzyl sulfide under the same experimental conditions afforded the corresponding sulfoxide in 15% yield and benzaldehyde in 37% yield. Likewise, dibenzothiophene and diphenyl sulfide provide the corresponding sulfoxides in 93% and 95% yields, respectively. Some selected examples of benzyl aryl sulfides and the oxidation reaction mechanism are depicted in Scheme 5.33. Persulfoxide radical cation is the key intermediate in the formation of sulfoxide, whereas the aryl aldehydes arise from the deprotonation of the sulfide radical cation, followed by the reaction with molecular oxygen and finally through a Pummerer rearrangement.

Photocatalytic oxidation of benzylaryl sulfides

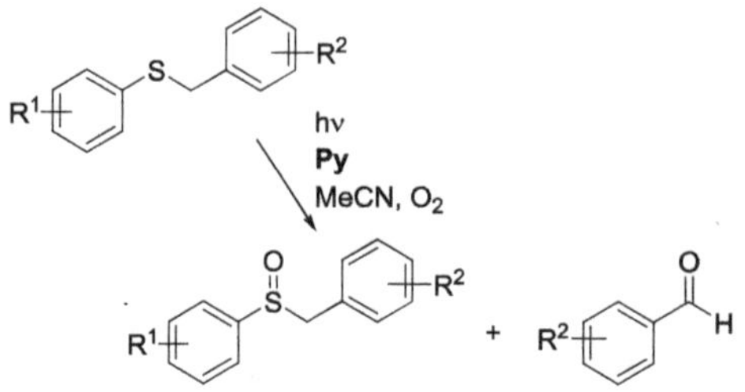

R^1	R^2	Yield (%) Sulfoxides	Aldehydes
H	H	14	38
	Br	15	40
	CH_3	3	43
	NO_2	3	46
CH_3	H	15	40
	Br	15	41
Cl	H	16	38
	Br	16	39
	NO_2	6	44

(a)

Proposed reaction mechanism

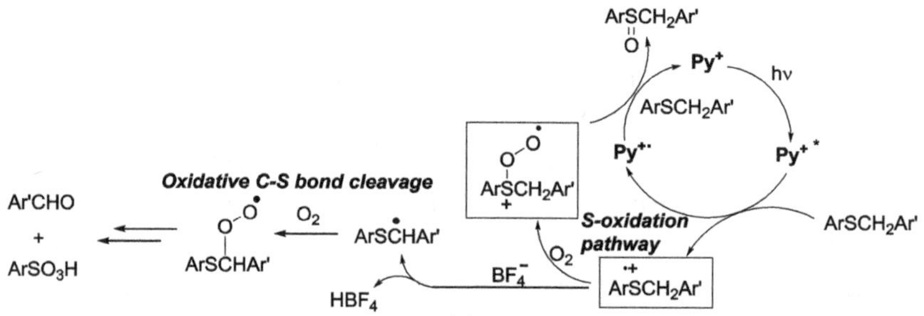

(b)

Scheme 5.33. (a) Photocatalyzed oxidation of some benzylaryl sulfides. (b) Proposed reaction mechanism.

Oxidation of a series of sulfides of the type R^1SR (where R^1 is dodecyl, phenethyl, benzyl, cumyl, benzhydryl; and R is an ethyl or a phenyl group) under ET conditions by using **Py** as the photocatalyst in the presence of molecular oxygen has been reported by Albini and co-workers [85]. The effect of the polarity of the solvent was studied. In fact, S-oxidation provides the sulfoxides with yields up to 75% when the solvent is MeOH, while the C–S bond cleavage gives the aldehydes/ketones in up to 85% yield in aprotic solvents.

The photocatalytic oxidation of a series of ethyl sulfides of the type RSEt (R = dodecyl; $PhCH_2CH_2$; $PhCH_2$; $PhCMe_2$; Ph_2CH) has been examined in the presence of **Py** in O_2-saturated MeCN [86]. The product distribution involves the formation of sulfoxides in up to 50% yield, as well as aldehydes and ketones and alkenes in yields up to 29% depending on the nature of the sulfides. On the other hand, cumyl ethyl and benzhydryl ethyl sulfides produce acetophenone and benzophenone, respectively, as the main photoproducts, but no S-oxidation occurs.

An interesting photocatalytic oxidation of alkyl and aryl sulfides has been reported by Albini and co-workers [87]. **Py** adsorbed on mesoporous silica, as well as encapsulated within zeolites Y (MCM), is an efficient solid photocatalyst in the oxidation of sulfides. Blue-light irradiation (410 nm) of alkyl and aryl sulfides in the presence of 6 mg of the solid photocatalyst in O_2-saturated solvents (MeOH, DCM, or MeCN), led to the formation of sulfoxides as the main photoproducts along with disulfides and sulfinic, sulfenic, and sulfonic acids. No significant degradation of the photocatalyst is observed in zeolites, whereas the photocatalyst is released when adsorbed on mesoporous silica.

Clennan and co-workers have investigated the photocatalytic oxidation of *5H,7H*-dibenzo[*b,g*][1,5]dithiocin (**24**) and 1,5-dithiacyclooctane with different photocatalysts in the presence of molecular oxygen in MeCN and chloroform [88]. Irradiation of **24** in the presence of **Py** or **2-(4-MeOPh)Py** under the aforementioned conditions with blue light, gave sulfoxide in a quantitative yield. Irradiation of 1,5-dithiacyclooctane in MeCN or in chloroform provided a mixture of 1,5-dithiacyclooctane-1-sulfoxide and 1,5-dithiacyclooctane-1,5-bissulfoxide. Examination of this reaction as a function of time clearly shows that the 1,5-bissulfoxide is a secondary product arising from the over-oxidation of 1,5-dithiacyclooctane-1-sulfoxide. Scheme 5.34 shows the photocatalytic oxidation of (**24**).

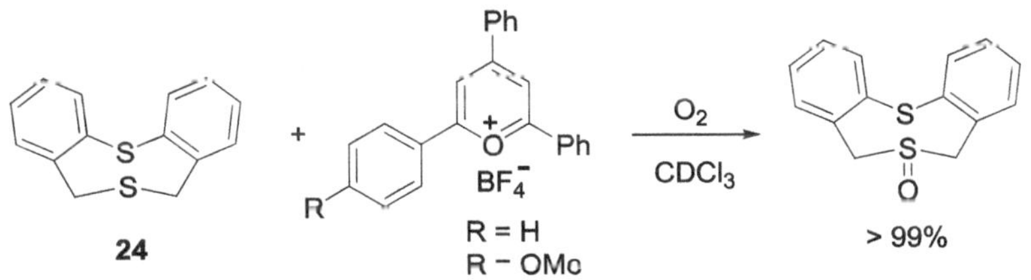

Scheme 5.34. Photocatalytic oxidation of *5H,7H*-dibenzo[*b,g*][1,5]dithiocin (**24**).

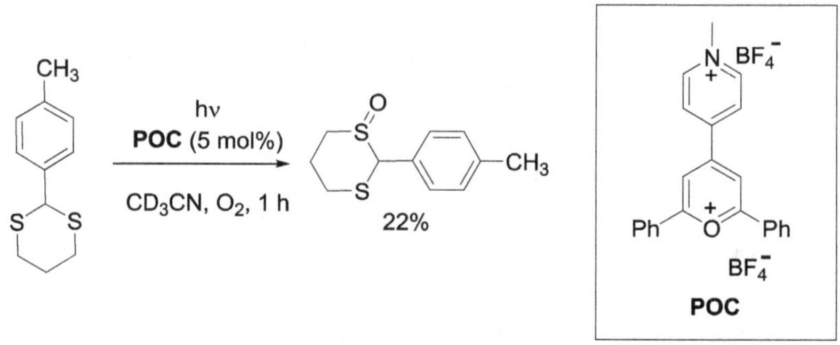

Scheme 5.35. Photoorganocatalyzed oxidation of 1,3-dithiane.

A series of new dicationic photocatalysts that are hybrids of pyrylium salts and viologens has been synthesized by Clennan and co-workers and they are called "pyrylogens" [89]. They have been shown to be excellent **POCs** from their singlet and triplet excited states, providing the radical cations of sulfides and alkenes. For example, in the presence of molecular oxygen, 1,3-dithiane derivatives afford the corresponding sulfoxides (Scheme 5.35).

Kamata and co-workers reported the photocatalytic oxidation of 2-(4-methylphenyl)- and 2,2-di(4-methylphenyl)-1,3-dithiane using **Py** as the photocatalyst in MeCN with $\lambda > 360$ nm. The corresponding carbonyl derivatives are formed in good to excellent yields [90]. In the case of the monosubstituted 1,3-dithianes, the sulfoxide was also formed in low yields. Photocatalytic ET is involved in the process. The photoreaction is depicted in Scheme 5.36. These examples pointed out the versatility of such photocatalytic reaction for the photorelease of carbonyl compounds.

PPh_3 has been converted to the corresponding phosphine oxide quantitatively when **Py** was used as the photocatalyst (10 mol%) in O_2-saturated solvents such as MeCN, methanol, DCM, and a mixture of aqueous MeCN (9:1) during 10 min [82]. However, in that period of time, the photocatalyst was partially consumed (up to 15%). LFP experiments account for the formation of the triphenyl phosphine radical cation, which in turn reacts with molecular oxygen to produce the triphenylphosphine peroxide radical cation intermediate that finally affords triphenylphosphine oxide.

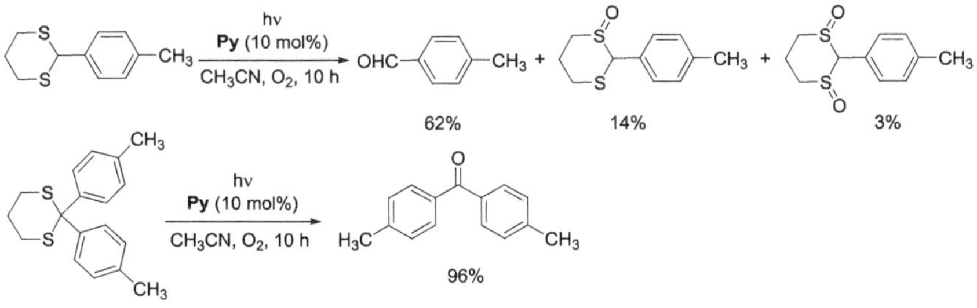

Scheme 5.36. Photocatalytic oxidation of 1,3-dithiane derivatives.

5.3.6 *Endoperoxydation reactions*

Pyrylium salts have been employed in the preparation of endoperoxides exhibiting interesting biological properties. For example, Posner and co-workers have reported that structurally simple and easily prepared 1,5-diaryl-6,7-dioxabicyclo[3.2.2]nonanes, where aryl is a phenyl group or a *p*-methoxyphenyl group, have potent antimalarial properties [91,92]. Later, Kamata and co-workers have applied the photooxygenation reaction of a series of 2,6-diaryl-1,6-heptadienes using **Py** as the photocatalyst (1 mol%) in MeCN and in the presence of oxygen to improve the synthesis of 1,5-diaryl-6,7-dioxabicyclo[3.2.2]nonanes [93]. In fact, the products are obtained in good yields with respect to those obtained when the photocatalyst is 9,10-dicyanoanthracene (**DCA**). A significant improvement is achieved when the aryl group is a *p*-methylphenyl or a phenyl group (see Scheme 5.37). However, no reaction is observed for the case of 2,6-di(*p*-methoxyphenyl)-1,6-heptadiene when **Py** is used as the **POC**. These valuable substrates are useful model compounds in the study of the mechanism of action of endoperoxides with Fe(II) and in the formation of a high-valent iron-oxo intermediate for high antimalarial activity of bicyclononane endoperoxides [94].

Triarylpyrylium salts have been employed as single electron photooxidants to catalyze a cyclization–endoperoxidation cascade of dienes. Nicewicz and co-workers have studied this reaction using 1,5-dienes in the presence of **4-MeOPy** as the photocatalyst in DCM

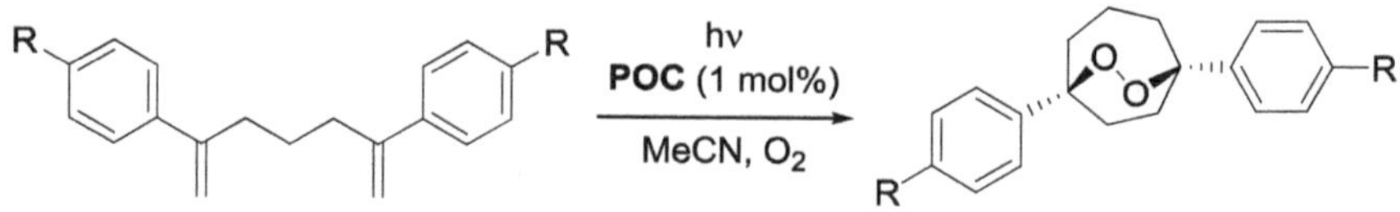

R	Irradiation time (min)	Isolated Yield (%) (**POC: Py**)	Isolated Yield (%) (**POC: DCA**)
F	50	33	----
H	65	44	0
Me	20	58	37
MeO	15	0	87

Scheme 5.37. Endoperoxidation of 2,6-diaryl-1,6-heptadienes.

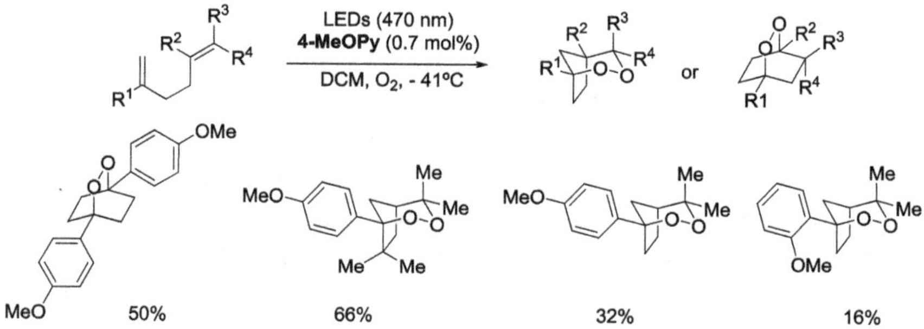

Scheme 5.38. Cyclization–endoperoxidation cascade reaction of dienes.

at −41°C affording the desired products in moderate to good yields (Scheme 5.38) [95]. Attempts to apply the photoreaction to less electron-rich dienes have failed to produce the desired endoperoxide products, while mainly oxidative cleavage products were observed.

The transformation is presumed to proceed *via* the intermediacy of a diene-localized radical cation that evolves to a stabilized distonic radical cation (Scheme 5.39) [96, 97].

Following excitation of pyrylium salt, one-electron oxidation of 1,5-diene produces a localized radical cation intermediate **A** and pyranyl radical. Cyclization of intermediate **A** forms a stabilized

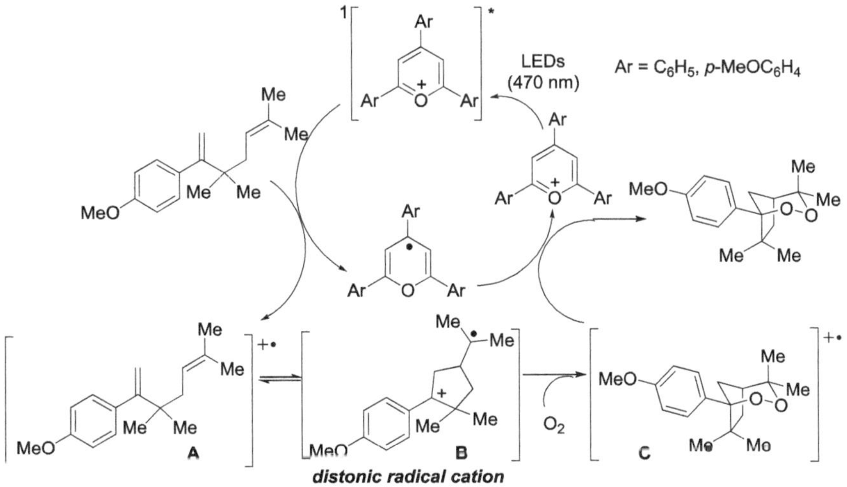

Scheme 5.39. Proposed mechanism for endoperoxide synthesis from tethered dienes.

distonic radical cation intermediate **B** which is trapped by molecular oxygen to form intermediate **C**. Single electron reduction forms the desired bicyclic endoperoxide.

Replacement of the isoprenyl substituent with a variety of other alkenes affords the desired endoperoxide adduct under the aforementioned experimental conditions. The products are formed in good yields with noticeable diastereocontrol (Scheme 5.40). However, the use of 1,2-, 1,1-dialkyls as well as monoalkyl-substituted alkenes appears to completely disfavor the endoperoxidation, while resulting in the unexpected isolation of homoallyl ketones (Scheme 5.40). Based on the Miyashi precedent [96], the formation of these adducts can be rationalized by invoking a formal [3,3]-rearrangement of the initial radical cation intermediate. The competing 6-*endo* cyclization mode and formation of distonic radical cation **D** ultimately provides access to the more stabilized radical cation **E**, which undergoes oxidative cleavage to the corresponding homoallyl ketones (Scheme 5.40). In the absence of the geminal dimethyl group, neither the Cope-like reactivity nor the endoperoxidation is observed.

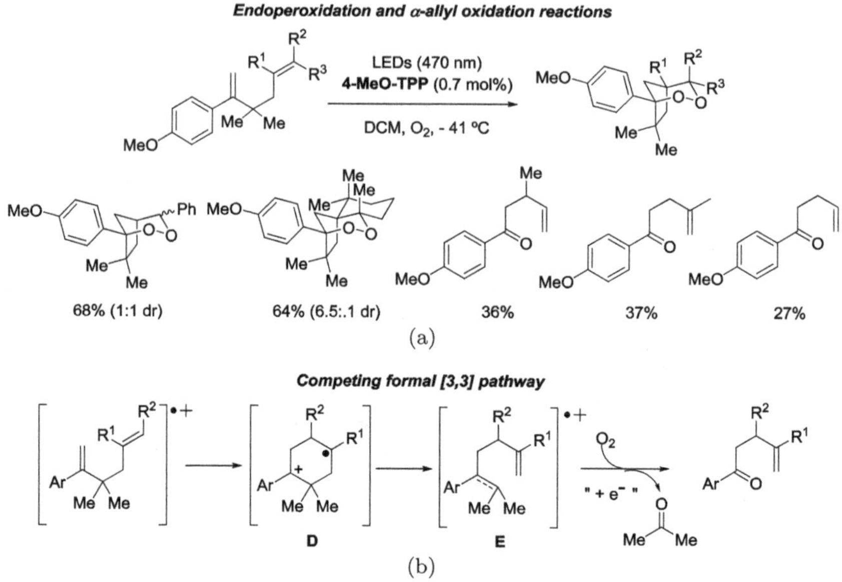

Scheme 5.40. Variation of the tethered alkene component in the endoperoxidation reaction.

Lastly, cyclization modes similar to Yoon's work [97] of dienes have also been reported [95]. Diene radical cations generated in the presence of **4-MeOPy** cyclize under a 6-*exo*-trig fashion due to the formation of a distonic radical cation intermediate. The expected trioxabicyclo[3.3.1]nonane product (**25**) is formed, albeit in low yields (16%), where oxidative degradation completes the mass balance (Scheme 5.41). Bis(styrene) **26** affords the fused 1,2-dioxane in 79% yield with significant diastereocontrol (5.7:1 dr).

5.3.7 *C–C and C–X bond cleavage*

Photocatalytic fragmentation of alkyl, benzyl, cumyl, and benzhydryl ethyl sulfides can be accomplished by using **Py** as the photocatalyst in MeCN under nitrogen atmosphere. Albini and co-workers have reported a systematic study on the C–S fragmentation of these sulfides [86]. Irradiation with blue light (410 nm) of the aforementioned sulfides affords aldehydes/ketones, alkenes, and alcohols in

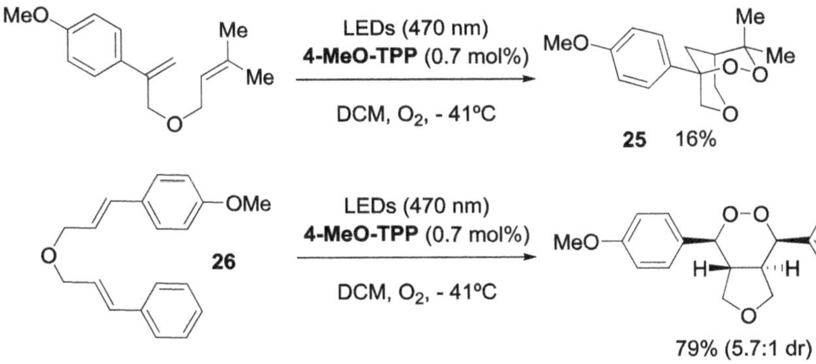

Scheme 5.41. Other cyclization modes and substrate designs.

fairly good yields along with photoproducts arising from the reaction of thiyl radicals formed during the irradiation. The photocatalytic fragmentation reaction is initiated through an exergonic ET process and provides the sulfide radical cation which is the key intermediate. From this, intermediate deprotonation of the Cα and asymmetric fragmentation afford the radical species that finally evolves to the photoproducts.

Irradiation of α-epoxyketones bearing electron-donating and electron-withdrawing substituents with **Py** as the photocatalyst in MeOH solutions has been reported [98]. An ET process occurs between the α-epoxyketones and the excited photocatalyst leading to the formation of the substrate radical cation followed by stereoselective fragmentation of the C–O bond. Whereas, irradiation of a mixture of α-epoxyketones bearing electron-rich substituents and the photocatalyst produces MeOH-adducts **27** and **28**, while those bearing a nitro group as substituent do not react under the same conditions (see Scheme 5.42).

The authors have also investigated the same photoreaction, but by using acetone as the solvent [99]. Under these conditions, C–O and C–C fragmentation occurs during the irradiation, and 1,3-dioxolanes (**29**, **30**) as well as products **31**, **32** are formed in moderate yields (Scheme 5.42).

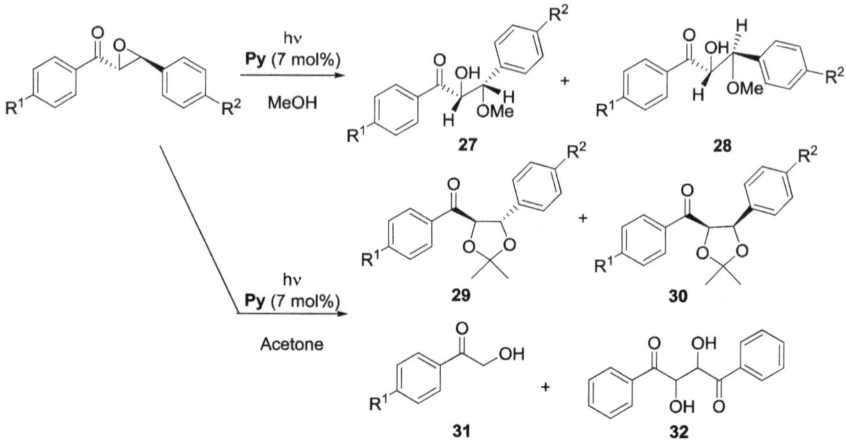

R¹	R²	MeOH: Yield (%) 27	28	Acetone: Yield (%) 29	30	31	32
H	H	29	71	14	23	—	60
H	Me	46	54	10	28	—	28
Me	H	14	86	11	24	41	—
H	OMe	52	48	11	29	—	21
OMe	H	25	75	8	27	42	—
OMe	OMe	50	50	—	23	—	38

Scheme 5.42. Photoorganocatalyzed reactions of α-epoxyketones.

Stereospecific formation of spirodioxolanes has been observed on ET-induced ring opening of α-epoxyketones by **Py** (10 mol%) in the presence of cyclohexanone [100]. The photoreaction is carried out in MeCN with λ_{exc} > 400 nm. Spirodioxolanes are formed in moderate yields (24%–33%) with high diastereoselectivity in favor of the *cis* diastereomer.

The Cα–O bond of chalcone epoxides is selectively cleaved by using **Py** as the **POC** under excitation with λ > 390 nm [101]. The product distribution depends on the solvent used. β-Ketoaldehydes are obtained in 80% yield when DCM is used as the solvent. When the photoreactions of chalcone epoxides are conducted in MeOH, MeOH adducts are formed instead of β-ketoaldehydes (Scheme 5.43). The photocatalytic ET reaction in DCM is proposed to follow the sequence: (i) formation of epoxide radical cation, (ii) Cα–O

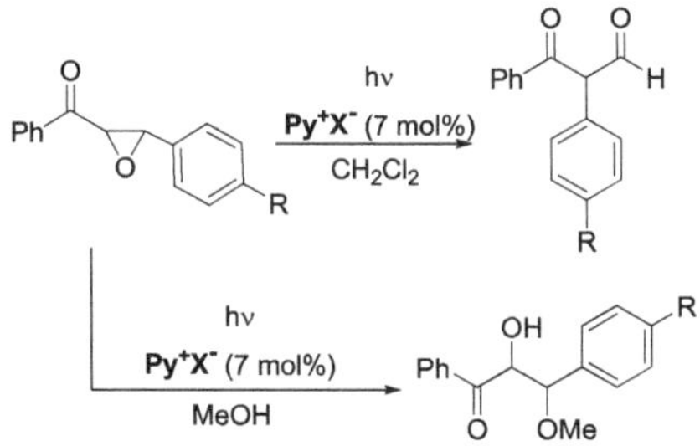

R	X^-	Yield (%)
Me	BF_4^-	80
	ClO_4^-	44
OMe	BF_4^-	51

R	X^-	Yield (%)
Me	BF_4^-	97
	ClO_4^-	90
OMe	BF_4^-	75

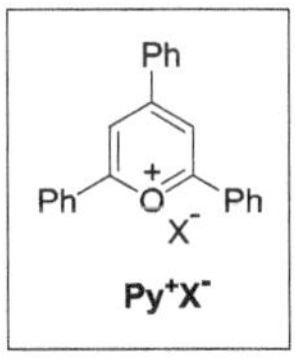

Scheme 5.43. Photocatalyzed ring opening of chalcone epoxides.

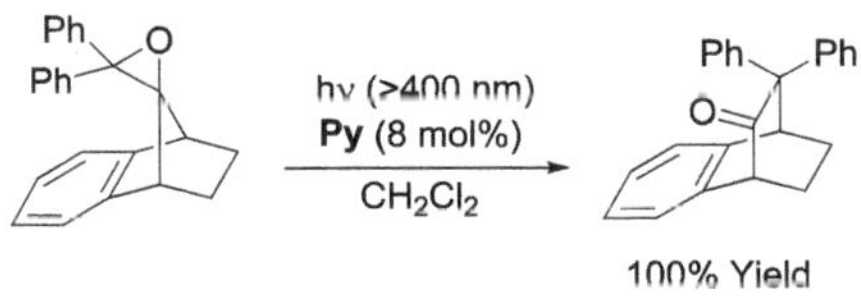

Scheme 5.44. Photocatalyzed epoxide ring opening.

cleavage of the radical cation, (iii) 1,2-benzoyl shift to give β-ketoaldehyde radical cation that finally collapses to the photoproduct. In MeOH, the Cα—O cleavage in this radical cation of chalcone epoxide is trapped by the nucleophilic solvent giving the methanol adduct.

Mukai and co-workers [102] reported the C–O cleavage of epoxides upon photocatalysis with **Py**, providing a rearranged ketone in quantitative yield (Scheme 5.44). The photoreaction is insensitive to the presence of molecular oxygen and occurs efficiently in halogenated solvents.

1,4-diaryl-2,3-dioxabicyclo[2.2.2]octanes in the presence of **Py** undergo a fragmentation (*via* radical cations) upon irradiation in the presence of 2,6-di-*tert*-butylpyridine, to give 1,4-diarylbutan-1, 4-diones as the sole photoproducts in good to excellent yields accompanied by the elimination of ethylene (Scheme 5.45) [103]. On the other hand, 4-aryl-cyclohex-3-en-1-ones, *p*-substituted phenols, and 4-aryl-4-aryloxycyclohexanones are produced through proton-catalyzed pathways when the photoreactions of the

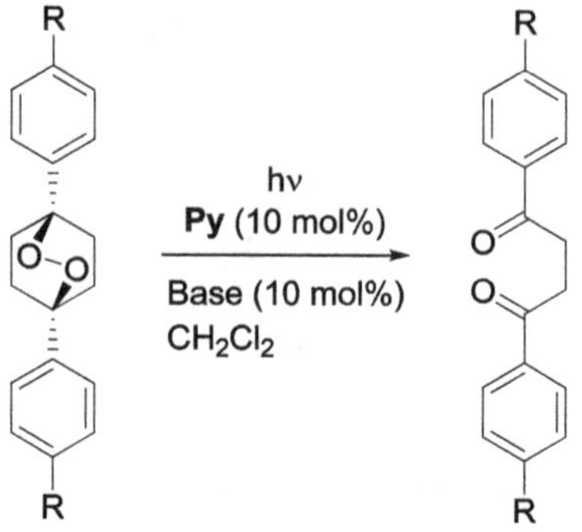

R	Yield (%)
OMe	83
Me	95
H	94

Scheme 5.45. Photocatalyzed fragmentation of 1,4-diaryl-2,3-dioxabicyclo[2.2.2] octanes.

endoperoxides are performed in the absence of the base. Particularly, the formation of 4-aryl-4-aryloxycyclohexanones is consistent with the novel cationic rearrangement involving nucleophilic *O*-1,2-aryl shifts and *C*-1,4-aryl shifts.

Several dithianes and dithiolanes produced carbonyl compounds in good to excellent yields under **Py** photocatalysis in oxygenated MeCN (Scheme 5.46) [104]. During irradiation of the disulfides, selective C–S bond cleavage in dithiane and dithiolane cation radicals occurs and molecular oxygen selectively traps the carbon radical intermediates, thereby generating the carbonyl-containing photoproducts. This photocatalytic reaction is an example of the photorelease of protecting groups such as dithianes and dithiolanes.

When 2-(*p*-methylphenyl)-1,3dithiane is subjected to the photocatalyzed oxygenation reaction with **Py** photocatalyst in MeCN, *p*-methylbenzaldehyde and 2-(*p*-methylphenyl)-1,3-dithiane-l-oxide are isolated in 62% and 14% yields, respectively [90]. Small amounts of 2-(*p*-methylphenyl)-1,3-dithiane-1,3-dioxide are also obtained (3% yield). Likewise, irradiation of 2,2-di-(*p*-methylphenyl)-1,3-dithiane under similar conditions affords quantitatively *p*-methylbenzophenone, and no sulfoxide nor sulfone is formed.

Photoreactions of various dithioacetals and ketals using **4-ClPy** perchlorate as the photocatalyst at $\lambda > 360$ nm in O_2-saturated DCM

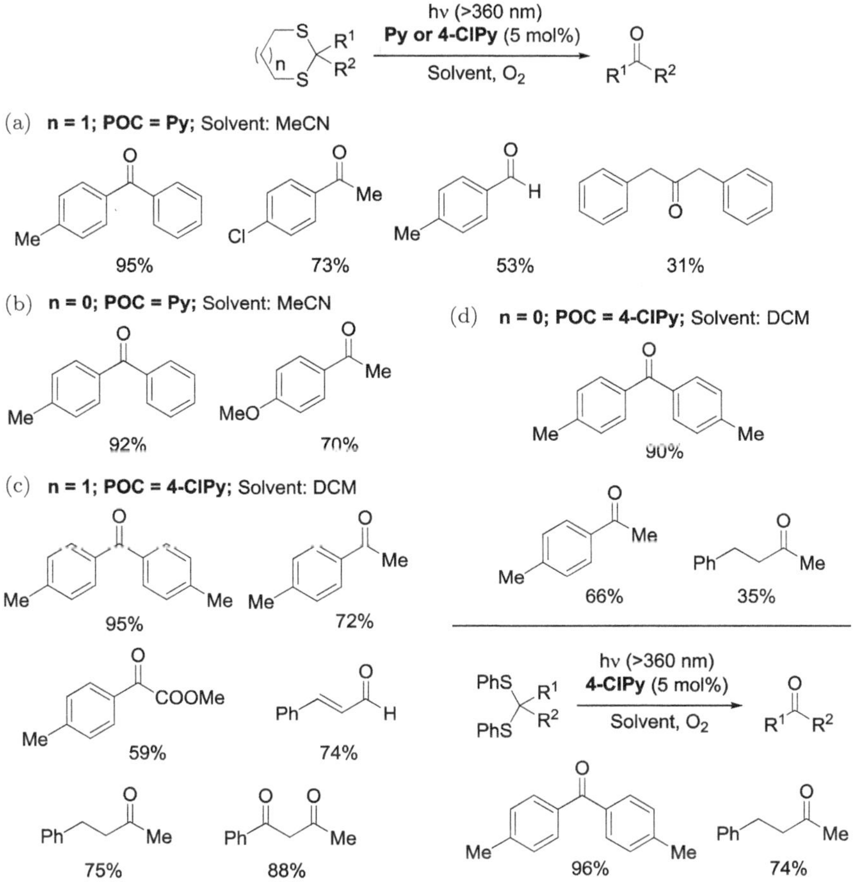

Scheme 5.46. Photoorganocatalyzed release of carbonyls from dithianes and dithiolanes.

afforded the corresponding carbonyl compounds in good to excellent yields (Scheme 5.46) [105].

Miranda and co-workers have reported the photocatalytic deprotection of 4-(*p*-isopropylphenyl)-2-nonyl-1,3-dioxolane with **Py** as the photocatalyst (20 mol%) in DCM, using the Pyrex-filtered light from a 400 W mercury lamp [106]. Under these conditions, the only absorbing species is the pyrylium salt acting as the photocatalyst, and *n*-decanal is formed in 48% yield. The reaction yield is enhanced in the presence of both water and oxygen.

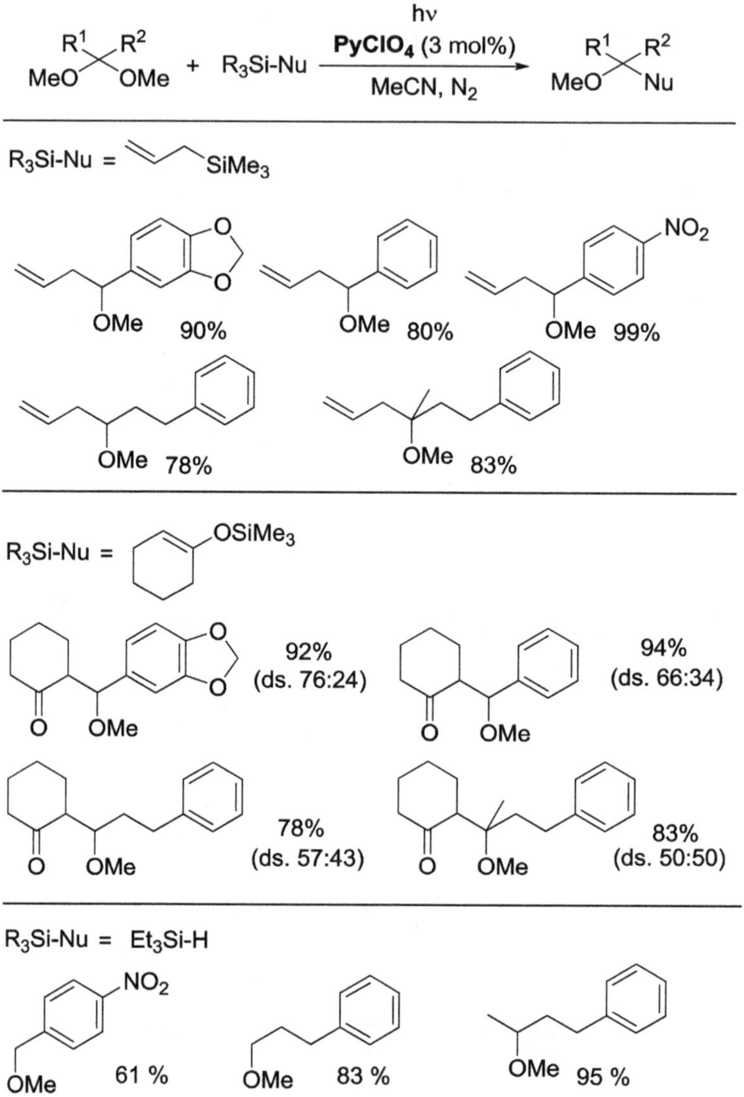

Scheme 5.47. Photoorganocatalyzed deprotection of acetals and ketals.

Catalytic amounts of **Py** perchlorate (3 mol%) photochemically ($\lambda > 360$ nm) promotes the substitution reactions of acetals with various silyl nucleophiles in N_2-saturated dry MeCN [107]. The photoproducts are obtained in excellent yields (up to 99% yield) and, in some cases, with good diastereoselectivity (Scheme 5.47).

The irradiation of a substituted oxasilacyclopropane in MeCN with 4-(*p*-butoxyphenyl)-2,6-diphenylpyrylium perchlorate as the photocatalyst produces a tetramethylindene in 77% yield together with dimesitylsilanediol [108]. The photoreaction occurs through an ET process and forms tetramethyl indenyl carbene intermediate, which after 1,2-methyl migration leads to the indene derivative.

Photocatalyzed ET oxygenation of several arylalkyl ketones and aldehydes such as 1,2,2-triphenylethanone and diphenyl and triphenylethanal has been carried out with **Py** (10 mol%) in DCM, affording the corresponding benzophenones in good yields (up to 80%) [109].

Oxygen-saturated MeCN solutions containing **Py** as the photocatalyst and 4-methoxybenzyl ethers as the substrates have been irradiated ($\lambda = 355$ nm) affording the corresponding ketones, alcohols, and esters in good yields depending on the nature of the substrate (see Scheme 5.48) [110]. LFP experiments demonstrate that the photoreaction occurs through an ET process, giving the corresponding radical cations.

The Newman–Kwart rearrangement is perhaps the quintessential method for the synthesis of thiophenols from the corresponding phenols. However, the harsh thermal conditions required for the rearrangement of the requisite *O*-aryl carbamothioates often leads to decomposition [111]. However, photoredox catalysis has been reported to improve the conversion of phenols to thiophenols at room temperature. Thus, irradiating

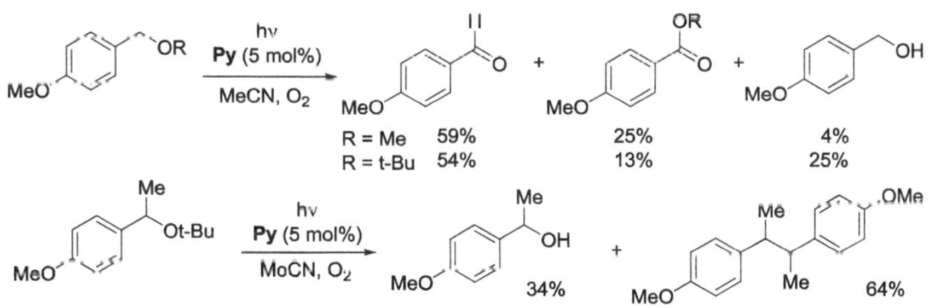

Scheme 5.48. Photocatalyzed reaction of benzyl ethers.

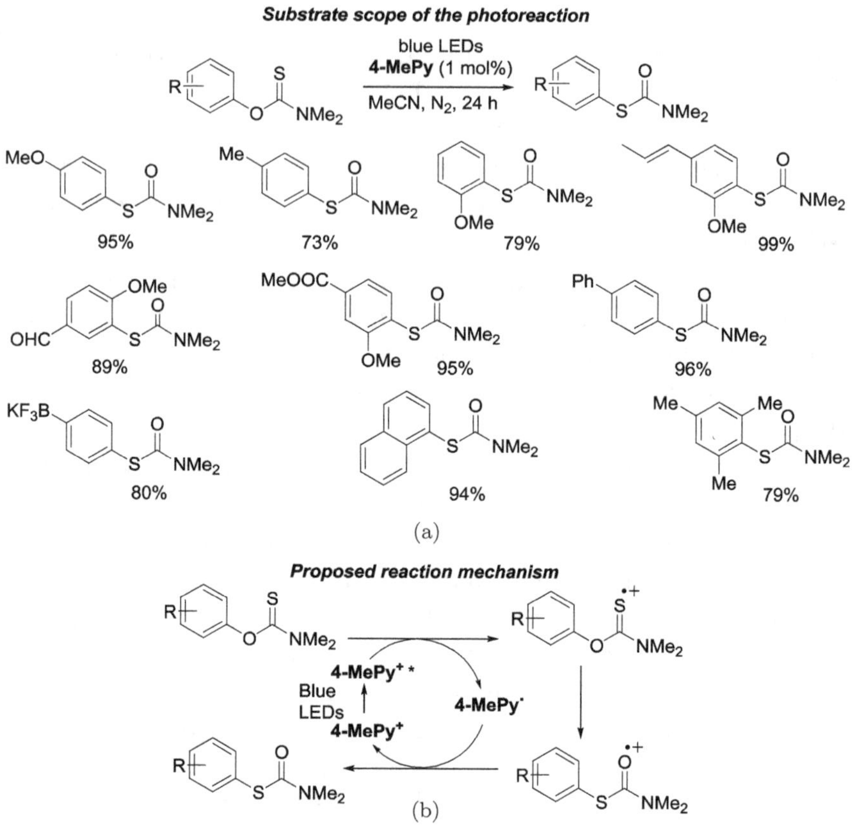

Scheme 5.49. Photoorganocatalyzed Newman–Kwart rearrangement of *O*-aryl carbamothioates.

solutions of *O*-aryl dimethylcarbamothioate derivatives in the presence of catalytic amounts of **4-MePy** using blue LEDs in N_2-saturated MeCN affords the thiol products in good to excellent yields in 24 h. Scheme 5.49 shows some examples and the proposed reaction mechanism.

This reaction tolerates a wide variety of acceptor and donor functional groups, and polycyclic substrates such as naphthyl moiety can also be adopted. Further, the suitability of this protocol using a flow apparatus has also been reported, affording the desired products in excellent yield under **Py**-photocatalyzed conditions.

5.4 Conclusions and Perspectives of Pyrylium Salt Photocatalysis

This chapter has reported an account of selected advances in the use of purely organic photooxidants such as pyrylium salts in synthetic transformations.

It has been demonstrated that 2,4,6-triarylpyrylium salts are aromatic compounds that can be used as excellent ET photocatalysts in organic synthesis. When they are irradiated with visible light ($\lambda > 400$ nm), and due to their moderate-to-high reduction potentials, they have become excellent oxidant species from the pyrylium singlet-excited-state manifold (E^*_{RED}). Accordingly, they are able to oxidize a wide variety of organic compounds such as alkenes, alkynes, (hetero)aromatic compounds, thioethers, ethers, and phosphines, among others.

These photocatalysts can react as good oxidant agents from their triplet excited states as well. Another feature that characterizes these photocatalysts is the fact that they are not able to photosensitize singlet oxygen through an energy transfer process, and neither produce a superoxide ion. This behavior is useful in the case of photooxygenation reactions of a variety of electron-rich substrates, such as alkenes, thioethers, and phosphines. In fact, the photooxygenation products come from pure ET processes instead of a singlet oxygen photosensitization.

Since such photocatalytic processes are metal- and/or organometal-free methods and use visible-light or low energy radiation sources, they can contribute significantly to the environmentally benign array of protocols available to the organic synthetic chemist to effect numerous organic transformations. An area for future development is the utilization of pyrylium salts and derivatives in aqueous media to promote photocatalytic processes on water-soluble or very polar substrates.

References

[1] (a) Prier, C. K.; Rankic, D. A.; MacMillan, D. W. C. *Chem. Rev.* **2013**, *113*, 5322–5363; (b) Yoon, T. P.; Ischay, M. A.; Du, J. *Nat. Chem.* **2010**, *2*, 527–532; (c) Narayanam, J. M. R.; Stephenson, C. R. J. *Chem. Soc. Rev.* **2011**, *40*, 102–113; (d) Skubi, K. L.; Blum, T. R.; Yoon, T. P. *Chem. Rev.* **2016**, *116*, 10035–10074; (e) Ravelli, R.; Dondi, D.; Fagnoni, M.; Albini, A. *Chem. Soc. Rev.* **2009**, *38*, 1999–2011.

[2] (a) Miranda, M. A.; García, H. *Chem. Rev.* **1994**, *94*, 1063–1089; (b) Fagnoni, M.; Dondi, D.; Ravelli, D.; Albini, A. *Chem. Rev.* **2007**, *107*, 2725–2756; (c) Ravelli, D.; Fagnoni, M. *ChemCatChem* **2012**, *4*, 169–171; (d) Marin, M. L.; Santos-Juanes, L.; Arques, A.; Amat, A. M.; Miranda, M. A. *Chem. Rev.* **2012**, *112*, 1710–1750; (e) Fukuzumi, S.; Ohkubo, K. *Chem. Sci.* **2013**, *4*, 561–574; (f) Ravelli, D.; Fagnoni, M.; Albini, A. *Chem. Soc. Rev.* **2013**, *42*, 97–113.

[3] (a) Che, Y.; Ma, W.; Ji, H.; Zhao, J.; Zang, L. *J. Phys. Chem. B* **2006**, *110*, 2942–2948; (b) Wei, D.; Liang, F. *Org. Lett.* **2016**, *18*, 5860–5864; (c) Clennan, E. L.; Liao, C. *Photochem. Photobiol.* **2014**, *90*, 344–357.

[4] Liang, S.; Zhu, L.; Hurst, J. K. *Langmuir* **2012**, *28*, 12171–12181.

[5] (a) Haucke, G.; Czerney, P.; Cebulla, F. *Ber. Bunsen-Ges. Phys. Chem.* **1992**, *96*, 880–886; (b) Tripathi, S.; Simalty, M.; Pouliquen, J.; Kwsanyi, J. *Bull. Soc. Chim. Fr.* **1986**, *4*, 600; (c) Wintgens, V.; Kossanyi, J.; Simalty, M. *Bull. Soc. Chim. Fr.* **1983**, *3–4*, 115; (d) Wintgens, V.; Pouliquen, J.; Valat, P.; Koseanyi, J.; Canonica, S.; Wild, U. P. *Chem. Phys. Lett.* **1986**, *123*, 282–288.

[6] Martiny, M.; Steckhan, E.; Esch, T. *Chem. Ber.* **1993**, *126*, 1671–1682.

[7] Mlcoch, J.; Steckhan, E. *Angew. Chem. Int. Ed.* **1985**, *24*, 412– 414.

[8] Wiest, O.; Steckhan, E. *Angew. Chem. Int. Ed.* **1993**, *32*, 901–903.

[9] Peglow, T.; Blechert, S.; Steckhan, E. *Chem. Commun.* **1999**, 433–434.

[10] Gunrtler, C. F.; Blechert, S.; Steckhan, E. *Synlett* **1994**, 14–142.

[11] Gunrtler, C. F.; Steckhan, E.; Blechert, S. *J. Org. Chem.* **1996**, *61*, 4136–4143.

[12] Haberl, U.; Steckhan, E.; Blechert, S.; Wiest, O. *Chem. Eur. J.* **1999**, *5*, 2859–2865.

[13] Wiest, O.; Steckhan, E. *Tetrahedron Lett.* **1993**, *34*, 6391–6394.

[14] Gieseler, A.; Steckhan, E.; Wiest, O.; Knoch, F. *J. Org. Chem.* **1991**, *56*, 1405–1411.

[15] Gieseler, A.; Steckhan, E.; Wiest, O. *Synlett* **1990**, *5*, 275–277.

[16] Ischay, M. A.; Yoon, T. P. *Eur. J. Org. Chem.* **2012**, 3359–3372.

[17] Gieseler, A.; Steckhan, E.; Wiest, O.; Knoch, F. *J. Org. Chem.* **1991**, *56*, 1405–1411.

[18] Galindo, F.; Miranda, M. A. *J. Photochem. Photobiol A Chem.* **1998**, *113*, 155–161.

[19] Arguello, J. E.; Pérez-Ruiz, R.; Miranda, M. A. *Org. Lett.* **2007**, *9*, 3587–3590.

[20] Pérez-Ruiz, R.; Domingo, L. R.; Consuelo Jiménez, M.; Miranda, M. A. *Org. Lett.* **2011**, *13*, 5116–5119.

[21] (a) Zhang, W.; Guo, Y., Liu, Zh.; Jin, X.; Yang, L.; Liu, Zh.-L. *Tetrahedron* **2005**, *61*, 1325–1333; (b) Zhang, W.; Jia, X.; Yang, L.; Liu, Z.-L. *Tetrahedron Lett.* **2002**, *43*, 9433–9436.

[22] Wang, K.; Meng, L.-G.; Wang, L. *Org. Lett.* **2017**, *19*, 1958–1961.

[23] Alfonzo, E.; Alfonso, F. S.; Beeler, A. B. *Org. Lett.* **2017**, *19*, 2989–2992.

[24] Riener, M.; Nicewicz, D. A. *Chem. Sci.* **2013**, *4*, 2625–2629.

[25] (a) Mahindru, R. N.; Taneja, S. C.; Dhar, K. L.; Brown, R. T. *Phytochemistry* **1999**, *32*, 1073–1075; (b) Malhorta, S.; Koul, S.; Taheja, S.; Pushpangadan, P.; Dhar, K. *Phytochemistry* **1990**, *29*, 2733–2734; (c) Kikuchi, T.; Kadota, S.; Yamada, K.; Tanaka, K.; Watanabe, K.; Yoshizaki, M.; Yokoi, T.; Shingu, T. *Chem. Pharm. Bull.* **1983**, *31*, 1112–1114.

[26] (a) Kuntzsch, D.; Bergann, T.; Dames, P.; Fromm, A.; Fromm, M.; Davis, R. A.; Melzig, M. F.; Schulzke, J. D. *PLoS One* **2012**, *7*, e49426, doi: 10.1371/journal.pone.0049426; (b) Ngadjui, B. T.; Lontsi, D.; Ayafor, J. F.; Sondengam, B. L. *Phytochemistry* **1989**, *28*, 231–234.

[27] Mattay, J.; Vondenhof, M.; Denig, R. *Chem. Ber.* **1989**, *122*, 951–958.

[28] Fornés, V.; García, H.; Miranda, M. A.; Mojarrad, F.; Sabater, M. J.; Suliman, N. N. E. *Tetrahedron* **1996**, *52*, 7755–7760.

[29] Mizuno, K.; Pac, C. *CRC Handbook of Organic Photochemistry and Photobiology*, Eds., Horspool, W. M. and Song, P. S., CRC Press, Boca Raton, Florida, **1995**, 359–365.

[30] Pérez-Ruiz, R.; Consuelo Jiménez, M., Miranda, M. A. *Acc. Chem. Res.* **2014**, *47*, 1359–1368.

[31] Izquierdo, M. A.; Domingo, L. R.; Miranda, M. A. *J. Phys. Chem. A* **2005**, *109*, 2602–2607.

[32] Miranda, M. A.; Izquierdo, M. A. *J. Am. Chem. Soc.* **2002**, *124*, 6532–6533.

[33] Miranda, M. A.; Izquierdo, M. A. *Chem. Commun.* **2003**, 364–365.

[34] Miranda, M. A.; Izquierdo, M. A.; Galindo, F. *Org. Lett.* **2001**, *3*, 1965–1967.

[35] Miranda, M. A.; Izquierdo, M. A.; Galindo, F. *J. Org. Chem.* **2002**, *67*, 4138–4142.

[36] Izquierdo, M. A.; Miranda, M. A. *Eur. J. Org. Chem.* **2004**, 1424–1431.

[37] Miranda, M. A.; Izquierdo, M. A.; Pérez-Ruiz, R. *J. Phys. Chem. A* **2003**, *107*, 2478–2482.

[38] Domingo, L. R.; Perez-Ruiz, R.; Arguello, J. E.; Miranda, M. A. *J. Phys. Chem. A.* **2011**, *115*, 5443–5448.

[39] Arguello, J. A.; Perez-Ruiz, R.; Miranda, M. A. *Org. Lett.* **2010**, *12*, 1884–1887.

[40] Perez-Ruiz, R.; Saez, J. A.; Domingo, L. R.; Jimenez, M. C.; Miranda, M. A. *Org. Lett.* **2012**, *14*, 5700–5703.

[41] (a) Okada, K.; Hisamitsu, K.; Mukai, T. *Tetrahedron Lett.* **1981**, *22*, 1251–1254; (b) Okada, K.; Hisamitsu, K.; Miyashi, T.; Mukai, T. *J. Chem. Soc. Chem. Commun.* **1982**, 974–976.

[42] Yamashita, Y.; Ikeda, H.; Mukai, T. *J. Am. Chem. Soc.* **1987**, *109*, 6682–6687.

[43] Yang, Z.; Li, H.; Zhang, L.; Zhang, M.-T.; Cheng, J.-P.; Luo, S. *Chem. Eur. J.* **2015**, *21*, 14723–14727.

[44] (a) Ozser, M. E.; Icil, H.; Makhynya, Y.; Demuth, M. *Eur. J. Org. Chem.* **2004**, 3686–3692; (b) Rosales, V.; Zambrano, J.; Demuth, M. *Eur. J. Org. Chem.* **2004**, 1798–1802.

[45] Warsecha, F. D.; Demuth, M.; Goerner, H. *J. Chem. Soc. Faraday Trans.* **1997**, *93*, 1523–1530.

[46] Jacobi, D.; Abraham, W.; Pischel, U.; Grubert, L.; Stoesser, R.; Schnabel, W. *J. Chem. Soc. Perkin Trans.* **1999**, *2*, 1695–1702.

[47] Mattes, S. L.; Farid, S. *J. Chem. Soc. Chem. Commun.* **1980**, 126.

[48] Becker, P.; Duhamel, T.; Stein, C. T.; Reiher, M.; Muñiz, K. *Angew. Chem. Int. Ed.* **2017**, *56*, 8004–8008.

[49] Jackl, M. K.; Legnani, L.; Morandi, B.; Bode, J. W. *Org. Lett.* **2017**, *19*, 4696–4699.

[50] Ortgies, S.; Rieger, R.; Rode, K.; Koszinowski, K.; Kind, J.; Thiele, C. M.; Rehbein, J.; Breder, B. *ACS Catal.* **2017**, *7*, 7578–7586.

[51] Amat, A. M.; Arques, A.; Miranda, M. A. *App. Cat. B Environ.* **1999**, *23*, 205–214.

[52] (a) Sanjuan, A.; Aguirre, G.; Alvaro, M.; Garcia, H. *App. Catal. B Environ.* **1998**, *15*, 247–257; (b) Sanjuan, A.; Alvaro, M.; Aguirre, G.; Garcia, H. J. C. Scaiano, *J. Am. Chem. Soc.* **1998**, *120*, 7351–7352; (c) Sanjuan, A.; Aguirre, G.; Alvaro, M.; Garcia, H. *Water Res.* **2000**, *34*, 320–326.

[53] Sanjuan, A.; Aguirre, G.; Alvaro, M.; Garcia, H.; Scaiano, J. C. *App. Catal. B Environ.* **2000**, *25*, 257–265.

[54] Srihdar, M.; Kumar, B. A. *Chem. Lett.* **1998**, *27*, 461–462.

[55] Ortgies, S.; Depken, C.; Breder, V. *Org. Lett.* **2016**, *18*, 2856–2859.

[56] Manoj, N.; Gopidas, K. R. *J. Photochem. Photobiol. A Chem.* **1995**, *85*, 53–61.

[57] Wei, D.; Liang, F. *Org. Lett.* **2016**, *18*, 5860– 863.

[58] Akaba, R.; Kamata, M.; Itoh, H.; Nakao, A.; Goto, S.; Saito, K.; Negishi, A.; Sakuragi, H.; Tokumarud, K. *Tetrahedron Lett.* **1992**, *33*, 7011–7014.

[59] (a) Akaba, R.; Aihara, S.; Sakuragi, H.; Tokumaru, K. *J. Chem. Soc. Chem. Commun.* **1987**, 1262–1263; (b) Akaba, R.; Sakuragi, H.; Tokumaru, K. *J. Chem. Soc. Perkin Trans.* 2, **1991**, 291–297.

[60] Fox, M. A.; Chanon, M. *Photoinduced Electron Transfer, Part C*, Elsevier, Amsterdam, 1988.

[61] Akaba, R.; Ohshima, K.; Eawai, Y.; Obuchi, Y.; Negishi, A.; Sakumg, H.; Tokumaru, E. *Tetrahedron Lett.* **1991**, *32*, 109–112.

[62] Garcia, H.; Miranda, M. A.; Mojarrad, F.; Sabater, M. *Tetrahedron* **1994**, *50*, 8773–8780.

[63] Del Giacco, V.; Lipparoni, L.; Ranchella, M.; Rol, C.; Sebastiani, G. V. *J. Chem. Soc. Perkin Trans.* **2001**, *2*, 1802–1807.

[64] Branchi, B.; Bietti, M.; Ercolani, G.; Izquierdo, M. A.; Miranda, M. A.; Stella, L. *J. Org. Chem.* **2004**, *69*, 8874–8885.

[65] Iijima, Y.; Obara, S.; Ohba, Y.; Kubo, K.; Sakurai, T. *J. Photochem. Photobiol. A Chem.* **2001**, *143*, 23–29.

[66] Ando, W. *Sulfur Rep.* **1981**, *1*, 147–213.

[67] Liang, J. J.; Gu, C. L.; Kacher, M. L.; Foote, C. S. *J. Am. Chem. Soc.* **1983**, *105*, 4717–4721.

[68] Clennan, E. L. *Sulfur Rep.* **1996**, *19*, 171–214.

[69] Clennan, E. L. *Acc. Chem. Res.* **2001**, *34*, 878–884.

[70] Eriksen, J.; Foote, C. S.; Parker, T. L. *J. Am. Chem. Soc.* **1977**, *99*, 6455–6456.

[71] Baciocchi, E.; Crescenzi, C.; Lanzalunga, O. *Tetrahedron* **1997**, *53*, 4469–4478.

[72] Shiraishi, Y.; Taki, Y.; Hirai, T.; Komasawa, I. *Chem. Commun.* **1998**, 2601–2602.

[73] Ando, W.; Nagashima, T.; Saito, K.; Kohmoto, S. *J. Chem. Soc. Chem. Commun.* **1979**, 154–156.

[74] Soggiu, N.; Cardy, H.; Habib-Jiwan, J. L.; Leray, I.; Soumillion, J. P.; Lacombe, S. *J. Photochem. Photobiol. A Chem.* **1999**, *124*, 1– 8.

[75] Miller, B. L.; Williams, T. D.; Schoneich, C. *J. Am. Chem. Soc.* **1996**, *118*, 11014–11025.

[76] Marciniak, B.; Bobrowski, V.; Hug, G. L. *J. Phys. Chem.* **1993**, *97*, 11937–11943.

[77] Guttenplan, J.; Cohen, S. G. *J. Chem. Soc. D Chem. Commun.* **1969**, 247–248.

[78] Inbar, S.; Linschitz, H.; Cohen, S. G. *J. Am. Chem. Soc.* **1982**, *104*, 1679–1682.

[79] Lacombe, S.; Cardy, H.; Simon, M.; Khoukh, A.; Soumillion, J. P.; Ayadim, M. *Photochem. Photobiol. Sci.* **2002**, *1*, 347– 354.

[80] Baciocchi, E.; Del Giacco, T.; Elisei, F.; Gerini, M.; Guerra, M.; Lapi, A.; Liberali, P. *J. Am. Chem. Soc.* **2003**, *125*, 16444–16454.

[81] Bonesi, S. M.; Manet, I.; Freccero, M.; Fagnoni, M.; Albini, A. *Chem. Eur. J.* **2006**, *12*, 4844–4857.

[82] Bonesi, S. M.; Protti, S.; Albini, A. *J. Org. Chem.* **2016**, *81*, 11678–11685.

[83] Che, J.; Ma, V.; Ji, H.; Zhao, J.; Zang, L. *J. Phys. Chem. B* **2006**, *110*, 2942–2948.

[84] Memarian, H. R.; Mohammadpoor-Baltork, I.; Bahrami, K. *Bull. Korean Chem. Soc.* **2006**, *27*, 106–110.

[85] Bonesi, S. M.; Fagnoni, M.; Albini, A. *J. Sulfur Chem.* **2008**, *29*, 367–376.

[86] Bonesi, S. M.; Fagnoni, M.; Albini, A. *Eur. J. Org. Chem.* **2008**, 2612–2620.
[87] Bonesi, S. M.; Carbonell, E.; Garcia, H.; Fagnoni, M.; Albini, A. *Appl. Catal. B Environ.* **2008**, *79*, 368–375.
[88] Clennan, E. L. *J. Am. Chem. Soc.* **2008**, *130*, 4057–4068.
[89] Clennan, E. L.; Liao, C. *Photochem. Photobiol.* **2014**, *90*, 344–357.
[90] Kamata, M.; Sato, M.; Hasewaga, E. *Tetrahedron Lett.* **1992**, *33*, 5085–5088.
[91] Posner, G. H.; Wang, D.; Gonzares, L.; Tao, X.; Cumming, J. N.; Klinedinst, D.; Shapiro, T. A. *Tetrahedron Lett.* **1996**, *37*, 815–818.
[92] Posner, G. H.; Gonzares, L.; Cumming, J. N.; Klinedinst, D.; Shapiro, T. A. *Tetrahedron* **1997**, *53*, 37–50.
[93] Kamata, M.; Ohta, M.; Komatsu, K.; Kim, H.; Wataya, Y. *Tetrahedron Lett.* **2002**, *43*, 2063–2067.
[94] Posner. G. H.; Cumming, J. N.; Ploypradith, P.; Oh, C. H. *J. Am. Chem. Soc.* **1995**, *117*, 5885–5886.
[95] Gesmundo, N. J.; Nicewicz, D. A. *Beilstein J. Org. Chem.* **2014**, *10*, 1272–1281.
[96] Miyashi, T.; Konno, A.; Takahashi, Y. *J. Am. Chem. Soc.* **1988**, *110*, 3676–3677.
[97] Parrish, J. D.; Ischay, M. A.; Lu, Z.; Gou, S.; Peters, N. R.; Yoon, T. P. *Org. Lett.* **2012**, *14*, 1640–1643.
[98] Memarian, H. R.; Hesami, A.; Nikpour, F.; Dopp, D. *Indian. J. Chem.* **2001**, *40B*, 662–666.
[99] Memarian, H. R.; Hesami, A.; Nikpour, F. *Monatsh. Chem.* **2002**, *133*, 1045.
[100] Memarian, H. R.; Nikpour, F. *Molecules* **2002**, *7*, 63–71.
[101] Hasegawa, E.; Ishiyama, K.; Kashiwasaki, H.; Horaguchi, T.; Shimizu, T. *Tetrahedron Lett.* **1990**, *31*, 4045–4048.
[102] Okada, K.; Hasegawa, E.; Mukai, T. *Chem. Lett.* **1983**, *12*, 305–308.
[103] Kamata, M.; Kaneko, J.; Hagiwara, J.; Akaba, R. *Tetrahedron Lett.* **2004**, *45*, 7423–7428.
[104] Kamata, M.; Kato, Y.; Hasewaga, E. *Tetrahedron Lett.* **1991**, *34*, 4349–4352.
[105] Kamata, M.; Murakami, Y.; Tamagawa, Y.; Kato, M.; Hasegawa, E. *Tetrahedron* **1994**, *50*, 12821–12828.

[106] Gaviña, P.; López Lavernia, N.; Mestres, R.; Miranda, M. A. *Tetrahedron* **1996**, *52*, 4911–4916.

[107] Kamata, V.; Nagai, S.; Kato, M.; Hasegawa, E. *Tetrahedron Lett.* **1996**, *37*, 7779–7782.

[108] Ando, W.; Hamada, Y.; Sekiguchi, A. *Tetrahedron Lett.* **1984**, *25*, 5057–5060.

[109] Akaba, R.; Niimura, Y.; Fukushima, T.; Kawai, Y.; Tajima, T.; Kuragami, T.; Negishi, A.; Kamata, M.; Sakuragi, H.; Tokumaru, K. *J. Am. Chem. Soc.* **1992**, *114*, 4460–4464.

[110] Carlotti, B.; Del Giacco, T.; Elisei, E. *Photochem. Photobiol. Sci.* **2013**, *12*, 489–499.

[111] Perkowski, A. J.; Cruz, C. L.; Nicewicz, D. A. *J. Am. Chem. Soc.* **2015**, *137*, 15684–15687.

Chapter 6

Oxygen Heterocycles: Eosin Derivatives

Wen-Jing Xiao*, Xiao-Qiang Hu, and Jia-Rong Chen†

CCNU-uOttawa Joint Research Center,
Key Laboratory of Pesticide & Chemical Biology,
Ministry of Education, College of Chemistry,
Central China Normal University,
152 Luoyu Road, Wuhan, Hubei 430079, China
**wxiao@mail.ccnu.edu.cn*
†chenjiarong@mail.ccnu.edu.cn

6.1 Introduction

Visible-light-induced photoredox catalysis has recently been established as a powerful technique for the activation of organic molecules, by which a wide variety of single electron transfer (SET) reactions and energy transfer processes can be achieved under very mild reaction conditions [1]. Intensive efforts from the synthetic community have been devoted to the important field of visible-light-induced organic synthesis for the efficient formation of various chemical bonds in a predictable and selective manner, which have yielded more than 70 reviews over the past few years. As demonstrated in numerous examples, the choice of the photocatalyst plays a pivotal role in the success of the exploration of new and practical visible-light-induced photocatalytic reactions [2]. Ruthenium- and iridium-based polypyridyl complexes are the most commonly employed photocatalysts in many photosynthetic reactions due to their excellent photophysical properties, such as the stable and long-lived photoexcited states [1, 3]. However, at the current stage, their relatively high cost, problematic recyclability, and potential toxicity might limit their general application, especially in large-scale synthesis.

Recently, organic dyes-mediated photocatalytic reactions have emerged as a cutting-edge research topic in organic synthesis, which provide attractive and metal-free alternatives to organometallic complexes due to their competitive advantages in terms of low cost, ready availability, as well as biological tolerance in some cases [4]. In this regard, Eosin Y and its derivatives found wide application in numerous photochemical transformations (Scheme 6.1) [5]. Eosin Y absorbs green light (absorption maximum at 539 nm in the UV–Vis spectrum), leading to the excited state ^{3}Eosin Y* with a

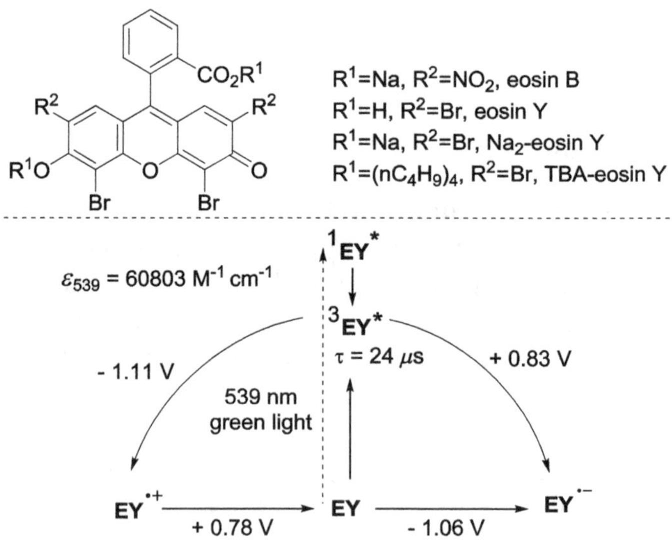

Scheme 6.1. Eosin derivatives and the corresponding redox potentials.

lifetime of 24 μs. This triplet excited state can act as a good oxidant or reductant, and the corresponding redox potentials are outlined in Scheme 6.1. In this chapter, we will highlight the recent advances on Eosin Y-promoted photocatalytic reactions, with an emphasis on their working models. To calibrate the reasonable scope of this topic, the chapter is mainly organized according to the different types of radical species (N-radical, S-radical, P-radical, C-radical, etc.) and their applications in the corresponding transformations.

6.2 Activation of N-Containing Molecules

6.2.1 *N-centered radical reactions*

Nitrogen-containing heterocyclic scaffolds are widely found in numerous natural products, functional materials, as well as medicinally relevant molecules [6]. In recent years, visible-light-induced N-radical reactions have been established as one of the most powerful methods for the formation of C–N bonds in a controllable and efficient fashion [7]. In 2014, Xiao and Chen reported the first example of visible-light-induced oxidative deprotonation electron transfer (ODET) for the direct conversion of N–H bonds to N-centered radicals, and further applied this strategy to the hydroamination reaction of β,γ-unsaturated hydrazone **1** (Scheme 6.2) [8]. Although only 43% yield of 4,5-dihydropyrazole **2** could be obtained by employing 2 mol% of Eosin Y as the photocatalyst in this reaction, the proposed method provides a direct, metal-free access to N-radical hydroamination reactions.

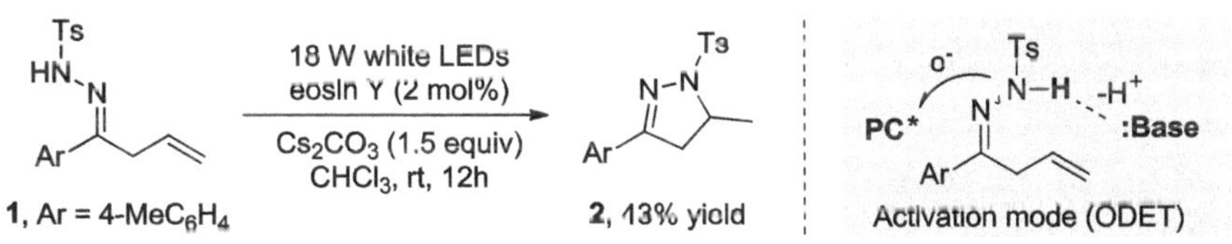

Scheme 6.2. Visible light photocatalytic hydroamination of alkenes by hydrazones.

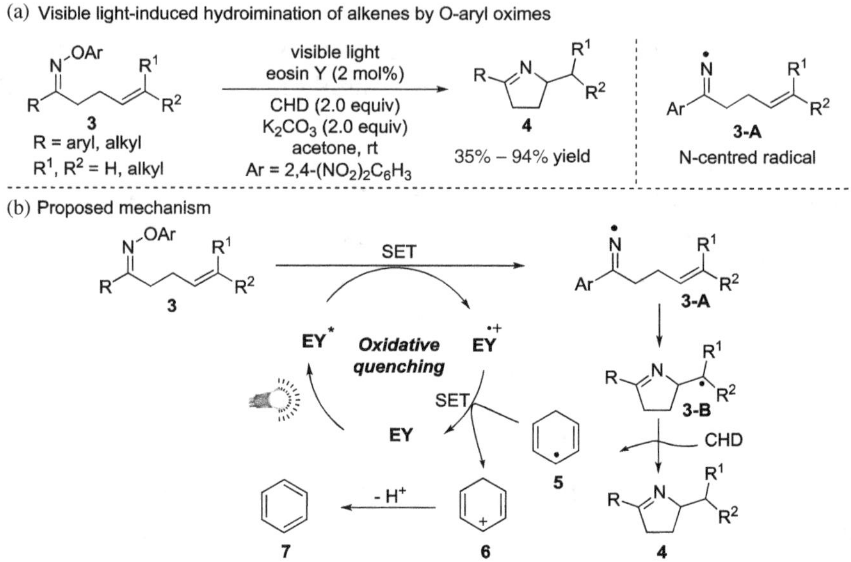

Scheme 6.3. Visible-light-induced hydroimination by O-aryl oximes.

In 2015, the group of Leonori disclosed an elegant Eosin Y-catalyzed photocatalytic hydroimination by O-aryl oximes with the use of 1,4-cyclohexadiene (CHD) as the H-source, furnishing a set of biologically important pyrrolines in generally good yields (Scheme 6.3(a)) [9]. The reduction potential profile of oximes has been carefully investigated by electrochemical studies to identify the most suitable activating groups. As a result, 2,4-dinitrophenyl-substituted oximes proved to be the most reactive species, which can be efficiently reduced by the excited state of Eosin Y under mild photocatalytic conditions.

Based on the results of mechanistic studies, it is proposed that the N–O bond of O-aryl oximes **3** is initially reduced by the excited photocatalyst Eosin Y* to generate iminyl radicals **3-A** *via* a SET process. Subsequently, a 5-*exo*-radical cyclization of **3-A** occurs to give the key C-centered radicals **3-B**, followed by a H-atom abstraction from CHD to deliver the final products **4** with the generation of C-radical **5** at the same time. Finally, the radical **5** undergoes a SET oxidation by the oxidized photocatalyst to form carbocation **6**

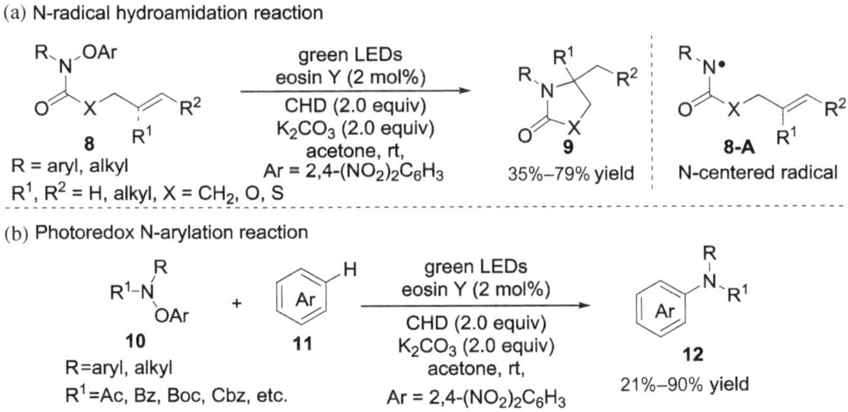

Scheme 6.4. Photocatalytic hydroamidation and N-arylation reactions.

and regenerates the ground state of the photocatalyst Eosin Y, closing the catalytic cycle (Scheme 6.3(b)).

As part of their work on N-radical reactions, in 2016, Leonori and co-workers further achieved a visible light photocatalytic generation of reactive amidyl radicals from various easy-to-make aryloxy-amides under metal-free conditions (Scheme 6.4) [10]. As aforementioned, the protecting group of the N-atom plays a core role in the formation of amidyl radicals, and 2,4-dinitrophenoxy-substituted amides were identified as the best substrates. Importantly, this methodology can be successfully applied in both N-radical hydroamidations and N-arylation reactions. These two transformations display a broad substrate scope and high functional group tolerance, which have been further utilized in the late-stage modification of terpenes, and some ergot derivatives and blockbuster drugs. Recently, by using a similar strategy, Wang *et al.* developed a visible-light-induced amidyl radical cyclization/addition cascade for the efficient construction of pyrroloindolines [11].

In addition, organic azides act as another important class of N-radical precursors with broad applications for the assembly of diverse N-containing compounds [12]. In 2015, Zhou and co-workers reported an interesting visible light photocatalytic N-radical cascade

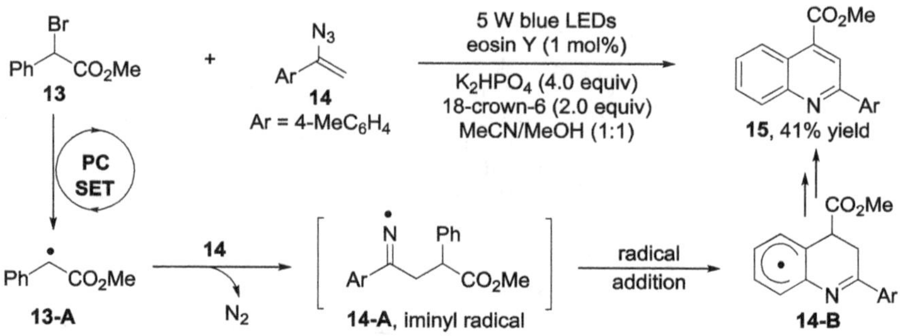

Scheme 6.5. N-radical cascade involving vinyl azides and benzyl bromides.

reaction of vinyl azides (e.g. **14**) with α-carbonyl benzyl bromides, such as **13**, furnishing the corresponding highly functionalized quinolines (**15**) in moderate to good yields (Scheme 6.5) [13]. In this reaction, Eosin Y proved to be a promising photocatalyst, but it was less efficient than $Ir(ppy)_3$ (41% vs 77% yield). The key iminyl radical **14-A** was generated from the radical addition of α-carbonyl benzyl radical **13-A** to vinyl azide **14** with the extrusion of N_2.

6.2.2 *α-functionalization of amines via iminium ions*

Iminium ions represent a class of highly reactive and versatile building blocks, and they have been widely utilized in visible-light-induced C(sp^3)–H bond functionalization and cyclization reactions [14]. In 2010, Stephenson *et al.* firstly developed a visible light photocatalytic *aza*-Henry reaction of *N*-aryl tetrahydroisoquinolines through a facile nucleophilic addition of nitromethane onto iminium ion intermediates, albeit with the use of external oxidants (TBHP, H_2O_2, etc.) [15]. Inspired by this work, in recent years, several reaction modes of amines, such as Mannich reaction, C(sp^3)–H bond cyanation, phosphonylation, and alkynylation have been successfully achieved by using these reactive intermediates under mild photocatalytic conditions (Scheme 6.6) [14].

In this field, Eosin Y proved to be a suitable and efficient photocatalyst for these types of reactions. In 2011, the group of König

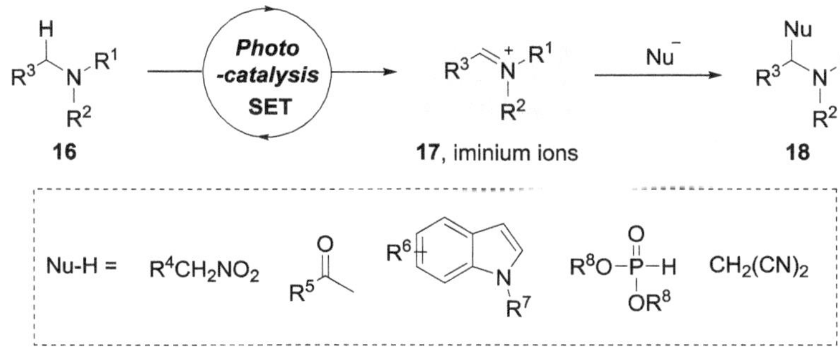

Scheme 6.6. Visible-light-induced C(sp^3)–H bond functionalizations.

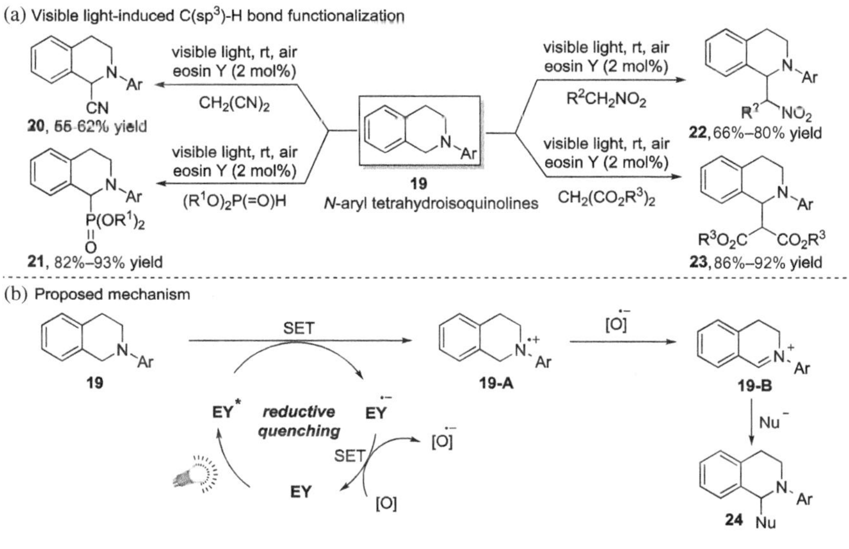

Scheme 6.7. Eosin Y-catalyzed oxidative functionalization of C(sp^3)–H bonds.

reported an elegant metal-free, Eosin Y-catalyzed oxidative coupling reaction of *N*-aryl tetrahydroisoquinolines through visible light-induced C(sp^3)–H bond functionalization. A wide range of nucleophiles, such as nitroalkanes, dialkyl phosphonates, dialkylmalonates, and malononitrile, proved to be suitable for this reaction, delivering the corresponding C–C and C–P bond coupling products in good to excellent yields (Scheme 6.7(a)) [16]. Although the detailed mechanism of this reaction has not been well-investigated at the current stage, a reasonable mechanism was proposed by the author, as shown in Scheme 6.7(b). Firstly, a rapid SET oxidation of *N*-aryl tetrahydroisoquinolines **19** by the excited state of Eosin Y upon irradiation with visible light, furnished the aminyl radical cation intermediates **19-A**. Subsequently, a H-atom abstraction from **19-A** occurred to generate the key iminium ion species **19-B**, which was easily trapped by various nucleophiles to give the corresponding products.

Inspired by König's work, Wu and co-workers developed a dehydrogenative coupling reaction of *N*-aryl tetrahydroisoquinolines and nitromethane enabled by TBA-Eosin Y-mediated photoredox catalysis at room temperature (Scheme 6.8(a)) [17]. To shed insight into the mechanism, a series of spectroscopic studies and product analysis have been conducted, which indicated a SET process between *N*-aryl tetrahydroisoquinoline and the excited state of Eosin Y. More importantly, the results of electron spin resonance (ESR) measurements suggested that the generation of the superoxide radical anion remarkably accelerated this oxidative coupling process. Subsequently, the same group reported a photocatalytic cross-coupling hydrogen evolution (CCHE) reaction by combining Eosin Y and a graphene-supported RuO_2 nanocomposite (G-RuO_2; Scheme 6.8(b)) [18]. Under the standard reaction conditions, the

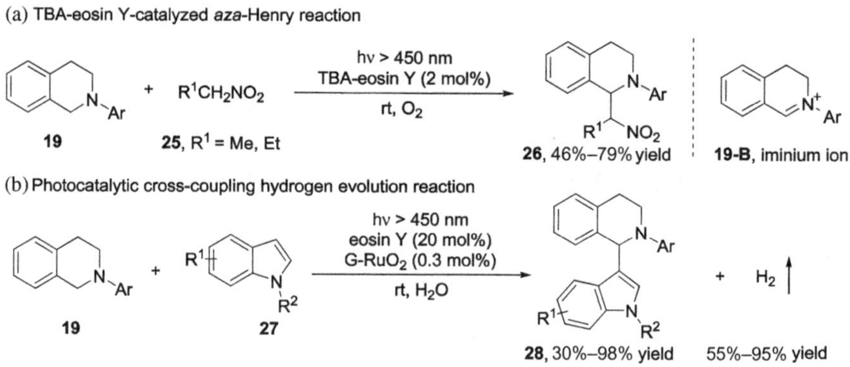

Scheme 6.8. Visible light photocatalytic cross-coupling reactions.

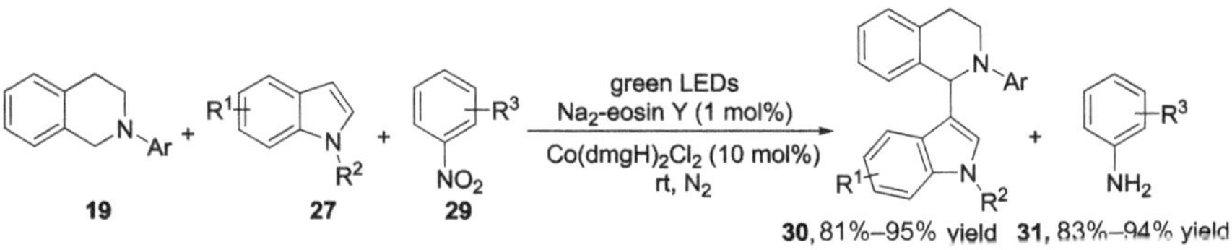

Scheme 6.9. Visible light photocatalytic cross-coupling reactions.

CCHE reaction proceeded very smoothly to give the desired coupling products in moderate to good yields with the concomitant generation of H_2 in quantitative yield. This work represents the first example of visible-light-induced CCHE reaction between two different C–H bonds, providing a clean and atom-economical approach to the formation of C–C bonds.

In 2014, Wu *et al.* developed an interesting visible-light-induced cross-coupling/*in situ* hydrogenation cascade reaction by rational combination of Na_2-Eosin Y (1 mol%) and the cobalt-complex $Co(dmgH)_2Cl_2$ (10 mol%; Scheme 6.9) [19]. In this work, Na_2-Eosin Y was used as a photoorganocatalyst for the SET activation of *N*-aryl tetrahydroisoquinolines **19** to generate the reactive iminium ion species **19-B**, while $Co(dmgH)_2Cl_2$ was employed as an efficient catalyst to trap the proton and electron species for the *in situ* reduction of nitrobenzene *via* a reactive Co(III)–H intermediate. Under the optimized conditions, both the desired cross-coupling products and hydrogenation products could be obtained in generally good yields.

Besides the cyanation, phosphonylation, alkynylation, and arylation of $C(sp^3)$–H bonds, in 2012, Xiao and co-workers reported one example of photocatalytic $C(sp^3)$–H bond amination reaction enabled by Na_2-Eosin Y (0.5 mol%) with the use of air as green oxidant (Scheme 6.10(a)) [20]. The reaction proceeds smoothly, affording the biologically important isoquino[2,1-a]pyrimidine derivative **33** in 85% yield. In addition, Yadav *et al.* demonstrated a successful photocatalytic thiocyanation of $C(sp^3)$–H bonds by employing readily available NH_4SCN as the nucleophile. Various tertiary amines can participate in this reaction, delivering the thiocyanation products **36** in good yields (Scheme 6.10 (b)) [21].

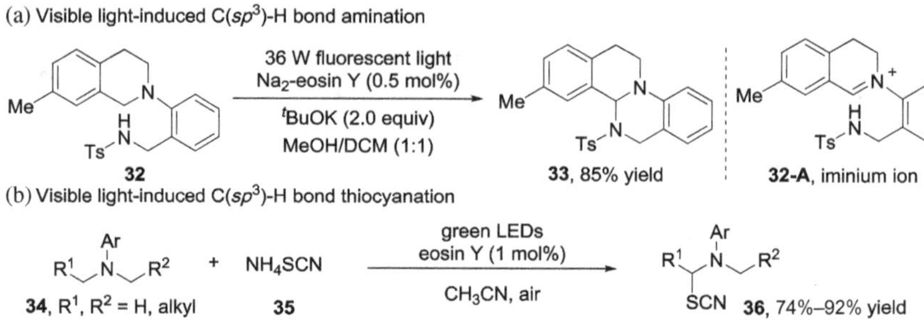

Scheme 6.10. C(sp^3)–H bond amination and thiocyanation reactions.

6.2.3 *α-functionalization of amines via α-amino radicals*

α-amino alkyl radicals have proven to be the reactive radical species that are useful in the construction of various biologically important *N*-heterocycles [14]. However, over the past decades, there has been a limited number of research reports dealing with these intermediates, because they tend to form iminium ions *via* a rapid SET oxidation process. In 2014, Zhou's group described an efficient intramolecular oxidation/cyclization sequence of *O*-alkynylated *N,N*-dialkylamines by visible light photocatalysis (Scheme 6.11) [22]. In this reaction, Eosin Y can be used as an alternative photocatalyst for Ir[(ppy)$_2$(dtbbpy)]PF$_6$, furnishing the desired 3-benzoylindole **39** in 50% yield. The reaction begins with a SET oxidation of tertiary amine **37** to give the key α-amino alkyl radical **37-A** under visible-light-induced photocatalytic conditions. Subsequently, an intramolecular addition of radical **37-A** to the C–C triple bond affords vinyl radical **37-B**, which can be easily trapped by O_2 to give the final product 3-benzoylindole **39** through the intermediate **38**.

In 2015, Zhang and co-workers developed a novel and practical strategy for the efficient construction of tetrahydroquinolines *via* a visible-light-induced, Eosin Y-promoted aerobic oxidation/radical cyclization reaction of *N,N*-dimethylanilines **40** with maleimides **41** (Scheme 6.12) [23]. This metal-free protocol features mild reaction conditions, broad substrate scope, and operational simplicity.

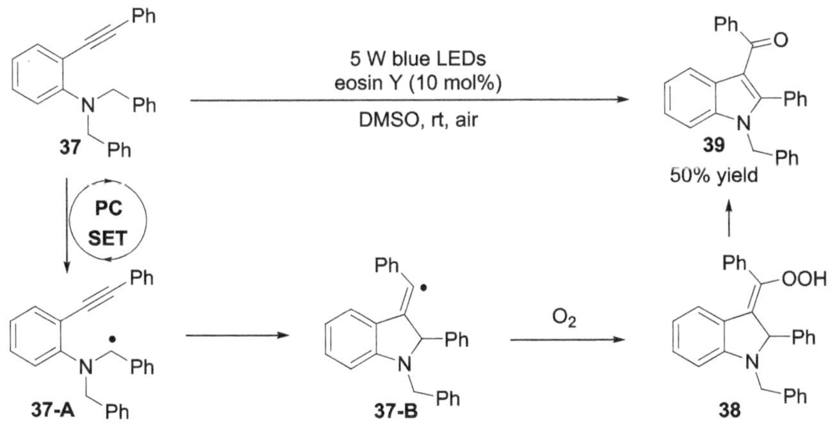

Scheme 6.11. Intramolecular oxidation/cyclization cascade of amines.

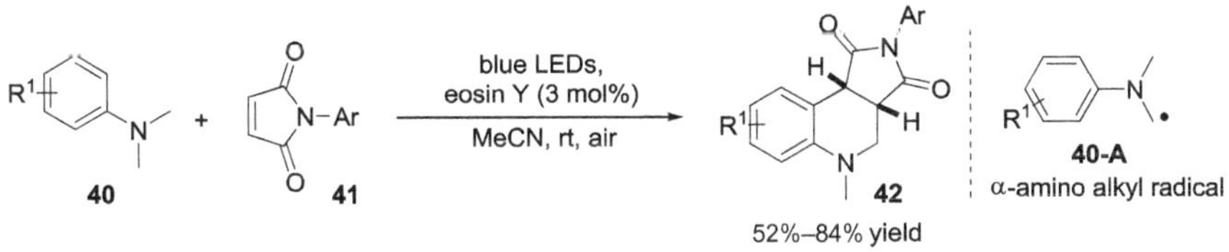

Scheme 6.12. Eosin Y-promoted oxidation/radical cyclization of *N,N*-dimethylanilines with maleimides.

In addition, visible-light-induced SET reduction of imines provides an alternative method to generate α-amino alkyl radicals [24]. In 2016, Dixon's group reported an Eosin Y-catalyzed photocatalytic three-component reductive coupling reaction of aromatic aldehydes **43**, amines **44**, and electron-poor sulfones **45**, under irradiation with green LEDs (Scheme 6.13) [25]. The results of crossover experiments suggested the intermediacy of imines **47**, which derived from the corresponding aromatic aldehydes and amines. In this reaction, the Hantzsch ester was employed as the terminal reductant, which played a critical role in the formation of α-amino alkyl radicals **47-A** *via* a SET reduction of *in situ-* generated imines **47**. Then, a nucleophilic addition of the α-amino radicals **47-A** to the electron-deficient sulfones **45** occurred to furnish the final amine products **46** with release of the sulfonyl radical **48**.

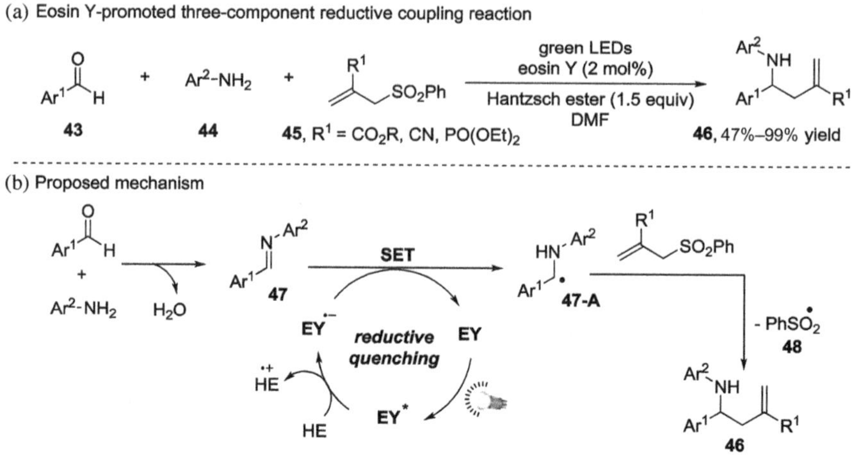

Scheme 6.13. Eosin Y-promoted three-component coupling of aldehydes anilines and sulfones.

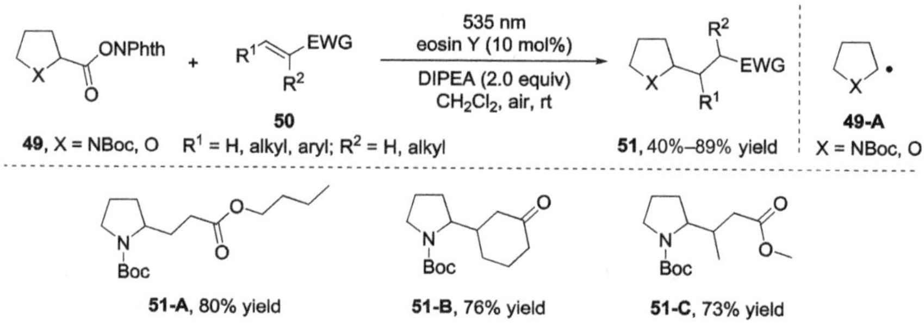

Scheme 6.14. Visible light photocatalytic alkylation reactions.

Visible-light-induced radical decarboxylation has been established as a powerful tool for the generation of alkyl radicals [26]. In 2016, the group of König demonstrated the first example of Eosin Y-mediated photocatalytic radical decarboxylative alkylation reaction of *N*-(acyloxy)phthalimides with electron-deficient alkenes (Scheme 6.14) [27]. As a result, a wide variety of *N*-(acyloxy)phthalimides, derived from readily available fatty acids, amino acids, and α-oxy acids, proved to be suitable for this reaction. It should be noted that the *N*-(acyloxy)phthalimide generated from primary amino acids showed slightly low efficiency due to the high reactivity of the corresponding α-aminoalkyl radical intermediate, which may lead to some by-products *via* radical–radical coupling or H-atom abstraction processes.

6.3 S-Centered Radical Reactions

6.3.1 *Sulfonyl radical-mediated reactions*

The selective and efficient construction of C–S bonds has been established as an important field of research in synthetic chemistry [28]. In recent years, the rapid development of visible-light-induced S-centered radical reactions provided a fruitful platform for the formation of C–S bonds in a controllable manner [29]. In this context, sulfonyl radicals represent a significant class of S-centered radicals that have been widely utilized in the efficient synthesis of various organosulfur compounds. Generally, sulfonyl radicals can be generated from sulfonic acids' derivatives, such as sulfonyl halides, or sodium sulfinates under visible light photocatalytic conditions.

In 2015, Wang and Mao reported a Na_2-Eosin Y-catalyzed, visible-light-photocatalytic-sulfonyl-radical-mediated addition/cyclization cascade occurring on *N*-aryl acrylamides, furnishing biologically important sulfonated oxindoles in generally good yields under mild reaction conditions (Scheme 6.15(a)) [30]. In this reaction, the addition of an oxidant played an important role in the generation of sulfonyl radicals, and *tert*-butyl hydroperoxide (TBHP, 70% aqueous) finally proved to be the best choice. More importantly, this reaction proceeded very well by employing water (H_2O) as the solvent, which further demonstrated the synthetic potential of this methodology.

Based on radical-trapping experiments and related literature reports, a visible-light-induced free-radical mechanism was proposed by the authors, as shown in Scheme 6.15(b). Initially, the formation of *tert*-butyloxy radical **58** was enabled through an oxidative quenching event of the photoexcited Na_2-Eosin Y by TBHP, followed by a H-atom abstraction from arylsulfinic acid **52** to give the key sulfonyl

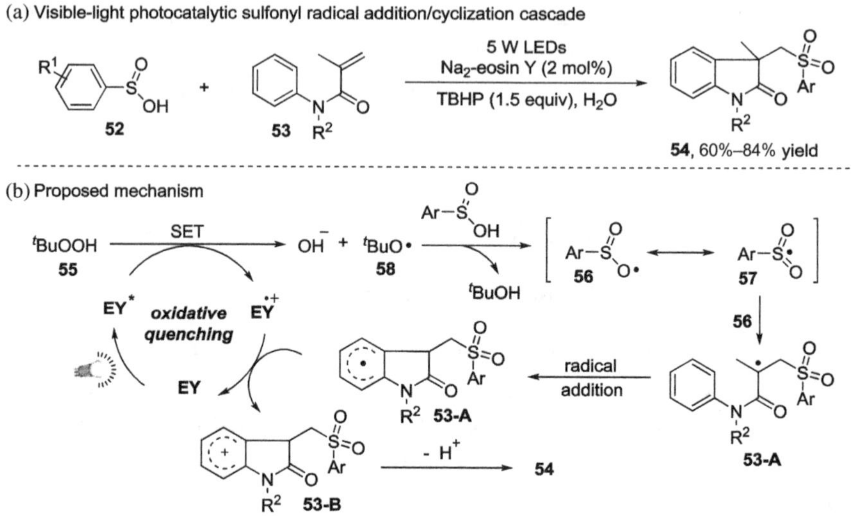

Scheme 6.15. Sulfonyl radical addition/cyclization cascade.

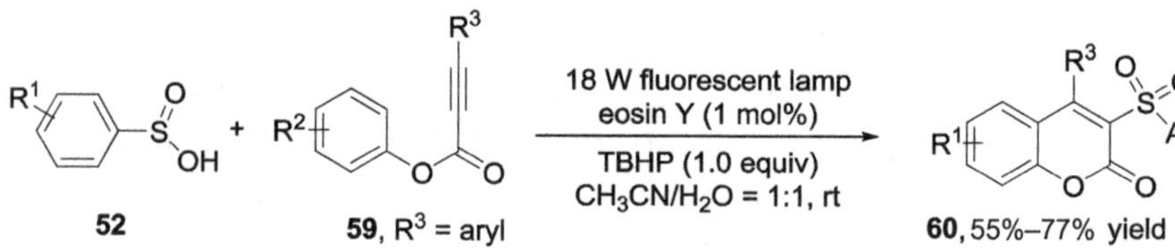

Scheme 6.16. Photocatalytic oxidative cyclization of phenyl propiolates.

radical **57**. Then, the rapid addition of **57** to acrylamide **53** formed a new C-centered radical intermediate **53-A**. Finally, the radical **53-A** underwent a radical cyclization/oxidation/deprotonation sequence to give the desired product **54**.

Shortly thereafter, by using the similar strategy, Wang and co-workers further developed an oxidative radical cyclization of phenyl propiolates with various sulfinic acids photoorganocatalyzed by Eosin Y (Scheme 6.16) [31]. Under the optimal conditions, a wide variety of phenyl propiolates reacted very well with various arysulfinic acids, leading to the formation of coumarin derivatives in moderate to good yields. In addition, Pan and co-workers reported a sulfonyl radical-mediated oxidative/reductive cyclization cascade reaction involving *N*-cyanamide alkenes for the preparation of sulfonated quinazolinones [32].

In 2015, the group of König described a novel, visible light photocatalytic synthesis of vinyl sulfones, such as **63**, from sodium sulfinates **61** and alkenes **62**, enabled by 10 mol% Eosin Y as the photocatalyst and nitrobenzene as the terminal oxidant (Scheme 6.17(a)) [33]. This strategy features metal-free reaction conditions and operational simplicity. A wide range of sodium aryl sulfinates reacted smoothly with styrene derivatives, delivering the corresponding vinyl sulfones **63** in generally good yields (up to 99%). However, α-methylstyrene was not well tolerated in this process and only afforded a 19% yield of the desired product **63-D**.

In 2016, König and co-workers extended the scope of this strategy to different alkyl and heteroaryl sulfinates, leading to the expected sulfones in moderate to good yields [34]. It should be noted that this methodology can be successfully applied in the construction of two important cyclic sulfones through a visible-light-induced radical sulfonylation and an intramolecular [2+2] cycloaddition cascade reaction.

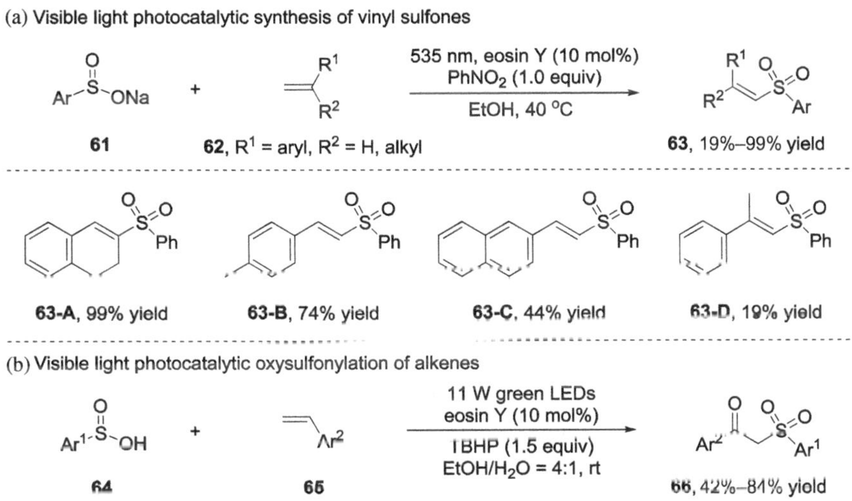

Scheme 6.17. Visible-light photocatalytic (oxy)sulfonylation of alkenes.

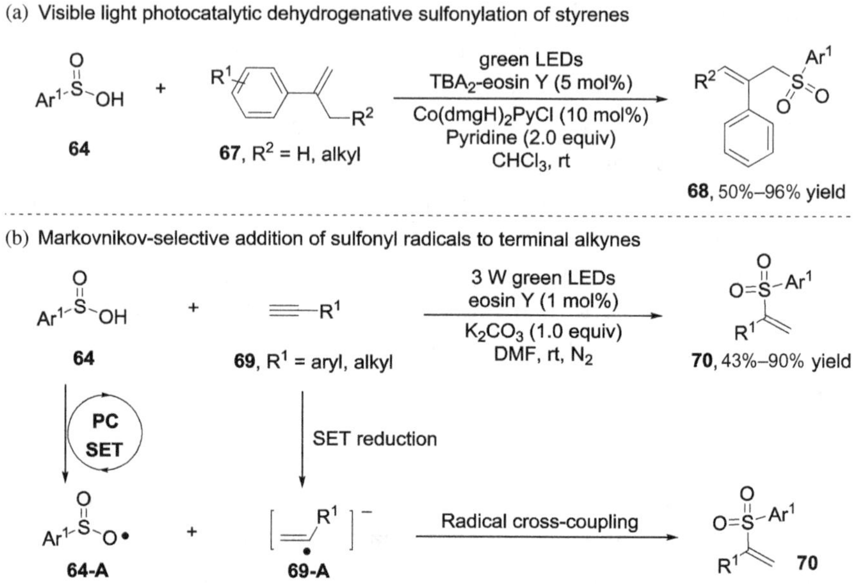

Scheme 6.18. Visible-light-induced sulfonylation of styrenes and alkynes.

In 2016, an unusual Eosin Y-mediated radical oxysulfonylation of alkenes has been achieved by Wang's group with the use of TBHP as an oxidant and EtOH/H_2O (4:1) as the solvent (Scheme 6.17(b)) [35]. The reaction provides an efficient and direct approach to various β-ketosulfones in moderate to excellent yields. Unfortunately, at the current stage, aliphatic alkenes failed to give the expected products.

In addition, Lei and Wu *et al.* developed a dual catalytic system by synergistic combination of TBA_2-Eosin Y-mediated visible light photocatalysis with a concomitant cobalt-catalyzed dehydrogenation process, to further extend the reaction of sulfonyl radicals with alkenes (Scheme 6.18(a)) [36]. In this work, an oxidant-free dehydrogenative sulfonylation of α-methylstyrenes has been achieved for the assembly of allylic sulfones with release of H_2 as the by-product. The radical sulfonylation reaction regioselectively took place at the terminal position of C=C bonds and allylic sulfones were obtained as the major products. The results of radical inhibition experiments, radical-clock experiments, and EPR measurements demonstrated a radical pathway for this transformation. The unusual Markovnikov-selective control of sulfonyl radical reactions provides a new opportunity for the discovery of other valuable transformations, but still remains a challenging goal for scientists due to the *Kharasch effect*. Very recently, the group of Lei developed a novel Markovnikov-selective addition of sulfonyl radicals to terminal alkynes, affording a wide array of vinyl sulfones with good yields and high regioselectivities (Scheme 6.18(b)) [37]. The key step of this reaction is the radical cross-coupling between a sulfonyl radical **64-A** and an α-vinyl carbon radical intermediate **69-A**, which has been demonstrated by a series of mechanistic studies, including EPR experiments and radical trapping experiments.

6.3.2 *Sulfenyl radical mediated reactions*

In recent years, Eosin Y-catalyzed, visible light photocatalytic sulfenyl radical reactions have also been widely employed in the construction of various highly functionalized *S*-containing compounds because of several attractive advantages, including mild reaction conditions and good functional group tolerance. In 2014, Yadav's group reported a sulfenyl radical-mediated difunctionalization of alkenes enabled by Eosin Y-visible-light photocatalysis (Scheme 6.19(a)) [38]. In this reaction, O_2 acted as the terminal oxidant as well as the oxygen source. Under the optimized conditions, a wide range of thiophenols **71** and alkenes **72**, including styrenes and aliphatic derivatives, participated in this reaction very well to give β-keto sulfoxides **73** in generally good yields. However, aliphatic thiols proved to be not suitable, probably due to the difficulty in the formation of sulfenyl radical species under the current catalytic system. As shown in Scheme 6.19(b), the postulated mechanism begins with the generation of a sulfenyl radical intermediate **71-B** *via* a SET oxidation/deprotonation sequence of thiophenol **71**. The subsequent addition of **71-B** to an alkene gives rise to a new C-radical species **74**, which can be trapped by O_2 to give the hydroperoxyl radical **75**. Finally, the radical **75** undergoes a H-abstraction/oxidation process to give the desired product **73**.

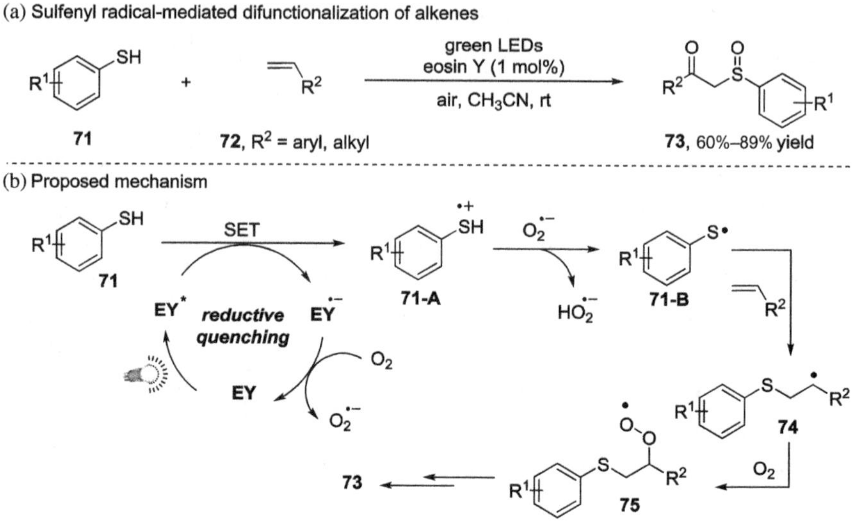

Scheme 6.19. Sulfenyl radical-mediated difunctionalization of alkenes.

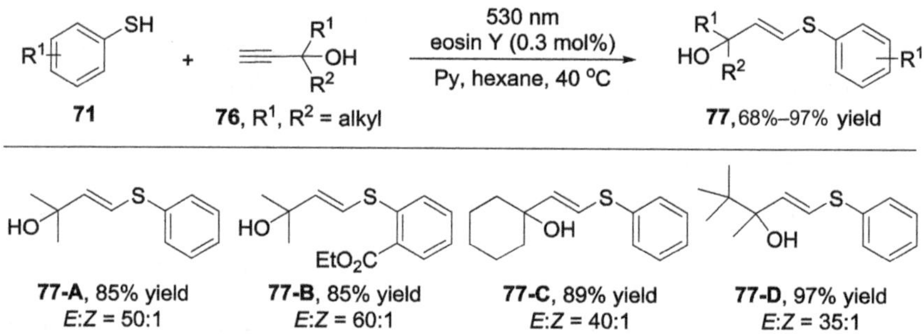

Scheme 6.20. Eosin Y-catalyzed photocatalytic thiol-yne click reaction.

Shortly after, Ananikov and co-workers designed an Eosin Y-promoted thiol-yne click reaction of aryl thiols with alkynes (Scheme 6.20) [39]. It was found that the solvent had a great influence on the yield and selectivity of the process. Hexane was finally identified as the optimal solvent, which largely slowed down the degradation of Eosin Y in this reaction. A wide range of C–S coupling products could be obtained in good yields and with high

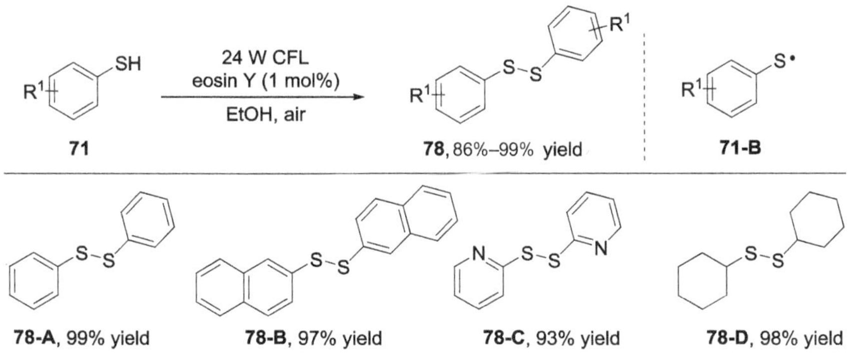

Scheme 6.21. Photocatalytic oxidation of thiols to organic disulfides.

selectivity (up to 96% yield and 60:1). It should be pointed out that low selectivities were observed when employing aliphatic thiols as substrates in this reaction (<10:1).

In 2015, a rapid continuous flow procedure was successfully applied to the visible-light-induced oxidation of thiols to organic disulfides by Noël's group (Scheme 6.21) [40]. A wide variety of disulfides can be obtained in generally good yields with the use of O_2 as green oxidant. More remarkably, the author expanded the scope of this strategy to produce the biologically active derivative oxytocin.

Recently, Hajra and co-workers reported a metal-free, oxidative radical thiocyanation of imidazole heterocycles enabled by Eosin Y-promoted photoredox catalysis (Scheme 6.22(a)) [41]. In this reaction, the stable, easy-to-handle, and inexpensive NH_4SCN was used as a thiocyanating agent and O_2 was the terminal oxidant. Notably, the SCN-containing products can be selectively transformed into the corresponding SCF_3 derivatives by employing $TMSCF_3$ as the trifluoromethylation reagent. In addition, Wang and co-workers developed a novel photocatalytic $C(sp^2)$–H sulfenylation of various imidazole heterocycles with sulfinic acids under metal-free conditions, employing Eosin B as the photoorganocatalyst [42].

On the basis of dedicated mechanistic studies, a pathway involving the visible-light-induced generation of a thiocyanate

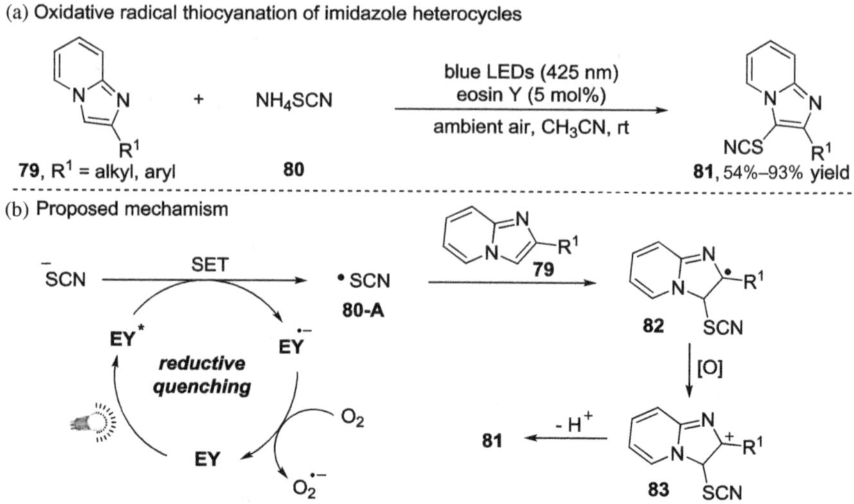

Scheme 6.22. Oxidative radical thiocyanation of imidazole heterocycles.

radical has been proposed (Scheme 6.22(b)). Upon irradiation with blue LEDs, a SET oxidation of the thiocyanate anion by the excited state of Eosin Y initially gives the thiocyanate radical **80-A**. Subsequently, radical **80-A** reacts with imidazole heterocycles **82** to afford C-centered radical intermediates **83**, followed by a SET oxidation and deprotonation process to produce the thiocyanation products **81**.

By using the same thiocyanate radical intermediates, the group of Yadav recently developed a visible light photocatalytic difunctionalization of alkenes for the efficient synthesis of biologically important 5-aryl-2-imino-1,3-oxathiolanes of type **85** (Scheme 6.23(a)) [43]. In most cases, the reaction proceeded well in the presence of 2 mol% Eosin Y as the photoredox catalyst and O_2 as the oxidant, giving the expected products in moderate to good yields (74%–96% yield). In addition, in 2016, Yadav and co-workers developed a direct photocatalytic approach for the synthesis of 2-aminobenzoxazoles through an Eosin Y-promoted cyclization/desulfurization cascade of *o*-phenolic thioureas **86** with the intermediacy of radical **86-A** (Scheme 6.23(b)) [44].

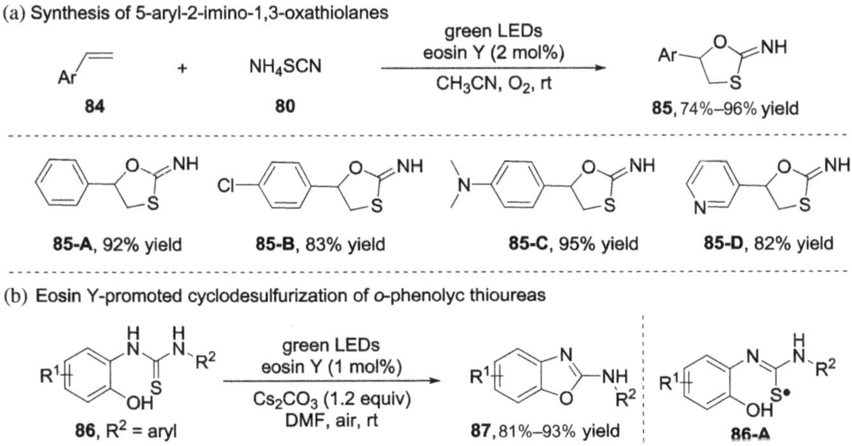

Scheme 6.23. Visible-light-induced synthesis of: (a) 5-aryl-2-imino-1,3-oxathiolanes and (b) 2-aminobenzoxazoles.

6.4 P-Centered Radical Reactions

Owing to the unique physical, biological, and chemical properties of organophosphorus compounds, the development of efficient methods for the incorporation of phosphorus-containing groups into a molecule has attracted much attention from the synthetic organic community [45]. In recent years, visible-light-induced P-centered radical reactions provided a rapid and reliable method for the construction of C–P bonds [46]. In 2016, Lei and co-workers developed a visible-light-photocatalytic C–H phosphonylation of benzothiazole derivatives by using O_2 as the oxidant (Scheme 6.24(a)) [47]. A wide array of diarylphosphine oxides reacted smoothly with benzothiazoles to give the corresponding 2-diarylphosphoryl benzothiazoles in medium to excellent yields under metal-free conditions. However, it was found that diethyl phosphite was not compatible for this reaction. In addition, EPR experiments indicated the intermediacy of phosphonyl radical **89-A** in this reaction system. Almost at the same time, the group of Wu independently demonstrated an Eosin B-promoted C–H phosphorylation of thiazole derivatives *via* visible-light-induced CCHE without the use of any oxidant

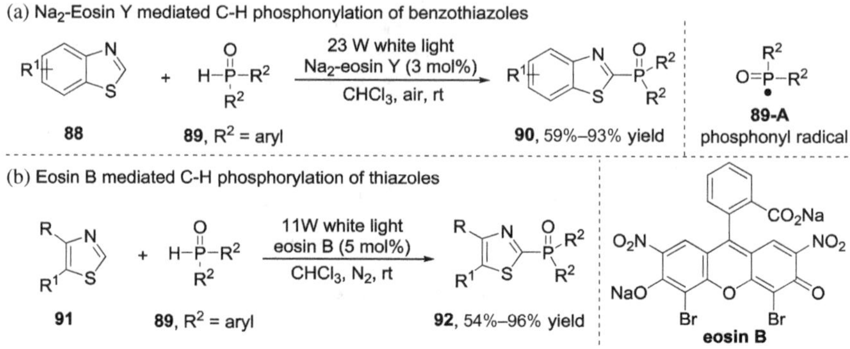

Scheme 6.24. Photocatalytic phosphonylation of thiazole derivatives.

(Scheme 6.24(b)) [48]. Notably, compared to Lei's work, this reaction shows a broader substrate scope with respect to thiazoles. In this work, both benzothiazoles and other simple thiazole derivatives can be well tolerated, furnishing the desired products in generally good yields (54%–96%).

In 2016, Lakhdar *et al.* disclosed an elegant photocatalytic reaction of arylphosphine oxides **95** with alkynes **93** for the synthesis of valuable benzo[*b*]phosphole oxides **96** promoted by Eosin Y (4 mol%) under mild, metal-free conditions (Scheme 6.25(a)) [49]. The oxidant proved to be very important for the success of this reaction, and *N*-ethoxy-2-methylpyridinium tetrafluoroborate was selected as the best oxidation agent. It was postulated that an electron donor–acceptor complex (EDA) **EY-94** was *in situ*-generated from Eosin Y and *N*-ethoxy-2-methylpyridinium salt **94** in this reaction system, which was confirmed by UV–Vis spectroscopy and X-ray diffraction analysis. In addition, EPR experiments suggested the key role of phosphonyl radicals and the ethoxy radical in this transformation. On the basis of these observations, a radical mechanism was proposed by authors as shown in Scheme 6.25(b). The reaction starts with the formation of an EDA complex **EY-94** from Eosin Y and *N*-ethoxy-2-methylpyridinium salt **94**. Upon irradiation with green light (525 nm), the **EY-94** complex undergoes a SET process to generate a reactive ethoxy radical **94-A**. Then, a

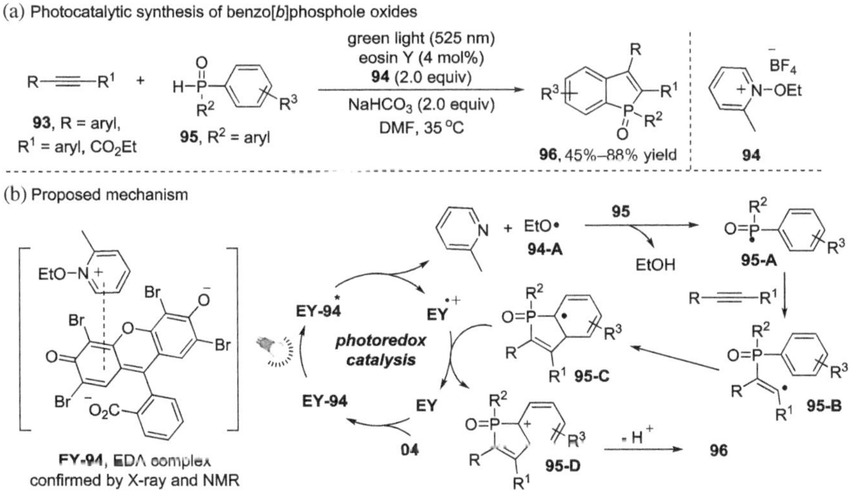

Scheme 6.25. Photocatalytic synthesis of benzo[*b*]phosphole oxides.

H-atom abstraction from arylphosphine oxide **95** by the ethoxy radical **94-A** gives the key phosphonyl radical **95-A**. Subsequently, addition of radical **95-A** to alkyne **93** affords a new alkenyl radical **95-B**, then a cyclization/oxidation/deprotonation sequence occurs to give the expected product **96** and finishes the catalytic cycle.

6.5 C-Centered Radical Reactions

6.5.1 *Aryl radical reactions*

Aryl radicals have been established as a class of well-known and useful radical species and have attracted the attention of synthetic chemists because of their wide applications in the formation of various C–C, C–B, C–N, and C–S bonds [50]. Generally speaking, aryl radicals can be generated from the corresponding aryl halides, aryl diazonium salts, aryl hydrazines, and arylsulfonyl chlorides by transition-metal catalysis. In recent years, Eosin Y-promoted visible-light photocatalysis has been considered as an ideal alternative for classic aryl radical reactions because of the mild, metal-free conditions, high efficiency, and good functional group tolerance [51]. A major

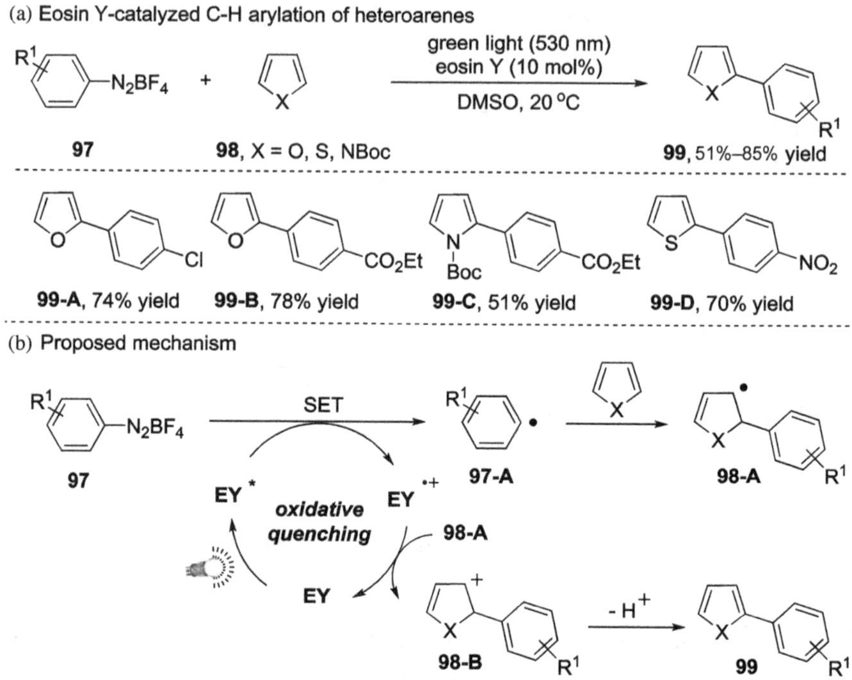

Scheme 6.26. Eosin Y-catalyzed C–H arylation of heteroarenes.

breakthrough in this field was achieved by the group of König in 2012. They reported an elegant Eosin Y-catalyzed C–H arylation of heteroarenes by using aryl diazonium salts as the aryl radical precursors at ambient temperature (Scheme 6.26(a)) [52]. The reaction showed a broad substrate scope with respect to aryl diazonium salts and heteroarenes, and the desired products were obtained in moderate yields (51%–85% yield). Control experiments confirmed that both the photocatalyst and visible light are critical to this transformation.

The plausible mechanism of this C–H arylation reaction is depicted in Scheme 6.26(b). Firstly, a SET reduction of aryl diazonium salt **97** by the photoexcited state of Eosin Y provides aryl radical **97-A**. Then, the addition of **97-A** to the heteroaromatic ring gives rise to a new radical intermediate **98-A**, followed by another SET oxidation to **98-B** and a deprotonation process to give the final product **99**.

Inspired by König's work, in 2015, Ranu and Kundu developed a visible-light-photoredox-catalyzed C–H heteroarylation procedure through an *in situ* diazotization of heteroarylamines under neutral conditions (Scheme 6.27(a)) [53]. The diazotizing agent played an important role in this reaction, and ^{t}BuONO proved to be the best choice, while $NaNO_2$/HCl failed to produce the desired products. This methodology provides a green and direct approach towards the synthesis of heterobiaryl compounds. Additionally, in 2016, König and co-workers reported an Eosin Y-catalyzed photocatalytic arylation of simple arenes **103** with the use of fluorinated aryl bromides **102** as the aryl radical precursors (Scheme 6.27(b)) [54]. Under the standard reaction conditions, a wide variety of polyfluorinated biaryl compounds **104** was obtained with high efficiency (up to 99% yield).

Visible-light photoredox-catalyzed aryl radical reactions have also been widely used in the arylation of alkenes and alkynes. In 2012, the group of König developed a visible-light-induced aryl radical addition/annulation process for the efficient synthesis of various benzothiophenes starting from diazonium salts **105** and alkynes **106** [55]. In this process, both terminal alkynes and dialkyl but-2-ynedioates were well tolerated and, generally, good yields were obtained for the corresponding benzothiophenes **107** (Scheme 6.28(a)). To further demonstrate the synthetic potential of

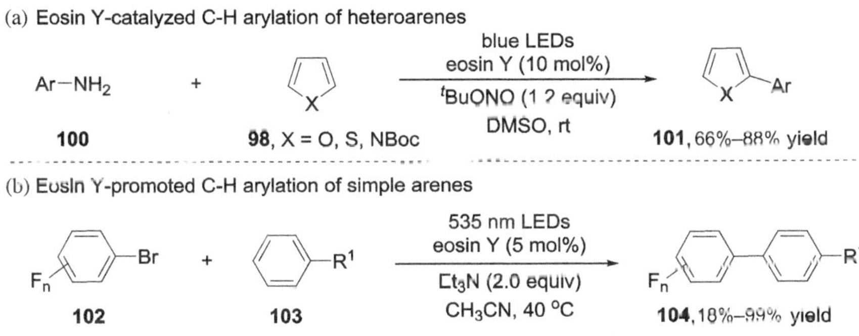

Scheme 6.27. Visible-light-induced C–H arylation of arenes.

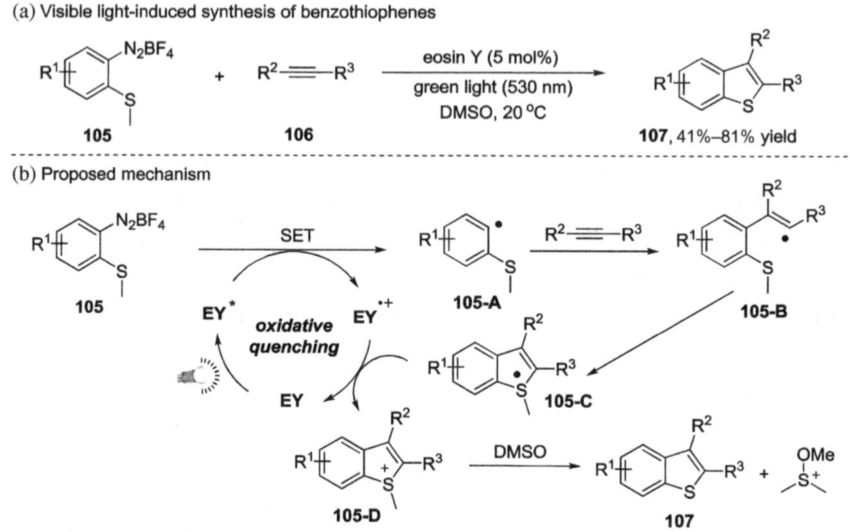

Scheme 6.28. Eosin Y-catalyzed aryl radical cascade reaction for the synthesis of benzothiophenes.

this reaction, the author utilized this protocol for the efficient synthesis of a key intermediate of the biologically active compound Raloxifene.

The plausible mechanism for this reaction is outlined in Scheme 6.28(b). The reaction begins with the generation of aryl radical **105-A** through an oxidative quenching of the excited photocatalyst Eosin Y* by diazonium salt **105**. Then, radical **105-A** reacts with alkyne **106** to produce vinyl radical intermediate **105-B**, followed by a cyclization process to generate sulfuranyl radical **105-C**. A SET oxidation of **105-C** gives the key cation intermediate **105-D**, which transfers a methyl group to DMSO to give the final product **107**.

In the same year, Zhou and co-workers reported an Eosin Y-mediated photocatalytic [4+2] benzannulation of biaryldiazonium salts with different alkynes for the synthesis of diverse substituted phenanthrenes under mild conditions [56]. In addition, an interesting visible-light-induced aryl radical addition/oxidation domino reaction has also been achieved by Wang's group [57]. This transformation provides a straightforward method towards the

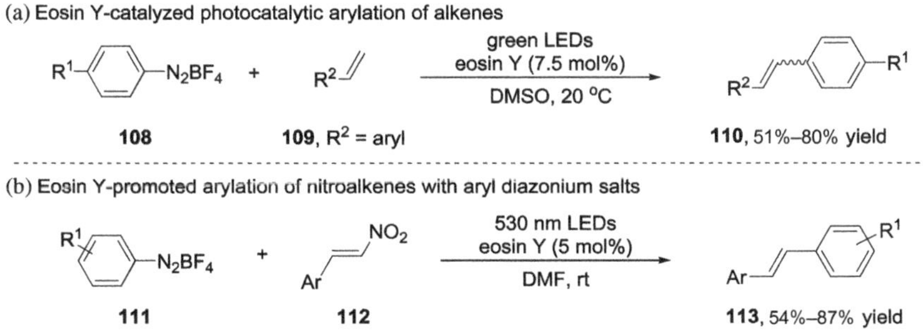

Scheme 6.29. Photocatalytic arylation of alkenes with diazonium salts.

construction of various α-chloro and α-alkoxy aryl ketones from readily available diazonium salts and alkynes.

In 2012, König and co-workers further significantly expanded the reaction scope of aryl radical addition to alkenes enabled by Eosin Y-mediated photoredox catalysis under mild, metal-free conditions (Scheme 6.29(a)) [58]. Recently, Wang and co-workers independently reported an Eosin Y-promoted photocatalytic arylation of nitroalkenes **112** with aryl diazonium salts **111** for the efficient synthesis of diverse stilbene derivatives **113** (Scheme 6.29(b)) [59].

In recent years, aryl radical alkoxycarbonylation reactions have been established as a powerful approach for the synthesis of aromatic esters. In 2015, the group of Xiao developed the first example of visible-light-induced photocatalytic alkoxycarbonylation of aryldiazonium salts **111** with CO (80 atm) catalyzed by 3 mol% fluorescein [60]. Almost at the same time, a similar reaction has been achieved by von Wangelin's group with the use of Eosin Y as the photoorganocatalyst (Scheme 6.30(a)) [61]. In this process, various alkyl benzoates **116** were obtained in good yields from the corresponding aryl diazonium salts **111**, alcohols **114**, and CO (50 bar) at room temperature.

The suggested mechanism for this type of reaction is described in Scheme 6.30(b). First, a SET event occurs between the aryl diazonium salt **111** and the photoexcited **EY*** to give the aryl radical intermediate **111-A** under irradiation with green LEDs. The reactive

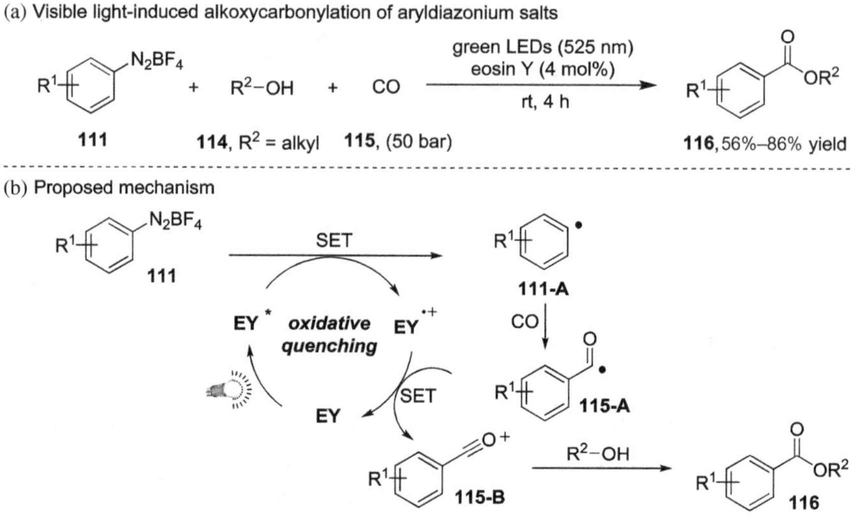

Scheme 6.30. Radical alkoxycarbonylation of aryldiazonium salts.

aryl radical species **111-A** rapidly reacts with CO to generate the key acyl radical **115-A**, followed by another SET oxidation by the radical cation of Eosin Y to give the electrophilic benzylidyneoxonium cation **115-B**, and regenerate the ground-state photocatalyst **EY**. Finally, **115-B** can be easily trapped by alcohols to give the expected aromatic esters **116**.

Inspired by these works, in 2015, a visible-light-induced, base-free radical carbonylation of (hetero)arenes with aryldiazonium salts and CO (70 atm) has been developed by Gu's group (Scheme 6.31(a)) [62]. As an extension of this work, in 2016, Liang and co-workers reported an Eosin Y-mediated photocatalytic carbonylation of indoles **120** with commercially available arylsulfonyl chlorides **119** as aryl radical precursors under CO atmosphere (80 atm; Scheme 6.31(b)) [63]. These works provide direct and efficient approaches to the synthesis of aryl ketones under very mild reaction conditions.

Arylboronates and arylsulfides represent two significant classes of versatile and robust synthetic building blocks and have found broad application in transition metal-catalyzed cross-coupling reactions. The

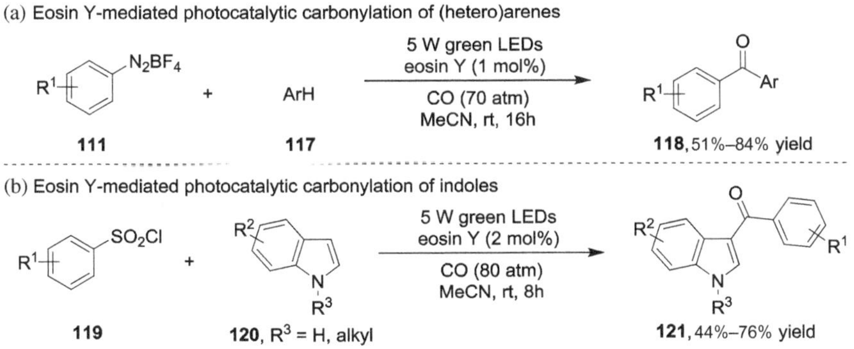

Scheme 6.31. Visible-light-induced carbonylation of arenes and indoles.

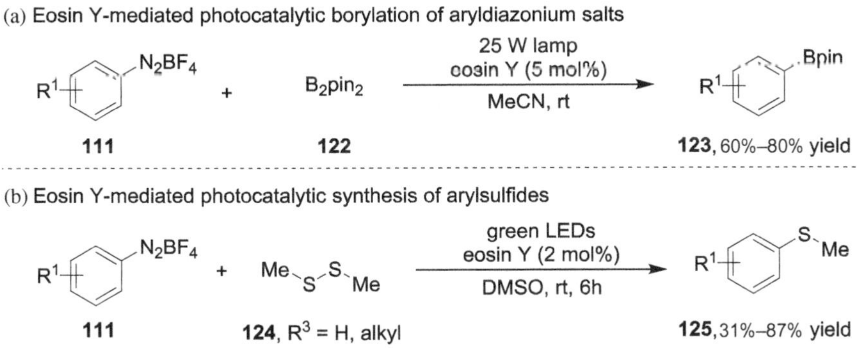

Scheme 6.32. Photocatalytic synthesis of arylboronates and arylsulfides.

development of efficient methods for the synthesis of these important compounds has attracted much attention from synthetic scientists. In 2012, Yan and co-workers documented a metal-free borylation of aryldiazonium salts **111** catalyzed by 5 mol% Eosin Y (Scheme 6.32(a)) [64]. Note that the choice of solvent was important for this reaction, and acetonitrile proved to be the most suitable solvent as it improved the reaction efficiency. Aryldiazonium salts bearing electron-poor or electron-rich substituents reacted well, allowing the facile synthesis of arylboronates in moderate to good yields (60%–80%).

Shortly after, a novel visible-light-induced thiolation reaction of aryldiazonium salts has been achieved by von Wangelin *et al.* in the presence of 2 mol% Eosin Y as the photocatalyst and dimethyl-disulfide (dmds; Scheme 6.32(b)) [65]. Recently, a photocatalytic batch and flow reaction has been developed by the group of Noël. Importantly, this protocol can be successfully applied to the synthesis of various arylated cysteine derivatives [66]. Additionally, Jiang and co-workers reported an Eosin Y-catalyzed sulfoxidation and sulfenylation with thiosulfate salts. Mechanistic studies indicated that this reaction occurs through tandem electron/energy-transfer and SET pathways [67].

6.5.2 *Functionalization of alkenes and alkynes*

Alkenes and alkynes are two classes of easily available feedstock that are amenable to a wide range of functionalization, given the presence of their unsaturated bonds. Not surprisingly, over the past decades, great efforts have been directed towards the development of mild, efficient, and atom-economic methods to synthesize diversely functionalized molecules using these compounds as starting materials. Visible-light photocatalytic C-radical reactions open a new way for the functionalization of alkenes and alkynes because of the mild conditions and high reaction efficiency [68].

In 2015, Qing and co-workers developed an Eosin Y-catalyzed hydrobromodifluoromethylation of alkenes using CF_2Br_2 (6.0 equiv) in the presence of $KHCO_3$ (1.0 equiv; Scheme 6.33(a)) [69]. Under visible-light photocatalytic conditions, a wide range of mono and disubstituted alkenes **126** can participate well in this transformation, delivering the desired products **128** in good yields (41%–90%). Notably, this transformation can be further applied in the late-stage hydrobromodifluoromethylation of other biologically active compounds, such as Vinclozolin and Rotenone.

Very recently, the group of Bolm developed a direct radical addition/elimination reaction of terminal olefins with perfluoroalkyl halides. Generally, in this reaction, a wide variety of perfluoroalkylated alkenes can be obtained in moderate to good yields (Scheme 6.33(b)) [70]. Xu and co-workers further demonstrated an Eosin Y-catalyzed decarboxylative fluoromethylation of cinnamic acids under mild photocatalytic conditions [71].

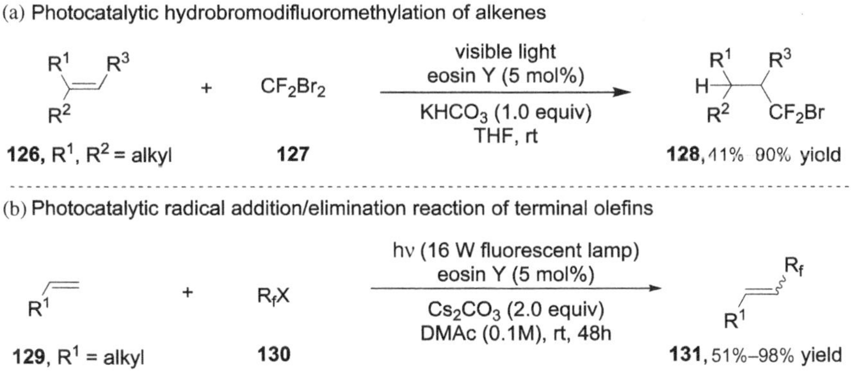

Scheme 6.33. Visible-light photocatalytic fluoroalkylation of alkenes.

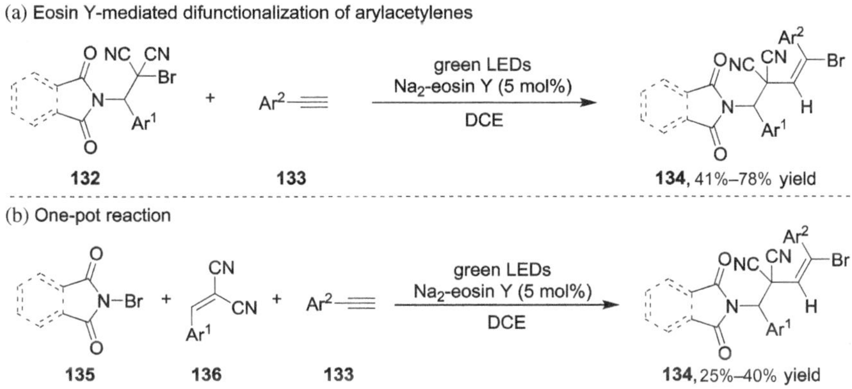

Scheme 6.34. Visible-light-induced difunctionalization of arylacetylenes.

In 2016, a direct and highly regioselective photocatalytic difunctionalization of arylacetylenes was developed by Wang *et al.* (Scheme 6.34(a)) [72]. In this reaction, substrates **132** were used as C-radical sources to react with a variety of alkynes **133**, furnishing the corresponding (*E*)-alkenyl bromides **134** in generally moderate to good yields (41%–78% yield). A simplified procedure, based on a one-pot reaction involving a dicyanoalkene and a *N*-bromo imide instead of substrate **132**, was likewise developed, albeit relatively lower yields (25%–40%) were obtained (Scheme 6.34(b)).

6.5.3 α-functionalization of aldehydes and ketones

Visible-light-induced one-electron activation has recently been applied in the α-functionalization of aldehydes and ketones, since the seminal works from MacMillan [73a, 73b]. In 2011, the group of Zeitler developed a dual synergistic catalytic system by combination of Eosin Y-mediated visible-light photocatalysis with an imidazolidinone-based organocatalyst to achieve the enantioselective α-alkylation of aliphatic aldehydes (Scheme 6.35) [74]. Compared with the $Ru(bpy)_3Cl_2$-catalyzed reaction from MacMillan [73a], the Eosin Y-promoted reaction requires somewhat longer reaction times. Generally, the desired alkylation products can be obtained in good yields with high enantioselectivities at −5°C. Importantly, this protocol can also be extended to the asymmetric α-perfluoroalkylation of aldehydes.

Similar to the mechanism proposed by MacMillan, a SET-based mechanism has been proposed (Scheme 6.36). Initially, a SET reduction of Eosin Y* by a catalytic amount of enamine intermediate **137-B** occurs to give the radical anion of Eosin Y, which can reduce the alkyl halide **138** to generate the alkyl radical species **138-A**. Then, the rapid addition of radical **138-A** to enamine **137-B** affords a new C-radical **137-C**, followed by a SET oxidation/hydrolysis sequence to produce the desired product **140**. In addition, Kappe and co-workers demonstrated an efficient continuous-flow

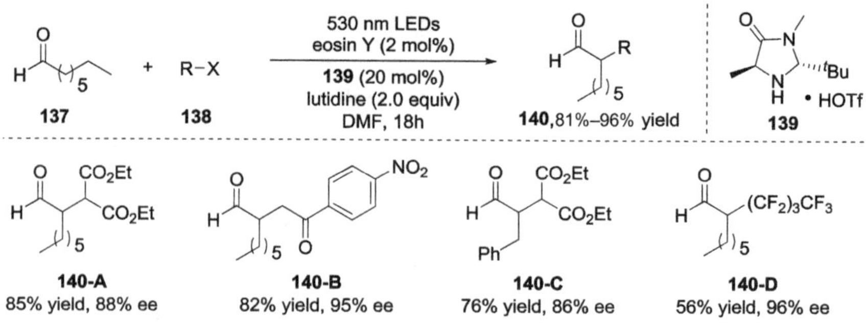

Scheme 6.35. Catalytic enantioselective α-alkylation of aldehydes.

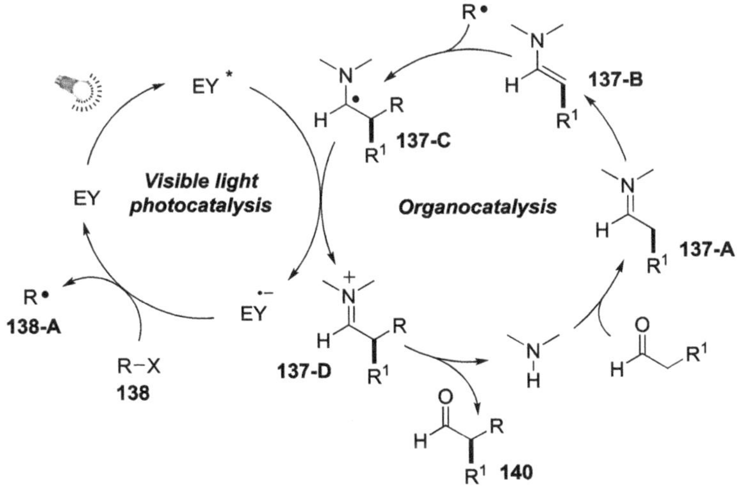

Scheme 6.36. Proposed mechanism for the α-alkylation of aldehydes.

procedure for the α-trifluoromethylation of ketones using CF_3SO_2Cl as the CF_3 radical source and Eosin Y as the photoredox catalyst [75].

In 2016, Yasuda and co-workers developed an Eosin Y-catalyzed allylation reaction of α-bromocarbonyl compounds **142** with potassium allyltrifluoroborates **141** accelerated by Bu_4NF or CsF (Scheme 6.37(a)) [76]. Control experiments indicated that the *in situ*-generated KF can significantly improve the yields of this reaction. The authors pointed out that KF may act as Lewis acid to accelerate the SET process between α-bromocarbonyls and the photoexcited Eosin Y*. By using this methodology, a broad array of α-allylated carbonyls **143** could be obtained in moderate yields, but with excellent selectivities.

The suggested mechanism for this reaction is outlined in Scheme 6.37(b). Firstly, the cation exchange between potassium allyltrifluoroborate **141** and Bu_4NF generates KF and the corresponding borate salt **141-A**. Then, KF acts as Lewis acid to activate α-halocarbonyl compound **142** to produce complex **142-A**, which can be reduced by the photoexcited state of Eosin Y to give rise to

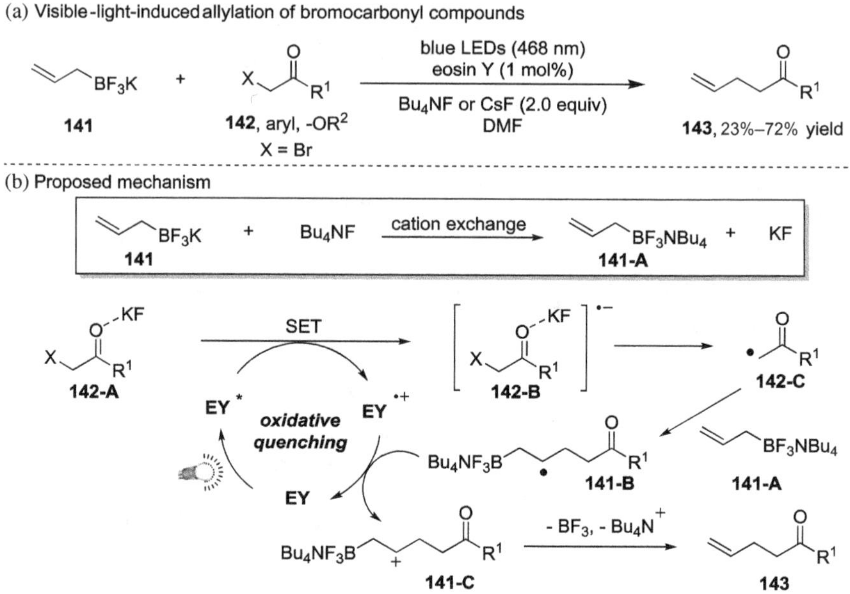

Scheme 6.37. Photocatalytic allylation of α-bromocarbonyl compounds.

the C-radical **142-C**. Then, **142-C** reacts with borate salt **141-A**, delivering a new radical intermediate **141-B**. The SET oxidation of radical **141-B** affords the cationic intermediate **141-C**, and then releases Bu_4N^+ and BF_3 to give the desired allylation product **143**. As an extension of this work, the same group reported a visible-light photocatalytic radical addition of α-bromocarbonyl compounds to silyl enol ethers, furnishing a variety of 1,4-dicarbonyls in moderate to good yields [77].

6.6 Oxidation and Reduction Reactions

In recent years, Eosin Y and its derivatives have also been used in many oxidation and reduction reactions. In 2012, Xiao and co-workers reported one example of Eosin Y-catalyzed photocatalytic oxidative hydroxylation of 4-methoxyphenylboronic acid **144** to 4-methoxyphenol **145** by using O_2 of air as the green oxidant (Scheme 6.38(a)) [78]. It was found that the use of Na_2-Eosin Y

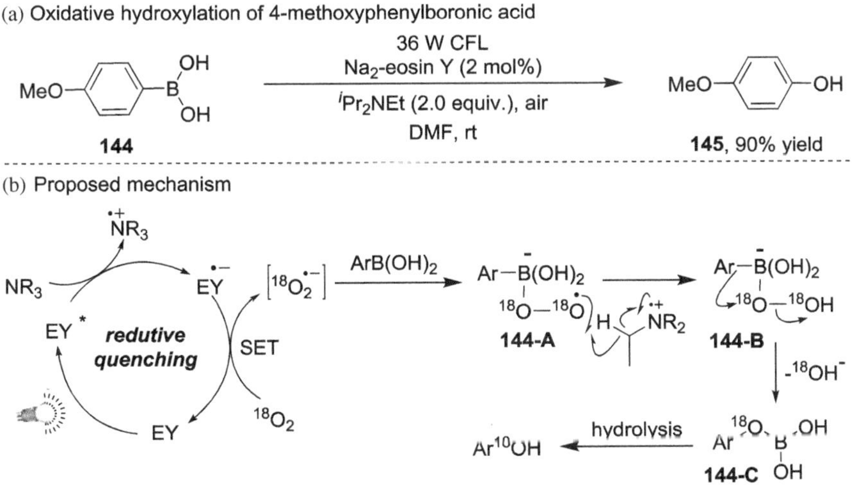

Scheme 6.38. Photocatalytic oxidation of arylboronic acids.

(2 mol%) as the photocatalyst and $^{i}Pr_2NEt$ (2.0 equiv) as the sacrificial reducing quencher under visible-light irradiation enabled a broad substrate scope, high functional group tolerance, and reaction efficiency (90% yield). Based on the results of calculation studies and ^{18}O-labeling experiments, the authors confirmed the two important intermediates **144-B** and **144-C** (Scheme 6.38(b)).

Inspired by this work, by employing $PhI(OAc)_2$ as the terminal oxidant, a similar transformation has been developed by the group of Yadav in 2015 (Scheme 6.39(a)) [79]. As part of their continuing work on visible-light-induced oxidation reactions, Yadav *et al.* reported an Eosin Y-catalyzed aerobic oxidation of styrenes **148** to aldehydes **149** at ambient temperature (Scheme 6.39(b)) [80]. Both electron-rich and electron-poor styrenes reacted smoothly in this reaction, delivering the corresponding aldehydes **149** in good yields. In addition, Yadav *et al.* have also achieved the direct transformation of primary amides and aldoximes into nitriles under mild conditions enabled by Eosin Y-mediated photoredox catalysis [81].

In 2016, Shah *et al.* described an Eosin Y-mediated oxidation of various benzylic alcohols **150** to the corresponding carbonyls **151** in the presence of TBHP (3.0 equiv; Scheme 6.40(a)) [82]. More

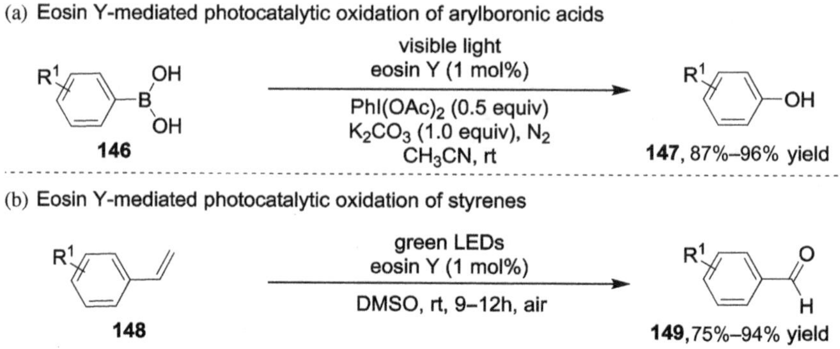

Scheme 6.39. Oxidative hydroxylation of arylboronic acids.

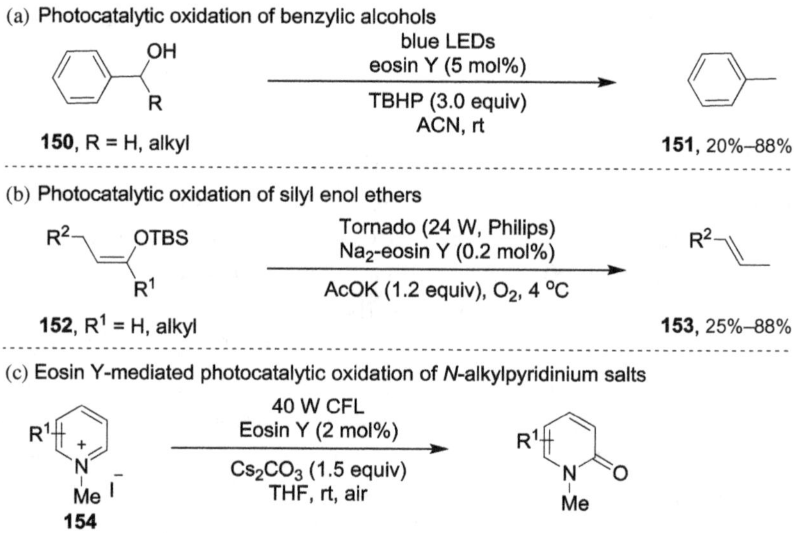

Scheme 6.40. Photocatalytic oxidation of alcohols, silyl enol ethers, and *N*-alkylpyridinium salts.

importantly, this strategy can be used for the chemoselective oxidation of alcohols of different types. Additionally, a visible-light-induced oxidation of silyl enol ethers to α,β-unsaturated aldehydes and ketones has been reported by Huang *et al.* by using Na_2-Eosin Y (5 mol%) as the photocatalyst and O_2 (1 atm) as the green oxidant (Scheme 6.40(b)) [83]. Through mechanistic studies, a single singlet oxygen $^1O_2$ process was proposed to elucidate the mechanism of this reaction. Recently, Fu and co-workers reported an elegant Eosin Y-catalyzed aerobic oxidation of *N*-alkylpyridinium salts under mild conditions (Scheme 6.40(c)) [84].

Since the pioneering work of Kellogg *et al.* on the photocatalytic reduction of sulfonium salts in 1978 [85], visible-light-induced reduction reactions have been applied in organic synthesis. In 2014, Wu *et al.* utilized triethanolamine (TEOA) as the reducing agent for the photocatalytic chemoselective reduction of nitrobenzenes to the corresponding anilines enabled by 1 mol% Eosin Y under green light irradiation (Scheme 6.41(a)) [86]. It was found that the pH value of the reaction mixture and the amount of TEOA are both crucial for the reaction efficiency. The mild reaction conditions tolerated various functional groups such as nitriles, bromine substituents, and carbonyls. Furthermore, Wu and co-workers have recently achieved the selective hydrogenation of alkynes to alkenes under visible-light photocatalytic conditions [87]. Recently, the group of von Wangelin developed an interesting Eosin B-mediated photocatalytic hydrodefunctionalization of aryldiazonium salts in the

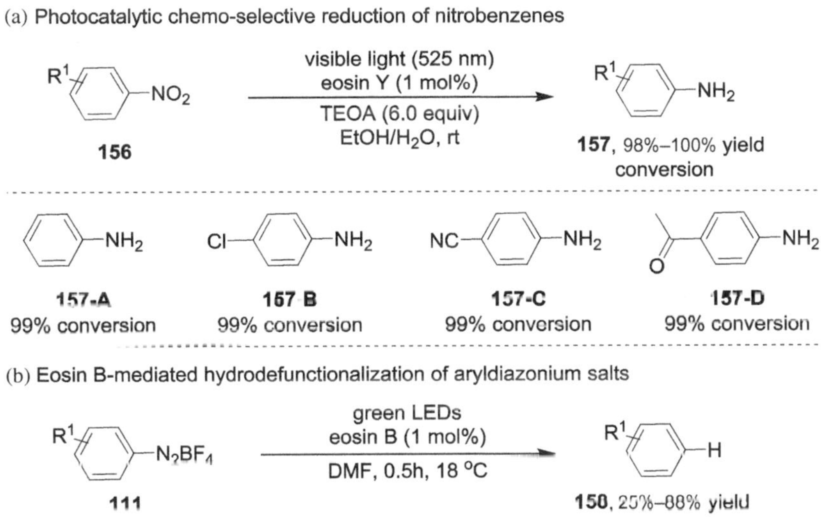

Scheme 6.41. Photocatalytic reduction of nitroaromatics and aryldiazonium salts.

presence of DMF (Scheme 6.41(b)) [88]. Notably, by employing $[D_7]$-DMF as the solvent, the corresponding deuterobenzenes can be successfully obtained in good yields with excellent isotopic purity.

6.7 Conclusions

Recently, organic dye Eosin Y (and its derivatives)-photoorganocatalyzed visible-light reactions have emerged as a cutting-edge research topic in organic chemistry, providing attractive alternatives to ruthenium- and iridium–polypyridyl complexes due to the competitive advantages in terms of low cost, wide availability, as well as biological tolerance in some cases. In this context, Eosin Y-mediated SET reactions and energy transfer processes have been applied in the generation of various reactive radical species (N-radical, S-radical, P-radical, C-radical, etc.) for the efficient formation of numerous C–C, C–B, C–N, C–P, and C–S bonds under very mild, metal-free conditions.

Despite these advances, however, some challenges in Eosin Y-catalyzed photocatalytic reactions still remain, and these are as follows: (1) the relatively low efficiency in large-scale reactions; (2) the asymmetric variants of different radical reactions are still lacking. We believe that the development of a more efficient catalytic system such as a dual catalytic system by combination of Eosin Y-mediated photoredox catalysis with other catalytic strategies (organocatalysis or transition-metal catalysis) would provide potential solutions to those challenging problems. It is hoped that, in the future, there will be more efforts from the synthetic community toward addressing these challenges.

Acknowledgments

We sincerely thank our collaborators and co-workers for their great contributions to our own work described. We are grateful to the National Science Foundation of China (Nos. 21472058, 21472057, 21622201, 21232003, and 21772053), CCNU (CCNU16JCZX02 and CCNU17TS0011), the Science Foundation of Hubei Province (2017AH047), and the Program of Introducing Talents of Discipline to Universities of China (111 Program, B17019).

Abbreviations

SET	single electron transfer
ODET	oxidative deprotonation electron transfer
CHD	1,4-cyclohexadiene
ESR	electron spin resonance
CCHE	cross-coupling hydrogen evolution
ppy	2-phenylpyridine
bpy	2,2′-bipyridine
THBP	*tert*-butyl hydroperoxide
EDA	electron donor-acceptor complex
dmds	dimethyldisulfide
TEOA	triethanolamine

References

[1] (a) Prier, C. K.; Rankic, D. A.; Macmillan, D. W. *Chem. Rev.* **2013**, *113*, 5322–5363; (b) Skubi, K. L.; Blum, T. R.; Yoon, T. P. *Chem. Rev.* **2016**, *116*, 10035–10074; (c) Romero, N. A.; Nicewicz, D. A. *Chem. Rev.* **2016**, *116*, 10075–10166; (d) Karkas, M. D.; Porco, J. A., Jr.; Stephenson, C. R. *Chem. Rev.* **2016**, *116*, 9683–9747; (e) Chen, J.-R.; Hu, X.-Q.; Lu, L.-Q.; Xiao, W.-J. *Acc. Chem. Res.* **2016**, *49*, 1911–1923; (f) Ravelli, D.; Protti, S.; Fagnoni, M. *Chem. Rev.* **2016**, *116*, 9850–9913.

[2] (a) Tucker, J. W.; Stephenson, C. R. *J. Org. Chem.* **2012**, *77*, 1617–1622; (b) Reckenthäler, M.; Griesbeck, A. G. *Adv. Synth. Catal.* **2013**, *355*, 2727–2744.

[3] Teplý, F. *Collect. Czech. Chem. Commun.* **2011**, *76*, 859–917.

[4] (a) Ravelli, D.; Fagnoni, M. *ChemCatChem* **2012**, *4*, 169–171; (b) Ravelli, D.; Fagnoni, M.; Albini, A. *Chem. Soc. Rev.* **2013**, *42*, 97–113; (c) Fukuzumi, S.; Ohkubo, K. *Org. Biomol. Chem.* **2014**, *12*, 6059–6071; (d) Nicewicz, D.; Hamilton, D. *Synlett* **2014**, *25*, 1191–1196; (e) Nicewicz, D. A.; Nguyen, T. M. *ACS Catal.* **2013**, *4*, 355–360; (f) Majek, M.; von Wangelin, A. J. *Acc. Chem. Res.* **2016**, *49*, 2316–2327; (g) Yan, G.; Huang, D.; Wu, X. *Adv. Synth. Catal.* **2018**, *360*, 1040–1053.

[5] (a) Hari, D. P.; Konig, B. *Chem. Commun.* **2014**, *50*, 6688–6699; (b) Srivastava, V.; Singh, P. P. *RSC Adv.* **2017**, *7*, 31377–31392.

[6] (a) Mailyan, A. K.; Eickhoff, J. A.; Minakova, A. S.; Gu, Z.; Lu, P.; Zakarian, A. *Chem. Rev.* **2016**, *116*, 4441–4557; (b) Blair, L. M.; Sperry, J. *J. Nat. Prod.* **2013**, *76*, 794–812; (c) Hili, R.; Yudin, A. K. *Nat. Chem. Biol.* **2006**, *2*, 284–287.

[7] (a) Zard, S. Z. *Chem. Soc. Rev.* **2008**, *37*, 1603–1618; (b) Xiong, T.; Zhang, Q. *Chem. Soc. Rev.* **2016**, *45*, 3069–3087; (c) Chen, J.-R.; Hu, X.-Q.; Lu, L.-Q.; Xiao, W.-J. *Chem. Soc. Rev.* **2016**, *45*, 2044–2056; (d) Kärkäs, M. D. *ACS Catal.* **2017**, *7*, 4999–5022.

[8] (a) Hu, X.-Q.; Chen, J.-R.; Wei, Q.; Liu, F.-L.; Deng, Q.-H.; Beauchemin, A. M.; Xiao, W.-J. *Angew. Chem. Int. Ed.* **2014**, *53*, 12163–12167; (b) Hu, X.-Q.; Qi, X.; Chen, J.-R.; Zhao, Q.-Q.; Wei, Q.; Lan, Y.; Xiao, W.-J. *Nat. Commun.* **2016**, *7*, 11188; (c) Hu, X.-Q.; Chen, J.; Chen, J.-R.; Yan, D.-M.; Xiao, W.-J. *Chem. Eur. J.* **2016**, *22*, 14141–14146.

[9] Davies, J.; Booth, S. G.; Essafi, S.; Dryfe, R. A.; Leonori, D. *Angew. Chem. Int. Ed.* **2015**, *54*, 14017–14021.

[10] Davies, J.; Svejstrup, T. D.; Reina, D. F.; Sheikh, N. S.; Leonori, D. *J. Am. Chem. Soc.* **2016**, *138*, 8092–8095.

[11] Minozzi, M.; Nanni, D.; Spagnolo, P. Wu, K.; Du, Y.; Wang, T. *Org. Lett.* **2017**, *19*, 5669–5672.

[12] Minozzi, M.; Nanni, D.; Spagnolo, P. *Chem. Eur. J.* **2009**, *15*, 7830–7840.

[13] Wang, Q.; Huang, J.; Zhou, L. *Adv. Synth. Catal.* **2015**, *357*, 2479–2484.

[14] Xie, J.; Jin, H.; Xu, P.; Zhu, C. *Tetrahedron Lett.* **2014**, *55*, 36–48.

[15] Condie, A. G.; Gonzalez-Gomez, J. C.; Stephenson, C. R. *J. Am. Chem. Soc.* **2010**, *132*, 1464–1465.

[16] Hari, D. P.; König, B. *Org. Lett.* **2011**, *13*, 3852–3855.

[17] Liu, Q.; Li, Y.-N.; Zhang, H.-H.; Chen, B.; Tung, C.-H.; Wu, L.-Z. *Chem. Eur. J.* **2012**, *18*, 620–627.

[18] Meng, Q.-Y.; Zhong, J.-J.; Liu, Q.; Gao, X.-W.; Zhang, H.-H.; Lei, T.; Li, Z.-J.; Feng, K.; Chen, B.; Tung, C.-H.; Wu, L.-Z. *J. Am. Chem. Soc.* **2013**, *135*, 19052–19055.

[19] Zhong, J.-J.; Wu, C.-J.; Meng, Q.-Y.; Gao, X.-W.; Lei, T.; Tung, C.-H.; Wu, L.-Z. *Adv. Synth. Catal.* **2014**, *356*, 2846–2852.

[20] Xuan, J.; Feng, Z.-J.; Duan, S.-W.; Xiao, W.-J. *RSC Adv.* **2012**, *2*, 4065–4068.

[21] Yadav, A. K.; Yadav, L. D. S. *Tetrahedron Lett.* **2015**, *56*, 6696–6699.

[22] Zhang, P.; Xiao, T.; Xiong, S.; Dong, X.; Zhou, L. *Org. Lett.* **2014**, *16*, 3264–3267.

[23] Liang, Z.; Xu, S.; Tian, W.; Zhang, R. *Beilstein J. Org. Chem.* **2015**, *11*, 425–430.

[24] (a) Fava, E.; Millet, A.; Nakajima, M.; Loescher, S.; Rueping, M. *Angew. Chem. Int. Ed.* **2016**, *55*, 6776–6779; (b) Uraguchi, D.; Kinoshita, N.; Kizu, T.; Ooi, T. *J. Am. Chem. Soc.* **2015**, *137*, 13768–13771.

[25] de Arriba, A. L. F.; Urbitsch, F.; Dixon, D. J. *Chem. Commun.* **2016**, *52*, 14434–14437.

[26] (a) Xuan, J.; Zhang, Z.-G.; Xiao, W.-J. *Angew. Chem. Int. Ed.* **2015**, *54*, 15632–15641; (b) Huang, H.; Jia, K.; Chen, Y. *ACS Catal.* **2016**, *6*, 4983–4988; (c) Jin, Y.; Fu, H.; *Asian J. Org. Chem.* **2017**, *6*, 368–385.

[27] Schwarz, J.; König, B. *Green. Chem.* **2016**, *18*, 4743–4749.

[28] (a) Matsumoto, K.; Sugiyama, H. *Acc. Chem. Res.* **2002**, *35*, 915–926; (b) Shen, C.; Zhang, P.; Sun, Q.; Bai, S.; Hor, T. S. A.; Liu, X. *Chem. Soc. Rev.* **2015**, *44*, 291–314; (c) Beletskaya, I. P.; Ananikov, V. P. *Chem. Rev.* **2011**, *111*, 1596–1636.

[29] Pan, X.-Q.; Zou, J.-P.; Yi, W.-B.; Zhang, W. *Tetrahedron* **2015**, *71*, 7481–7529.

[30] Xia, D.; Miao, T.; Li, P.; Wang, L. *Chem. Asian J.* **2015**, *10*, 1919–1925.

[31] Yang, W.; Yang, S.; Li, P.; Wang, L. *Chem. Commun.* **2015**, *51*, 7520–7523.

[32] Qian, P.; Deng, Y.; Mei, H.; Han, J.; Zhou, J.; Pan, Y. *Org. Lett.* **2017**, *19*, 4798–4801.

[33] Meyer, A. U.; Jäger, S.; Hari, D. P.; König, B. *Adv. Synth. Catal.* **2015**, *357*, 2050–2054.

[34] Meyer, A. U.; Strakova, K.; Slanina, T.; König, B. *Chem. Eur. J.* **2016**, *22*, 8694–8699.

[35] Yang, D.; Huang, B.; Wei, W.; Li, J.; Liu, G.; Liu, Y; Ding, J.; Sun, P.; Wang, H. *Green Chem.* **2016**, *18*, 5630–5634.

[36] Zhang, C.; Zhang, L.; Yi, H.; Luo, Y.; Qi, X.; Tung, C.-H.; Wu, L.-Z.; Lei, A. *Chem. Commun.* **2016**, *52*, 10407–10410.

[37] Wang, H.; Lu, Q.; Chiang, C. W.; Luo, Y.; Zhou, J.; Wang, G.; Lei, A. *Angew. Chem. Int. Ed.* **2017**, *56*, 595–599.

[38] Keshari, T., Yadav, V. K.; Srivastava, V. P.; Yadav, L. D. S. *Green Chem.* **2014**, *16*, 3986–3992.

[39] Zalesskiy, S. S.; Shlapakov, N. S.; Ananikov, V. P. *Chem. Sci.* **2016**, *7*, 6740–6745.

[40] Talla, A.; Driessen, B.; Straathof, N. J. W.; Milroy, L.-G.; Brunsveld, L.; Hessel, V.; Noël, T. *Adv. Synth. Catal.* **2015**, *357*, 2180–2186.

[41] Mitra, S.; Ghosh, M.; Mishra, S.; Hajra, A. *J. Org. Chem.* **2015**, *80*, 8275–8281.

[42] Sun, P.; Yang, D.; Wei, W.; Jiang, M.; Wang, Z.; Zhang, L.; Zhang, H.; Zhang, Z.; Wang, Y.; Wang, H. *Green Chem.* **2017**, *19*, 4785–4791.

[43] Yadav, A. K.; Yadav, L. D. S. *Green Chem.* **2015**, *17*, 3515–3520.

[44] Yadav, V. K.; Srivastava, V. P.; Yadav, L. D. S. *Tetrahedron Lett.* **2016**, *57*, 155–158.

[45] Horsman, G. P.; Zechel, D. L. *Chem. Rev.* **2017**, *117*, 5704–5783.

[46] (a) Luo, K.; Yang, W. C.; Wu, L. *Chem. Asian J.* **2017**, *6*, 350–367; (b) Shaikh, R. S.; Düsel, S. J. S.; König, B. *ACS Catal.* **2016**, *6*, 8410–8414; (c) Leca, D.; Fensterbank, L.; Lacote, E.; Malacria, M. *Chem. Soc. Rev.* **2005**, *34*, 858–865; (d) Liu, D.; Chen, J.-Q.; Wang, X.-Z.; Xu, P.-F. *Adv. Synth. Catal.* **2017**, *359*, 2773–2777.

[47] Peng, P.; Peng, L.; Wang, G.; Wang, F.; Luo, Y.; Lei, A. *Org. Chem. Front.* **2016**, *3*, 749–752.

[48] Luo, K.; Chen, Y. Z.; Yang, W. C.; Zhu, J.; Wu, L. *Org. Lett.* **2016**, *18*, 452–455.

[49] Quint, V.; Morlet-Savary, F.; Lohier, J. F.; Lalevee, J.; Gaumont, A. C.; Lakhdar, S. *J. Am. Chem. Soc.* **2016**, *138*, 7436–7441.

[50] Hari, D. P.; König, B. *Angew. Chem. Int. Ed.* **2013**, *52*, 4734–4743.

[51] (a) Majek, M.; Filace, F.; von Wangelin, A. J. *Beilstein J. Org. Chem.* **2014**, *10*, 981–989; (b) Gu, L.; Jin, C.; Liu, J.; Ding, H.; Fan, B. *Chem. Commun.* **2014**, *50*, 4643–4645; (c) Shaaban, S.; Roller, A.; Maulide, N. *Eur. J. Org. Chem.* **2015**, *2015*, 7643–7647; (d) Zhao, D.; Xie, Z. *Angew. Chem. Int. Ed.* **2016**, *55*, 3166–3170.

[52] Hari, D. P.; Schroll, P.; König, B. *J. Am. Chem. Soc.* **2012**, *134*, 2958–2961.

[53] Maity, P.; Kundu, D.; Ranu, B. C. *Eur. J. Org. Chem.* **2015**, *2015*, 1727–1734.

[54] Meyer, A. U.; Slanina, T.; Yao, C.-J.; König, B. *ACS Catal.* **2015**, *6*, 369–375.

[55] Hari, D. P.; Hering, T.; König, B. *Org. Lett.* **2012**, *14*, 5334–5337.

[56] Xiao, T.; Dong, X.; Tang, Y.; Zhou, L. *Adv. Synth. Catal.* **2012**, *354*, 3195–3199.
[57] Niu, T. F.; Jiang, D. Y.; Li, S. Y.; Ni, B. Q.; Wang, L. *Chem. Commun.* **2016**, *52*, 13105–13108.
[58] Schroll, P.; Hari, D. P.; König, B. *ChemistryOpen* **2012**, *1*, 130–133.
[59] Zhang, N.; Quan, Z.-J.; Zhang, Z.; Da, Y.-X.; Wang, X.-C. *Chem. Commun.* **2016**, *52*, 14234–14237.
[60] Guo, W.; Lu, L.-Q.; Wang, Y.; Wang, Y.-N.; Chen, J.-R.; Xiao, W.-J. *Angew. Chem. Int. Ed.* **2015**, *54*, 2265–2269.
[61] Majek, M.; von Wangelin, A. J. *Angew. Chem. Int. Ed.* **2015**, *54*, 2270–2274.
[62] Gu, L.; Jin, C.; Liu, J. *Green Chem.* **2015**, *17*, 3733–3736.
[63] Li, X.; Liang, D.; Huang, W.; Zhou, H.; Li, Z.; Wang, B.; Ma, Y.; Wang, H. *Tetrahedron* **2016**, *72*, 8442–8448.
[64] Yu, J.; Zhang, L.; Yan, G. *Adv. Synth. Catal.* **2012**, *354*, 2625–2628.
[65] Majek, M.; von Wangelin, A. J. *Chem. Commun.* **2013**, *49*, 5507–5509.
[66] Bottecchia, C.; Rubens, M.; Gunnoo, S. B.; Hessel, V.; Madder, A.; Noël, T. *Angew. Chem. Int. Ed.* **2017**, *56*, 12702–12707.
[67] Li, Y.; Wang, M.; Jiang, X. *ACS Catal.* **2017**, *7*, 7587–7592.
[68] (a) Cao, M.-Y.; Ren, X.; Lu, Z. *Tetrahedron Lett.* **2015**, *56*, 3732–3742; (b) Angnes, R. A.; Li, Z.; Correia, C. R. D.; Hammond, G. B. *Org. Biomol. Chem.* **2015**, *13*, 9152–9167; (c) Heitz, D. R.; Rizwan, K.; Molander, G. A. *J. Org. Chem.* **2016**, *81*, 7308–7313.
[69] Lin, Q.-Y.; Xu, X.-H.; Qing, F.-L. *Org. Biomol. Chem.* **2015**, *13*, 8740–8749.
[70] Tiwari, D. P.; Dabral, S.; Wen, J.; Wiesenthal, J.; Terhorst, S.; Bolm, C. *Org. Lett.* **2017**, *19*, 4295–4298.
[71] Tang, W. K.; Feng, Y. S.; Xu, Z. W.; Cheng, Z. F.; Xu, J.; Dai, J. J.; Xu, H. J. *Org. Lett.* **2017**, *19*, 5501–5504.
[72] Wang, K.; Meng, L. G.; Wang, L. *J. Org. Chem.* **2016**, *81*, 7080–7087.
[73] (a) Nicewicz, D. A.; MacMillan, D. W. *Science* **2008**, *322*, 77–80; (b) Nagib, D. A.; Scott, M. E.; MacMillan, D. W. *J. Am. Chem. Soc.* **2009**, *131*, 10875–10877.
[74] Neumann, M.; Füldner, S.; König, B.; Zeitler, K. *Angew. Chem. Int. Ed.* **2011**, *50*, 951–954.

[75] Cantillo, D.; de Frutos, O.; Rincón, J. A.; Mateos, C.; Kappe, C. O. *Org. Lett.* **2014**, *16*, 896–899.

[76] Suzuki, I.; Esumi, N.; Yasuda, M. *Asian J. Org. Chem.* **2016**, *5*, 179–182.

[77] Esumi, N.; Suzuki, K.; Nishimoto, Y.; Yasuda, M. *Org. Lett.* **2016**, *18*, 5704–5707.

[78] Zou, Y.-Q.; Chen, J.-R.; Liu, X.-P.; Lu, L.-Q.; Davis, R. L.; Jorgensen, K. A.; Xiao, W.-J. *Angew. Chem. Int. Ed.* **2012**, *51*, 784–788.

[79] Paul, A.; Chatterjee, D.; Rajkamal; Halder, T.; Banerjee, S.; Yadav, S. *Tetrahedron Lett.* **2015**, *56*, 2496–2499.

[80] Singh, A. K.; Chawla, R.; Yadav, L. D. S. *Tetrahedron Lett.* **2015**, *56*, 653–656.

[81] Yadav, A. K.; Srivastava, V. P.; Yadav, L. D. S. *RSC Adv.* **2014**, *4*, 4181–4186.

[82] Devari, S.; Rizvi, M. A.; Shah, B. A. *Tetrahedron Lett.* **2016**, *57*, 3294–3297.

[83] Zhang, J.; Wang, L.; Liu, Q.; Yang, Z.; Huang, Y. *Chem. Commun.* **2013**, *49*, 11662–11664.

[84] Jin, Y.; Ou, L.; Yang, H.; Fu, H. *J. Am. Chem. Soc.* **2017**, *139*, 14237–14243.

[85] Hedstrand, D. M.; Kruizinga, W. H.; Kellogg, R. M. *Tetrahedron Lett.* **1978**, *19*, 1255–1258.

[86] Yang, X.-J.; Chen, B.; Zheng, L.-Q.; Wu, L.-Z.; Tung, C.-H. *Green Chem.* **2014**, *16*, 1082–1086.

[87] Zhong, J.-J.; Liu, Q.; Wu, C.-J.; Meng, Q.-Y.; Gao, X.-W.; Li, Z.-J.; Chen, B.; Tung, C.-H.; Wu, L.-Z. *Chem. Commun.* **2016**, *52*, 1800–1803.

[88] Majek, M.; Filace, F.; von Wangelin, A. J. *Chem. Eur. J.* **2015**, *21*, 4518–4522.

Chapter 7

Oxygen Heterocycles: Fluorescein, Rhodamines, Rose Bengal

Daniele Leonori* and Fabio Juliá

School of Chemistry, University of Manchester, Manchester M13 9PL, UK

**daniele.leonori@manchester.ac.uk*

7.1 Introduction

Xanthene dyes have been well known in the literature since the first synthesis of fluorescein (**1**) by von Baeyer in 1871 [1]. These compounds are characterized by strong absorptions in the visible range with a bright fluorescence in solution in many cases [2]. These peculiar properties have been responsible for their broad applications spanning biological stains, tracing agents as well as laser dyes, among others. More recently, some of these dyes have witnessed rejuvenated interest by the organic chemistry community owing to their ability to serve as powerful visible-light-absorbing photocatalysts [3–6].

In this chapter, we will discuss the synthetic applications of fluorescein (**1**), rhodamine B (**2**), rhodamine 6G (**3**), and Rose Bengal (**4**) in visible-light-mediated photocatalytic transformations. Readers are directed to Chapter 6 for a discussion on the application of related xanthene dyes belonging to the family of Eosin derivatives.

7.2 Photophysical and Electrochemical Properties

In general, dyes **1–4** present two main structural similarities (Scheme 7.1):

(i) they are all characterized by a xanthene core difunctionalized in the meta position with respect to the central oxygen atom;
(ii) they all contain a 2-benzoate group at the C-9 position.

An important feature associated with the photochemistry of **1–4** is that these compounds undergo pH-dependent structural changes which dramatically impact their photophysical properties and, as a result, their ability to act as photocatalysts [7]. Such a pH-dependent effect is described below for **1** and can be extended to other xanthene dyes (Scheme 7.2). Neutral **1** exists in three forms: quinonoid (**1**-*q*), lactonoid (**1**-*l*), and zwitterionic (**1**-*z*). These species are all in equilibrium in solution and their distribution is pH and solvent dependent.

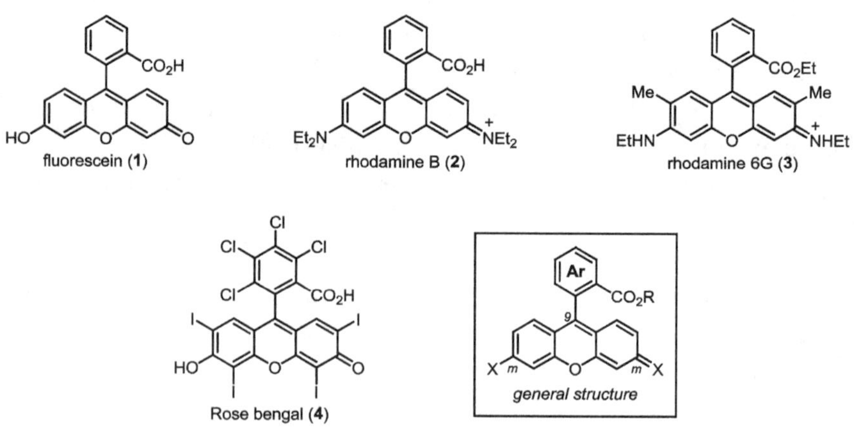

Scheme 7.1. Structures of xanthene dyes **1–4**.

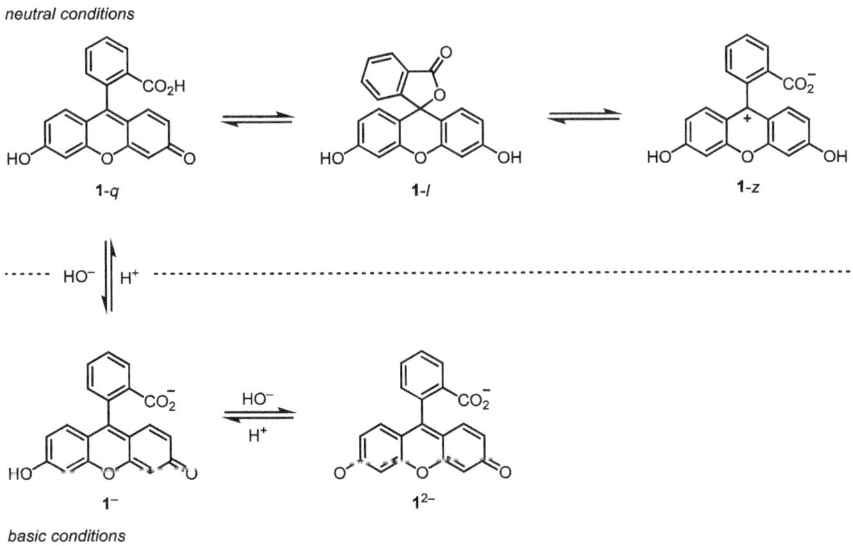

Scheme 7.2. pH-dependent structural changes of **1**.

Under neutral conditions, **1**-*l* is the predominant form. As this species does not absorb in the visible range and is weakly fluorescent, a poor photochemical activity is normally observed when using **1** in neutral form [8].

Under basic conditions, **1** is deprotonated to the yellow dianionic species $\mathbf{1}^{2-}$, which displays strong fluorescence and excited state lifetime long enough to engage in bimolecular reactions and therefore serve as a competent photocatalyst.

In view of the above considerations, it is strongly recommended to specify whether the neutral, monoanionic, or dianionic species is used in a particular reaction, as some xanthene dyes are commercially available in all these three forms.

Table 7.1 outlines the photophysical and electrochemical properties of **1–4** in both the ground and the excited state (**POC***) [3]. In general, **1–4** absorb very well in the visible range at wavelengths tailing in the green. This makes them attractive photocatalysts as low energy light can be used for their excitation, which minimizes any unwanted reactivity arising for the direct light-absorbance from other reagents.

Table 7.1. Selected photophysical and electrochemical data for xanthene dyes [1].[a]

	1	**2**	**3**	**4**
λ_{abs} (nm)[b]	491	550	530	549
Φ_F	0.93	0.58	0.90	0.09
τ_F (ns)	4.73	2.45	4.13	0.50
Φ_{ISC}	0.03	0.12	0.002	0.77
E(**POC**/**POC**$^+$)[c,d]	+0.83	+0.91	+1.23	+0.84
E(**POC**/**POC**$^-$)[c,d]	−1.22	−0.96	−1.14	−0.96
E(**POC***/**POC**$^+$)[c,d]	−1.55	−1.31	−1.09	−0.96[e]
E(**POC***/**POC**$^-$)[c,d]	+1.25	+1.26	+1.18	+0.81[e]

Notes: [a]Data obtained under basic conditions. [b]Wavelength at the maximum of the lowest-energy band. [c]V vs SCE. [d]**POC**, **POC**$^+$, and **POC**$^-$ refer to the ground state, oxidized and reduced forms of the photocatalyst, respectively. [e]Triplet excited state is considered.

Dianionic **1** ($\mathbf{1}^{2-}$) absorbs in the blue region and exhibits a bright green fluorescent emission ($\Phi_F = 0.93$, $\tau = 4.73$ ns). As the population of its triplet state (T_1) by intersystem crossing (ISC) is not efficient ($\Phi_{ISC} = 0.03$), any photochemistry is expected to arise from the excited singlet state (S_1). Upon blue light irradiation, **1*** undergoes an amplification of its redox properties both as an oxidant and a reductant with E^*_{red} and E^*_{ox} values similar to the ones of the widely-adopted Ru/Ir-based photocatalysts.

Dyes **2** and **3** are structurally related to **1**, but they bear (di)-ethylamino substituents instead of hydroxyl groups at the xanthene core. This substitution pattern is responsible for their batochromic absorption (green region instead of blue) and emission (orange region instead of green) bands [9, 10]. The photophysical properties of **2** are highly dependent on the solvent and pH as discussed for **1** [11]. In this regard, **3** includes some structural features that considerably improve its fluorescent quantum yield ($\Phi_F = 0.90$):

(i) Esterification of the C-9 benzoate group prevents pH- dependent lactonization which would generate a photoinactive species.

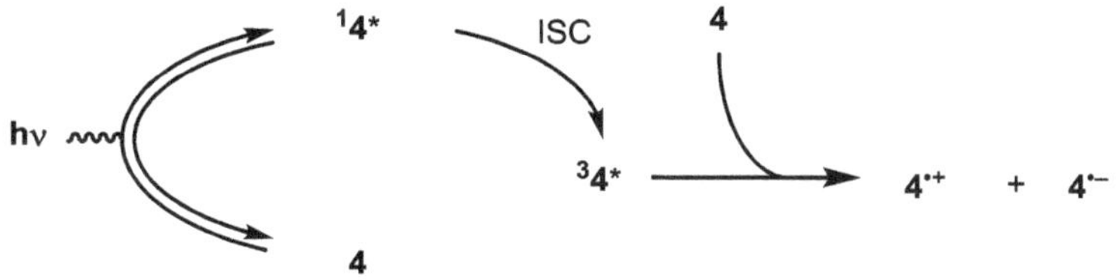

Scheme 7.3. Mechanism of self-quenching for **4** following photoexcitation.

(ii) The presence of an ethylamine instead of diethylamine substitution at the xanthene core minimizes non-radiative processes resulting from the population of intramolecular charge-transfer states [9, 12].

These two structural modifications significantly reduce the sensitivity of **3**'s photophysical properties to pH and solvent factors.

In view of the low quantum yields for T_1 population by ISC, the active excited states involved in bimolecular reactions with **2** and **3** are expected to be predominantly singlet S_1. Also, in this case, the corresponding excited states have enhanced redox properties.

Compound **4** absorbs in the green region and exhibits fluorescence in the orange region. However, it is considerably weaker in comparison with **1**, mainly owing to the efficient population of T_1 ($\Phi_{ISC} = 0.77$). The iodine atoms are responsible for the efficient ISC, due to the well-known "heavy-atom effect". As a result, upon visible-light irradiation, the photochemistry of **4** is dominated by the reactivity of its T_1. Thus, it is not surprising that one of the main uses of **4** is the photochemical generation of singlet oxygen ($^1O_2$) by Dexter energy transfer, which has found many applications in organic synthesis [13, 14]. In addition, an interesting self-quenching process has been characterized for this dye [15], consisting in the generation of both radical cation and radical anion species by reaction between the triplet excited state and ground state molecules (Scheme 7.3).

7.3 Fluorescein

In synthetic settings, **1** has found application as a photoredox catalyst in the generation of aryl radicals by single electron reduction of aryl diazonium salts.

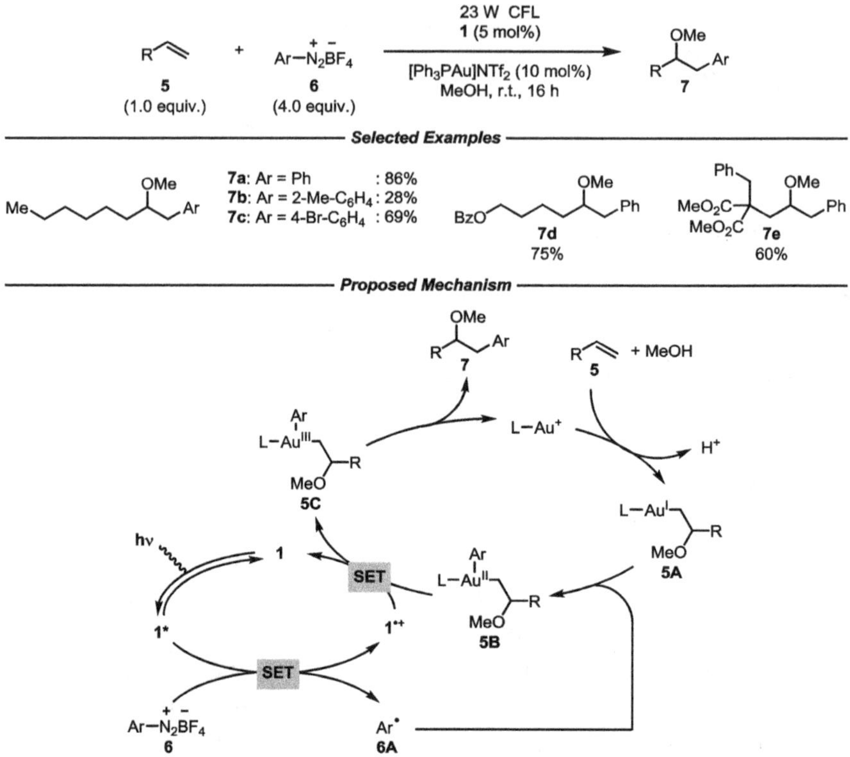

Scheme 7.4. Oxyarylation of olefins promoted by **1** *via* dual Au-photoredox catalysis.

In 2014, Glorius and co-workers demonstrated that **1** could be used as a competent light-absorbing catalyst in a dual photoredox–Au catalytic strategy for the three-component oxyarylation of olefins (**5**) using aryl diazonium salts (**6**) (Scheme 7.4) [16]. This work was based on previous results obtained by the same group that identified a related intramolecular-transformation using $Ru(bpy)_3Cl_2$ as the photoredox catalyst [17]. The process has been only applied to terminal and monosubstituted olefins, but several functional groups were shown to be compatible (**7a–7e**).

Mechanistically, upon visible-light irradiation, **1*** was supposed to generate aryl radicals **6A** from **6** by SET reduction and N_2 extrusion. At the same time, the $[Ph_3PAu]NTf_2$ co-catalyst would promote

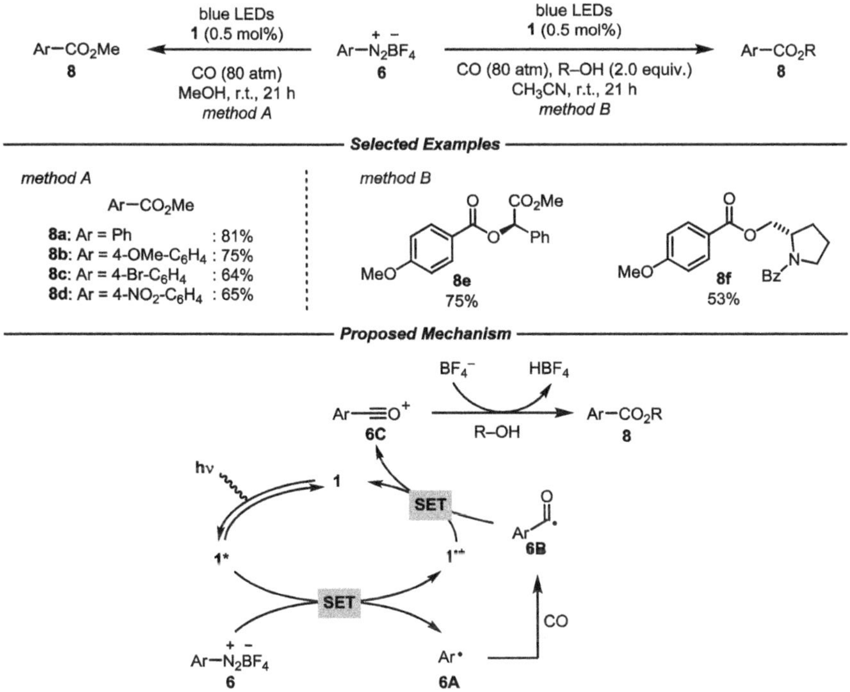

Scheme 7.5. Photoredox esterification of aryldiazonium salts with CO.

olefin methoxylation delivering the Au(I)-species **5A**. Aryl radical addition at Au(I) was proposed to generate a highly reactive diorganyl Au(II) complex **5B**. At this point, the authors suggested that the redox potential of $\mathbf{1}^{\bullet+}$ and the Au(II) metal could be adjusted by SET. This event would close the photoredox cycle and generate a diorganyl-Au(III) species that, upon facile reductive elimination, would form the product **7** while regenerating the catalytically active Au(I) cationic catalyst.

Xiao and co-workers have also harnessed the ability of **1** to generate aryl radicals as part of a strategy for the preparation of aromatic esters using CO as reactive coupling partner (Scheme 7.5) [18]. This process involved the direct reaction of aryl diazonium salts **6** with **1** and various alcohols under an atmosphere of CO and blue LEDs' irradiation.

Depending on the nature of the alcohol, the authors developed two methods whereby cheap and readily available alcohols (e.g. MeOH) can be conveniently used as reaction media (method A), while more complex and difficult-to-make substrates can be employed in stoichiometric amounts (method B). As aryl radicals are very reactive and not highly affected by the aromatic substitution pattern, a broad range of aromatic precursors was successfully engaged (**8a–8d**) as well as several complex and chiral alcohols (**8e**, **8f**). The authors proposed a catalytic cycle based on the generation of aryl radical **6A** from **6** by reductive SET and N_2 extrusion. Following CO capture would deliver an acyl radical **6B** that was proposed to undergo SET oxidation with $\mathbf{1}^{\bullet+}$ closing the photoredox cycle and generating the acylium ion **6C**. Nucleophilic trapping by the alcohol gave the ester **8**. The feasibility of the final SET ($\mathbf{1}^{\bullet+}$ + **6B** → **1** + **6C**) was supported by DFT studies, which showed it to be extremely exergonic ($\Delta G° = -37$ kcal mol^{-1}).

7.4 Rhodamine

As discussed in Sec. 7.1, **2** and **3** have ground- and excited-state redox potentials similar to the ones of transition metal-based photoredox catalysts. As a result, they can be employed in numerous visible-light-mediated transformations based on both oxidative and reductive quenching cycles.

7.4.1 *Oxidative quenching cycle*

One of the first applications of **2** as photoredox catalyst has been disclosed by Zeitler that developed a metal-free asymmetric α-alkylation of aldehydes [19]. This study was aimed at replacing $Ru(bpy)_3Cl_2$ as the photoredox catalyst in the dual organocatalytic–photoredox strategy reported by MacMillan in 2008 [20]. In this study, the authors not only identified Eosin Y as the optimum catalyst but also reported a single example showing that **2** could efficiently promote the reductive dehalogenation of Br-acetophenone **9** using Hantzsch ester **10** as the H-donor and $(i\text{-Pr})_2NEt$ (DIPEA) as the stoichiometric reductant upon green LED irradiation (Scheme 7.6).

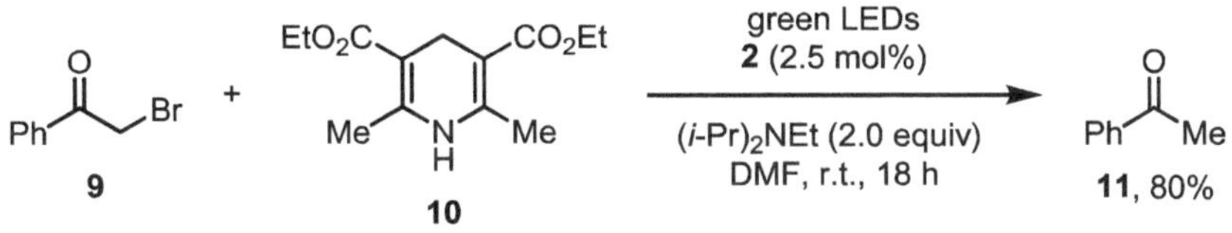

Scheme 7.6. Photo-organocatalytic reduction of a α-bromoacetophenone.

More recently, Kohtani and Miyabe have shown that **2** was a useful catalyst in radical cascade reactions involving diolefin acceptors (**12**), perfluoro-alkyl iodides (**13**), and (*i*-Pr)$_2$NEt under white LEDs' irradiation (Scheme 7.7) [11, 21]. The process was found effective for the assembly of polyfluorinated iodo-containing pyrrolidines (**14a**, **14c**, **14d**) and cyclopentanes (**14b**). Interestingly, the chemistry could also be extended to the use of simple alkyl iodides, albeit in lower yields (**14e**). The mechanism likely involves the generation of persistent perfluoro-alkyl radicals (**13A**) *via* SET reduction of **13** by **2***. This event would then trigger a cascade of intermolecular addition to **12**, intramolecular cyclization (**13B**→**13C**), and final iodine-atom transfer from **12**. While the authors proposed (*i*-Pr)$_2$NEt to serve as the stoichiometric electron donor required to close the photoredox cycle, a radical chain propagation process is also feasible. In this case, the photoredox manifold would serve as the radical initiation process.

A photoredox catalytic method employing **2** accomplished the direct C-3 arylation of indole **15** using aryl diazonium salts **6** (Scheme 7.8) [22]. This protocol worked well with electron-poor, halogen-containing, and polyaromatic diazonium salts (**16a–d**) and could also be successfully extended to functionalized indoles (**16e–g**). Mechanistically, upon generation of an aryl radical by reductive SET (see also Scheme 7.4), C-3 radical addition to indole **15** and final oxidation delivered the products **16**.

7.4.2 *Reductive quenching cycle*

The use of **2** in reductive quenching cycles is significantly less developed. However, a successful example can be found in the direct

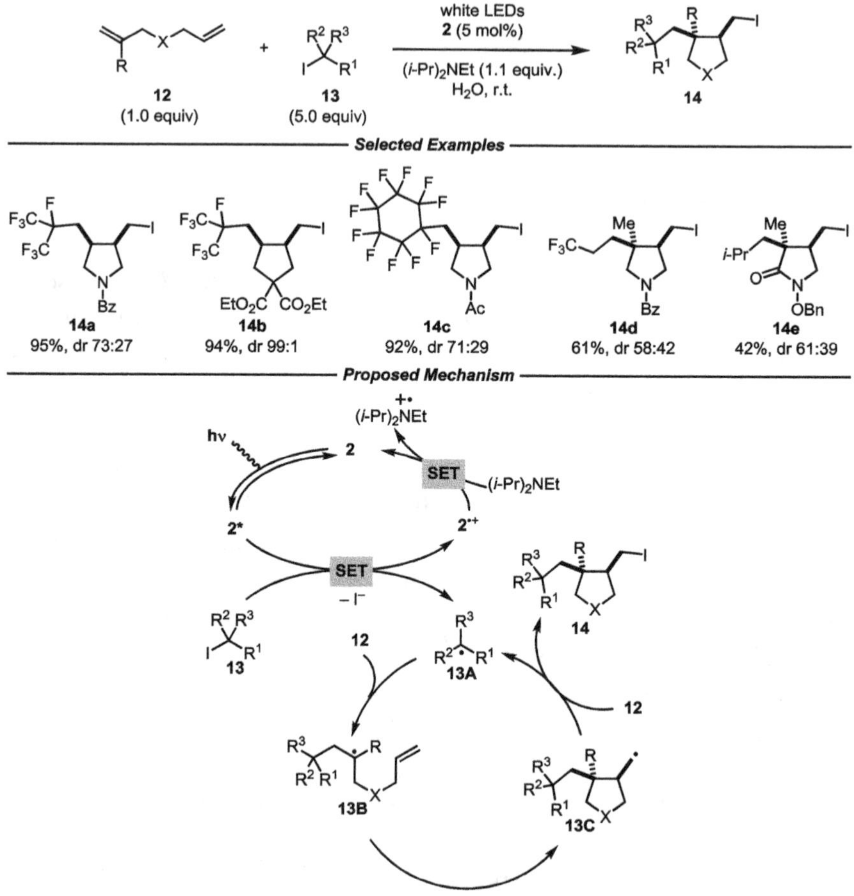

Scheme 7.7. Cascade process leading to polyfluorinated iodo-containing cycloalkanes/heterocycles.

hydrophosphinylation of olefins reported by Kobayashi and co-workers (Scheme 7.9) [23]. This process enables the anti-Markovnikov addition of disubstituted phosphine oxides (**17**) to alkenes (**18**) using **2** as the photoredox catalyst under white **LED** irradiation. Although terminal olefins gave the highest yields (**19a**), several electron-rich disubstituted olefins were included in the substrate scope (**19b**). In terms of the phosphine oxides, electron-rich diaryl substrates could be used (**19c, 19d**), but electron-poor diaryl as well as dialkyl substrates were not found compatible (**19e, 19f**). This

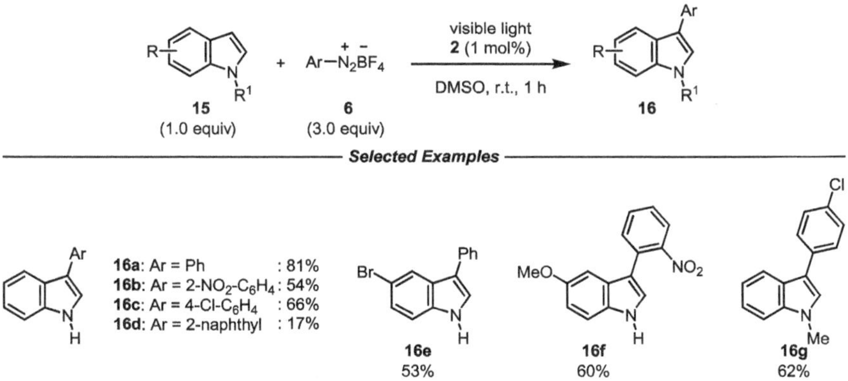

Scheme 7.8. Photoredox C-3 arylation of indoles.

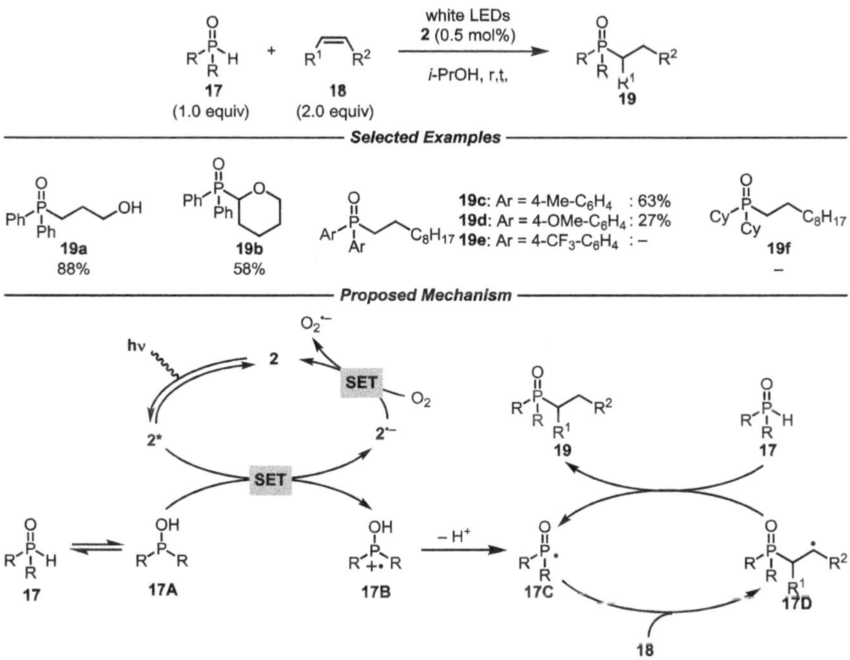

Scheme 7.9. Photoredox reaction of phosphine oxides with olefins.

reaction was thought to start by the SET oxidation of the diaryl phosphine oxides (**17** or **17A**) by **2*** delivering the P-centred radical cation **17B**. *In situ* deprotonation would give access to the phosphinoyl radical **17C** that, upon anti-Markovnikov addition to olefin **18**, would furnish the C-radical intermediate **17D**. At this stage, the authors suggested a radical chain propagation by H-atom abstraction from either **17** or **17A** to form **19** and regenerate the P-radical **17C**.

7.4.3 *Oxidative quenching cycle upon excitation of rhodamine radical anion*

The highly negative reduction potential of aryl halides precludes their direct photoredox SET reduction-fragmentation *en route* to aryl radicals. While aryl diazonium salts have been proven powerful aryl radical precursors [16, 18], the possibility of engaging readily available building-blocks in related transformations can facilitate the preparation of many high-value molecules. A strategy to overcome this challenge has been recently introduced by König and co-workers, that took advantage of the unique photophysical properties of **3** [24, 25]. As mentioned in Sec. 7.1, **3** can easily be photoexcited upon either blue or green light irradiation. In the presence of a good electron donor (e.g. $(i\text{-Pr})_2\text{NEt}$), a facile SET delivers its radical anion $\mathbf{3}^{\bullet-}$ (Scheme 7.10). Interestingly, $\mathbf{3}^{\bullet-}$ displays a significant absorption in the blue region and can be further photoexcited to deliver $\mathbf{3}^{\bullet-*}$. This species has been shown to have a remarkably high reductive power reaching almost −2.4 V vs SCE. In this way, organic

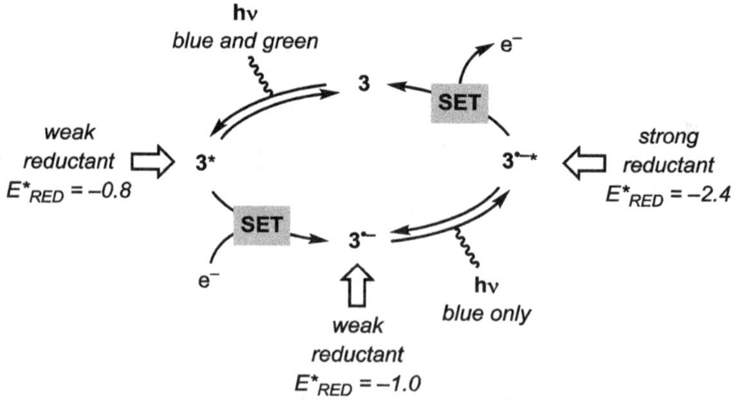

Scheme 7.10. Photoexcitation of **3** and $\mathbf{3}^{\bullet-}$.

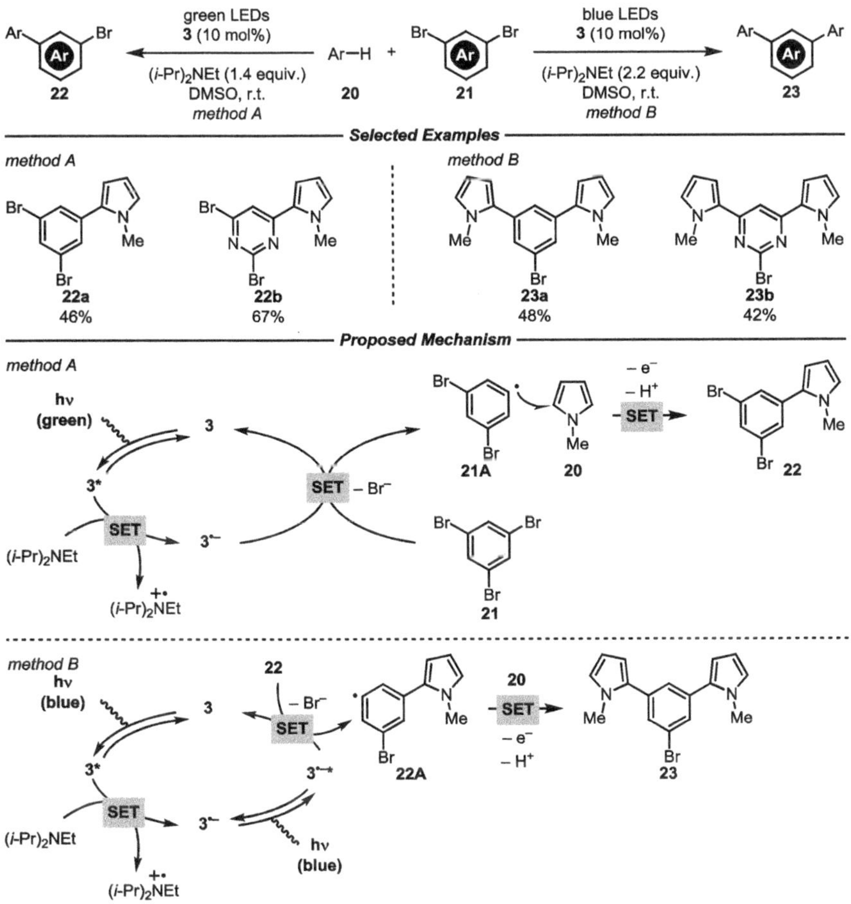

Scheme 7.11. Reductive arylation of aryl bromides with pyrrole.

molecules that would be normally very difficult to reduce *via* photoredox catalysis can be engaged in such transformations.

The power of this approach has been demonstrated by König in a unique "light-color-guided" process that enables the chemoselective coupling of electron-rich (**20**) and polybrominated (**21**) aromatics using **3** as the photocatalyst and (i-Pr)$_2$NEt as the stoichiometric reductant (Scheme 7.11) [24]. The authors demonstrated that, depending on the light color used, two different products could be selectively obtained. Upon green light irradiation, compound **21** underwent selective monoarylation (product **22**, method A), while under blue light irradiation the diarylated product **23** was selectively formed (method B). The process was found feasible only with highly electron-rich aromatics and highly activated aryl bromides. This remarkable difference in selectivity stems from the ability of different light colors to modulate the redox properties of **3**. In both cases, upon visible-light irradiation (either blue or green), **3*** underwent reductive quenching with (i-Pr)$_2$Net, giving **3**$^{\bullet-}$ which is reactive enough to promote SET reduction and fragmentation of the highly activated bromide **21**. This gave access to aryl radical **21A** that underwent addition to the electron-rich pyrrole **20**. Radical oxidation–deprotonation furnished the product **22**. As the reduction potential of **22** is out of the range for SET from ground state **3**$^{\bullet-}$, under green light irradiation, the process does not provide further arylation. However, when blue light was used, a second photoexcitation took place delivering **3**$^{\bullet-}$* which was now able to engage **22** in a reductive SET, thus triggering the second arylation process.

In two following reports, König and co-workers have expanded this strategy to the arylation of bromine-containing 5-membered heterocycles [26], as well as chloro-pyrimidines [27].

The ability of **3**$^{\bullet-}$* to generate aryl radicals by SET reduction and fragmentation of aryl halides has been used by König and co-workers to develop a direct photoArbuzov reaction using trialkyl phosphites to deliver aryl phosphonates **26** (Scheme 7.12) [28]. The protocol

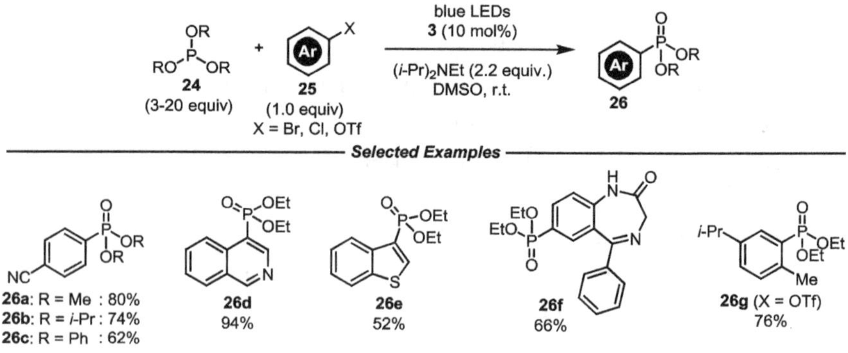

Scheme 7.12. Synthesis of aryl phosphonates by a photoArbuzov reaction.

was generally applicable to electron-poor (**26a–d**) and electron-rich (**26e**) aryl bromides and chlorides, and it was used in the modification of bioactive molecules (**26f**). Sometimes, however, aryl triflates had to be used like in the generation of cymene aryl-radical leading to product **26g**.

7.5 Rose Bengal

Rose Bengal (**4**) has been extensively adopted as photoredox catalyst and photosensitizer. As discussed in Sec. 7.1, **4** has, upon visible-light excitation, a very high ISC quantum yield (Φ_{ISC}) and its triplet excited state ^{3}**4*** (T_1) can act as a competent reductant, oxidant, and energy-transfer sensitizer. All these reaction pathways will be described in the following sections.

7.5.1 *Oxidative quenching cycle*

Compound **4** was utilized by Tan and co-workers in the a-oxyamination of 1,3-dicarbonyl compounds **27** through their SET reduction and further reaction with TEMPO **28** (Scheme 7.13) under visible-light irradiation [29]. The process tolerated several functionalities

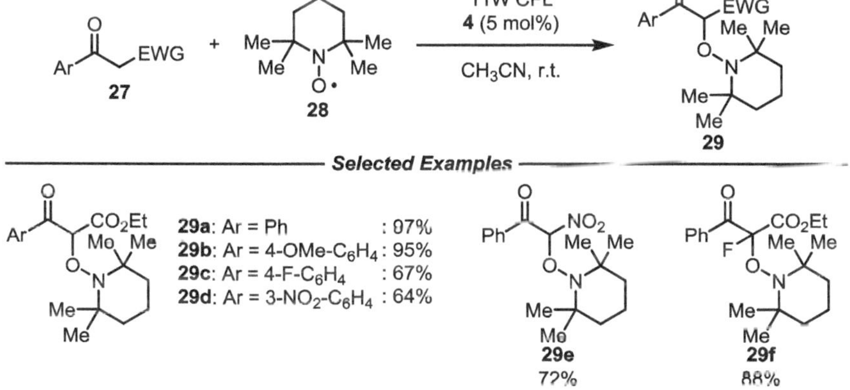

Scheme 7.13. α-Hydroxylation of β-keto-esters using TEMPO.

(**29a–d**) and could also be applied to the functionalization of α-nitro-acetophenones (**29e**). Furthermore, the authors showed that α-F-1,3-dicarbonyl compounds could be efficiently employed in the process delivering quaternary fluorinated compounds in good yields (**29f**). Remarkably, the use of a CFL as the light source was not essential and, upon simple sunlight irradiation, similar conversions were obtained.

The introduction of fluorine in organic molecules is important for the development of molecules of both pharmaceutical and agrochemical relevance. Postigo and co-workers explored the use of **4** in the addition of perfluoro-alkyl radicals generated from the corresponding iodides (**31**) to anilines (**30**) (Scheme 7.14) [30]. High

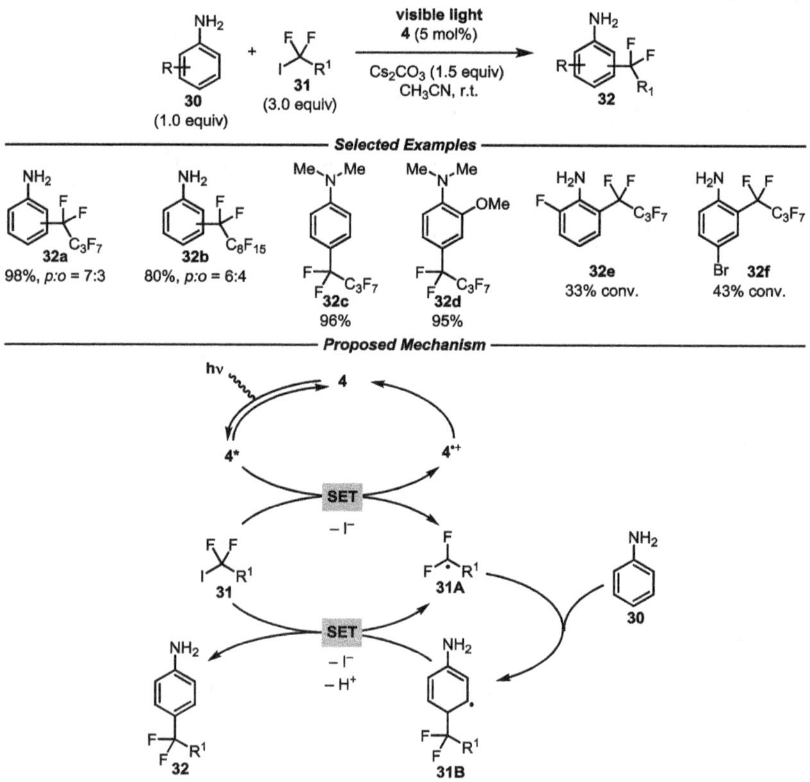

Scheme 7.14. Reductive perfluoroalkylation of anilines.

yields and good-to-moderate selectivity favoring the *para*-addition products could be achieved on a number of electron-rich aromatics (**32a–d**). Somewhat lower conversions were, however, obtained when electron-withdrawing groups were incorporated in the aromatic ring (**32e**, **32f**). This transformation is mechanistically related to other perfluoro-functionalizations as, for example, described in Sec. 7.4.1 (Scheme 7.7). Upon visible-light irradiation, **4*** promotes the SET reduction and fragmentation of **31** that generates the stable perfluoro-alkyl radicals **31A**. As **31A** is an electrophilic species, it undergoes facile reaction with electron-rich aniline **30** to give the C-radical **31B**. At this point, direct SET between **31B** and the starting iodo-perfluoroalkane **31** completes a radical chain propagation and delivers **32**. In this way, **4** would act as photoinitiator.

7.5.2 *Reductive quenching cycle*

One of the most exploited reaction manifold in photoredox catalysis is represented by the α-functionalization of tertiary amines *via* tandem two-electron oxidation and nucleophilic trapping. These reactions are generally classified as "dehydrogenative couplings" [31] and share a very similar mechanism. Organic dyes have played a prominent role here proving their chemical power and selectivity when compared to classical strategies relying on the use of Cu(I) and stoichiometric oxidants (e.g. *t*-BuOOH) as well as transition metal-based photoredox catalysts [32]. In general, the most reactive system is represented by *N*-aryl tetrahydroisoquinolines (**33**) and several authors have demonstrated the ability of **4** to enable their selective α-functionalization (Scheme 7.15). Tan and co-workers reported the dehydrogenative coupling with TMSCN, to give α-CN-amines **34** [33], and reactions with nitroalkanes and ketones, furnishing **35** and **36** [34]. On the other hand, Zhou has successfully developed a α-diazotization strategy giving **37** using simple diazocarbonyls as nucleophilic partners [35]. These reaction pathways are not restricted to the functionalization of tetrahydroisoquinolines and can be applied as a general strategy to the α-modification of many cyclic and acyclic tertiary amines, as showcased by the

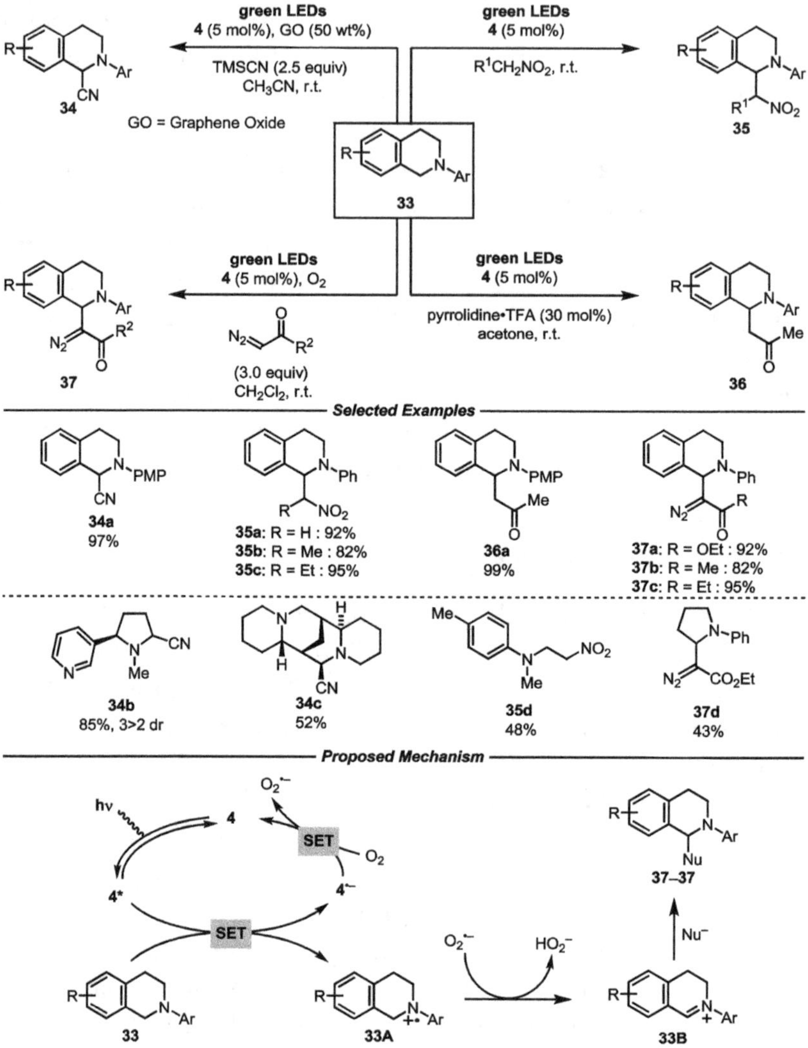

Scheme 7.15. Oxidation of *N*-Aryl tetrahydroisoquinolines and their following nucleophilic trapping.

selective modification of many alkaloids such as (+)-nicotine (**34b**) and (–)-sparteine (**34c**) [36–38]. Mechanistically, all of these reactions are believed to start with the SET oxidation of the amine **33** by the visible-light-excited **4***, which is a competent oxidant ($E^*(\mathbf{4}^*/\mathbf{4}^{\bullet-}) = +0.81$ V vs SCE). This step delivers an amine radical cation **33A** that can be converted into an iminium ion **33B** by H-atom abstraction. As this sequence involves the overall removal of two electron and one proton, it equates to a net oxidation. Nucleophilic trapping then delivers the reaction products. As these reactions are normally run under aerobic conditions, O_2 is believed to serve as the terminal oxidant enabling regeneration of ground state **4** and delivering $O_2^{\bullet-}$ which is generally considered to be involved in the formation of the iminium ion **33B**.

The ability of **4*** to promote the SET oxidation of amines has been used by Li and co-workers as part of a strategy for the selective C-3 formylation of indoles (**38**; Scheme 7.16) [39]. The process uses *N,N,N′,N′*-tetramethylethylenediamine (TMEDA) as the one-carbon source upon a cascade of oxidation–nucleophilic trapping followed by a second SET oxidation and hydrolysis *en route* to **40**. This process tolerated several functional groups across the indole ring. During the optimization of the process, the authors serendipitously identified KI as a beneficial additive, but unfortunately its role in the reaction pathways has not been elucidated yet.

Indoles were also found to be competent coupling partners that underwent selective C-3 thiocyanation using NH_4SCN, **4** as the photoredox catalyst and air as the terminal oxidant (Scheme 7.17) [40]. As the SCN-functionality can be converted into an array of other groups including 5-membered ring heterocycles, this approach can facilitate the synthesis of S-containing molecules. This photoredox process has been shown to tolerate functionalized indoles (**42b–42c**) as well as *N*-alkyl derivatives (**42d**). However, the use of *N*-Boc protected indoles (**42e**) completely suppressed the reactivity. The authors suggested this transformation to involve the SET oxidation

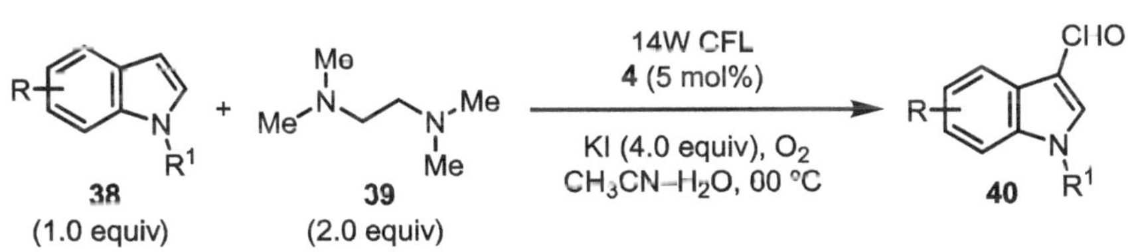

Scheme 7.16. Preparation of 3-formyl-indoles.

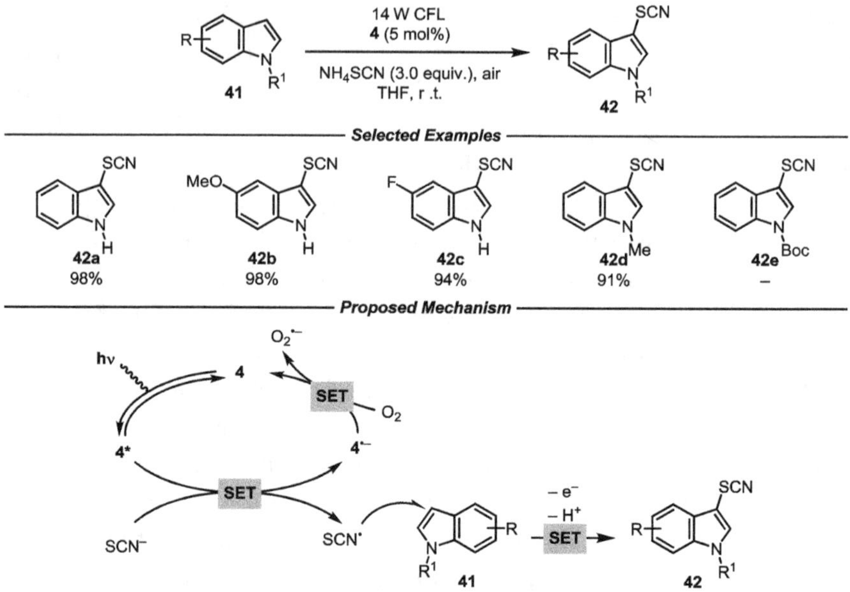

Scheme 7.17. C-3 Addition of SCN• radical to indoles.

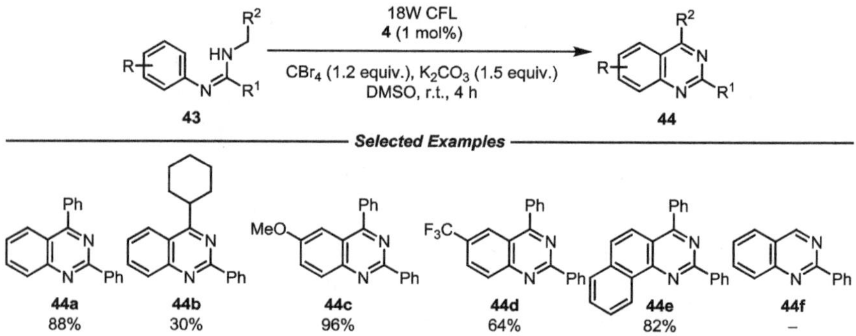

Scheme 7.18. Photoorganocatalyzed preparation of quinazolines.

of SCN^- to SCN• by **4***. Following radical addition to the C-3 position of indole **41**, a final oxidation-deprotonation gave the product **42**.

Tang and co-workers also used the ability of **4*** to promote the SET oxidation of amines as part of a strategy for the synthesis of quinazolines (**44**) *via* annulation of arylamidines (**43**; Scheme 7.18) [41].

This process differs from other oxidative protocols, because CBr_4 had to be used as the terminal oxidant in place of air or O_2.

7.5.3 *Energy transfer cycle via the generation of $^1O_2$*

The high triplet state quantum yield (Φ_{ISC}) yield and energy of **4** enables ^{3}**4*** (T_1) to easily participate in the triplet energy transfer processes (see Table 7.1, Sec. 7.2). The most prominent application of this reactivity is certainly represented by the generation of singlet oxygen ($^1O_2$), which has been extensively used in organic synthesis (Scheme 7.19) [42–44]. Only the more recent applications will be illustrated here and the reader is remanded to several authoritative reviews [13, 45, 46].

The use of **4** for the photocatalytic generation of $^1O_2$ has been generally applied to perform chemoselective oxidation reactions under mild reaction conditions. Barbas has used this approach as a tool to develop light-responsive prodrugs (Scheme 7.20(a)) [47]. In their report, the authors showed that thioketals (**45**), which are commonly used as carbonyl protecting groups, can be easily cleaved back to the ketone by irradiation in the presence of O_2 and **4** in a buffered system. As thioketals are stable to other reactive oxygen species (e.g. metal-oxo species), the authors envisaged this approach as a tool to enable the light activation of fluorescent sensors and prodrugs. In terms of substrate scope, the process enabled the formation of many ketones (**46a–d**), but could not be extended to thioketals derived from aldehydes (**46e**).

Wan and Wen have harnessed the ability of **4** to generate $^1O_2$ for the preparation of 1,2-diketones (**48**) by the oxidative cleavage of

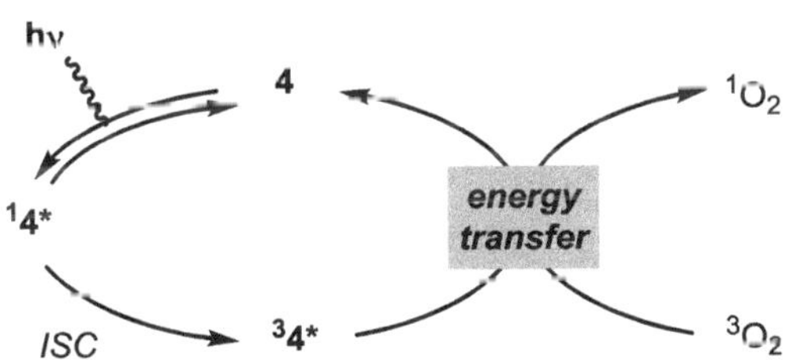

Scheme 7.19. Generation of $^1O_2$ by photoinduced energy transfer.

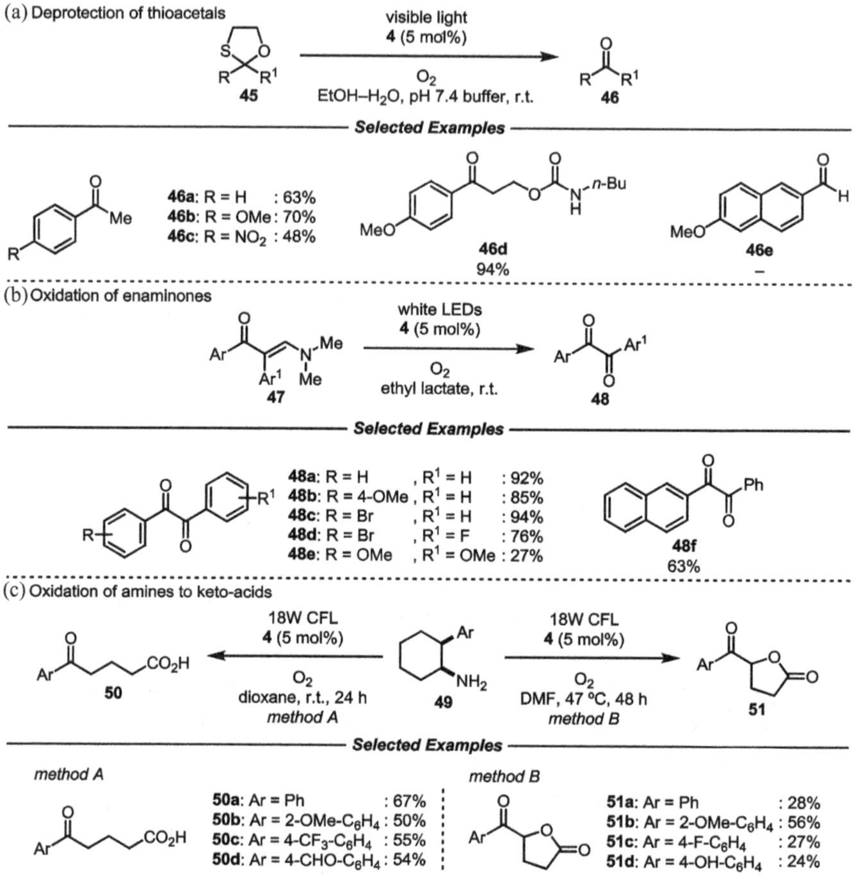

Scheme 7.20. Photochemical oxidations and rearrangements using $^1O_2$.

the C=C double bonds of enaminones (**47**) (Scheme 7.20(b)) [48]. This process has only been applied to the preparation of diaryl derivatives that were then used as part of a telescoped approach towards quinoxalines.

An interesting application of $^1O_2$ in the oxidative structural rearrangement of amines has been recently described by Lu and co-workers (Scheme 7.20(c)) [49]. Here, the authors showed that 2-aryl-cyclohexyl-1-amines (**49**) can be selectively converted to keto-acids (**50**) or lactones (**51**) depending on the reaction solvent, using

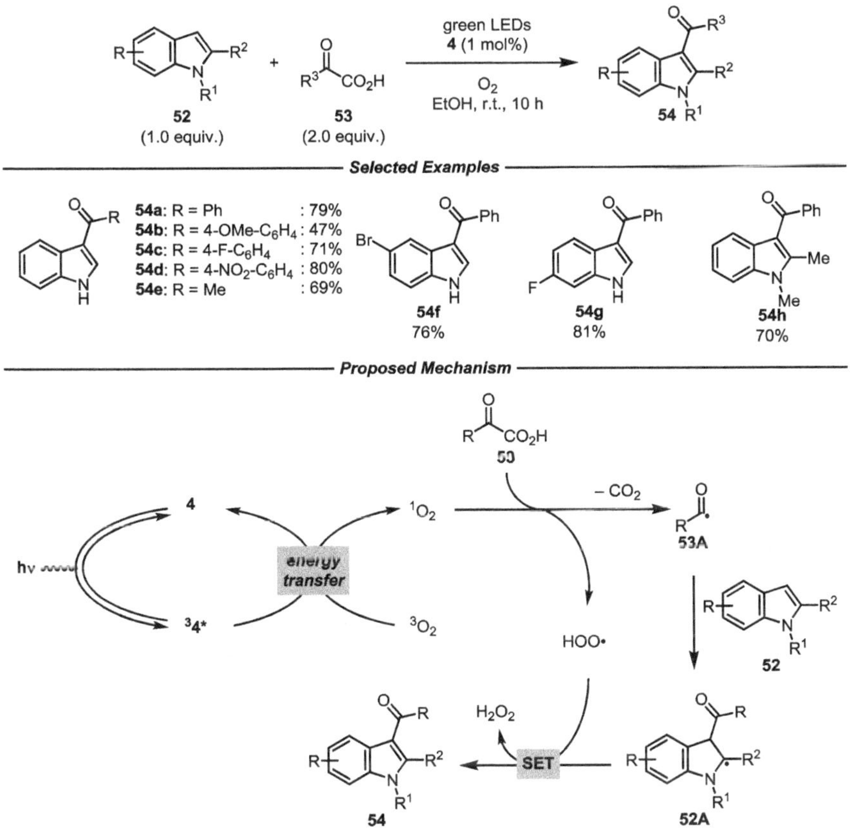

Scheme 7.21. C-3 addition of acyl radicals to indoles.

4 under visible-light irradiation. The authors have proposed a reaction mechanism where **4** acts sequentially as a sensitizer (for the formation of $^1O_2$) and photoredox catalyst in order to readjust the oxidation level of the reaction products.

Selective C-3 acylation of indoles (**52**) using α-keto-acids (**53**) can also be triggered by the **4**-promoted photochemical formation of $^1O_2$ (Scheme 7.21).

In particular, Wang and co-workers have shown that **4** promoted the formation of **54** by simple green LEDs' irradiation of **52** and **53** in EtOH under an atmosphere of O_2 [50]. In this case, α-keto-acids

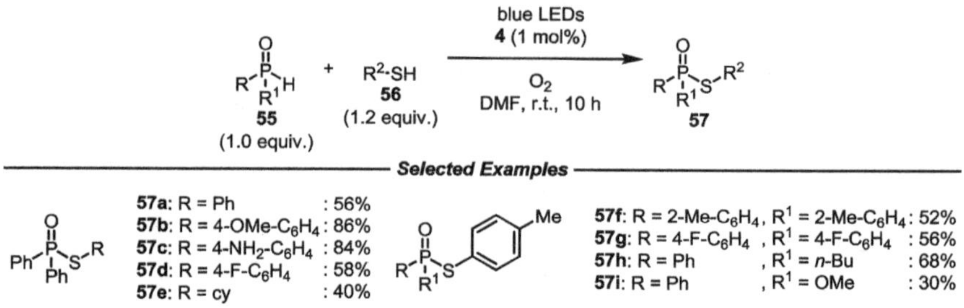

Scheme 7.22. Photochemical (O)P–S coupling between phosphine oxides and thiols.

with both electron-rich and electron-poor aromatic groups (**54a–d**) as well as alkyl groups (**54e**) could be used. The mechanism for this transformation differs somewhat from the usual reactivity of $^1O_2$ as this reactive species was proposed to undergo selective H-atom abstraction from the carboxylic acid functionality of **53** [51]. This gave the peroxyl radical and the acyl radical **53A** upon exergonic decarboxylation. The authors suggested a C-3 selective radical reaction between **53A** and **52** to give the C-radical **52A**. The reaction product **54** would then be generated upon SET between **52A** and the peroxyl radical followed by deprotonation.

Dye **4** was also found successful for the direct (O)P–S coupling between disubstituted phosphine oxides (**55**) and thiols (**56**; Scheme 7.22) [52]. This process enables the fast generation of products **57** with good functional group tolerance. This process most likely starts from the generation of a S-radical *via* H-atom abstraction from **56** by $^1O_2$.

An area of particular interest has been the use of **4** for the photochemical oxidation of small electron-rich heterocycles like pyrroles and furans. Bissember and Smith have reported the scalable preparation of 3-pyrrolin-2-ones (**59** and **60**) by **4**-sensitized oxidation of pyrroles (**58**) (Scheme 7.23) [53]. This process could be implemented in a flow system enabling fast scale-up and has the advantage of bypassing some of the hurdles present in the direct photooxidation of pyrrole by UV light, which undergoes uncontrolled polymerization and decomposition pathways.

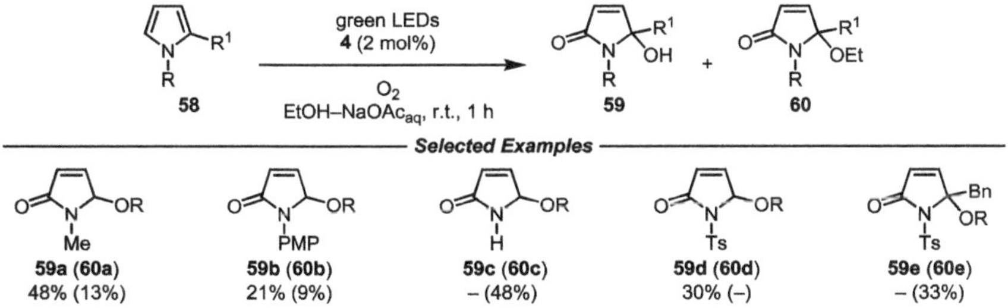

Scheme 7.23. Oxidation of pyrroles with $^1O_2$.

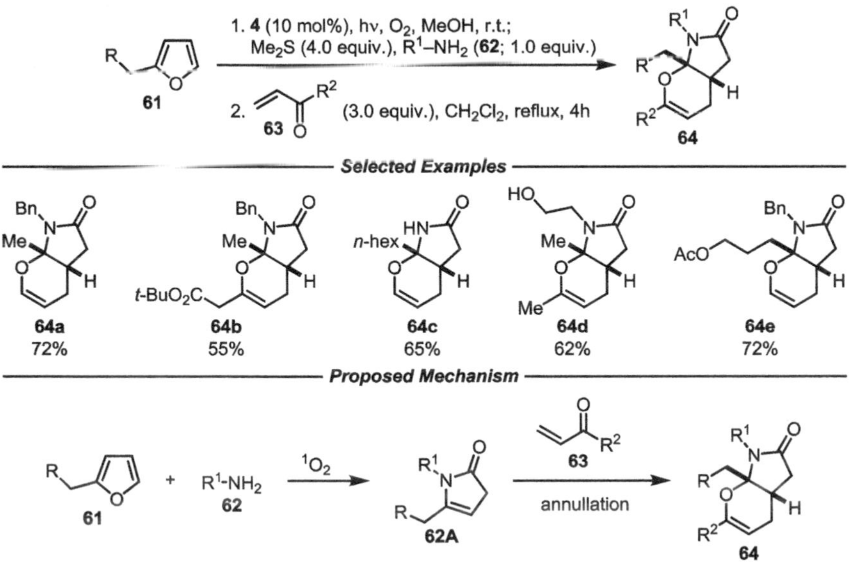

Scheme 7.24. Preparation of complex *N*-heterocycles from furans.

An elegant and diastereoselective approach for the fast conversion of simple furans (**61**) into complex tetrahydropyranopyrrolone scaffolds (**64**) has been recently developed by Vassilikogiannakis and co-workers (Scheme 7.24) [54]. Irradiation of **61** using **4** as the sensitizer in MeOH under an atmosphere of O_2, followed by treatment with Me_2S and a primary alkyl amine (**62**), gave the 2-pyrrolinone **62A** in almost quantitative conversion. This intermediate was then

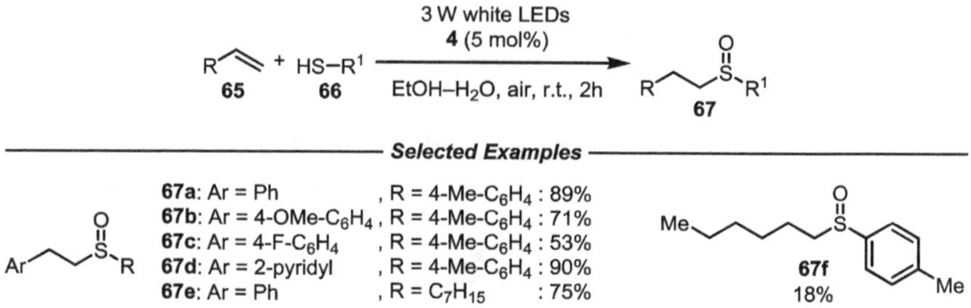

Scheme 7.25. Synthesis of sulfoxides from thiols and olefins using $^1O_2$.

directly engaged in a telescoped annulation reaction with the Michael acceptor **63**. This process forged the C–O and the C–C bonds in **64**, and various functional groups and substitution patterns were tolerated. The reaction products could be further elaborated giving access to valuable building blocks for the synthesis of alkaloids.

Very recently, Wang and co-workers have developed an approach for the generation of sulfoxides **67** from simple olefins **65** and thiols **66** using air as the oxidant and **4** as the photocatalyst (Scheme 7.25) [55]. The process required the use of an aromatic olefin in order to provide high yields, but both aromatic (**67a–d**) and aliphatic thiols (**67e**) could be used.

7.6 Conclusions

Organic photocatalysts are attractive alternatives to their transition metal-analogs because of their lower cost, ease of handling, and low environmental toxicity. In this chapter, we have reviewed the application of the xanthene dyes fluorescein, rhodamine B and 6G, and Rose Bengal. Their good absorbance in the visible range and the strong redox properties of their excited states make these dyes a powerful tool for the development of photochemical transformations. In some cases, their use has led to the development of processes that cannot be mirrored by other systems.

References

[1] von Baeyer, A. *Chem. Ber.* **1871**, *4*, 555–558.

[2] Xanthene dyes, Van Nostrand's Scientific Encyclopedia, American Cancer Society, 2006.

[3] Romero, N. A.; Nicewicz, D. A. *Chem. Rev.* **2016**, *116*, 10075–10166.

[4] Hari, D. P.; König, B. *Chem. Commun.* **2014**, *50*, 6688–6699.

[5] Fukuzumi, S.; Ohkubo, K. *Org. Biomol. Chem.* **2014**, *12*, 6059–6071.

[6] Ravelli, D.; Fagnoni, M.; Albini, A. *Chem. Soc. Rev.* **2013**, *42*, 97–113.

[7] Majek, M.; von Wangelin, A. J. *Acc. Chem. Res.* **2016**, *49*, 2316–2327.

[8] Klonis, N.; Sawyer, W. H. *J. Fluoresc.* **1996**, *6*, 147–157.

[9] Beija, M.; Afonso, C. A. M.; Martinho, J. M. G. *Chem. Soc. Rev.* **2009**, *38*, 2410–2433.

[10] Neckers, D. C.; Valdes-Aguilera, O. M. Photochemistry of the Xanthene dyes, *Advances in Photochemistry*, Vol. 18, Eds D. H. Volman, G. S. Hammond and D. C. Neckers, John Wiley & Sons, Inc., Hoboken, NJ, USA, **2007**, 315–394.

[11] Yoshiokaa, E.; Kohtani, S.; Jichua, T.; Fukazawaa, T.; Nagaia, T.; Takemotob, Y.; Miyabe, H. *Synlett* **2015**, *26*, 265–270.

[12] Magde, D.; Rojas, G. E.; Seybold, P. G. *Photochem. Photobiol.* **1999**, *70*, 737–744.

[13] Ghogare, A. A.; Greer, A. *Chem. Rev.* **2016**, *116*, 9994–10034.

[14] Ogilby, P. R. *Chem. Soc. Rev.* **2010**, *39*, 3181–3209.

[15] Wintgens, V.; Scaiano, J. C.; Linden, S. M.; Neckers, D. C. *J. Org. Chem.* **1989**, *54*, 5242–5246.

[16] Hopkinson, M. N.; Sahoo, B.; Glorius, F. *Adv. Synth. Catal.* **2014**, *356*, 2794–2800.

[17] Hopkinson, M. N.; Sahoo, B.; Li, J.-L.; Glorius, F. *Chem. Eur. J.* **2014**, *20*, 3874–3886.

[18] Guo, W.; Lu, L. Q.; Wang, Y.; Wang, Y.-N.; Chen, J.-R.; Xiao, W.-J. *Angew. Chem. Int. Ed.* **2015**, *54*, 2265–2269.

[19] Neumann, M.; Fuldner, S.; König, B.; Zeitler, K. *Angew. Chem. Int. Ed.* **2010**, *50*, 951–954.

[20] Nicewicz, D.; MacMillan, D. W. C. *Science* **2008**, *322*, 77–80.

[21] Yoshioka, E.; Kohtani, S.; Jichu, T.; Fukazawa, T.; Nagai, T.; Kawashima, A.; Takemoto, Y.; Miyabe, H. *J. Org. Chem.* **2016**, *81*, 7217–7229.

[22] Zhang, Y.-P.; Feng, X.-L.; Yang, Y.-S.; Cao, B.-X. *Tetrahedron Lett.* **2016**, *57*, 2298–2302.

[23] Yoo, W.-J.; Kobayashi, S. *Green Chem.* **2013**, *15*, 1844–1848.

[24] Gosh, I.; König, B. *Angew. Chem. Int. Ed.* **2016**, *55*, 7676–7679.

[25] Ghosh, I.; Ghosh, T.; Bardagi, J. I.; König, B. *Science* **2014**, *346*, 725–728.

[26] Marzo, L.; Ghosh, I.; Esteban, F.; König, B. *ACS Catal.* **2016**, *6*, 6780–6784.

[27] Graml, A.; Ghosh, I.; König, B. *J. Org. Chem.* **2017**, *82*, 3552–3560.

[28] Shaikh, R. S.; Düsel, S. J. S.; König, B. *ACS Catal.* **2016**, *6*, 8410–8414.

[29] Liu, H.; Feng, W.; Kee, C. W.; Zhao, Y.; Leow, D.; Pan, Y.; Tan, C.-H. *Green Chem.* **2010**, *12*, 953–956.

[30] Barata-Vallejo, S.; Yerien, D. E.; Postigo, A. *Eur. J. Org. Chem.* **2015**, 7869–7875.

[31] Li, C.-J. *Acc. Chem. Res.* **2009**, *42*, 335–344.

[32] Beatty, J. W.; Stephenson, C. R. J. *Acc. Chem. Res.* **2015**, *48*, 1474–1484.

[33] Pan, Y.; Wang, S.; Kee, C. W.; Dubuisson, E.; Yang, Y.; Loh, K. P.; Tan, C.-H. *Green Chem.* **2011**, *13*, 3341–3344.

[34] Pan, Y.; Kee, C. W.; Chen, L.; Tan, C.-H. *Green Chem.* **2011**, *13*, 2682–2685.

[35] Xiao, T.; Li, L.; Lin, G.; Mao, Z.-W.; Zhou, L. *Org. Lett.* **2014**, *16*, 4232–4235.

[36] Pacheco, J. C. O.; Lipp, A.; Nauth, A. M.; Acke, F.; Dietz, J.-P.; Opatz, T. *Chem. Eur. J.* **2016**, *22*, 5409–5415.

[37] Chandrasekhar, D.; Borra, S.; Kapure, J. S.; Shivaji, G. S.; Srinivasulu, G.; Maurya, R. A. *Org. Chem. Front.* **2015**, *2*, 1308–1312.

[38] Gandy, M. N.; Raston, C. L.; Stubbs, K. A. *Chem. Commun.* **2015**, *51*, 11041–11044.

[39] Li, X.; Gu, X.; Li, Y.; Li, P. *ACS Catal.* **2014**, *4*, 1897–1900.

[40] Fang, W.; Yang, Q.; Xu, F.; Li, P. *J. Org. Chem.* **2014**, *79*, 10588–10592.

[41] Shen, Z.-C.; Yang, P.; Tang, Y. *J. Org. Chem.* **2016**, *81*, 309–317.

[42] Levésque, F.; Seeberger, P. H. *Angew. Chem. Int. Ed.* **2012**, *51*, 1706–1709.

[43] Nicolaou, K. C.; Gunzner, J. L.; Shi, G.-Q.; Agrios, K. A.; Gartner, P.; Yang, Z. *Chem. Eur. J.* **1999**, *5*, 646–658.

[44] Jung, M.; Ham, J.; Song, J. *Org. Lett.* **2002**, *4*, 2763–2765.

[45] Montagnon, T.; Kalaitzakis, D.; Triantafyllakis, M.; Stratakis, M.; Vassilikogiannakis, G. *Chem. Commun.* **2014**, *50*, 15480–15498.

[46] Montagnon, T.; Kalaitzakis, D.; Sofiadis, M.; Vassilikogiannakis, G. *Org. Biomol. Chem.* **2016**, *14*, 8636–8640.

[47] Lamb, B. M.; Barbas III, C. F. *Chem. Commun.* **2015**, *51*, 3196–3199.

[48] Cao, S.; Zhong, S.; Xin, L.; Wan, J.-P.; Wen, C. *ChemCatChem* **2015**, *7*, 1478–1482.

[49] Cheng, X.; Yang, B.; Hu, X.; Xu, Q.; Lu, Z. *Chem. Eur. J.* **2016**, *22*, 17566–17570.

[50] Shi, Q.; Li, P.; Zhu, Z.; Wang, L. *Green Chem.* **2016**, *18*, 4916–4923.

[51] Mondal, R. R.; Khamarui, S.; Maiti, D. K. *Org. Lett.* **2017**, *19*, 5964–5967.

[52] Sun, J.-G.; Yang, H.; Li, P.; Zhang, B. *Org. Lett.* **2016**, *18*, 5114–5117.

[53] Howard, J. K.; Rihak, K. J.; Hyland, C. J. T.; Bissember, A. C.; Smith, J. A. *Org. Biomol. Chem.* **2016**, *14*, 8873–8880.

[54] Kalaitzakis, D.; Triantafyllakis, M.; Ioannou, G. I.; Vassilikogiannakis, G. *Angew. Chem. Int. Ed.* **2017**, *56*, 4020–4023.

[55] Cui, H.; Wei, W.; Yang, D.; Zhang, Y.; Zhao, H.; Wang, L.; Wang, H. *Green Chem.* **2017**, 19, 3520–3524.

Chapter 8

Nitrogen Heterocycles: Porphyrins

Hermenegildo García* and Sonia Remiro-Buenamañana

Instituto Universitario de Tecnología Química CSIC-UPV,
Universitat Politècnica de València,
Av. De los Naranjos s/n, 46022 Valencia, Spain
**hgarcia@qim.upv.es*

8.1 Introduction

Porphyrins are a class of naturally occurring colorful pigments that have been used in many applications due to their versatile and tunable photophysical properties [1]. They play a key role in nature, for instance, in photosynthesis or for transporting oxygen [1–3]. Two of the most relevant examples are protoporphyrin-IX in the heme group of hemoglobin, an iron-complexed porphyrin that is responsible for the red color of blood, and the magnesium-related porphyrinoids present in chlorophyll in plants [4]. Porphyrins present some advantages with respect to other photo and electroactive compounds, arising from their highly conjugated π-electron system and its ability to bind metal ions.

The structure of porphyrins was first suggested by Küster in 1912, but the structure was not accepted until 1929, when Fisher

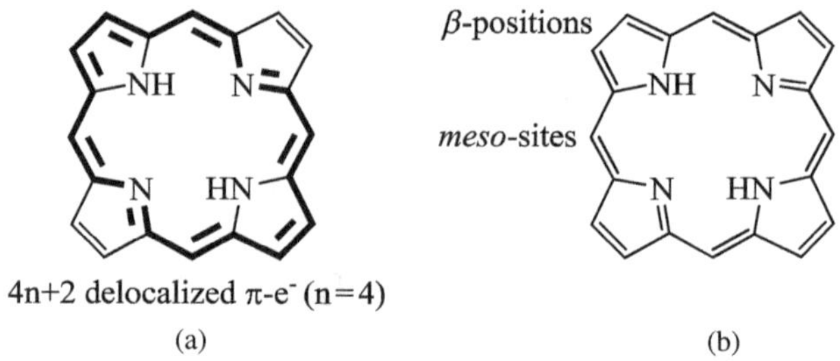

Figure 8.1. (a) π-e^{-} delocalization in the macrocyclic ring. (b) Planar structure and sites for possible functionalization.

and co-workers accomplished the total synthesis of heme [5]. Porphyrins are formed by four pyrrolic units linked by methine bridges forming a planar macrocycle, known as free base porphyrin when the central hydrogen atoms are present. The macrocycle contains 22 π-electrons, however, only 18 are delocalized according to Huckel's rule of aromaticity (see Figure 8.1). They can be functionalized either at the β-pyrrolic positions or at the *meso*methine carbon atoms by electrophilic substitution reactions. Aza-substitution at the methine bridge to form azoporphyrinoids is another possible ring modification [6]. Moreover, the four nitrogen atoms in the inner core can act as a dianionic ligand to coordinate with almost every metal ion in the periodic table to form *metalloporphyrins.*

The porphyrin name originated from the Greek word *porphura* which means purple. They are very colorful materials widely employed as pigments and dyes. Absorption spectra of porphyrins show two main components; a very intense peak in the near-ultraviolet (UV) region called Soret or B band, ranging from 390 to 425 nm, that involves an electronic transition from the ground state to the second excited state ($S_0 \rightarrow S_2$), and weaker bands known as Q bands corresponding to the transitions from the ground state to the first excited state ($S_0 \rightarrow S_1$) in the range between 480–700 nm [7]. UV–visible (UV–Vis) spectroscopy provides an easy way of monitoring the metal binding process as the number of bands will diminish from four in free base porphyrin (D_{2h} molecular symmetry) to two when the porphyrin is metallated

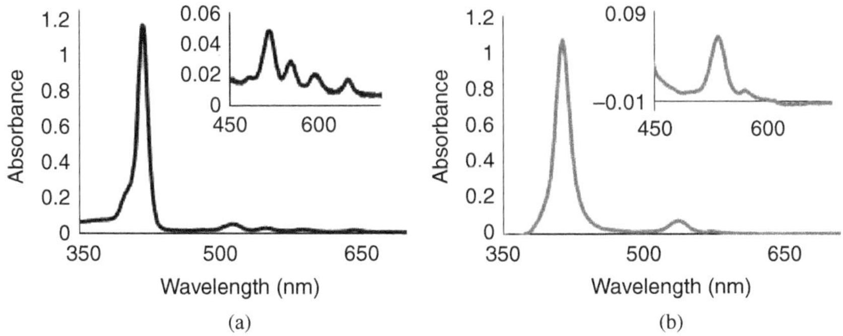

Figure 8.2. General example of the absorption spectrum of a free porphyrin (a) and a porphyrin hosting a metal (b) in dichloromethane.

(D_{4h}) due to its higher symmetry. Figure 8.2 shows the two types of absorption spectra found for free and metallated porphyrins.

Porphyrin spectra can be explained according to Gouterman's four orbitals model [8]. The theory postulates that the absorption bands observed in the UV–Vis spectrum are transitions occurring from two HOMOs to two LUMOs based on the direct LCAO-MO calculations for the porphyrin ring. Hence, HOMO orbitals are designated as a_{1u} and a_{2u}, whilst the degenerate e_g correspond to the LUMO orbitals (see Figure 8.3). Two excited states result from these transitions, both exhibiting a ^1E_u symmetry but a different energy, where the lower energy band would give rise to the Q band while the transition to the higher energy level relates to the Soret band (Figure 8.3(b)).

Due to the light-absorbing capacity of porphyrins, they have been used as light-harvesting systems for optoelectronic applications [9]. For instance, tetraphenylporphyrin (**TPP**) (see structure in Figure 8.4) can harvest around 46% of the energy in the solar spectrum [10].

Porphyrins present high molar extinction coefficients in the range of 10^5 M^{-1} cm^{-1} for the Soret band [11]. Moreover, the molar absorptivity of the series of Q bands is in the range of 10^4 M^{-1} cm^{-1}. It has been reported that the substitution in the *para* position in the phenyl rings of **TPP** has little effect on the shape

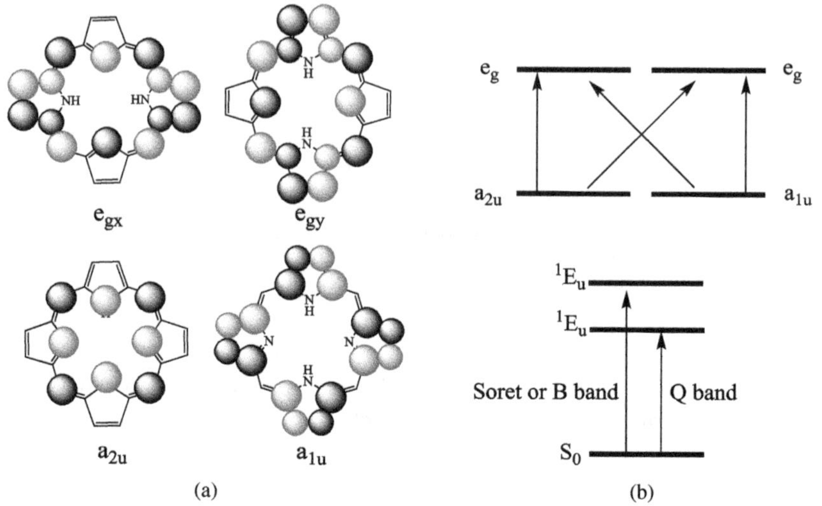

Figure 8.3. (a) Representation of the four molecular orbitals of porphyrins in Gouterman's model (atomic orbital coefficients are drawn proportional to the size of the spheres, color indicates different sign). (b) Energy levels and transitions corresponding to the B and Q bands.

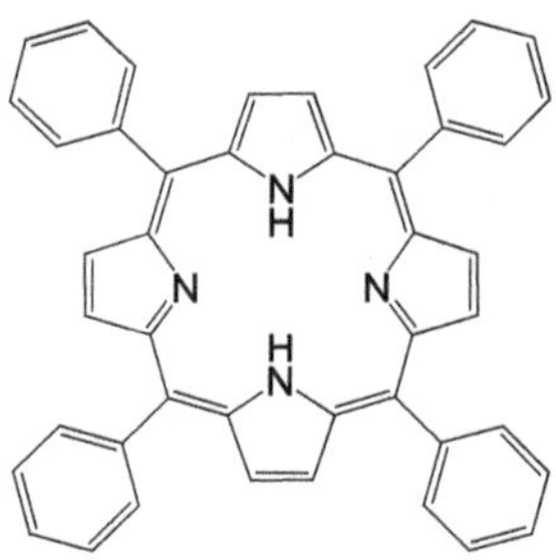

TPP

Figure 8.4. Structure of **TPP**.

of the bands, however, a red shift of the absorption bands can be observed for the substitutions, independent of the nature of the functional group [12]. Peripheral functionalization resulted in a blue shift of the Soret bands when moving to withdrawing substituents [13]. All absorption bands also exhibit blue shifts when the polarity of the solvent used for recording the spectrum is reduced, possibly owing to the hydrogen-bonding interactions between the inner pyrrole NH protons with solvent molecules [13–14].

There have been extensive studies reporting different electronic configuration of porphyrins [15], including fluorescence quantum yields (Φ_F) [16, 17] that range from 0.11 to 0.98. Several porphyrins also exhibit emission in the red region of the spectrum, more specifically, from 600 to 800 nm. Triplet quantum yields have also been reported to range from approximately 0.3 to 0.81, and to be highly dependent on the substituents at the *meso* position of the porphyrin ring [18]. Photoexcited singlet state lifetimes are typically around 60 ns. Phosphorescence quantum yield range from 10^{-4} to 0.2, phosphorescence lifetimes are typically less than 3.0 ms at room temperature [9].

8.2 Porphyrins as Singlet Oxygen Photosensitizers

Formation of singlet molecular oxygen $^1O_2$ plays a crucial role in applications such as photodynamic therapy or photocatalysis [19]. Porphyrins have proven to be effective singlet oxygen generators (Scheme 8.1) [20, 21] and they are among the most widely used photosensitizers for this purpose. A porphyrin is excited to an electronically singlet excited state (^{1}**P***) by means of photon absorption at the Soret or Q band, while undergoing intersystem crossing (ISC) to the lowest-lying triplet excited state (^{3}**P***), which, in the presence of ground-state oxygen, forms singlet oxygen *via* energy transfer. Interestingly, $^1O_2$ has the ability to oxidize electron-rich substrates in a process denoted as quenching, while most of these substrates are stable in the presence of ground-state oxygen (O_2).

Porphyrins exhibit high singlet oxygen quantum yields, for instance, Φ_Δ for **TPP** in benzene is 0.78 ± 0.04 (λ_{exc} = 355 nm). Metallated porphyrins also show high efficiency values for $^1O_2$ generation, such as 0.81 for Zn-**TPP** or 0.78 for Mg-**TPP** [22]. Porphyrins have also served to dope materials in order to increase singlet oxygen production [23].

$$\text{Porphyrin (P)} \xrightarrow{h\nu} {}^{1}\text{P}^{*} \xrightarrow[\text{crossing}]{\text{intersystem}} {}^{3}\text{P}^{*}$$

$${}^{3}\text{P}^{*} + \text{O}_2 \xrightarrow{\text{energy transfer}} \text{P} + {}^{1}\text{O}_2$$

$${}^{1}\text{O}_2 + \text{Acceptor (A)} \xrightarrow{\text{quenching}} \text{AO}_2$$

Scheme 8.1. Mechanism of formation of molecular singlet oxygen upon photosensitization by porphyrins (**P**) and subsequent chemical quenching.

As commented above, porphyrins are highly efficient singlet oxygen photosensitizers and, therefore, visible-light irradiation of solutions, mainly organic, containing porphyrins under oxygenated conditions opens the possibility to promote all the reactions characteristic of singlet oxygen [24–26].

Photocatalytic transformations and, more specifically, those that involve singlet oxygen, are one of the most difficult reaction types to induce chirality. While the use of the Sharpless titanium catalyst for the enantioselective epoxidation of allylic alcohols and the use of chiral manganese salen complexes for the enantioselective epoxidation of alkenes have constituted significant achievements in asymmetric catalysis, most of the attempts to promote asymmetric induction using singlet oxygen have resulted in very low enantiomeric excesses. One exception in this area is the work of Gryko and co-workers who reported the enantioselective α-oxygenation of aldehydes through the formation of the corresponding enamines in the presence of a large excess of secondary pentacyclic chiral amines, such as proline and some of its derivatives, using **TPP** as photosensitizer [27]. Scheme 8.2 illustrates the process of the enantioselective photooxygenation of aldehydes *via* enamine. The available experimental evidence shows that the corresponding enamine is formed by the condensation of the secondary amine and the aldehyde (the key intermediate undergoing oxygenation), under the reaction conditions. In CCl_4 or DMF upon irradiation with white LED light, low to moderate yields of diols were obtained after ensuing the reduction of the aldehyde with sodium borohydride. Furthermore, by using the MacMillan imidazolidinone

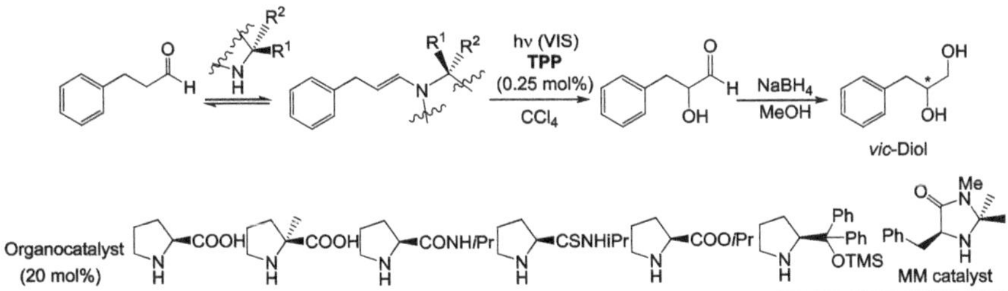

Organocatalyst (20 mol%)							MM catalyst
Yield of *vic*-Diol (%)[a]	23[b]	16[b]	52[c]	20[c]	37[c]	11	62
ee (%)	14 (S)	50 (S)	74 (R)	62 (R)	0	62 (S)	80 (S)

Notes: [a]Reaction conditions: aldehyde (1.5 mmol), organocatalyst (20 mol%), **TPP** (0.25 mol%) in 6 mL of CCl_4, 3 hours. [b]In DMF. [c]**TPP** (0.04 mol%).

Scheme 8.2. Enantioselective α-hydroxylation of aldehydes, followed by their $NaBH_4$ reduction to the corresponding *vic*-diol. The structure of the amines tested as organocatalysts has been indicated.

(**MM catalyst**) or the propyl amide of L-proline as the secondary amine, enantiomeric excesses of 80% and 74%, respectively, could be achieved for the diol derived from 3-phenylpropionaldehyde.

Theoretical calculations suggest that the enantioselectivity depends on the *E* or *Z* configuration of the C=C double bond and the *s-trans* or *s-cis* conformation of the C–N bond. These calculations predict that the free energy difference between the transition states corresponding to the singlet oxygen attack to the enamine is small (in the range of 1–1.8 kcal/mol) for the most favorable case of the MacMillan imidazolidinone derivative. These small differences in energy between the transition states involved justify the low enantiomeric excesses that are generally observed in many cases. It has to be considered that singlet oxygen is a highly reactive species that, in addition to its high reactivity, has a small molecular size and, therefore, its attack is less influenced by the steric encumbrance around the reactive site.

Extensive studies changing solvents, secondary amine, and temperature have led to the conclusion that the optimal irradiation conditions are those using the **MM catalyst** as the secondary cyclic amine and CCl_4 as solvent at 1°C or below. Under these conditions, a yield to 3-phenylpropane-1,2-diol of 67% with an enantiomeric excess towards the (*S*)-enantiomer of 78% has been reached.

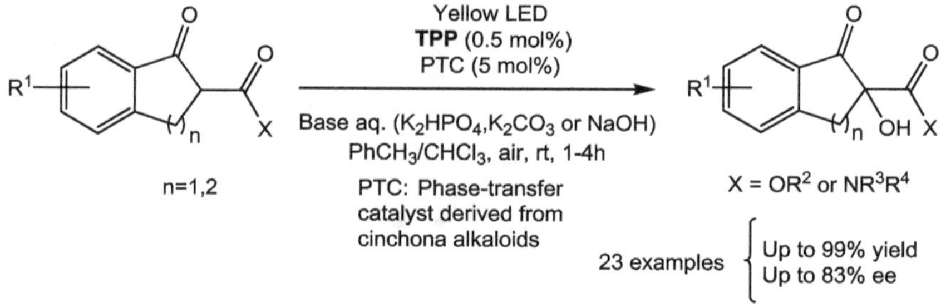

Scheme 8.3. Reaction conditions for the enantioselective α-hydroxylation of β-ketoesters or β-ketoamides by $^1O_2$ generated by irradiation of **TPP** under aerobic conditions.

The scope of the process was explored by including aliphatic, linear aldehydes such as decanal that also gave high yields and enantiomeric excesses. In contrast α,α-disubstituted 2-phenylpropanal failed to give an adequate product yield.

In another recent example, Meng and coworkers have enantiomerically used pure cinchona-derived *N*-oxides as the chiral inductor for the biphasic enantioselective α-hydroxylation of β-keto esters and amides by singlet oxygen [28], as illustrated in Scheme 8.3, where visible light and **TPP** (0.5 mol%) as the $^1O_2$ photosensitizer were used. Apparently, the key step of the asymmetric induction is the association between the chiral cinchona *N*-oxide and the enolate of the β-keto ester or amide.

The proposed reaction mechanism involves an electrophilic addition of singlet oxygen to electron-rich enolate that should initially lead to an anionic peroxide. This species can either cyclize to the corresponding dioxetane or, in the cases that an allylic aldehyde is available, form the allylic hydroperoxide. Hydroperoxides would be easily obtained from protonation of the anionic peroxide by the aqueous phase. Scheme 8.4 summarizes the reaction mechanism.

The inorganic base is present in the aqueous phase, where the conjugate base of the starting diketonic reagent is, and migrates to the organic phase as ion pair associated to the chiral cinchona *N*-oxide as the counter cation. Evidence in support of the initial formation of α-hydroperoxide was obtained by mass spectrometry,

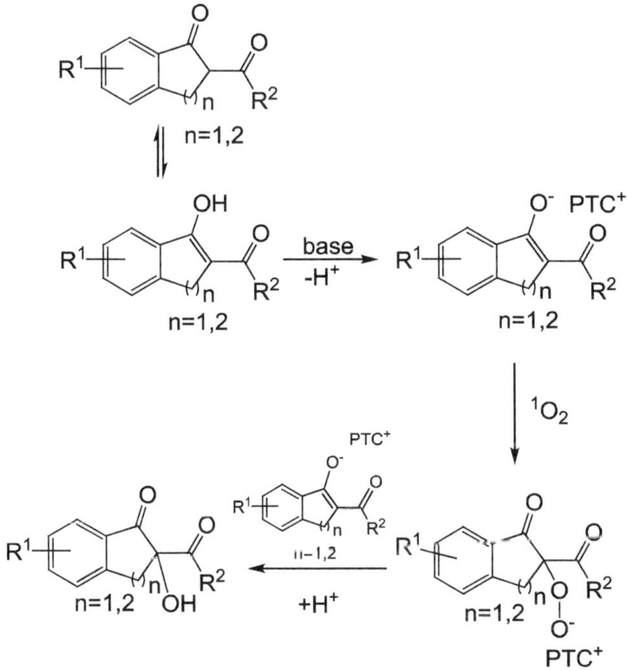

Scheme 8.4. Proposed mechanism for the photooxygenation reaction.

by performing the reaction at –20°C instead of room temperature in order to avoid its spontaneous decomposition. This hydroperoxide, due to the presence of the two carbonyls in vicinal positions, should be very reactive and could even promote the dark α-hydroxylation of an enolate, resulting in the formation of a second α-hydroxy diketonic product.

The scope of this enantioselective reaction appears to be general for indanone-substituted esters and amides, while the optimization of the cinchona PTC as chiral inductor has shown that the resulting enantioselectivity strongly depends on the structure of the natural alkaloid derivative. The current results constitute an important contribution to the field of enantioselective oxidations by singlet oxygen, considering the high enantiomeric excess values obtained. It would be important to expand this methodology based on the use of an enantiomerically pure phase transfer agent to other singlet oxygen oxygenation reactions, because in this way, using porphyrins as Photoorganocatalysts (**POCs**) and ambient oxygen, the use of hydroperoxides or other oxidizing reagents in these types of reactions will be avoided.

As mentioned earlier, one of the key features resulting in high enantiomeric excess in the α-hydroxylation promoted by $^1O_2$ is the structure of the chiral cinchona derivative that pairs with the enolate and favors the attack of $^1O_2$ selectively to only one of the two faces. High enantiomeric excess values (close to 90%) can be achieved, when suitably substituted cinchona derivatives (see the structure in Scheme 8.5) as chiral phase-transfer catalysts are used at 10 mol% working at 50°C [29].

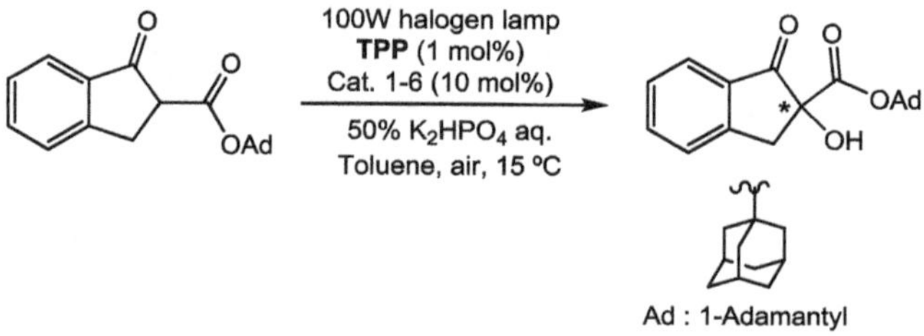

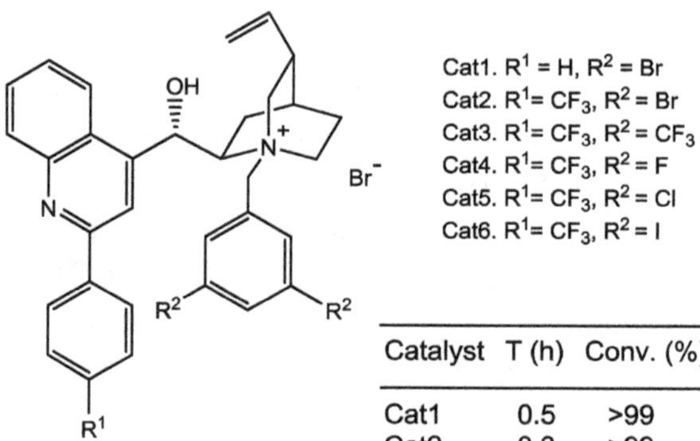

Catalyst	T (h)	Conv. (%)	ee (%)
Cat1	0.5	>99	85
Cat2	0.3	>99	86
Cat3	0.5	95	78
Cat4	0.5	>99	80
Cat5	0.5	>99	84
Cat6	0.5	>99	80

Scheme 8.5. Influence of the substitution at the cinchona-derivative phase-transfer agent on the asymmetric induction achieved in the α-hydroxylation.

Among the structural factors controlling asymmetric induction, it was proposed that the stereocontrol was highly sensitive to the substitution at the C(2) position of the cinchona quinoline ring. In contrast, asymmetric induction was not much influenced by the presence of different groups at the benzylic positions of cinchona. These studies on the correlation between the structure of the homochiral cinchona phase-transfer catalyst and asymmetric induction are of significant interest to understand the main steric factors that control enantioselective hydroxylation and could eventually lead to optimize enantioselective transfer reactions, beyond $^1O_2$ hydroxylation. The key species, from which asymmetry arises, was the anionic enolate strongly associated to the positive cinchona phase transfer catalyst. This association derives from three types of interactions, namely, Coulombic attraction, hydrogen bonding, and π–π stacking that should keep the enolate in a fixed geometry, exposing to the attack of $^1O_2$ only one face, in turn necessary for achieving high asymmetric induction.

This process was proposed to occur through singlet oxygen, rather than photoinduced electron-transfer oxygenation, and it exhibits a wide scope, particularly for β-keto esters, even if it can also be applied to β-keto amides. Involvement of $^1O_2$ is compatible with the results of the quenching experiments. For instance, DABCO is an efficient quencher of the reaction, in agreement with the key-role of $^1O_2$ in the photooxygenation. On the contrary, the addition of 10 mol% of *p*-benzoquinone, a typical $O_2^{\bullet-}$ quencher, does not have a significant effect on the reaction rate, indicating that the superoxide anion is not likely to be involved in the process.

An additional feature of the asymmetric α-hydroxylation process of β-keto carboxylic acid derivatives is that the reaction can be carried out on gram scale with a relatively minor decrease in the conversion and enantioselectivity of the final α-hydroxyl derivative [30].

As mentioned earlier, the organic hydroperoxides proposed as intermediates are highly reactive and difficult to be isolated and characterized, while evidence of their presence has been obtained by mass spectrometry. Therefore, all the experimental evidences are compatible with a mechanism involving a α-hydroperoxide that

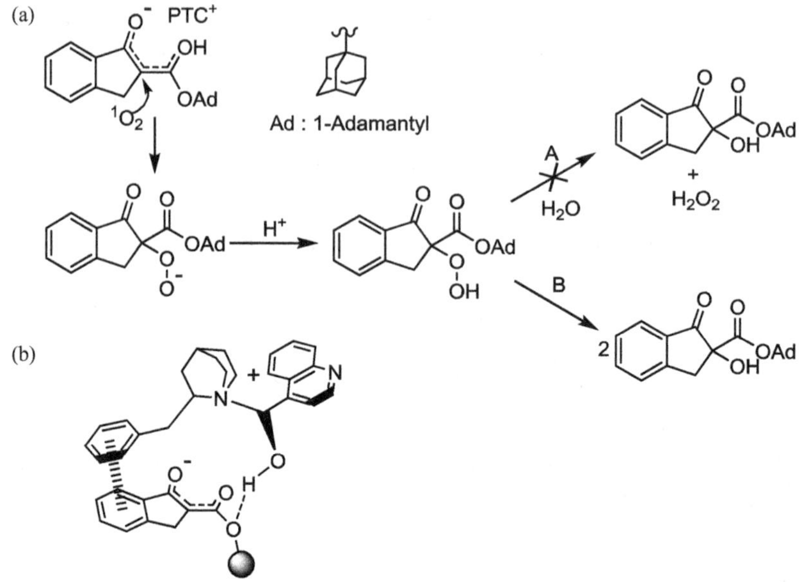

Scheme 8.6. (a) Proposed generation of a reactive hydroperoxide intermediate that, rather than decomposing and forming H_2O_2, reacts with another enolate substrate to form two molecules of α-hydroxylated β-keto ester. (b) Proposed enolate-PTC complex.

reacts with the enolate in a dark reaction, resulting in two molecules of a α-hydroxylated derivative, rather than a possible decomposition of this hydroperoxide by nucleophilic water attack that results in the generation of hydrogen peroxide. Scheme 8.6 completes the last steps of the α-hydroxylation process by considering the decomposition pathways of the hydroperoxide intermediate.

Singlet oxygen has been used for the preparation of valuable chemicals such as fragrances and therapeutical drugs [31, 32]. In one example, using **TPP** as a singlet oxygen photosensitizer aimed at the semisynthesis of artemisinin (**ART**) (see Figure 8.5) from artemisinic acid, Poliakoff and co-workers developed a strategy that prepared the antimalarial drug in a greener way [33]. The method involves the immobilization of **TPP** on an acidic sulfonated cross-linked polystyrene resin (Amberlyst-15) by protonation of the free nitrogen of the macrocyclic ring. The resulting Amberlyst resin

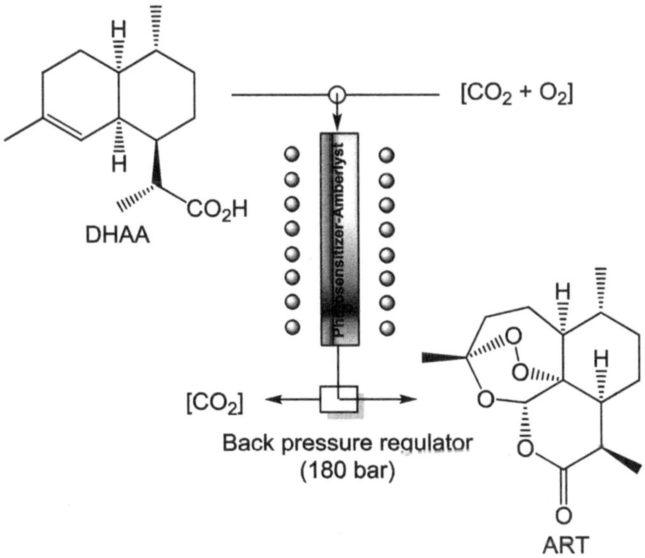

Figure 8.5. Conversion of DHAA into **ART** based on the immobilization of **TPP** photosensitizer in Amberlyst resin using a fixed-bed photoreactor.

containing **TPP** acts as a dual catalyst, since this material contains the photosensitizer for singlet oxygen generation and the Brønsted acid necessary for the ensuing hydroperoxide decomposition, ring opening, and rearrangement of the intermediate generated in the first singlet oxygen addition on dihydroartemisinic acid (DHAA). The process was reported to occur in liquid CO_2 as the solvent and the calculated E-factor, measuring the kg of waste generated per kg of products, was somewhat lower than that of the current Sanofi commercial process which, although also based on **TPP** singlet oxygen generation, requires trifluoroacetic acid as a soluble Brønsted acid and employs dichloromethane as the solvent [33].

A further step to improve the commercial production of **ART** based on DHAA oxidation by singlet oxygen is to develop a continuous flow process for this cascade process. In this regard, Seeberger and co-workers have reported the production of **ART** under continuous flow operation in 65% yield [34]. In the photoreactor, an array of high power LEDs (72 W) emitting at 420 nm, and matching

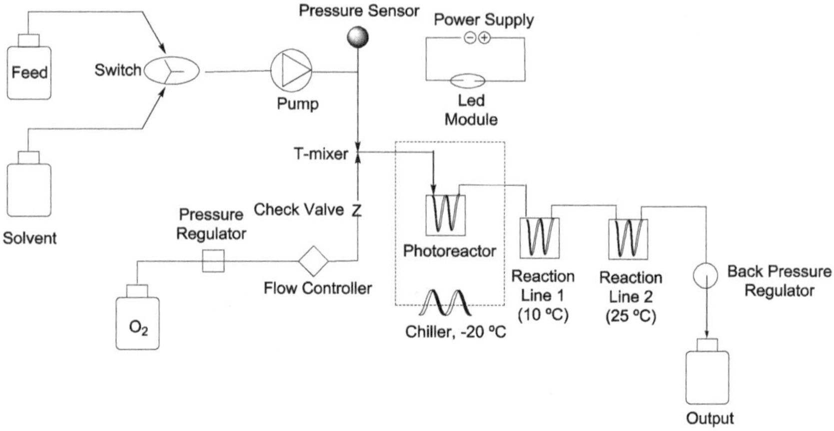

Figure 8.6. Schematic representation of the continuous-flow reactor to produce **ART** starting from DHAA through a cascade process including a singlet oxygen photooxygenation step and subsequent acid-catalyzed rearrangements.

with the Soret band of **TPP**, was used as the light source. The heat of the LEDs was dissipated by air cooling (see Figure 8.6). The photoreactor was constructed by wrapping transparent tubing around a transparent glass plate in two layers. Among the different porphyrins employed in the continuous flow process, Zn porphyrin exhibited the largest quantum yield, but did not perform well in the continuous flow photoreactor due to photobleaching. 9,10-Dicyanoanthracene (**DCA**) exhibited higher quantum yields than porphyrins, but the lower absorption coefficient made it necessary to use much higher concentrations of this photosensitizer with respect to 1 mM concentration of **TPP**. It was found that trifluoroacetic acid was the most convenient reagent in low polarity solvents, with toluene being the best choice among the various alternatives. The design of the continuous flow reactor including the photoreactor unit as well as two additional reactors connected in line, allowed sufficient flexibility to perform the photochemical step at a lower temperature (−10°C), while the subsequent acid-catalyzed reaction was performed at 25°C.

Besides oxygenations, porphyrin photosensitization under aerobic conditions can also be employed to achieve other

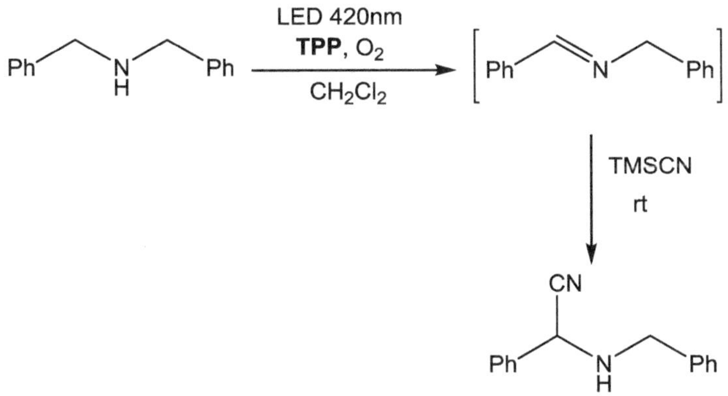

Scheme 8.7. Example of oxidative cyanation of amines.

transformations. Thus, following previous work from Che *et al.* [35, 36], Seeberger and co-workers developed a highly efficient method for the aerobic oxidative cyanation of amines using a variable-temperature flow photoreactor [37]. Singlet oxygen was generated *via* **TPP** photosensitization. In the first step, dibenzylamine was oxidized to the corresponding *N*-benzylidene-1-benzylamine and, interestingly, the by-products that are usually commonly observed in these processes, such nitriles, amides, or carbonyl compounds, were not observed in this case. The *in situ*-formed imine underwent a nucleophilic addition using TMSCN at room temperature, leading very rapidly to the quantitative formation of 2-(benzylamino)-phenylacetonitrile (see Scheme 8.7).

Pyrrolidine and piperidine were successfully converted into the corresponding α-cyano heterocycles in high yields. Primary amines behave somewhat different and undergo oxidative coupling to the corresponding *N*-substituted imines, before the subsequent cyanation. Furthermore, extensive screening of the reaction conditions allowed to determine optimal conditions to obtain an effective control over the dimeric or monomeric products. In this way, lowering the reaction temperature to –50°C and adding small amounts of tetrabutylammonium fluoride (TBAF, 4 mol% regarding TMSCN), the quantitative conversion of primary amines to the corresponding

Table 8.1. Results of the α-cyanation of amines promoted by **TPP** [37].

Entry	Substrate	Product	Yield (%)
1		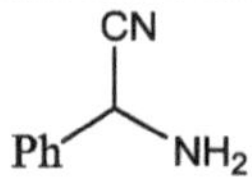	90
2	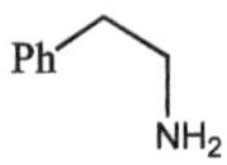	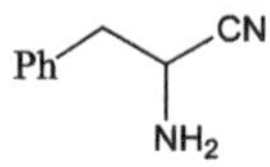	73
3	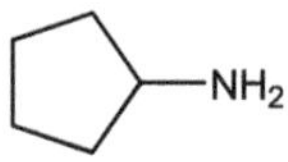	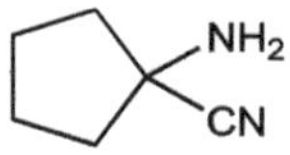	81
4	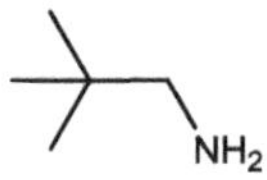	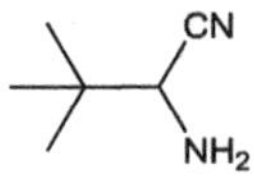	92

primary α-aminonitriles could be achieved without the oxidative coupling preferentially observed under other conditions for the first time. Table 8.1 shows some examples reported and the reaction yields obtained [37].

Cyano groups can be hydrolyzed to carboxylic acids and, in this way α-cyanation of amines can serve to synthesize α-amino acids. Taking advantage of the mild conditions of this photosensitized α-cyanation, Seeberger and co-workers also reported an effective method for the synthesis of racemic α-amino acids from the fluorinated amines [38]. Fluorinated biomolecules are of large interest in activity studies and therapeutical applications, since the positions occupied by fluorine are not susceptible to hydroxylation during the metabolization of the compound and, therefore, fluorinated compounds are more resistant to biotransformation, while preserving most of the biological activity of the non-fluorinated analog [39, 40]. Using the optimized conditions for the direct oxidation of primary amines to imines, these authors have developed a continuous flow process and demonstrated that a subsequent acid-mediated hydrolysis of α-cyano amines gives access to a range of fluorinated α-amino acids. Scheme 8.8 shows the scope of this singlet oxygen-driven photooxidative cyanation mediated by **TPP** as sensitizer.

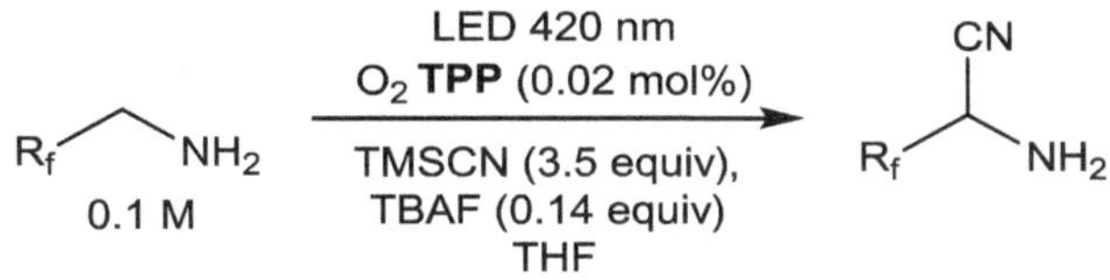

R_f	Yield %[a]
4-FC_6H_4	70
3-FC_6H_4	88
2-FC_6H_4	61
4-$CF_3C_6H_4$[b]	27
3-$CF_3C_6H_4$[b]	52
4-$CF_3C_6H_4$[b]	36
3,4-$F_2C_6H_3$	55
3,5-$F_2C_6H_3$	40
4-$CF_2(OH)C_6H_4$	55
4-$FC_6H_4CH_2$	89
3-$FC_6H_4CH_2$	79
2-$FC_6H_4CH_2$	75
$CF_3(CH_2)_2$	Decomposition

Notes: [a]Isolated yield. [b]Reaction temperature = −60°C.

Scheme 8.8. Synthesis of fluorinated α-amino nitriles.

After testing the influence of the temperature and residence time, an optimized semicontinuous process was developed, combining the photooxidative cyanation with the subsequent hydrolysis under acid conditions to achieve the desired benzylic and homobenzylic racemic α-amino acids. This process exhibits the advantage of transforming an aliphatic fluorinated compound straight away to the corresponding α-amino acid. Unlike its unstable intermediate, the product can be isolated in good yield compared to those obtained for aromatic derivatives, and it also does not require chromatographic techniques to isolate the final products. Scheme 8.9 shows the steps of this semicontinuous synthesis and some relevant examples.

One of the main problems of using porphyrins and, more specifically, **TTP**, as singlet oxygen photosensitizer is the tendency of these macrocycles to undergo oxidative degradation which results in

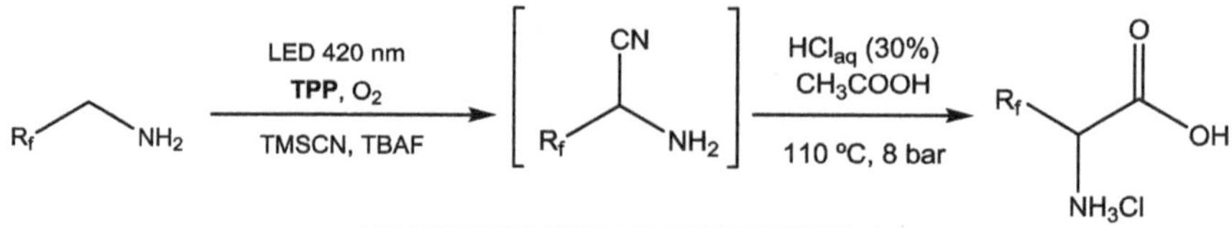

R_f	Yield (%)
4-$FC_6H_4CH_2$	64
4-FC_6H_4	67
3-$CF_3C_6H_4$	50
3,4-$F_2C_6H_3$	60
$CF_3(CH_2)_2$	63

Scheme 8.9. Synthesis of α-amino acids.

the decomposition of the **POC** during the course of the process. Accordingly, one important issue in this area is increasing the photostability of these porphyrins in the presence of molecular oxygen upon irradiation. This problem has been addressed by applying different strategies, and one of them involves the introduction of substituents and groups that increase, in general, the thermal stability of this derivative. Perfluorination of porphyrin is one case that illustrates this general methodology in order to increase the stability [41–43]. In a different approach, it has been reported that **TPP** forms a 1-to-2 intermolecular aggregation complex with 2,3-dichloro-5,6-dicyano-1,4-benzoquinone (**DDQ**) [44]. This strongly electron-poor quinone associates above and below the porphyrin ring, and the combination of charge transfer and steric encumbrance seem to increase the photostability of porphyrin in the aerobic photooxidation of cycloalkenes. The higher photostability of **TPP**, when **DDQ** is present forming the complex, is reflected in more than double cyclooctene conversion at much shorter irradiation times, resulting in a turnover number of about 800. Particularly under acidic conditions, in which the protonated form of **TPP** is present as determined by the shifts in the absorption spectrum, porphyrin degradation was completed in 24 h. In contrast, the adduct **TPP-(DDQ)**$_2$ (see Figure 8.7) does not undergo protonation under the same conditions and, as a result, it is photostable for at least 72 h.

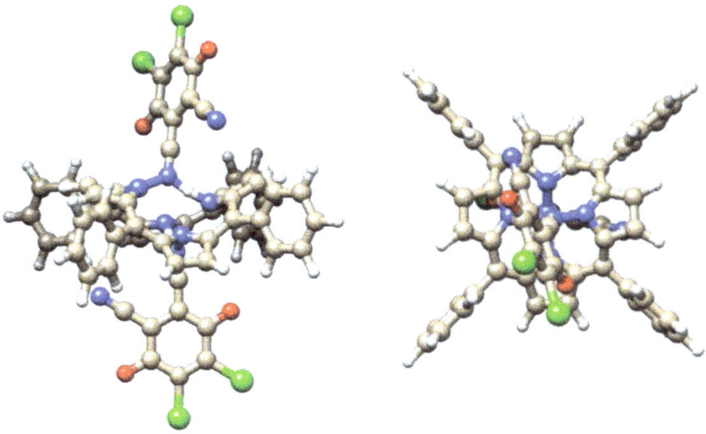

Figure 8.7. Proposed structure of the **TPP-(DDQ)$_2$** complex.

It could be also possible, however, that this higher conversion derives from a higher quantum yield of singlet oxygen generation for the **TPP-(DDQ)$_2$** adduct with respect to **TPP**, or even, from the observed red shifts of the Soret band for the association complex that makes absorption of the adduct to match better the blue LED emission of the lamp used in the photochemical experiments. It was indeed observed that blue light exciting the Soret band was more efficient than red light emitted from a different LED that was able to excite the Q_{0-0} absorption of the **TPP-(DDQ)$_2$** complex. Quantum yield of the photocatalytic cyclooctene oxidation for the **TPP-(DDQ)$_2$** complex upon irradiation with blue LEDs was 0.62, higher than the one obtained in the presence of **TPP** ($\Phi = 0.25$). By means of singlet oxygen scavengers, it was assessed that the photocatalytic oxidation by the association complex involves the generation of singlet oxygen.

Other examples on the use of porphyrins as singlet oxygen photosensitizers for the synthesis or modification of natural products include the photooxidation of isocolchicine to a mixture of endoperoxide diastereomers derived from tropolone [26, 45] and the conversion of cholesterol into 5*R*-hydroperoxycholesterol using, in this case, immobilized tetraaryl porphyrin (see Scheme 8.10) [46].

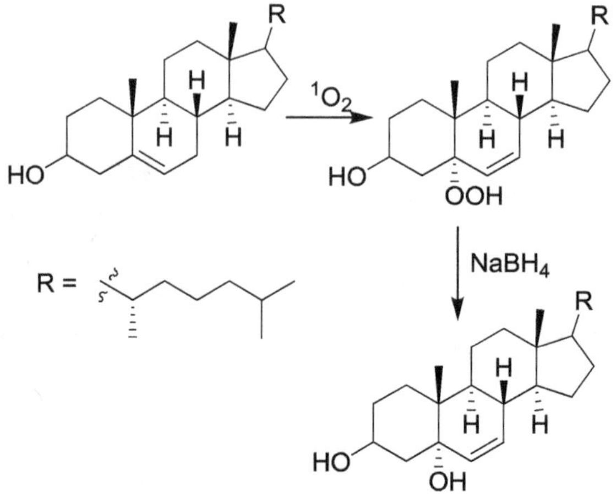

Scheme 8.10. Photooxidation of cholesterol.

8.3 Porphyrins as Electron-Transfer POCs

Porphyrins can act as oxidants as well as reductants. As a generic example, oxidation and reduction potentials of **TPP** and Zn**TPP** have been measured [47] by means of cyclic voltammetry in DMSO and a mixture of DMSO/buffered H_2O at pH = 4 in the presence of tetrabutylammonium perchlorate (TBAP) as the supporting electrolyte. Besides the redox processes in the ground state, the photoexcited porphyrins are simultaneously more oxidizing and reducing agents than the corresponding ground state. Table 8.2 summarizes some redox data for **TPP** and Zn**TPP**, depending on the singlet or triplet multiplicity of the excited state.

Redox properties of porphyrins can be tuned, both in the ground and the excited states, by introducing substituents onto the aromatic macrocycle, as well as by changing the metal in the core center [48–50], which makes this versatile family of compounds promising candidates for photoredox catalytic reactions.

In photoredox catalysis, a **POC** absorbs light, generally in the visible range, reaching an electronic excited state, either of singlet or triplet multiplicity. These electronic excited states are oxidizing and reducing species, due to the presence of an electron vacancy in

Table 8.2. Redox potentials of the free **TPP** and zinc metallated porphyrin (Zn**TPP**) in the singlet and triplet state, values are given in Volts and referenced vs the saturated calomel electrode (SCE).

	Oxidative Quenching		Reductive Quenching	
	$TPP^{\bullet+}/TPP^*$	$ZnTPP^{\bullet+}/ZnTPP^*$	TPP^*/TPP	$ZnTPP^*/ZnTPP^{\bullet-}$
Singlet state	−0.91	−1.18	0.91	0.79
Triplet state	−0.42	−0.73	0.42	0.27

the highest occupied molecular orbital and an electron in the lowest unoccupied molecular orbital, respectively. For this reason, excited states can be involved in oxidations and reductions when the electronically excited molecule is in the presence of electron donors or electron acceptors, respectively.

Although the number of examples is scarce, porphyrins can be used as photocatalysts (**PCs**) to promote photoinduced electron transfer. Thus, for instance, in the presence of suitable organic compounds with low oxidation potential, generally electron donors such as aliphatic amines, in the excited state, the photoexcited porphyrin can abstract one electron from an electron-rich organic compound, leading to the corresponding radical cation that may in turn activate a series of consecutive pathways. When this electron donor is an alkene, the overall photoredox process (Scheme 8.11) can lead to the formation of a product that formally corresponds to the anti-Markovnikov adduct in the reaction with a nucleophile [51].

The main prerequisite for this process is that the oxidation potential of the alkene is lower than the reduction potential of the **POC** in the excited state. The result is the hydrofunctionalization of the alkene. This process is particularly interesting as it promotes the intramolecular cyclization of unsaturated alcohols to the corresponding cyclic anti Markovnikov ethers. Examples are indicated in the Scheme 8.12. In this case, the presence of hydrogen atom donors in catalytic amounts appears to be necessary to achieve high product yields. Suitable hydrogen atom donors include α-phenylmalonodinitrile and 2-cyanoacetate as well as diethyl phenylmalonate.

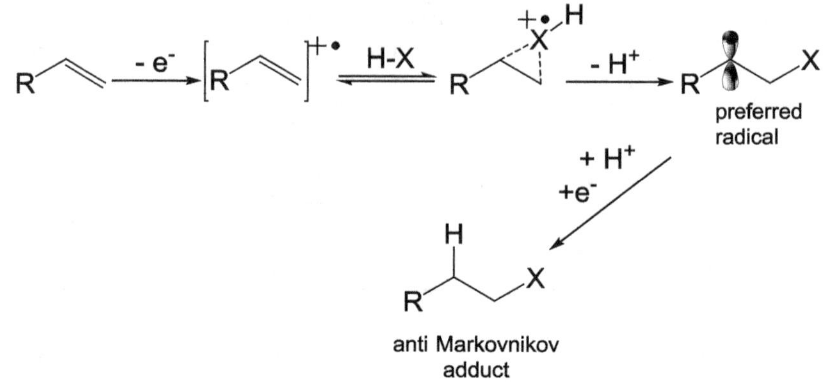

Scheme 8.11. Nucleophilic addition leading to anti Markovnikov product formation.

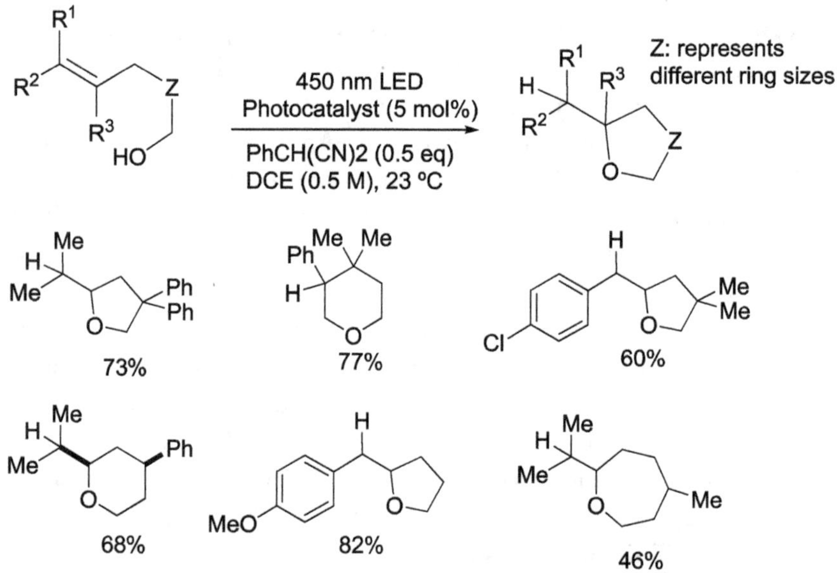

Scheme 8.12. Scope of the hydroalkoxylation reaction mediated by **POC**.

Most of the reports on the use of porphyrins as **POCs** have so far appeared to correspond to the oxygenations and oxidations, generally involving singlet oxygen generation. It is clear, however, that photocatalysis is a much broader phenomenon and has a

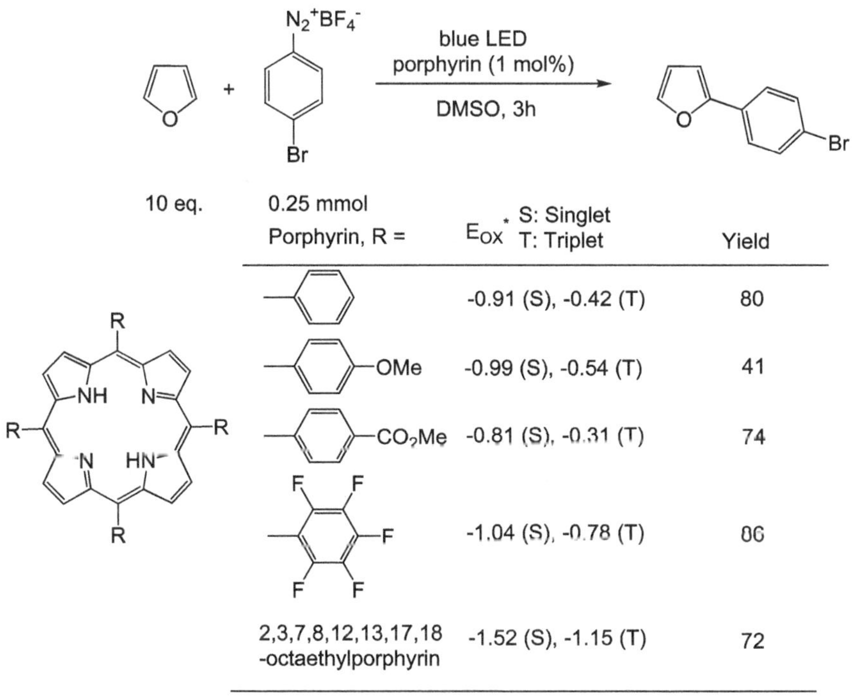

Scheme 8.13. C–H arylation of furan with diazonium salts.

considerable potential, when taking into account the photophysical properties of porphyrins in applying these dyes to other reaction types. Thus, the use of porphyrins as **POCs** for processes other than photooxidations is still underdeveloped, and there are only a few examples compared to their use as singlet oxygen photosensitizers. In one of the cases, wherein porphyrins have been applied as **POCs** for the synthesis of organic compounds, Gryko and co-workers reported the C–H arylation of heterocycles *via* the photocatalytic decomposition of an aryl diazonium salt (Scheme 8.13) [52].

Aryl diazonium decomposition is promoted by photoinduced single electron reduction operated by the excited catalyst. Among the different *meso*tetraarylporphyrins evaluated as **POCs**, those that exhibit an electron-poor character are more efficient in terms of the final product yields, than the **TPP** or tetraarylporphyrins substituted with electron donating groups. In an attempt to correlate the photocatalytic behavior of aryl porphyrins, for the C–H arylation by aryl diazonium salts, with the oxidation potential of the porphyrin's excited state and photoluminescence quenching by the electron donor, it was found that there are optimal porphyrin properties in the excited state to achieve the highest yield. Among the various solvents screened for the process, including methanol, dichloromethane, DMF, and DMSO, the latter led to the highest product yield. Optimized photocatalytic conditions used only 1 mol% of perfluoroaryl porphyrin with 10 equiv excess furan with respect to 4-bromobenzenediazonium tetrafluoroborate and blue LED (455 nm) as the light source. Although further screening of the scope of this interesting photocatalytic arylation is needed, it seems that the reaction can tolerate different electron donor and acceptor groups on the diazonium salts, although higher yields are obtained with the electron-withdrawing substituents. With regard to the heterocycle, besides furan, examples employing thiophene and *N*-substituted pyrrole were also reported. In the only example involving *N*-substituted indole as substrate, the yield of the arylation product was negligible.

Besides five-membered-ring heterocycles, the C–H arylation by aryl diazonium salts promoted by porphyrins can also be applied to the derivatization at the C(3) position of coumarins (Scheme 8.14). Besides perfluorophenyl porphyrin used by Gryko, an earlier report described the use of 5,10,15,20-tetrakis(4-*N,N*-diethylaminophenyl)-porphyrin as the **POC** for coumarin arylation, following the lead of earlier studies on the photocatalytic C(3)–H coumarin arylation using $Ru(bpy)_3^{2+}$ [53, 54].

As the scope, it was found that the presence of electron-donating groups on the coumarin favors the formation of the corresponding arylated product in good yield. The reaction can also proceed in the presence of electron-deficient substituents, but in this case the yields are lower. Moreover, C(4)-substituted coumarins can also undergo C(3)–H arylation in spite of steric encumbrance. The photocatalytic reaction also exhibits a wide scope with respect to the substituents on the aryl diazonium tetrafluoroborate, and it appears that the

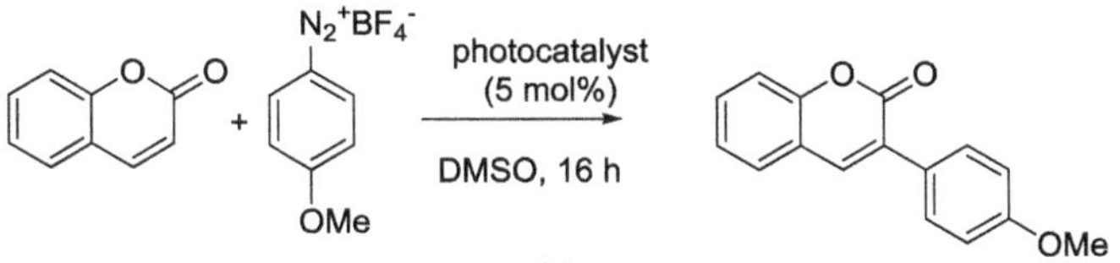

(a)

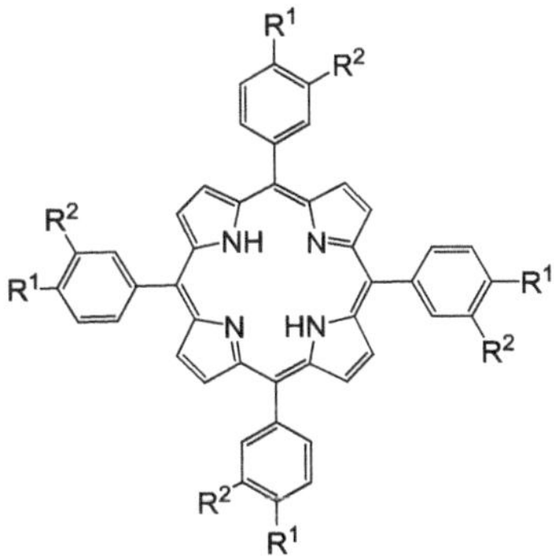

Photocatalyst	Yields[a] (%)
$R^1 = R^2 = H$	0
$R^1 = OMe$, $R^2 = H$	0
$R^1 = NH_2$, $R^2 = H$	17
$R^1 = NMe_2$, $R^2 = H$	0
$R^1 = NEt_2$, $R^2 = H$	24 (68)[b]
$R^1 = NH_2$, $R^2 = OMe$	20

(b)

Notes: [a]All the reactions shown here were carried out using 5 mol% of the porphyrin at room temperature. Yields determined by ^{1}H-NMR spectroscopy. [b]Yield using 10 mol% at 40°C.

Scheme 8.14. (a) C-3 arylation of coumarins promoted by porphyrins as **POCs**. (b) Yields that have been achieved.

photocatalytic arylation can be applied to halo-substituted substrates that are not suitable reagents under conventional conditions for the Heck reaction involving Pd catalysis.

As in the case of the C–H arylation of five-membered-ring heterocycles, the proposed reaction mechanism also involves the initial generation of aryl radicals from aryl diazonium salts by photoinduced electron transfer. Reduction of the aryl diazonium cation leads to the instantaneous decomposition of this labile reagent to form nitrogen and an aryl radical intermediate that can attack the electron-rich heterocycle. Scheme 8.15 depicts the proposed elementary steps of the mechanism. According to this scenario, limitations of the photocatalytic reaction are mainly due to alternative quenching pathways of porphyrin excited state by the heterocycle, rather than by the diazonium salt. For instance, indole, which is a good electron donor, could compete favorably in the quenching or

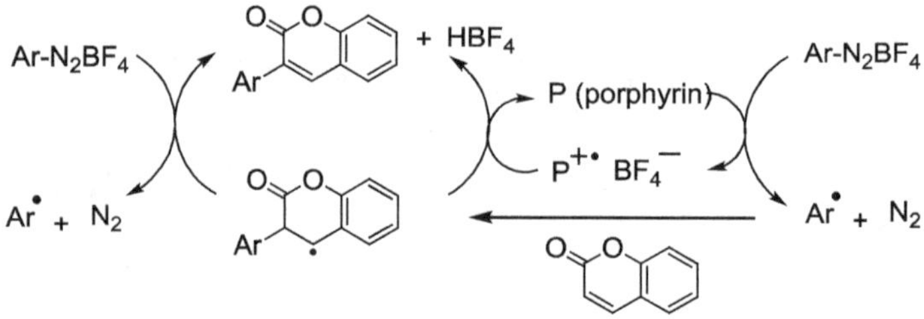

Scheme 8.15. Proposed reaction mechanism.

porphyrin excited state with the reaction of the aryl diazonium salt, and in this way aryl radical formation may be inhibited.

Photoredox organocatalysis is currently attracting much interest due to the large diversity of the processes for organic synthesis that have been uncovered recently, the very mild conditions in which they can be carried out, and the high efficiencies that can be achieved. These photoredox processes typically involve an amine as the sacrificial electron donor agent and the use of a transition metal complex, generally Ir, but also Ru^{2+} polypyridyl, to promote the decomposition of the reagent by photoinduced electron transfer. Among the photoredox catalytic reactions, there are some that involved the derivatization of aldehydes by functionalization with trifluoromethyl groups [55], benzylations [56], and even alkylations [57].

There is an interest in expanding the use of organic dyes such as methylene blue, Eosin Y, and fluorescein as photoredox catalysts in order to reduce the use of valuable transition metals. Due to the already mentioned electrochemical properties, porphyrins have been recently employed as **POC**s in the α-alkylation of aldehydes with diazo esters (Scheme 8.16) [47]. The optimized reaction conditions employ 3-phenylpropanal (0.5 mmol), 1 equiv of ethyl diazoacetate, morpholine (as the electron donor, 0.4 equiv), and four white LEDs as the light source. Finally, free **TPP** and Zn-**TPP** were used for this reaction (1 mol%). Remarkably, both porphyrins promoted the light-induced redox α-alkylation to give the desired product in 84% and 88% yields, respectively, in 5 h. The drop in yield when the

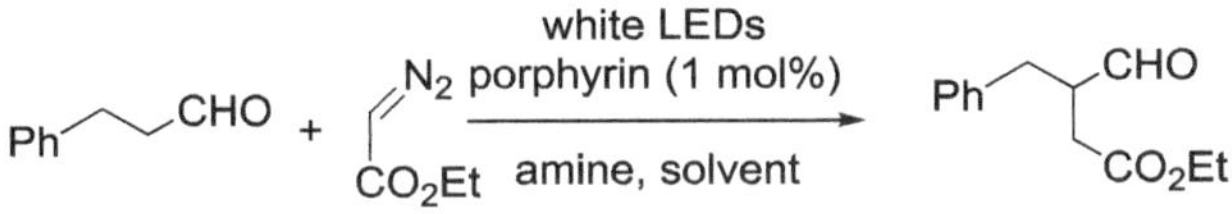

Scheme 8.16. C–C bond formation reaction between an aldehyde and ethyl diazoacetate catalyzed by porphyrins as **POCs**.

reaction was carried out open to air or under O_2 is noteworthy; this effect is due to the generation of singlet oxygen that competes with the photoredox mechanism and decreases the product formation.

In order to maximize product yields, the authors tested a variety of functionalized free-base and zinc-metallated **TPPs**, optimized the catalyst loading, and determined the influence of the selected amine. Their experiments led to the conclusion that the most efficient photoredox catalysts were in fact, **TPP** and Zn-**TPP**. As the decorating groups became more electron rich, the catalytic performance improves, and a reduced loading (0.1 mol%) could be achieved using Zn-**TPP** due to its good solubility, showing the promising use of porphyrins for this reaction. Scheme 8.17 illustrates some examples highlighting the scope of this reaction.

In these reactions, **TPP** acts as the photosensitizer and photoredox catalyst. Scheme 8.18 illustrates the reaction mechanism for this reaction of aldehydes with alkyl diazo derivatives induced by light. Such a mechanism involves two cycles; first, **TPP** is excited from the ground state to its excited state under light irradiation and transfers energy to the diazo compound. The excited diazo compound undergoes a fast decomposition and is transformed into the corresponding triplet state carbene. The second cycle starts with the oxidation of the enamine by the excited **TPP** to generate a radical that subsequently reacts with the carbene to form another radical. **TPP** transfers an electron, gives a zwitterion which after protonation and hydrolysis releases the desired aldehyde.

In another example of photoorganocatalysis, it has been reported that α-halocarbonyls can undergo photodehalogenation promoted by thiaporphyrins upon irradiation with visible light in the presence of *N,N* diisopropylethylamine, and with the Hantzsch dihydropyridine ester (HEH, see structure below) as the hydrogen donor [58].

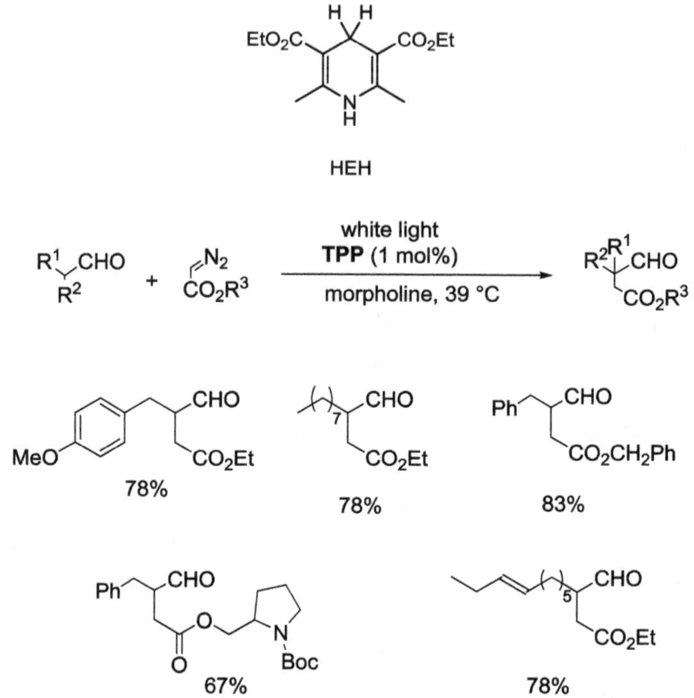

Scheme 8.17. α-functionalization of aldehydes.

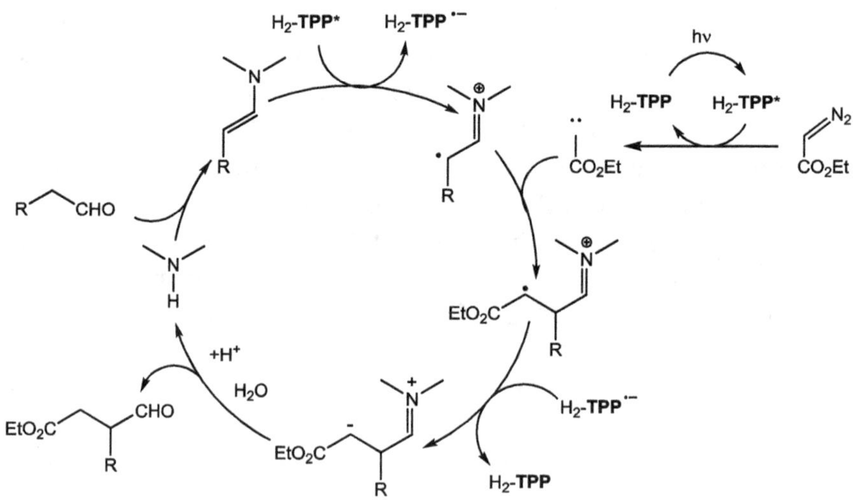

Scheme 8.18. Proposed catalytic cycles for the α-functionalization of aldehydes.

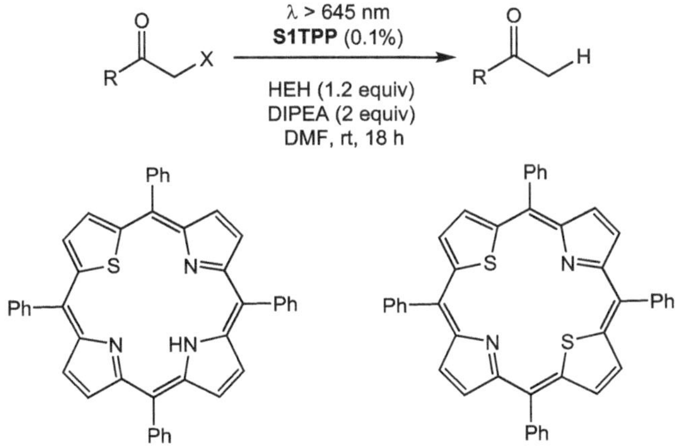

S1TPP **S2TPP**

Entry	Substrate	Product	% Yield[a]	Entry	Substrate	Product	% Yield[a]
1	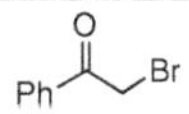	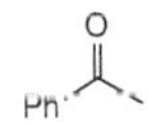	90 00[b]	5	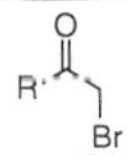	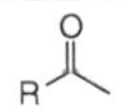	NR (R= -OBn) NR (R= -NHBn) 16[c] (R= -*t*-Bu) 40 (R= -adam)
2	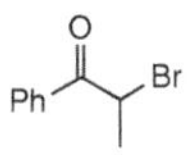	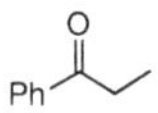	78 60[b]	6	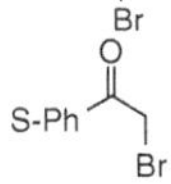	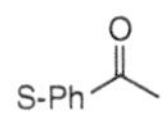	86 (X= -*p*-NO_2) 85 (X= -*p*-OMe) 76 (X= -*p*-Br) 80 (X= -*p*-CN) 92 (X= -*m*-OMe)
3	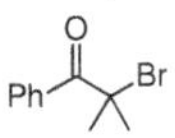	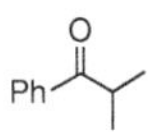	75 40[b]	7	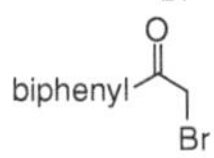	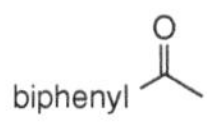	98
4	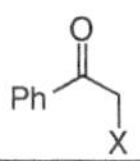	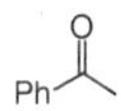	12 (X=Cl) 100 (X=I)				

Notes: [a] Isolated yield on 1 mmol scale. [b] **S2TPP** used as catalyst. [c] Yield measured by means of GC-MS analysis. NR = no reaction, **adam** = 1-adamantyl (1-(1,1′-biphenyl)-4-yl), **S1TPP** = meso5,10,15,20-tetraphenyl-21-monothiaporphyrin, **S2TPP** = 5,10,15,20-tetraphenyl-21,23-dithiaporphyrin.

Scheme 8.19. Reaction scope and yields obtained for the photoredox organocatalytic dehalogenation of α-halo ketones.

One of the main characteristics of S-substituted porphyrins is the long wavelength absorption (beyond 650 nm) and the moderate excited state reduction potentials (about 0.6 V to 0.7 V vs SCE). Analogous irradiations using other **PCs** such as $Ru(bpy)_3Cl_2$, result in only poor dehalogenation product yields.

Scheme 8.19 summarizes the scope of the reaction and some of the yields that have been reported for this photocatalytic process.

Besides the use of porphyrins dissolved in organic solvents as **POCs**, one possibility is to fabricate an insoluble solid containing the photoactive material. The main problem with this approach is that the interaction with the photoactive component and the substrates may become much disfavored as a consequence of the inefficient contact and mass transfer limitations. This drawback can be minimized by developing porous materials where the dye is attached to a porous support with large surface area. In this regard, metal organic frameworks (MOFs), where the structure contains a bi- or multi-podal rigid organic linker and metallic nodes, have attracted considerable attention due to the remarkably high surface area, very frequently above 1,000 m^2/g and even reaching, in some cases, 3,000 m^2/g, as well as a large porosity, site accessibility, and a considerable percentage of empty space. Typical linkers for MOFs are aromatic polycarboxylic acids, but the use of porphyrins substituted with phenyl carboxylic groups has been widely adopted, starting from the seminal synthesis of porphyrin MOFs by Chen and co-workers [59].

One of the examples of MOFs containing porphyrins as linkers is MOF-525. MOF-525 is constituted by zirconium nodes and tetrakis (4-carboxyphenyl)porphyrin (**TCPP**) as linker. MOF-525 was prepared and used as a solid and recyclable **PC** for the oxidative hydroxylation of aryl boronic acids [60], as indicated in Scheme 8.20.

The photocatalytic reactions are carried out under oxygen in the presence of an excess of triethylamine as an electron donor upon irradiation with green light at room temperature. Interestingly, control experiments showed that the solid porphyrin MOF exhibits a higher catalytic activity than the linker dissolved in MeCN/H_2O under the same operating conditions (see Figure 8.8).

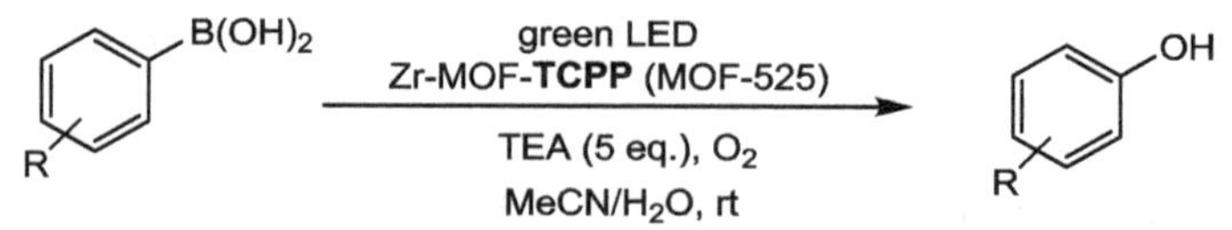

Scheme 8.20. Hydroxylation of arylboronic acids using MOF-525 (Zr-MOF-**TCPP**).

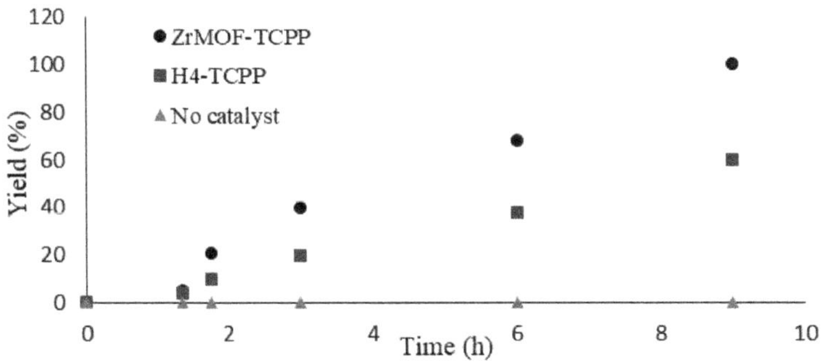

Figure 8.8. Time-conversion plots for the oxidation of phenylboronic acids to the corresponding phenols in the absence or in the presence of *meso*-tetra(4-carboxyphenyl) porphyrin dissolved in CH_3CN/H_2O, or by a zirconium MOF containing this tetracarboxylic porphyrin as linker.

A TON of 70 was calculated for the reaction based on the porphyrin present in the solid, firmly establishing the role of the porphyrin as the photoredox catalyst. The reaction has a wide scope regarding the substrate, including arylboronic acids containing electron donor and acceptor substituents and also sterically hindered 2,6-dimethylphenol (Scheme 8.21). The yield in all cases was quantitative, although the time required to achieve complete conversion depended on the substrate.

Besides photochemical oxygenation by singlet oxygen generated through an energy transfer mechanism, photooxygenation of alkenes by water through a photoinduced electron-transfer mechanism has also been reported for Al^{3+} and Si^{4+} metallated porphyrins [61]. In the case of silicon porphyrin, the Si^{4+} central atom exhibits a distorted octahedral coordination, with four N atoms in equatorial positions and two additional OH groups in the axial coordination. Upon irradiation with *quasi* monochromatic light from a xenon lamp using pass band filters that selectively allow to irradiate the Soret (420 nm) or the Q band (560 nm), the silicon porphyrin undergoes photoexcitation in the presence of K_2PtCl_4 as the sacrificial electron acceptor. In this way, it is possible to promote the activation of a H_2O molecule that can result in oxygenation of cyclohexene. Scheme 8.22 illustrates the photocatalytic reaction occurring upon

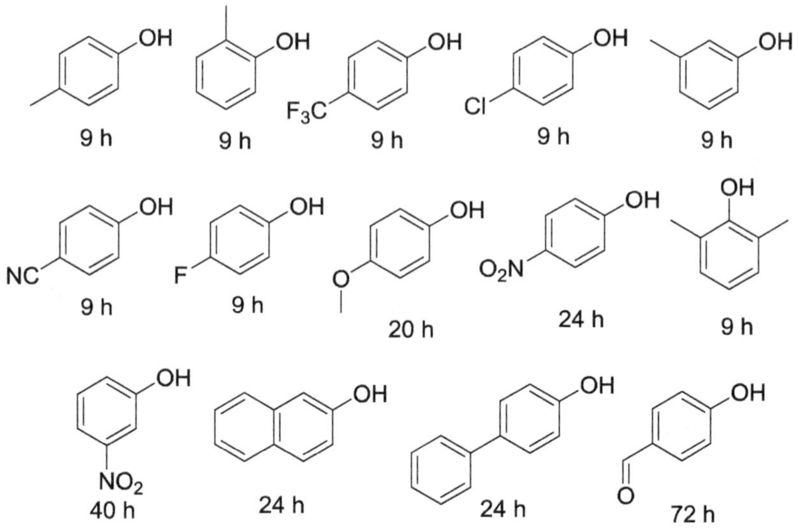

Scheme 8.21. Examples of phenols obtained through oxidative hydroxylation of arylboronic acids using MOF-525. Reaction conditions: Zr-MOF-**TCPP** (0.002 mmol of **TCPP**), arylboronic acid (0.6 mmol), TEA (3.0 mmol), MeCN/H_2O (4:1), green LED, rt. Product yields were, in all cases, quantitative at the indicated time.

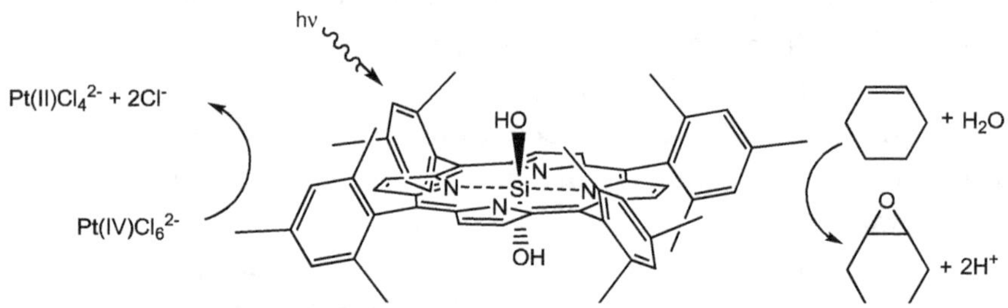

Scheme 8.22. Photooxygenation with water promoted by tetra-((2,4,6-trimethyl) phenyl)porphyrinatosilicon (**SiTMP**) in the presence of K_2PtCl_6 as the electron acceptor.

excitation of the Si porphyrin, while Table 8.3 summarizes some of the results that can be obtained under these conditions.

Other alkenes such as norbornene and styrenes can also undergo photochemical oxygenation by water under these conditions. It is interesting to note that the experimental procedure excludes the presence of molecular oxygen from the ambient and

Table 8.3. Photochemical oxygenation of cyclohexene by H_2O under visible-light irradiation ($\lambda = 420$ nm) using **SiTMP** as **POC** and reported yields [61]. Reaction conditions: $\lambda_{IRR} = 420$ nm, [**SiTMP**] $= 1 \times 10^{-5}$ M, [K_2PtCl_6] $= 5 \times 10^{-4}$ M, [Cyclohexene] $= 1 \times 10^{-2}$ M.

	Yield (μM)						
KOH (mM)		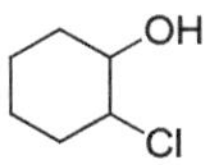	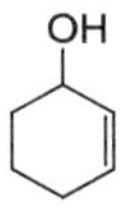		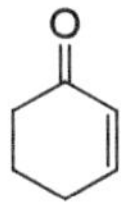	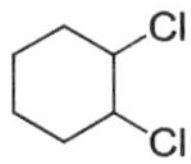	Total Yield
0	1.0	16.5	6.7	0.0	0.0	312.9	337.2
0.5	0.7	11.9	5.9	1.3	1.3	141.8	162.6
1.0	3.4	2.8	4.2	1.8	2.1	33.4	47.7
1.5	2.1	0.5	5.9	2.7	4.8	10.9	26.9
3.0	0.1	0.1	1.6	0.1	4.2	0.0	6.1
5.0	0.1	0.1	0.8	0.0	2.7	0.0	3.7

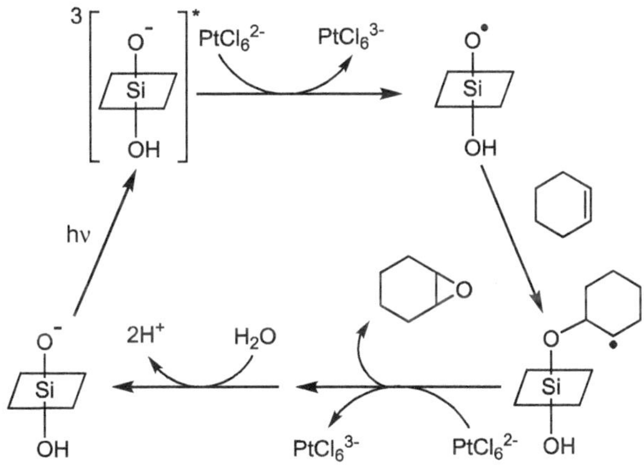

SiTMP (O⁻)(OH)

Scheme 8.23. Proposed mechanism for the photooxygenation of cyclohexene using visible light and **SiTMP** as **POC**.

the occurrence of water activation was conclusively demonstrated by using isotopically-labeled $H_2{}^{18}O$ as reagent, whereby a certain degree of ^{18}O incorporation was observed. Accordingly, the proposed reaction mechanism is indicated in Scheme 8.23, where an oxide ligand axially bound to silicon transfers an electron at the

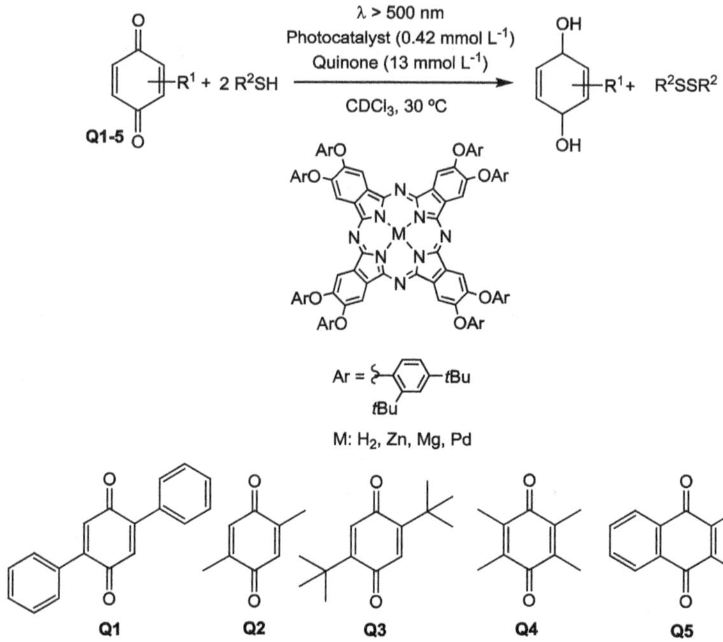

Scheme 8.24. General reaction scheme of the photocatalyzed reduction of quinones by thiols.

triplet excited state stage by photoinduced electron transfer to $PtCl_6^{2-}$, resulting in an oxygen-centered radical that is sufficiently electrophilic to attack the C=C double bond, the initial complex being restored by subsequent reaction with water. The estimated TON for photooxygenation by H_2O was 30.

In addition to porphyrins, structurally-related phthalocyanines can also be potentially used as **PCs** under similar conditions as those of porphyrins. However, in spite of the initial advantage derived from the presumably higher photocatalytic stability of phthalocyanines, their low solubility in most common solvents constitutes a severe limitation for these compounds. In one of the so-far rare examples about the use of phthalocyanines as **POCs**, quinones have been reduced to the corresponding hydroquinones by thiols using red light (690 nm) excitation (see Scheme 8.24 for reaction conditions and Table 8.4 for reported rates) [62]. It is also interesting to

Table 8.4. Reported initial rates for the photocatalyzed reduction of some quinones.

Entry	Q	Thiol	Sensitizer	Relative Rate[a]
1	Q1	4-*t*-BuC$_6$H$_4$SH	ZnPc	12
2	Q2	4-*t*-BuC$_6$H$_4$SH	ZnPc	3.3
3	Q3	4-*t*-BuC$_6$H$_4$SH	ZnPc	1
4	Q4	4-*t*-BuC$_6$H$_4$SH	ZnPc	0.6
5	Q5	4-*t*-BuC$_6$H$_4$SH	ZnPc	0
6	Q3	C$_6$F$_5$SH	ZnPc	14
7	Q3	1-C$_8$H$_{17}$SH	ZnPc	0.003
8[b]	Q1	4-*t*-BuC$_6$H$_4$SH	ZnPc	310
9[b]	Q2	4-*t*-BuC$_6$H$_4$SH	ZnPc	82
10[b]	Q3	4-*t*-BuC$_6$H$_4$SH	ZnPc	11
11[b]	Q4	4-*t*-BuC$_6$H$_4$SH	ZnPc	2.5
12	Q3	C$_6$F$_5$SH	H$_2$Pc	0.1
13	Q3	C$_6$F$_5$SH	MgPc	1.8
14	Q3	C$_6$F$_5$SH	PdPc	0.1

Notes: [a]Rate relative to entry 3. [b]Additive: acetic acid (42 mmol L^{-1}).

remark that photocatalysis at such a long wavelength is not very common due to the scarce number of **PCs** absorbing in this region and also due to the lower energy of red photons compared to UV or blue photons.

Reduction of quinones to hydroquinones is proposed to occur through photoinduced electron transfer from phthalocyanines in the triplet excited state as electron donors to quinones, which are known to be strong electron acceptors in their ground state (see Scheme 8.25). This elementary step would generate a semiquinone anion radical and protonation of this intermediate would constitute the rate determining step. For this reason, the presence of acetic acid as a Brønsted acid, together with an excess of thiol exerted a positive influence on the reaction rate. Stoichiometric amounts of disulfide (resulting from the oxidation of the thiol) were simultaneously formed in the process.

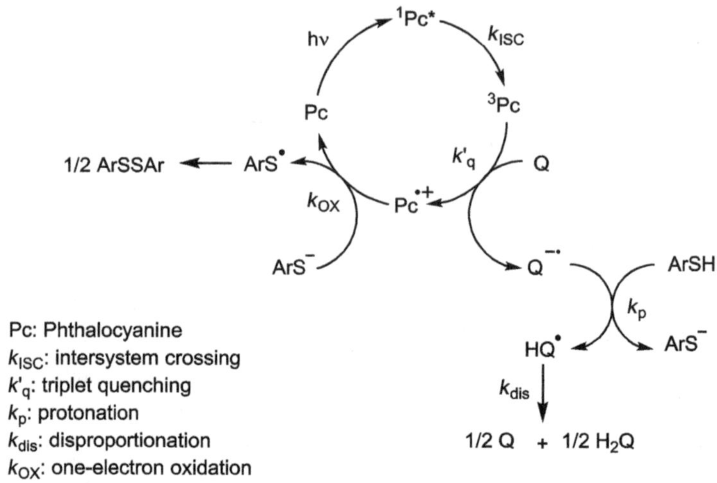

Scheme 8.25. Proposed reaction mechanism.

Among the different phthalocyanines tested, Zn^{2+}-metallated phthalocyanine is the one that exhibits the highest efficiency. Concerning the reaction scope, four different 2,5-disubstituted quinones and one antraquinone were tested, while, on the other hand, aryl thiols appeared to be more suited than alkyl thiols.

8.4 Conclusions and Future Prospects

Porphyrins are key pigments in natural photosynthesis; nature, over millions of years of evolution, has selected porphyrins to harvest solar light among other alternatives and, based on them, life has been sustained. Photocatalysis aims at mimicking natural photosynthesis using the energy of light to synthesize a diversity of organic compounds and chemicals. From the current state of the art that we have briefly summarize in the previous sections, it is clear that porphyrins are underutilized in photocatalysis. The vast majority of the current reported examples use porphyrins, mainly **TPP**, to generate singlet oxygen and porphyrins and have shown their excellent performance in this area. Regarding singlet oxygen, the present state-of-the-art focuses on the immobilization of these dyes on supports in order to develop continuous flow processes, showing that these types of photocatalytic reactions can reach industrial applications. Also remarkable are the achievements in the area of asymmetric induction in photooxygenations, and it can be easily foreseen that there will be an effort in expanding the scope of enantioselective hydroxylations beyond β-keto carboxylic acid derivatives and carbonyls.

Porphyrins are also used to promote electron-transfer reactions, and the area is currently moving toward the development of photoredox organocatalysis using porphyrins as the photoresponsive components. In these reactions, the excited state of porphyrins generates radicals by transferring electrons to (or from) the organic substrate.

However, in spite of the two types of general photocatalytic transformations that have been explained, porphyrins offer many other possibilities that will have to be developed in the future, including photocatalytic processes based on energy transfer to suitable acceptors, and application of porphyrins in the area of solar light conversion for the development of novel generations of sustainable fuels. Key reactions in this area include hydrogen evolution from water and photocatalytic-selective CO_2 reduction. For these applications, it will be necessary to control the redox potential of the excited states by appropriate substitution, to introduce catalytic sites at the periphery of the porphyrin chromophores and, more importantly, to increase the photostability of porphyrins. Spatial designing, inspired by the structure of the natural photosynthetic centers and ultrafast quenching of excited states by quinones or alternative electron acceptors are complement methodologies. These are based on the use of chemically-robust porphyrins less suitable to undergo self-degradation, particularly under oxidative conditions. Therefore, there are still plenty of opportunities for the development of new photocatalytic processes based on these dyes.

Acknowledgments

Sonia Remiro-Buenamañana would like to acknowledge the Research Executive Agency (REA) and the European Commission, for the funding received under the Marie Skłodowska-Curie actions (H2020-MSCA-IF-2015/Grant Agreement Number 709023/ZESMO).

References

[1] Jori, G.; Spikes, J. D., Photobiochemistry of Porphyrins, In *Topics in photomedicine*, Plenum Press, New York, 1984, pp. 183–318.

[2] Riess, J. G. *Chem. Rev.* **2001**, *101*, 2797–2920.

[3] van der Rest, M.; Gingras, G. *J. Biol. Chem.* **1974**, *249*, 6446–6453.

[4] Milgrom, L. R. *The Colours of Life: An Introduction to the Chemistry of Porphyrins and Related Compounds*, Oxford University Press, Oxford, 1997.

[5] Fischer, H.; Zeile, K. *Justus Liebigs Ann. Chem.* **1929**, *468*(1), 98–116.

[6] Cammidge, A. N.; Chambrier, I.; Cook, M. J.; Hughes, D. L.; Rahman, M.; Sosa-Vargas, L. *Chem. Eur. J.* **2011**, *17*, 3136–3146.

[7] Kadish, K. M., *Porphyrin Science: With Applications to Chemistry, Physics, Materials Science, Engineering, Biology and Medicine*, World Scientific Publishing, Company, 2010.

[8] Gouterman, M. *J. Chem. Phys.* **1959**, *30*, 1139–1161.

[9] Okura, I., *Photosensitization of Porphyrins and Phthalocyanines*, CRC Press, 2017.

[10] Darwent, J. R.; Douglas, P.; Harriman, A.; Porter, G.; Richoux, M.-C. *Coord. Chem. Rev.* **1982**, *44*, 83–126.

[11] Rimington, C. *Biochem. J.* **1960**, *75*, 620–623.

[12] Meot-Ner, M.; Adler, A. D. *J. Am. Chem. Soc.* **1975**, *97*, 5107–5111.

[13] Li, M.; Xiao, Z.; Xiang, H.; Lu, Z. *Spectrochim. Acta A* **1997**, *53*, 1691–1695.

[14] Bhyrappa, P.; Sankar, M. *Spectrochim. Acta A* **2018**, *189*, 80–85.

[15] Nakamura, M.; Ohgo, Y.; Ikezaki, A. Electronic and magnetic structures of iron porphyrin complexes, In *Handbook of Porphyrin Science*, Vol. 7, World Scientific Publishing Company, 2010, pp. 1–146.

[16] Barker, C. A.; Zeng, X.; Bettington, S.; Batsanov, A. S.; Bryce, M. R.; Beeby, A. *Chem. Eur. J.* **2007**, *13*, 6710–6717.

[17] Kimura, M.; Shiba, T.; Yamazaki, M.; Hanabusa, K.; Shirai, H.; Kobayashi, N. *J. Am. Chem. Soc.* **2001**, *123*, 5636–5642.

[18] de Souza, T. G. B.; Vivas, M. G.; Mendonça, C. R.; Plunkett, S.; Filatov, M. A.; Senge, M. O.; De Boni, L. *J. Porphyr. Phthalocyanines* **2016**, *20*, 282–291.

[19] Dougherty, T. J.; Gomer, C. J.; Henderson, B. W.; Jori, G.; Kessel, D.; Korbelik, M.; Moan, J.; Peng, Q. *J. Natl. Cancer Inst.* **1998**, *90*, 889–905.

[20] Mojarrad, A. G.; Zakavi, S. *Catal. Sci. Technol.* **2018**, *8*, 768–781.

[21] Lanzilotto, A.; Kyropoulou, M.; Constable, E. C.; Housecroft, C. E.; Meier, W. P.; Palivan, C. G. *J. Biol. Inorg. Chem.* **2018**, *23*, 109–122.

[22] Schmidt, R.; Tanielian, C. *J. Phys. Chem. A* **2000**, *104*, 3177–3180.

[23] Spada, R. M.; Macor, L. P.; Hernández, L. I.; Ponzio, R. A.; Ibarra, L. E.; Lorente, C.; Chesta, C. A.; Palacios, R. E. *Dyes Pigm.* **2018**, *149*, 212–223.

[24] Rånby, B. G.; Rabek, J. F. *Singlet Oxygen Reactions with Organic Compounds and Polymers*, John Wiley & Sons, 1978.

[25] DeRosa, M. C.; Crutchley, R. J. *Coord. Chem. Rev.* **2002**, *233*, 351–371.

[26] Ghogare, A. A.; Greer, A. *Chem. Rev.* **2016**, *116*, 9994–10034.

[27] Walaszek, D. J.; Rybicka-Jasińska, K.; Smoleń, S.; Karczewski, M.; Gryko, D. *Adv. Synth. Catal.* **2015**, *357*, 2061–2070.

[28] Wang, Y.; Yin, H.; Tang, X.; Wu, Y.; Meng, Q.; Gao, Z. *J. Org. Chem.* **2016**, *81*, 7042–7050.

[29] Wang, Y.; Zheng, Z.; Lian, M.; Yin, H.; Zhao, J.; Meng, Q.; Gao, Z. *Green Chem.* **2016**, *18*, 5493–5499.

[30] Lian, M.; Li, Z.; Cai, Y.; Meng, Q.; Gao, Z. *Chem. Asian J.* **2012**, *7*, 2019–2023.

[31] Wasserman, H. H.; Ives, J. L. *Tetrahedron* **1981**, *37*, 1825–1852.

[32] Ohloff, G. *Pure Appl. Chem.* **1975**, *43*, 481–502.

[33] Amara, Z.; Bellamy, J. F. B.; Horvath, R.; Miller, S. J.; Beeby, A.; Burgard, A.; Rossen, K.; Poliakoff, M.; George, M. W. *Nat. Chem.* **2015**, *7*, 489.

[34] Kopetzki, D.; Lévesque, F.; Seeberger, P. H. *Chem. Eur. J.* **2013**, *19*, 5450–5456.

[35] To, W.-P.; Tong, G. S.-M.; Lu, W.; Ma, C.; Liu, J.; Chow, A. L.-F.; Che, C.-M. *Angew. Chem.* **2012**, *124*, 2708–2711.

[36] Jiang, G.; Chen, J.; Huang, J.-S.; Che, C.-M. *Org. Lett.* **2009**, *11*, 4568–4571.

[37] Ushakov, D. B.; Gilmore, K.; Kopetzki, D.; McQuade, D. T.; Seeberger, P. H. *Angew. Chem. Int. Ed.* **2014**, *53*, 557–561.

[38] Vukelić, S.; Ushakov, D. B.; Gilmore, K.; Koksch, B.; Seeberger, P. H. *Eur. J. Org. Chem.* **2015**, *2015*, 3036–3039.

[39] Smart, B. E. *J. Fluorine Chem.* **2001**, *109*, 3–11.

[40] Ojima, I. *Fluorine in Medicinal Chemistry and Chemical Biology*, John Wiley & Sons, 2009.

[41] Shimizu, S.; Shin, J. Y.; Furuta, H.; Ismael, R.; Osuka, A. *Angew. Chem. Int. Ed.* **2003**, *42*, 78–82.

[42] Lai, S.-W.; Hou, Y.-J.; Che, C.-M.; Pang, H.-L.; Wong, K.-Y.; Chang, C. K.; Zhu, N. *Inorg. Chem.* **2004**, *43*, 3724–3732.

[43] Balkus Jr, K. J.; Eissa, M.; Levado, R. *J. Am. Chem. Soc.* **1995**, *117*, 10753–10754.

[44] Mojarrad, A. G.; Zakavi, S. *RSC Adv.* **2016**, *6*, 100931–100938.

[45] Brecht, R.; Büttner, F.; Böhm, M.; Seitz, G.; Frenzen, G.; Pilz, A.; Massa, W. *J. Org. Chem.* **2001**, *66*, 2911–2917.

[46] Lumley, E. K.; Dyer, C. E.; Pamme, N.; Boyle, R. W. *Org. Lett.* **2012**, *14*, 5724–5727.

[47] Rybicka-Jasińska, K.; Shan, W.; Zawada, K.; Kadish, K. M.; Gryko, D. *J. Am. Chem. Soc.* **2016**, *138*, 15451–15458.

[48] Reekie, T. A.; Sekita, M.; Urner, L. M.; Bauroth, S.; Ruhlmann, L.; Gisselbrecht, J.-P.; Boudon, C.; Trapp, N.; Clark, T.; Guldi, D. M.; Diederich, F. *Chem. Eur. J.* **2017**, *23*, 6357–6369.

[49] Lu, G.; Jiang, X.; Ou, Z.; Yan, S.; Kadish, K. M. *J. Porphyr. Phthalocyanines* **2017**, *21*, 465–475.

[50] Ke, X.; Kumar, R.; Sankar, M.; Kadish, K. M. *Inorg. Chem.* **2018**, *57*, 1490–1503.

[51] Nicewicz, D. A.; Hamilton, D. S. *Synlett* **2014**, *25*, 1191–1196.

[52] Rybicka-Jasińska, K.; König, B.; Gryko, D. *Eur. J. Org. Chem.* **2017**, *2017*, 2104–2107.

[53] Schroll, P.; Hari, D. P.; König, B. *ChemistryOpen* **2012**, *1*, 130–133.

[54] Schroll, P.; Fehl, C.; Dankesreiter, S.; Konig, B. *Org. Biomol. Chem.* **2013**, *11*, 6510–6514.

[55] Nagib, D. A.; Scott, M. E.; MacMillan, D. W. C. *J. Am. Chem. Soc.* **2009**, *131*, 10875–10877.

[56] Shih, H.-W.; Vander Wal, M. N.; Grange, R. L.; MacMillan, D. W. C. *J. Am. Chem. Soc.* **2010**, *132*, 13600–13603.

[57] Nicewicz, D. A.; MacMillan, D. W. C. *Science* **2008**, *322*, 77.

[58] Lee, J.; Papatzimas, J. W.; Bromby, A. D.; Gorobets, E.; Derksen, D. J. *RSC Adv.* **2016**, *6*, 59269–59272.

[59] Yang, X.-L.; Xie, M.-H.; Zou, C.; He, Y.; Chen, B.; O'Keeffe, M.; Wu, C.-D. *J. Am. Chem. Soc.* **2012**, *134*, 10638–10645.

[60] Toyao, T.; Ueno, N.; Miyahara, K.; Matsui, Y.; Kim, T.-H.; Horiuchi, Y.; Ikeda, H.; Matsuoka, M. *Chem. Commun.* **2015**, *51*, 16103–16106.

[61] Remello, S. N.; Hirano, T.; Kuttassery, F.; Nabetani, Y.; Yamamoto, D.; Onuki, S.; Tachibana, H.; Inoue, H. *J. Photochem. Photobiol. A* **2015**, *313*, 176–183.

[62] Yusa, M.; Nagata, T. *Photochem. Photobiol. Sci.* **2017**, *16*, 1043–1048.

Chapter 9

Nitrogen Heterocycles: Acridinium Salts

Peter Fodran, Leticia Monjas, and Carl-Johan Wallentin*

Department of Chemistry and Molecular Biology,
University of Gothenburg, Sweden
**carl.wallentin@chem.gu.se*

9.1 Introduction

Due to their straightforward synthesis and unique properties, acridinium salts are perhaps the most widely used photoorganocatalysts (**POCs**). Typically, acridinium cations are prepared by 1,2-addition of Grignard reagents or aryllithiums to acridanones, followed by elimination of the so-formed hydroxyl group under acidic conditions. The choice of acid determines the counterion in the final salt. Most commonly, $HClO_4$ or HBF_4 are used (Figure 9.1(a)) [1].

The facile and versatile synthesis of acridinium salts together with the ability of these salts to absorb light in the blue region of the visible part of the electromagnetic spectrum, have stimulated the development of a wide variety of redox catalysis competent structures, from which one stands out — **Mes-Acr-Me**$^+$ [2, 3]. This compound is a very strong oxidant in the excited state (E^{*}_{RED}(**POC**$^{+*}$/**POC**$^{\bullet}$) = +2.06 V vs SCE in MeCN) and is a moderate reductant in the ground state (E_{RED} = −0.5 V vs SCE in MeCN). As such, it can

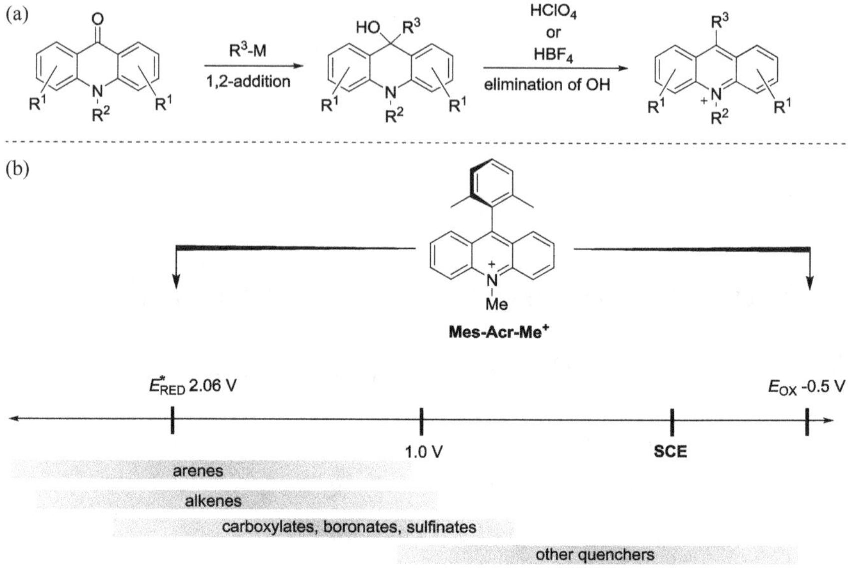

Figure 9.1. (a) General synthesis of acridinium salts. (b) Electrochemical range of **Mes-Acr-Me**$^+$ and oxidation potentials of quenchers in this chapter.

engage substrates that are usually inert towards oxidation (Figure 9.1(b)). The strong oxidizing properties of **Mes-Acr-Me**$^+$ stem from a long-lived charge-shift state in which an electron has been transferred from the mesityl group to the acridinium moiety [4]. The excited state lifetime is 6 ns, which, in a synthetic context, is sufficient for productive encounters with appropriate quenchers. Several derivatives of **Mes-Acr-Me**$^+$ with slightly altered excited-state redox properties and lifetimes have been developed [5]. The most important aspect of these derivatives is perhaps their increased stability, which addresses the typical bleaching seen for **Mes-Acr-Me**$^+$ under prolonged irradiation.

The typical substrates that can be utilized as quenchers include arenes, alkenes, carboxylates, sulfinates, boronates, 2H-azirines [6–8]. Upon encounter with the excited **POC**, the quenchers are oxidized and the resulting intermediates further undergo subsequent chemical transformations that typically follow free radical or radical-polar crossover mechanistic pathways. More details on the photophysical properties of acridinium salts will be found in Ref. [8].

This chapter is divided into segments that focus on the structural nature of the substrates and their ability to act as quenchers of the excited state of the catalysts. Furthermore, the formation of different covalent bonds serves as another systematic subdivision of this chapter.

9.2 Arenes as Quenchers of the Photoexcited State

9.2.1 *Formation of C–C and C–heteroatom bonds*

Due to their high oxidation potential, arenes are challenging substrates in photoredox catalysis relying on pathways that involve the reductive quenching of the catalyst's excited state. In order to employ these substrates in an exergonic photoinduced electron transfer, the reduction potential of the catalyst has to lie above E_{OX} = +1.0 V vs SCE (which corresponds to the oxidation potential of commonly encountered electron-rich arenes), leaving only a few photocatalysts as competent oxidants. **Mes-Acr-Me**$^+$ has shown to exhibit singlet excited state reduction potential E^*_{RED}(**POC***/**POC**$^\bullet$) = +2.06 V vs SCE, thus fulfilling this criterion. Single electron oxidation of arenes results in positively charged cation radicals (Figure 9.2) that are prone to be attacked by carbon-, heteroatom-, and halogen-centered nucleophiles. Consequently, this allows for

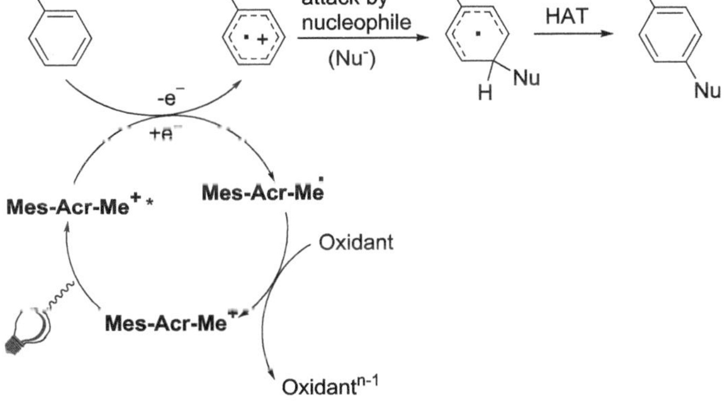

Figure 9.2. Working principle of photoredox derivatizations of arenes.

the formation of C–C, C–heteroatom, and C–halogen bonds. The so-formed radical undergoes a subsequent hydrogen atom transfer (HAT) or oxidation followed by deprotonation, which restores aromaticity in the substrate. The original oxidation state of the photocatalyst is restored by oxidation. This working principle has been exploited in several creative ways.

Seminal work employing this strategy was reported by the group of Fukuzumi in 2011 [9], in which **Mes-Acr-Me$^+$** was used as the photocatalyst in the bromination of electron-rich arenes and heteroarenes in an overall oxidative process (Figure 9.3(a)). A clean monobromination was observed with the majority of the examined substrates. Given that the oxidation potentials of arenes and their brominated analogs are close and both can undergo exergonic electron transfer, the authors concluded that the selective monobromination stems from the different reactivities of the transient radical cations. The observed regioselectivity could be explained by nucleophile trapping occurring at the most electron-depleted carbon atom of the conjugated system. The authors suggested the following mechanism (Figure 9.3(b)). After reductive quenching, the generated radical cation reacts with a nucleophilic bromide anion resulting in a brominated cyclohexadienyl radical. A following HAT process by hydroperoxyl radical, which was generated by the oxidation of the **Mes-Acr-Me$^•$** radical as a turnover event of the catalytic cycle, restores aromaticity and provides the desired products.

Another important contribution to this field of research was published by the group of Nicewicz in 2015 [10]. In a very detailed study, the authors developed a site-selective C–H aminating protocol of arenes (Figure 9.4). The reaction displays a relatively broad scope with preference for *para*-selectivity and can be utilized for late-stage amination of biologically active compounds. A regioselectively challenging situation arose when the nitrogens of the nitrogen-based nucleophile were chemically different. For example, when $(Boc)_2$-protected histidine was used, the amination took place with 1.2/1 *p*/*o* selectivity. The main isomer was obtained as a 5/1 mixture of isomers obtained by N1 vs N3 attack. The authors

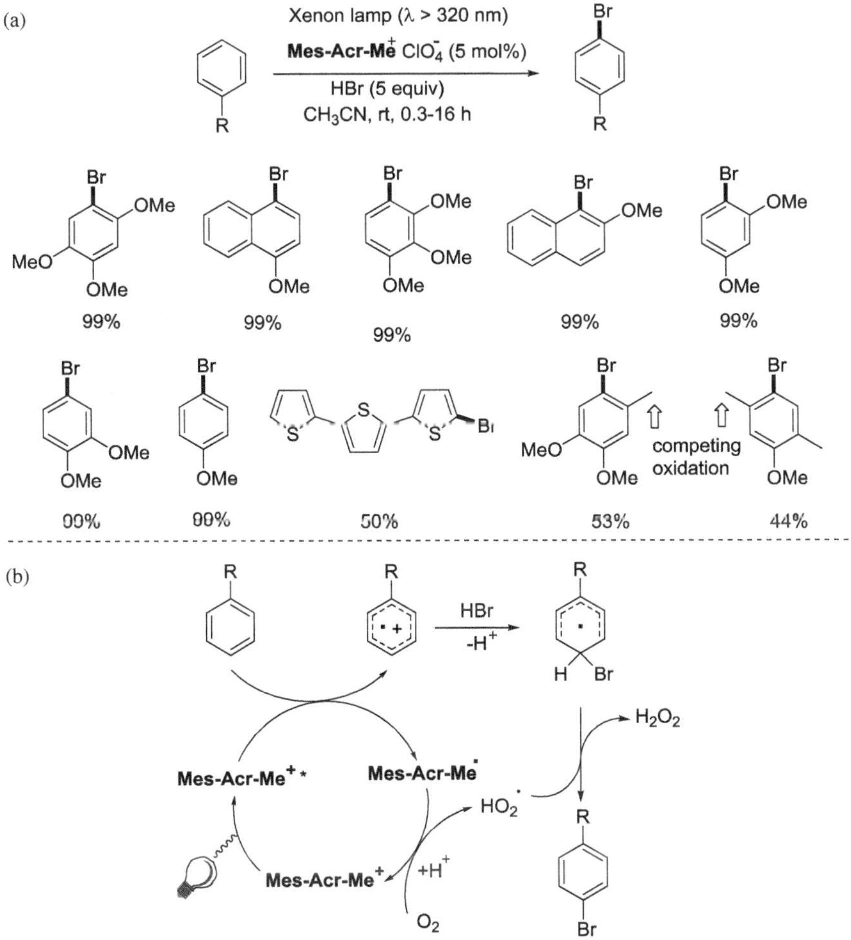

Figure 9.3. (a) Photoredox bromination of arenes; selected products. Formed bond is indicated in bold. (b) Proposed mechanism.

noticed that the **Mes-Acr-Ph**$^+$ catalyst deteriorates during the reaction, presumably by an oxidative process mediated by oxygen-centered radicals. In order to make the catalyst chemically more robust, the authors substituted the 3 and 6 positions of the acridinium moiety with bulky *t*Bu groups, which decreased the reactivity towards radicals or nucleophiles. Additionally, they included TEMPO that acts as a scavenger of reactive oxygen radicals. Using

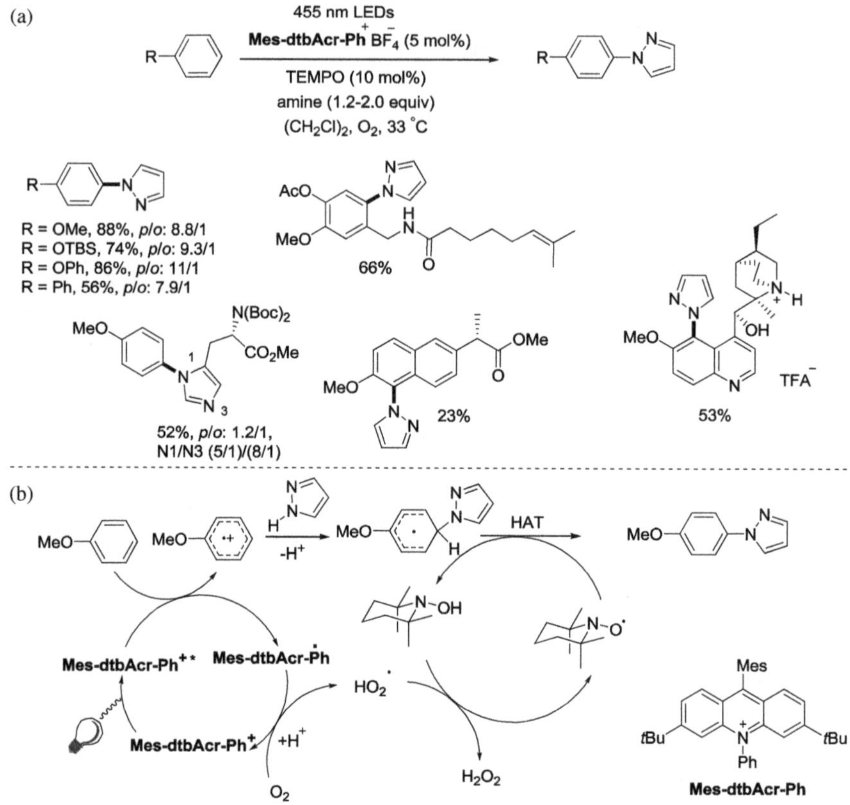

Figure 9.4. (a) Photoredox amination of arenes; selected products. Formed bond is indicated in bold. (b) Proposed mechanism.

10 mol% of TEMPO completely suppresses the formation of phenyl formate, a product of the uncatalyzed oxidation of the methoxy groups.

Following this publication, the authors reported a predictive model and conditions allowing for a selective and more general amination of several classes of aromatic and heteroaromatic compounds [11].

A shortcoming of this protocol is the potential oxidation of benzylic positions. In order to circumvent this, the group of Lei developed a protocol that allows for the corresponding amination

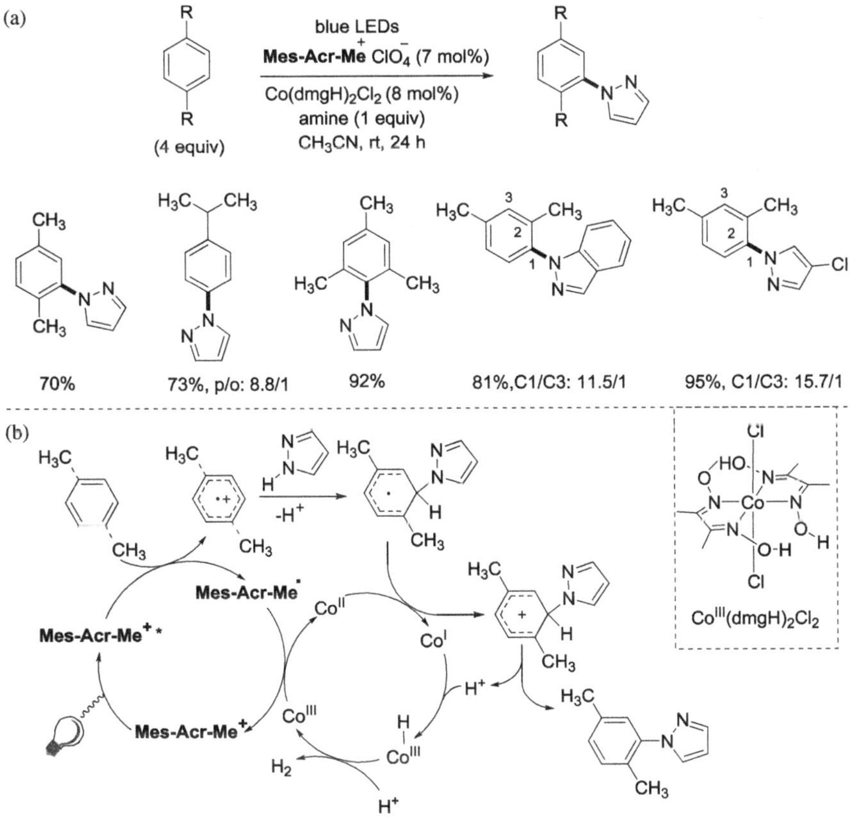

Figure 9.5. (a) Photoredox formation of Ar-N bonds; selected products. Formed bond is indicated in bold. (b) Proposed mechanism.

under oxidant-free conditions, exploiting a cobalt-based dual catalytic system (Figure 9.5(a)) [12]. The protocol showed a relatively broad scope in substrates with the sensitive benzylic position. Some representative examples are depicted in Figure 9.5(a). After single electron oxidation (Figure 9.5(b)) of an arene, the acridinium radical is a strong enough reductant to reduce $Co^{III}(dmgH)_2Cl_2$ to $Co^{II}(dmgH)_2Cl$ (E_{RED} ($Co^{III}/Co^{II} = -0.42$ V vs SCE) [13]. Subsequently, the Co^{II} complex acts as an oxidant for the neutral, aminated radical intermediate resulting in a cyclohexadienyl cation and a Co^{I} complex. The proton obtained from aromatization of the cation intermediate is reduced to molecular hydrogen by the Co^{I} complex.

Cation radicals accessible by single electron oxidation of arenes by acridinium catalysts have been utilized also in C–C bond forming reactions. In a report by McManus and Nicewicz, the authors used the knowledge obtained from the amination of arenes (Figure 9.4) in the development of direct C–H cyanation [14]. The generated cation radical was trapped with a CN^-, generated by slow hydrolysis of TMSCN with a pH = 9 buffer. Oxidation of the so-formed radical using molecular oxygen furnished the aryl nitrile. Compared to the previous report on the amination of arenes, the cyanation displays lower *o*/*p* selectivity. For example, while the amination of diphenylether afforded the corresponding products in an 11:1 ratio, the analogous products of cyanation were obtained only in a 1.6:1 *p*/*o* ratio (Figure 9.6). In addition to a relatively broad scope of substrates, the authors illustrated several examples in which the cyanation protocol could be used for the late-stage modification of few biologically active compounds, like Naproxen-methyl ester, *N*-Boc-melatonin, and Gemfibrozil-methyl ester.

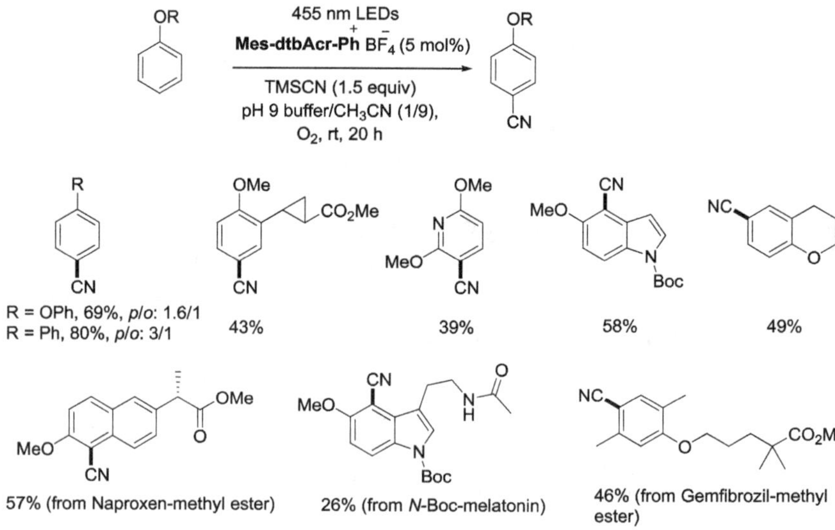

Figure 9.6. Photoredox cyanation of electron-rich substrates; selected substrate scope. Formed bond is indicated in bold.

An interesting feature of the radical cations obtained by single electron oxidation of arenes is that the benzylic position becomes superacidic. For example, the methyl group in toluene has a pK_a = 43 (in DMSO). After single electron oxidation, the methyl group of the corresponding cation radical will have a pK_a = −20 [15–17a]. For comparison, H_2SO_4 has a pK_a = −2.5 (in DCE), TfOH has a pK_a = −11.4 (in DCE), and $HClO_4$ as the strongest commonly used acid in synthetic organic chemistry has the pK_a = −13. Consequently, upon single electron oxidation of arenes with methyl, methylene, or methine substituents, these can easily be deprotonated. This deprotonation constitutes a rearomatization of the arene and provides, *via* radical translocation, a benzyl-centered radical. This working principle has been utilized in some C–C bond forming reactions reported by Wu and co-workers [17b], who engaged isochromans in functionalization with a co-catalytic system analogous to Lei (Figure 9.7) [12].

The isochroman is oxidized by the excited acridinium catalyst. After deprotonation, the generated benzylic radical is oxidized by Co^{II} to afford a stabilized oxocarbenium cation, which in turn reacts with $copper^{II}$ ethyl acetoacetate to afford the corresponding alkylated product. The two generated protons are reduced to H_2 by Co^{I}, thereby turning over the catalytic system. In terms of substrates, the authors showed a variety of aromatic and non-aromatic β-ketoesters being compatible with a few examples of isochromans giving rather high yields (76%–89%). A particularly interesting case is the reported example using 2-nitroacetophenone, which delivers the product in 43% yield.

9.2.2 *Oxidations*

The above-mentioned working principle in which photoredox catalysis can be utilized in the generation of benzylic radicals *via* deprotonation-mediated radical translocation provides a great promise for clean, environmentally benign oxidative transformations of benzylic positions. This has also been achieved in several different ways. In a very mechanistic-oriented publication, the group

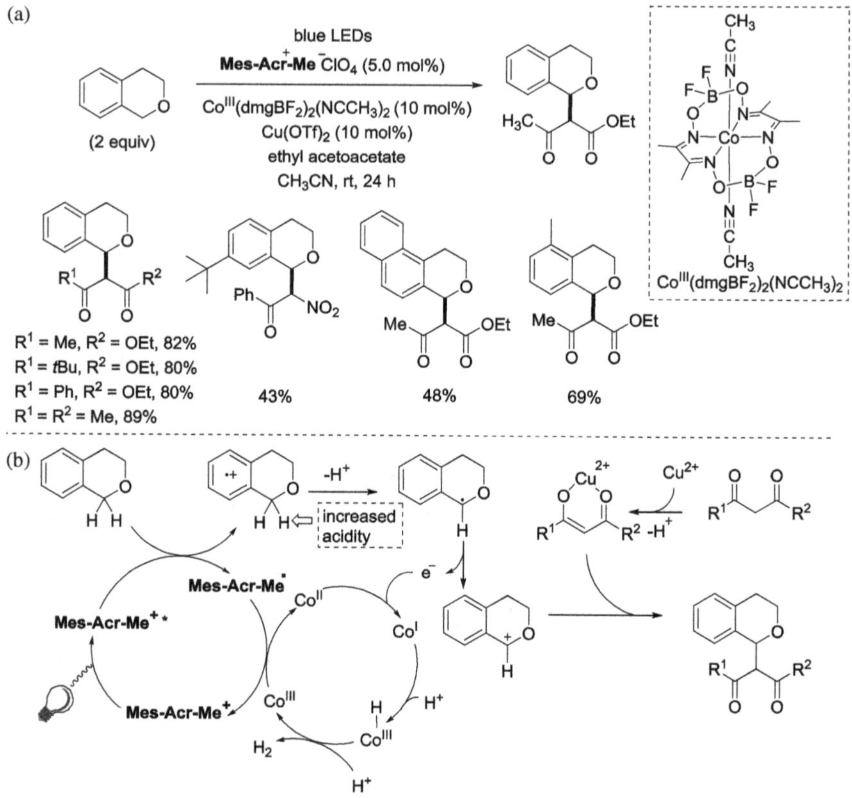

Figure 9.7. (a) Photoredox functionalization of isochromans; selected products. Formed bond is indicated in bold. (b) Proposed mechanism.

of Fukuzumi reported that benzylalcohol can be oxidized to benzaldehyde using molecular oxygen as the stoichiometric oxidant without any observed over-oxidation (Figure 9.8) [18]. The obtained aldehyde has a significantly increased oxidation potential and cannot be further oxidized by the photoexcited **Ph-Acr-Me**$^+$. The reaction was performed neat and the only observed side-product was H_2O_2. Subsequently, several variations on this publication were reported [19–23].

Fu and co-workers demonstrated that when CH_3CN is used as solvent, **Acr-Me**$^+$ and **Ph-Acr-Me**$^+$ are able to catalyze the oxidation of primary and secondary benzylic alcohols to the corresponding

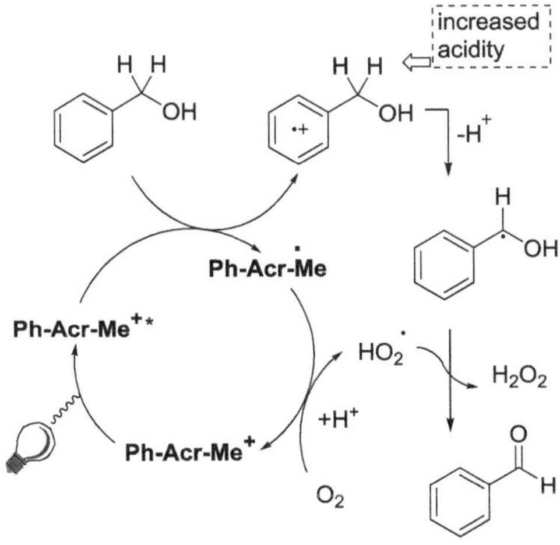

Figure 9.8. Oxidation of benzyl alcohol to benzaldehyde by phenylacridinium catalyst and O_2.

aldehydes and ketones [19]. Another step forward came with the subsequent publication of the group of Fukuzumi [20], which reported the oxidation of methyl-substituted arenes with various acridinium catalysts under oxygen atmosphere (Figure 9.9(a)). When varying the structure of the catalyst, the authors observed that with certain substitution patterns of both the acridinium and the phenyl moieties, the performance of the catalyst could be improved. This increased efficiency is probably caused by the increased chemical stability of the modified acridinium cations.

The system of Fukuzumi was utilized by the group of Lei for the oxidation of various diphenyl methanes [21]. Additional contributions by the Lei group include a coupled catalytic system where **Mes-Acr-Me**$^+$ and a Brønsted acid catalyst working in concert were able to convert various methyl-substituted aromatics to methyl esters in moderate to good yields (Figure 9.9(b)) [22]. When the reaction was performed in EtOH or *i*PrOH, the corresponding ethyl and isopropyl esters were obtained in somehow humble yields (15%–26%). In terms of substrate scope and functional group compatibility, the authors reported alkyl-, aryl-, methoxy-, and halo-substituted methyl aromatic compounds. Subsequently, the Lei group also showed that

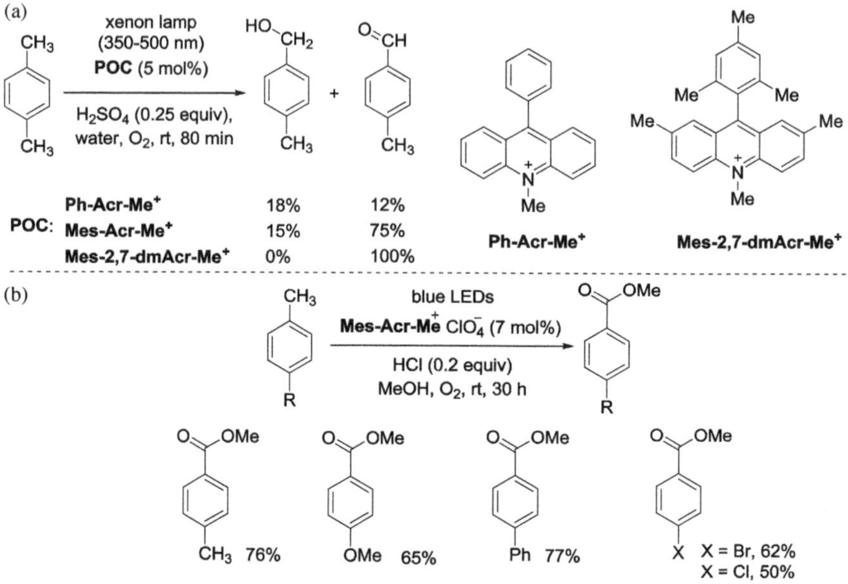

Figure 9.9. Variation of structure of the catalyst has a direct impact on the performance of the catalyst.

esters could be obtained from benzylic alcohols upon exchanging the Brønsted acid to $Cu(OTf)_2$ [23].

In addition to the oxidative formation of C–O bonds, the intriguing concept of photoredox-based catalytic acceptorless dehydrogenation (CAD) as a type of oxidative aromatization reaction has been reported by Kanai [24]. While CAD typically requires temperatures exceeding 100°C, the authors reported that photoredox catalysis enables CAD to take place at room temperature. The reaction has a relatively broad functional group tolerance, in which halides, esters, amides, ketones, and alkyl substituents were tolerated. This concept could be applied to the aromatization of tetrahydroisoquinolines, indolines, tetrahydronaphthalenes, and even cyclohexenes (Figure 9.10(a)). The authors suggested that 1-phenyltetrahydroisoquinoline undergoes oxidation by photoexcited **Mes-Acr-Me**$^+$, and upon deprotonation gives a transient *N*-centered radical. In a coupled fashion with metal catalysis, referred to as a binary hybrid catalysis, the *N*-centered radical

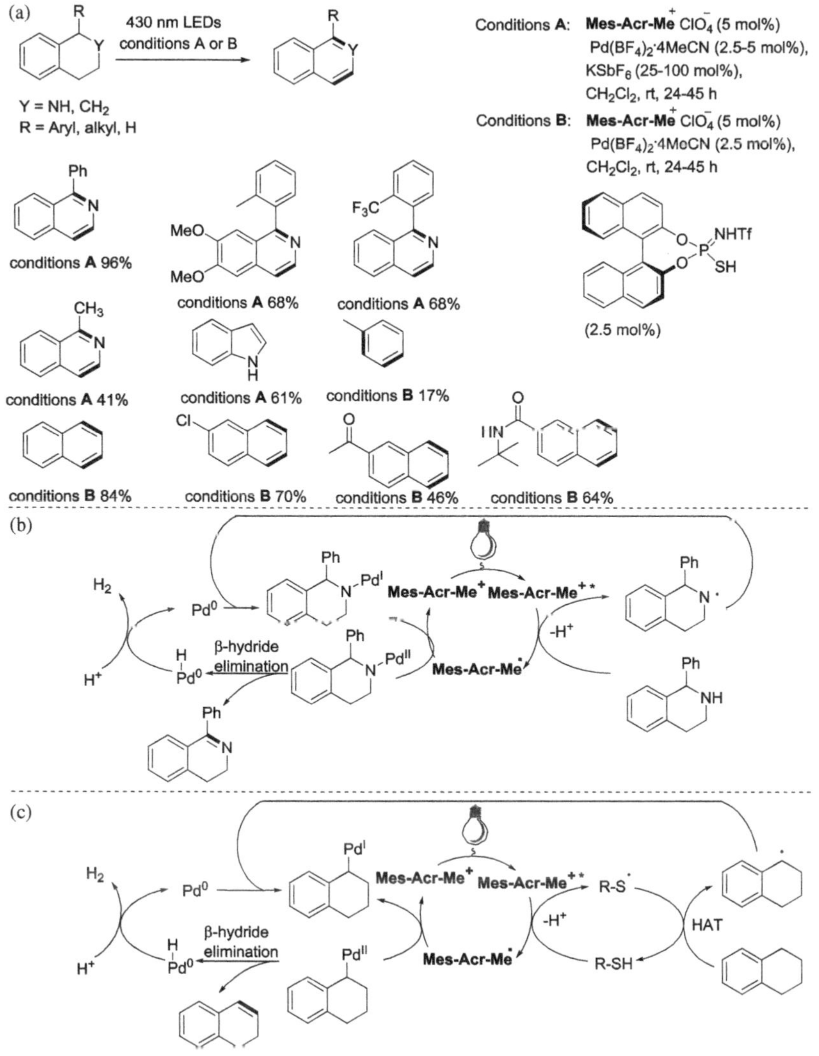

Figure 9.10. (a) Catalytic photoredox acceptorless dehydrogenation of tetrahydroisoquinolines, selected products. Formed bond is indicated in bold. (b) Proposed mechanism. (c) Inclusion of thiols enables CAD of tetrahydronaphthalenes.

species could be further oxidized to 1-phenylisoquinoline (Figure 9.10(b)). As expected, tetrahydronaphthalene derivatives were more challenging substrates due to their higher oxidation potentials. Here, the authors introduced one more catalytic cycle, thus achieving a ternary hybrid catalytic system (Figure 9.10(c)). Under these conditions, the photoexcited acridinium catalyst was quenched by a thiol co-catalyst resulting in the generation of a thiyl radical, which is known to be an efficient hydrogen atom abstractor [25, 26]. The thiyl radical accordingly abstracts a benzylic hydrogen resulting in a benzylic radical that can undergo the same process as in the case of tetrahydroisoquinolines interception by metal, β-hydride elimination, and further aromatization.

9.3 Alkenes as Quenchers of the Photoexcited State

Alkenes are abundant, cheap, and highly utilized building blocks in synthetic chemistry. Given their relatively high redox potential, alkenes require a potent oxidant in order to generate radical cations. The Fukuzumi catalyst has an E^*_{RED}(**POC***/**POC**$^{\bullet}$) = +2.06 vs SCE, which is achieved by visible-light irradiation ($\lambda > 450$ nm), making it a suitable catalyst to obtain radical cations from alkenes [2, 20]. The chemistry of these radical cations differs in many regards from those radical cations obtained from aromatic systems covered in the previous section. In this section, hydrofunctionalization of alkenes that encompasses the formation of new C–C, C–O, C–N, and C–halogen bonds will be covered in some detail. Special attention will be given to the dual functionalization of alkenes that encompass the formation of two new C–C or C–heteroatom bonds.

9.3.1 *Formation of C–C bonds*

One of the best-known methodologies to construct all carbon six-membered rings is the Diels–Alder reaction [27, 28]. Seminal work by Yoon and co-workers in 2011 provided a visible-light-mediated protocol to achieve this type of reactivity between styrene derivatives and aliphatic dienes by means of radical cation generation [29]. Similar to this transformation, Lei and co-workers developed a method for cyclodimerization-oxygenation of styrenes, which includes a formal [4+2] cyclization, giving access to 4-aryl tetralones. The overall cyclodimerization-oxygenation process

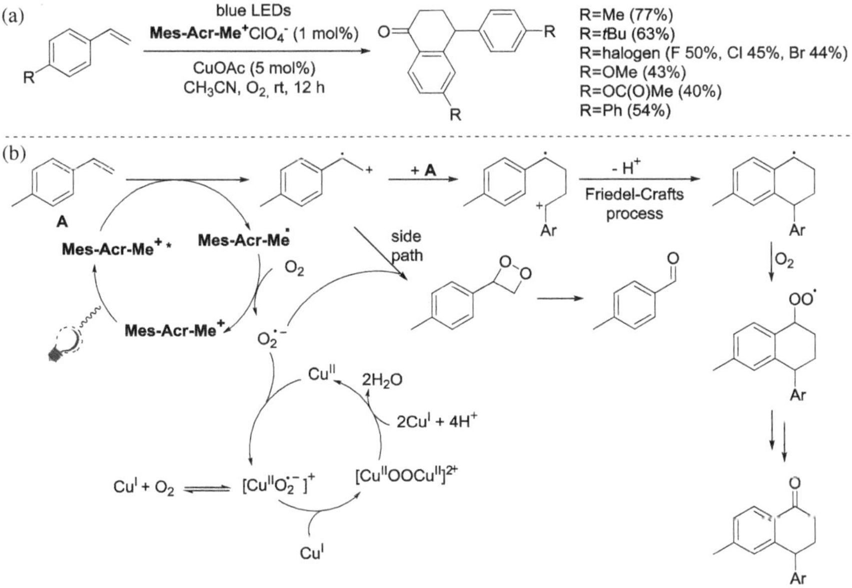

Figure 9.11. (a) Cyclodimerization-oxygenation of styrene for the synthesis of 4-aryl tetralones; selected products. (b) Proposed mechanism.

utilizes **Mes-Acr-Me**$^+$ as photoredox catalyst and O_2 as oxidant (Figure 9.11(a)) [30]. The proposed mechanism involves a single electron transfer (SET) between the styrene derivative and the photoexcited **Mes-Acr-Me**$^+$ catalyst (Figure 9.11(b)). The so-formed radical cation reacts with another molecule of styrene *via* intermolecular electrophilic addition producing a new radical cation, which subsequently undergoes a Friedel–Crafts process providing the tetralone scaffold containing a benzylic radical. This radical reacts rapidly with O_2 yielding the desired product after elimination of water. The turnover event in this process involves the formation of a superoxide radical anion ($O_2^{\bullet-}$), which is a reactive-oxygen species (ROS) that can be used to incorporate oxygen into the target structure. A catalytic amount of CuOAc serves in this system to control the reactivity of $O_2^{\bullet-}$, and thereby decreasing a deleterious side reaction that leads to the formation of benzaldehyde.

The same reaction was reported by Liang and co-workers, but using 2-phenylmalononitrile (PMN) as co-catalyst [31]. In the same

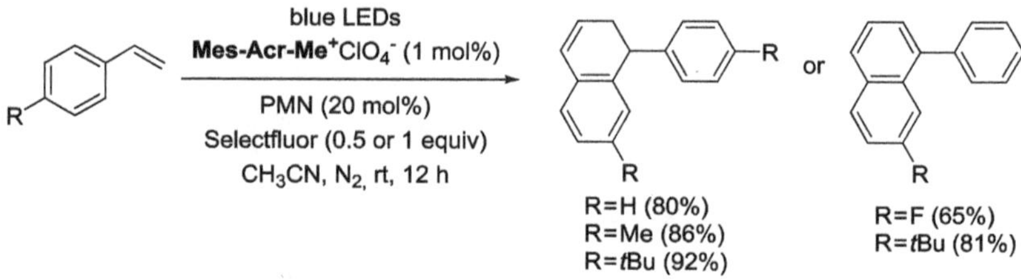

Figure 9.12. Radical cation [4+2] annulation for the synthesis of (dihydro)naphthalenes; selected products.

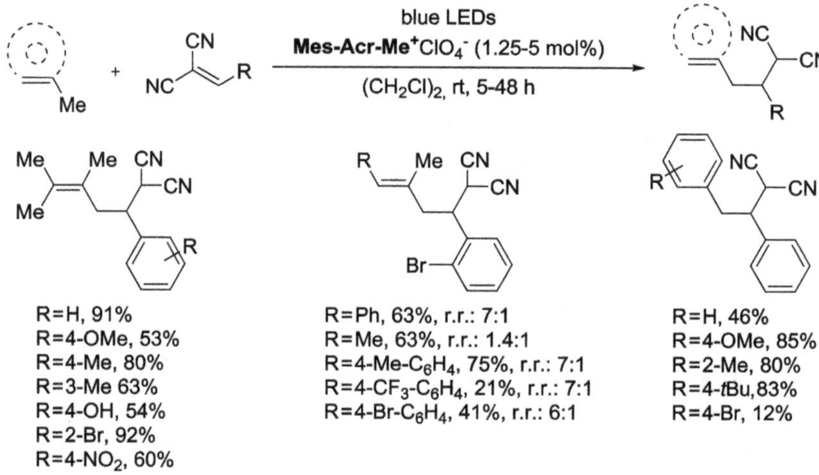

Figure 9.13. Alkylation of allylic and benzylic C(sp^3)–H bonds; selected products.

report, an oxidative [4+2] cycloaddition reaction led to dihydronaphthalenes under inert atmosphere using substoichiometric amount of Selectfluor as oxidant. Increasing the amounts of Selectfluor promoted a dehydrogenative aromatization of the 1,2-dihydronaphthalenes giving naphthalenes as final products (Figure 9.12).

Wu and co-workers described the direct alkylation of unactivated allylic and benzylic C(sp^3)–H bonds with electron-deficient alkenes, using the **Mes-Acr-Me⁺** catalyst (Figure 9.13) [32]. The products of alkylation can be converted into γ,δ-unsaturated or α,β-diaryl-acids, -esters, -amides, -pyrazoles, -isoxazoles, or -lactones, important intermediates in the synthesis of pharmaceuticals.

9.3.2 *Formation of C–O bonds*

The first example of alkene oxygenation catalyzed by an acridinium catalyst was reported by Fukuzumi in 2003 [33]. In this mechanistic study, the authors reported that styrenes are reductive quenchers of the photoexcited acridiniums and the corresponding radical cations undergo oxygenation, yielding products analogous to ozonolysis. Together with the subsequent studies (Figure 9.14) [3, 34], the authors very carefully characterized all key intermediates and provided the knowledge necessary for further developments in this area.

Nicewicz and co-workers have developed a general approach for the anti-Markovnikov hydrofunctionalization of alkenes using acridinium-type photocatalysts, first reported by Fukuzumi [3], in combination with redox-active H atom donors, such as phenyl malononitrile (PMN) or aryl thiols [8–36]. This method relies on the single electron oxidation of alkenes to reverse their polarity, which results in anti-Markovnikov regioselective functionalization.

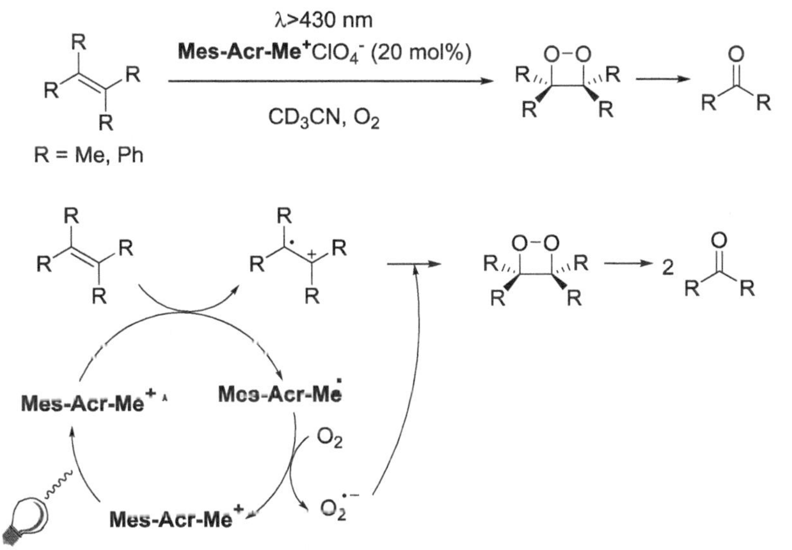

Figure 9.14. Photocatalytic oxygenation of tetrasubstituted alkenes with O_2 *via* photoinduced electron transfer.

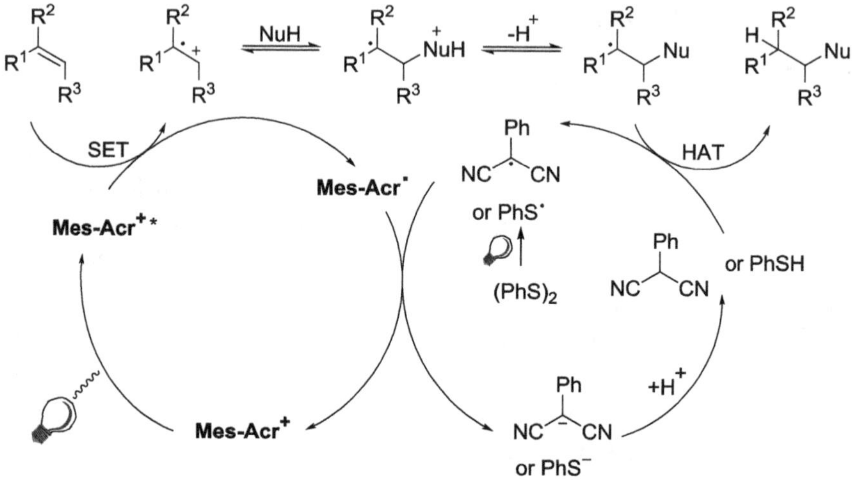

Figure 9.15. Proposed mechanism for the anti-Markovnikov alkene hydrofunctionalization reactions.

The proposed mechanism for these transformations states that electron transfer from the alkene to the excited state of **Mes-Acr-Me**$^+$ produces the alkenyl radical cation (Figure 9.15). Nucleophilic addition to the cation radical and subsequent HAT affords the desired product. After the HAT, a radical (PhS$^•$ or PMN$^•$) is formed, which, in an oxidative process, regenerates **Mes-Acr-Me**$^+$. Protonation of the so-formed PhS$^-$ or PMN$^-$ is coupled to the deprotonation of the nucleophile-cation radical adduct and turns over the catalytic cycle of the HAT donor [37]. Initially, PMN was used as HAT donor, but eventually superior performance was realized using aryl thiols, such as thiophenol and disulfides. Phenyl disulfide is more convenient to use because it is an odorless bench-stable solid, in contrast to thiophenol, which is a pungent toxic liquid that must be stored under nitrogen. Phenyl disulfide forms thiyl radicals *via* an energy transfer event with the photoexcited catalyst. Consequently, it serves as a thiyl radical reservoir in the coupled catalytic system.

The first report of alkene hydrofunctionalization using an acridinium-type catalyst involved an intramolecular hydroetherification of alkenols (oxidation potentials for the reported substrates (E_{OX} = +1.22 V to 2.06 V vs SCE, Figure 9.16(a)) [38]. PMN was

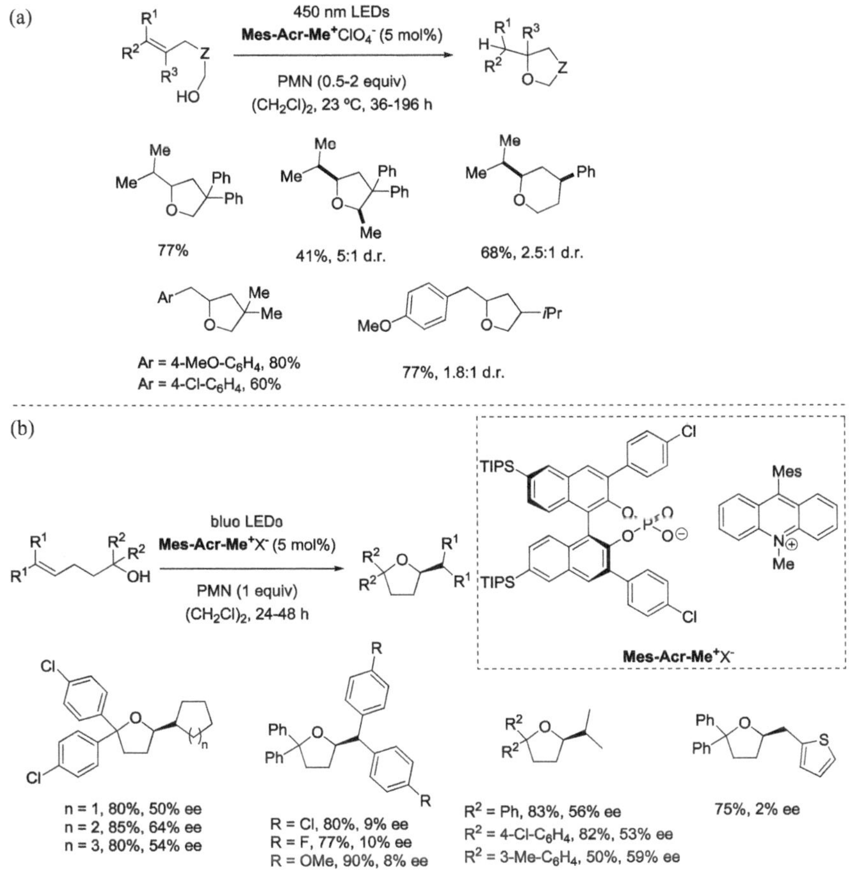

Figure 9.16. (a) Anti-Markovnikov hydroetherification of alkenols; selected products. (b) Enantioselective version of the reaction catalyzed by a chiral ion-pair **POC**.

used as a redox-cycling source of the hydrogen atom, typically in stoichiometric amounts (some substrates required more than one equivalent for synthetically useful yields). Aliphatic alkenes and electronically distinct styrenes were used to demonstrate the efficient formation of five- to seven-membered cyclic ethers. The same report showed one example with a carboxylic acid acting as the nucleophile, which yielded the corresponding lactone, as well as one example with MeOH, showing that the reaction could be applied to intermolecular couplings (*vide infra*). Notably, all reactions proceeded with complete regioselectivity, without any trace of the undesired Markovnikov regioisomer. Control reactions comparing the alkene reactivity with cation radicals or Brønsted acids confirmed that the regioselectivity was due to the reactivity pattern of alkene cation radicals and not the result of acid activation.

Luo reported an enantioselective version of this reaction [39]. Using a chiral ion-pair acridinium photoredox catalyst (**Mes-Acr-Me**$^+$X$^-$), accessible by anion exchange with a chiral phosphate ion, cyclic ethers could be obtained with an enantiomeric excess and with an overall increased efficiency compared to its parent Fukuzumi catalyst (Figure 9.16(b)). Although the achieved enantioselectivity is rather low, this report shows that the chiral ion-pair strategy can be applied to achieve enantioselective control of the photoinduced radical cation reactions of alkenes. The transformation follows the same mechanism proposed by Nicewicz [37]. However, the chiral phosphate anion endows a longer lifetime of the chiral photocatalyst's triplet state, induces enantioselectivity, and assists the H-shift.

In a subsequent report, Nicewicz used the same catalytic system to construct oxazolines by using amides as the intramolecular nucleophiles (Figure 9.17) [40]. The reaction worked well with a variety of aromatic and aliphatic amides. However, highly acidic amides

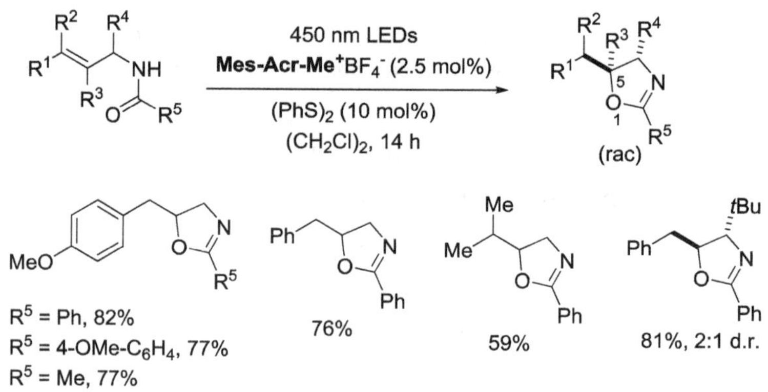

Figure 9.17. Intramolecular anti-Markovnikov addition of amides to alkenes; selected products.

were not compatible with the reaction conditions (e.g. trifluoromethyl and 2-pyridinyl amides). A variety of substituents on the alkene were tolerated, including trisubstituted derivatives. Alkenes bearing a pre-existing stereogenic center cyclized with modest levels of diastereoselectivity. Overall, the strategy is useful to obtain oxazolines substituted at the 5-position [41].

Similarly, thioamides undergo an intramolecular cyclization to form thiazolines *via* a different mechanism (Figure 9.18(a)), in which oxidation of the thioamide occurs instead of oxidation of the alkene (Figure 9.18(b)) [40]. This allows for the use of terminal and disubstituted alkenes as substrates, which would otherwise not react as a consequence of their high oxidation potentials: for example, in Figure 9.18(c), the thioamide has an oxidation

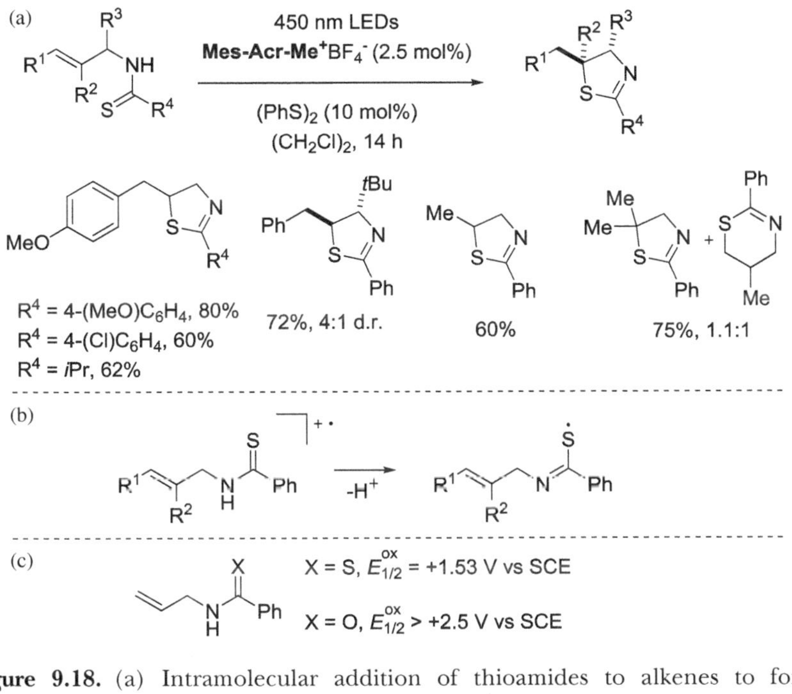

Figure 9.18. (a) Intramolecular addition of thioamides to alkenes to form thiazolines; selected products. (b) Oxidation of the thioamide according to the proposed mechanism. (c) Difference in oxidation potential of a thioamide and its corresponding amide.

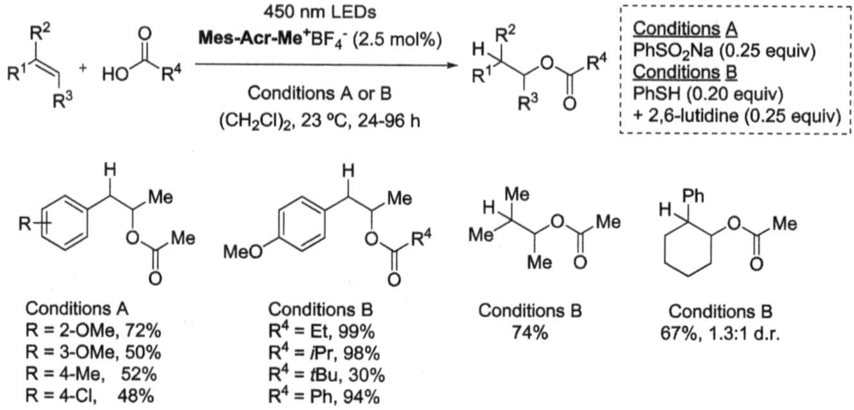

Figure 9.19. Anti-Markovnikov addition of carboxylic acids to alkenes; selected products.

potential of +1.53 V vs SCE, whereas the corresponding amide has an oxidation potential greater than +2.5 V vs SCE, significantly higher than the excited-state reduction potential of **Mes-Acr-Me**$^+$ (E^*_{RED}(**POC*/POC**$^\bullet$) = +2.06 V vs SCE) [38]. Typically, 5-exo-trig cyclization is favored, but for substrates giving highly stabilized radicals upon 6-endo-trig cyclization the selectivity is poor. Thus, regioisomeric control is highly dependent on the nature of the substrate.

Nicewicz extended this method to the intermolecular anti-Markovnikov addition of carboxylic acids to alkenes (Figure 9.19) [42].

In this case, a base was necessary to promote the reaction. β-Methyl styrenes bearing a variety of substituents, trisubstituted aliphatic alkenes, and enamides reacted smoothly with the alkyl carboxylic acids and benzoic acid to form the corresponding esters, using either sodium benzenesulfinate or thiophenol as the HAT co-catalyst.

Further development of this chemistry includes the addition of strong Brønsted acids, such as phosphates and sulfonates, to styrenes (Figure 9.20) [43]. The phosphoric acid substrates were initially used with only a catalytic amount of base (2,6-lutidine), in combination with **Mes-2,7-dmAcr-Ph**$^+$ BF_4^- as the catalyst and diphenyl disulfide as the H-atom donor. These conditions (Conditions A in

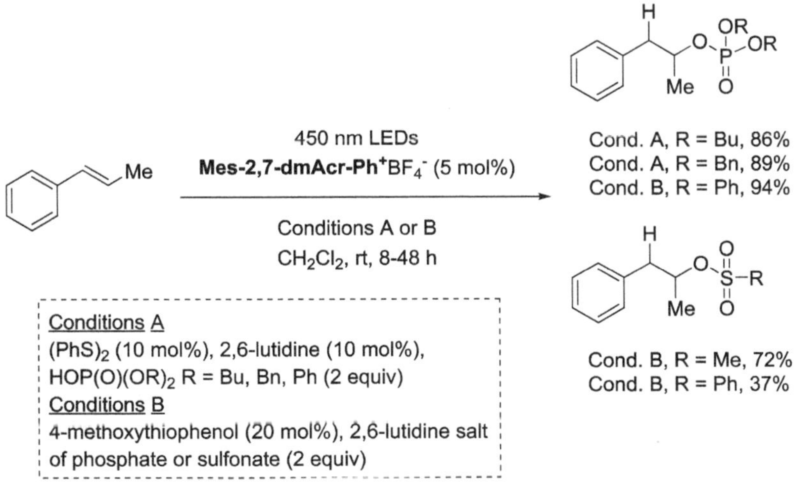

Figure 9.20. Anti-Markovnikov addition of phosphoric and sulfonic acids to β-methylstyrene; selected products.

Figure 9.20) provided good yields for dibutyl and dibenzyl phosphoric acids (86% and 89% yield, respectively). However, diphenyl phosphate gave a much lower yield (12%) when subjected to these reaction conditions; but it could be improved by employing the 2,6-lutidine salt of diphenyl phosphate (94% yield). For sulfonic acid substrates, the 2,6-lutidine salt in combination with 4-methoxythiophenol as the H-atom donor was needed for achieving acceptable yields. In the same report, the method was also used to form carbon–halogen bonds (see Sec. 9.3.4).

Nicewicz further expanded the anti-Markovnikov hydrofunctionalization to also encompass the intermolecular cycloaddition reactions, by taking advantage of a key feature of this method: the simultaneous polar and radical nature of the radical cation species. The new strategy, referred to as catalytic polar-radical-crossover cycloaddition (PRCC), can be viewed as an anti-Markovnikov hydrofunctionalization interrupted by a radical cyclization onto pendant unsaturation tethered to the nucleophile (Figure 9.21) [44, 8]. The PRCC strategy allowed for the synthesis of complex tetrahydrofurans bearing a variety of substituents, including free alcohols,

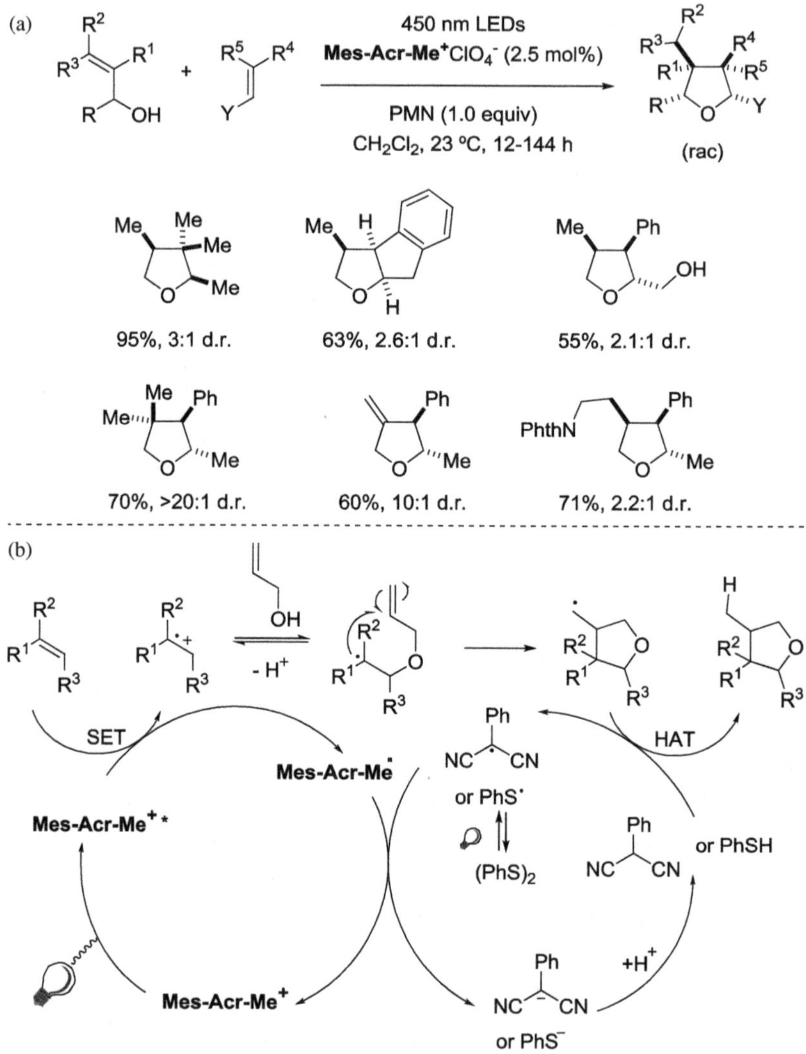

Figure 9.21. (a) PRCC reaction between alkenes and unsaturated alcohols; selected products. (b) Proposed mechanism.

esters, protected amines, and aryl halides, from readily available allylic alcohols and electron-rich alkenes (Figure 9.21(b)) [44].

In a subsequent report, the PRCC method was used for the synthesis of γ-butyrolactones utilizing styrenes or trisubstituted aliphatic

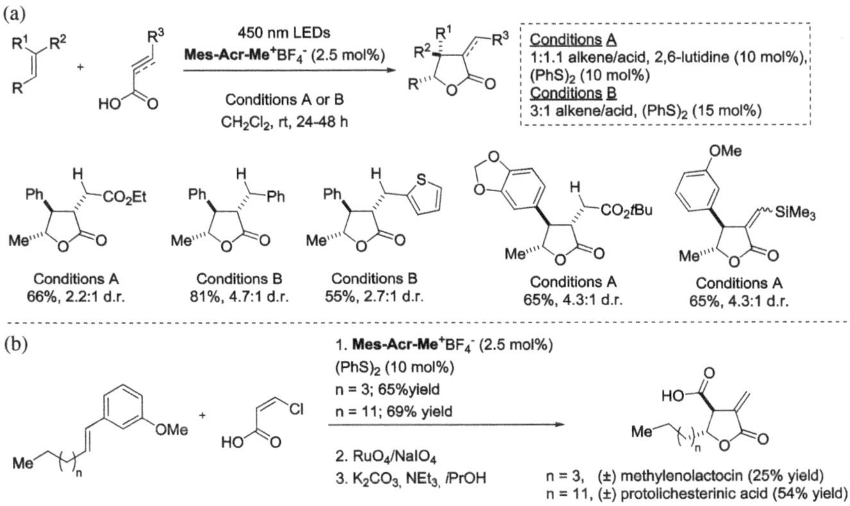

Figure 9.22. (a) PRCC reaction between alkenes and unsaturated carboxylic acids; selected products. (b) Synthesis of methylenolactocin and protolichesterinic acid.

alkenes together with α,β-unsaturated carboxylic acids (Figure 9.22) [45]. Unlike the previous studies, in which stoichiometric amounts of the H-atom donor (PMN) were required, only catalytic amounts of $(PhS)_2$ were necessary, together with 2,6-lutidine. This method was applied to the diastereoselective synthesis of two bioactive paraconic acids: methylenolactocin and protolichesterinic acid, in which the PRCC procedure was followed by arene oxidation and elimination of chloride (Figure 9.22(b)). Furthermore, *O*-benzyloxime acids could be engaged in a similar fashion with alkenes to form α-benzyloxyamino-γ-butyrolactones [46]. The scope of the transformation exhibited both a steric and electronic tolerance as well as a broad functional group compatibility.

Using water as nucleophile opens up for the possibility to convert alkenes to ketones, a process that could be viewed as a Wacker-type transformation with inversed regioselectivity [47]. Lei reported such an anti-Markovnikov oxidation of β-alkyl styrenes in the presence of water under external oxidant-free conditions using a photoredox-metal dual catalytic system [48]. This system utilizes **Mes-Acr-Me**$^+$ in

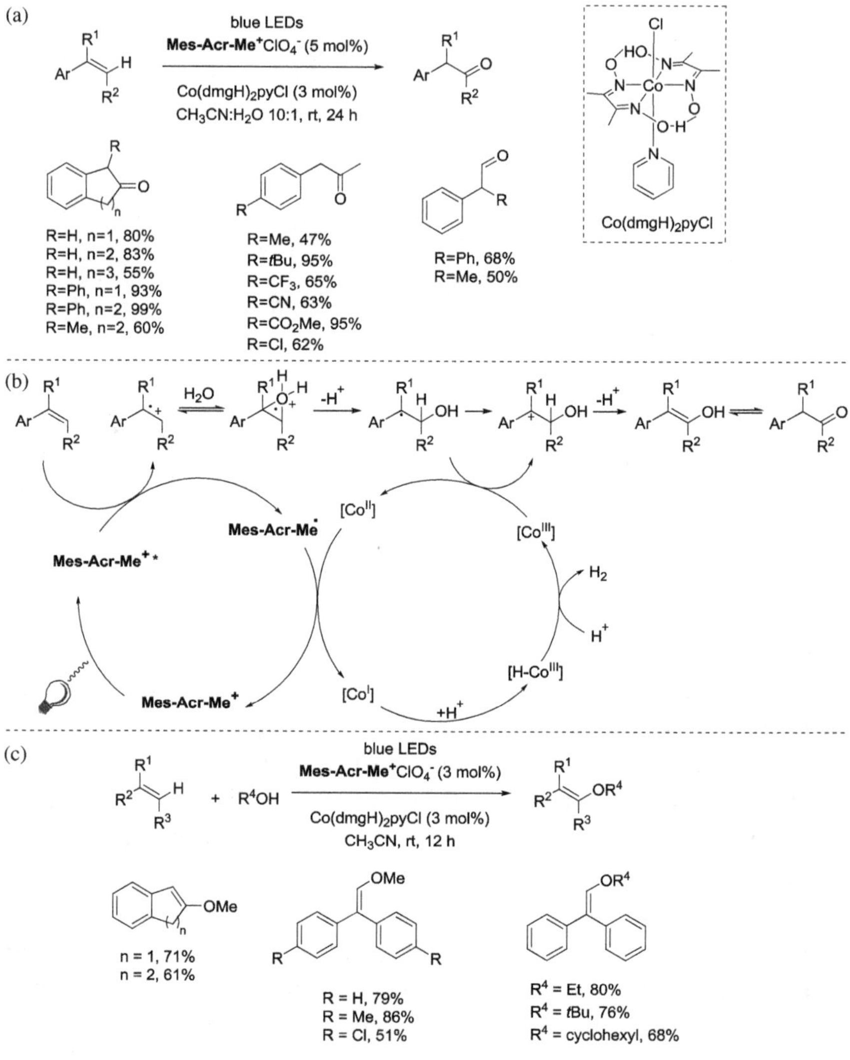

Figure 9.23. (a) Anti-Markovnikov oxidation of β-alkyl styrenes with water; selected products. (b) Proposed mechanism. (c) Coupling of alkenes with alcohols for the synthesis of enol ethers.

concert with the cobalt complex Co(dmgH)$_2$pyCl (dimethylglyoximate monoanion (**dmgH**), pyridine (**py**)) (Figure 9.23(a)). According to the proposed mechanism, the radical cation resulting from the oxidation of the alkene by the excited state of the acridinium catalyst reacts with water (anti-Markovnikov addition), and after deprotonation, generates a C-centered radical. This radical is further oxidized by the cobalt catalyst to form a cation, which after deprotonation and keto-enol tautomerism, furnishes the desired product. The acridinium radical may be oxidized by Co^{II} to reinitiate the photocatalytic reaction and generate a Co^{I} species, which after protonation provides a Co^{III}–H hydride. This species reacts with a proton to release H_2 (Figure 9.23(b)). Using the same catalytic system, a subsequent report describes the dehydrogenative cross-coupling of alkenes with alcohols (Figure 9.23(c)) [49]. This method provides a mild and robust synthesis of enol ethers from readily available alkenes and alcohols.

Lei also reported the anti-Markovnikov addition of water to alkenes for the rapid synthesis of a number of alcohols in good to excellent yields [50]. Based on their earlier studies related to the anti-Markovnikov hydration of styrenes [48], and the work of Nicewicz, in which a H-atom donor is used in combination with the acridinium catalyst for the anti-Markovnikov hydrofunctionalization of alkenes [8–36], Lei achieved the hydration of alkenes exchanging the cobaloxime catalyst with a HAT catalyst (1,2-diphenyldisulfide). The method works for terminal and internal alkenes (Figure 9.24). Moreover, the anti-Markovnikov selective addition of water to a steroid derivative was successfully achieved. The proposed mechanism is analogous to that proposed by Nicewicz for the hydrofunctionalizations of alkenes [37].

Yang and co-workers described the dihydroxylation of styrenes with water and molecular oxygen for the construction of vicinal alcohols, using **Mes-Acr-Me**$^+$ and a base, followed by reductive workup with triphenylphosphine (Figure 9.25) [51]. The styrene is oxidized to its radical cation by the excited state of **Mes-Acr-Me**$^+$. This radical cation is oxygenated by atmospheric O_2 and subsequently attacked by H_2O to generate a peroxy radical, which could undergo two possible pathways: reduction by **Mes-Acr-Me**$^\bullet$, which regenerates the catalyst through a SET process, or reduction by another molecule of styrene, while the catalyst would be regenerated by O_2. In both cases, after protonation and reduction, the product is obtained.

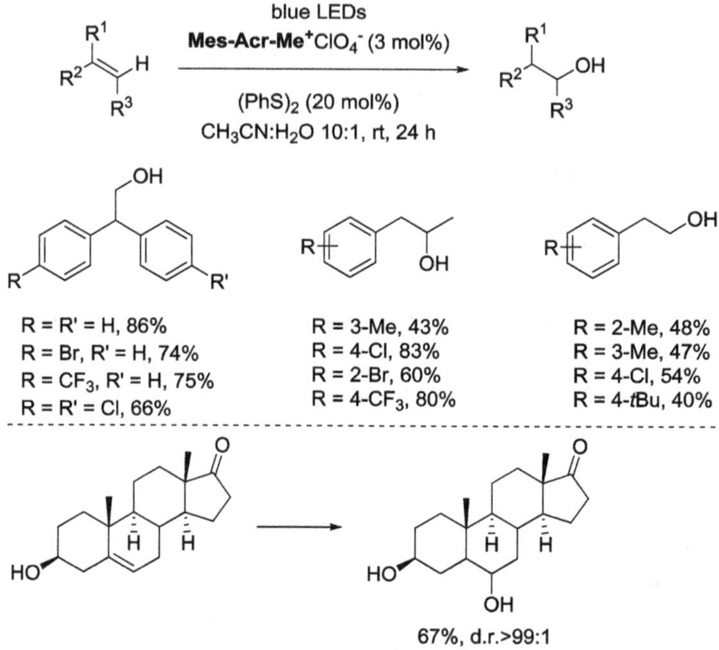

Figure 9.24. Anti-Markovnikov hydration of alkenes; selected products.

The formal [3+2] cycloaddition chemistry discussed above (PRCC) is one example of such a powerful transformation. Combining nucleophilic trapping of the radical cation obtained upon oxidation of an alkene with a secondary C–C or C–X bond forming reaction presents numerous possibilities to achieve structural complexity from alkenes. In accordance with this logic, Liu and co-workers reported an oxo-acyloxylation of olefins with carboxylic acids and molecular oxygen for the synthesis of α-acyloxy ketones [52]. The method relies on **Mes-Acr-Me**$^+$ as the photoredox catalyst, an organic base, and molecular sieves (4 Å). Both, intra- and inter-molecular reactions were studied (Figures 9.26(a) and 9.26(b)). Electron transfer from the alkene to the excited state of **Mes-Acr-Me**$^+$ generates a radical cation of the alkene, which reacts with the carboxylic acid with the aid of the base. The resulting radical is captured by O_2, forming a peroxyl radical, which further transforms into the corresponding peroxide through

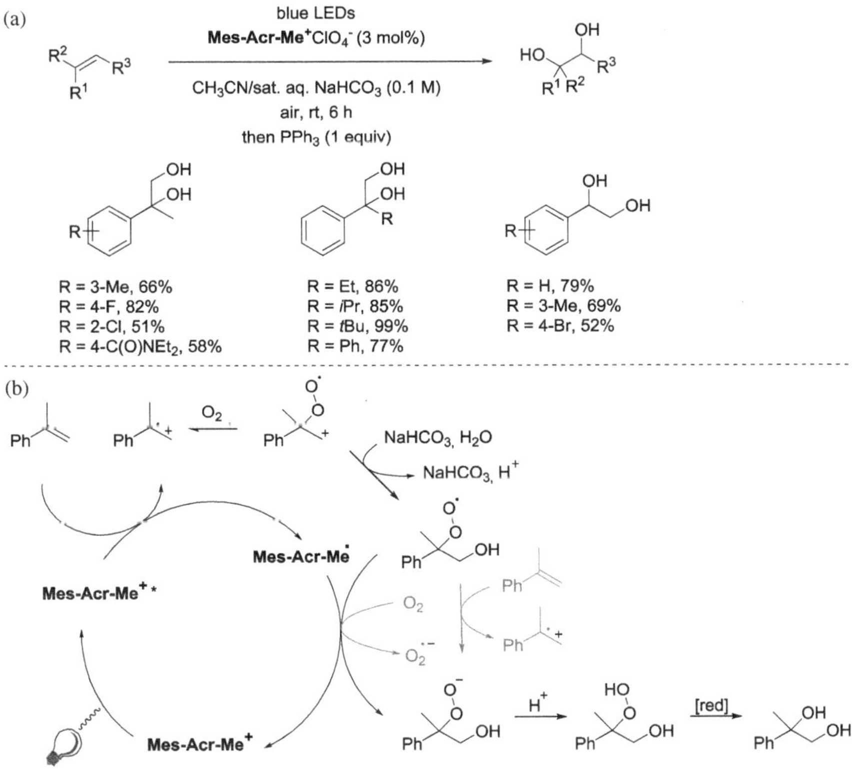

Figure 9.25. (a) Dihydroxylation of styrenes with water and molecular oxygen; selected products. (b) Proposed mechanism. Alternative pathway indicated in gray.

hydrogen-atom abstraction or electron transfer followed by proton transfer. Elimination of H_2O generates the desired α-acyloxy ketone. The reduced acridinium catalyst can be regenerated by O_2 or the peroxyl radical intermediate (Figure 9.26(c)).

Nicewicz and co-workers developed a halofunctionalization methodology of alkenes, through cooperative photoredox and copper catalysis [53]. The halofunctionalized products could be obtained by intercepting the radical intermediate with a halogen source (Figure 9.27). For the synthesis of chlorolactons, *N*-chlorosuccinimide (**NCS**) or *N*-chlorophthalimide (**NCP**) together with $CuCl_2$ were used. In the absence of copper, only the undesired Markovnikov regioisomer was obtained. $CuCl_2$ undergoes an

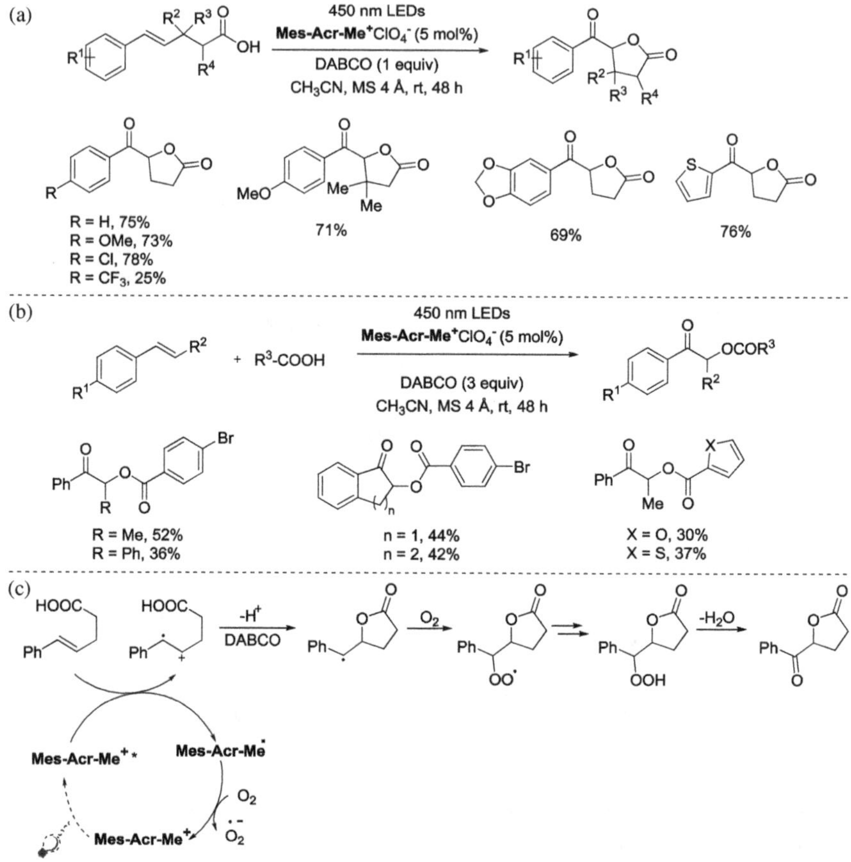

Figure 9.26. (a) Intramolecular oxylactonization of alkenoic acids; selected products. (b) Intermolecular oxo-acyloxylation of alkenes with carboxylic acids. (c) Proposed mechanism for the oxo-acylation of olefins with carboxylic acids.

irreversible chlorine-atom transfer with C-centered radicals, which makes it a suitable chlorine radical source for this reaction [54–56]. Better yields and diastereoselectivities were obtained when using phenanthroline (**phen**) as ligand. For bromolactonization, the combination of diethyl bromomalonate (**DEBM**) and $CuBr_2$ afforded the desired products. Depending on the substrate, the authors could obtain either five- or six-membered lactones. The proposed mechanism shares a common starting point with the anti-Markovnikov

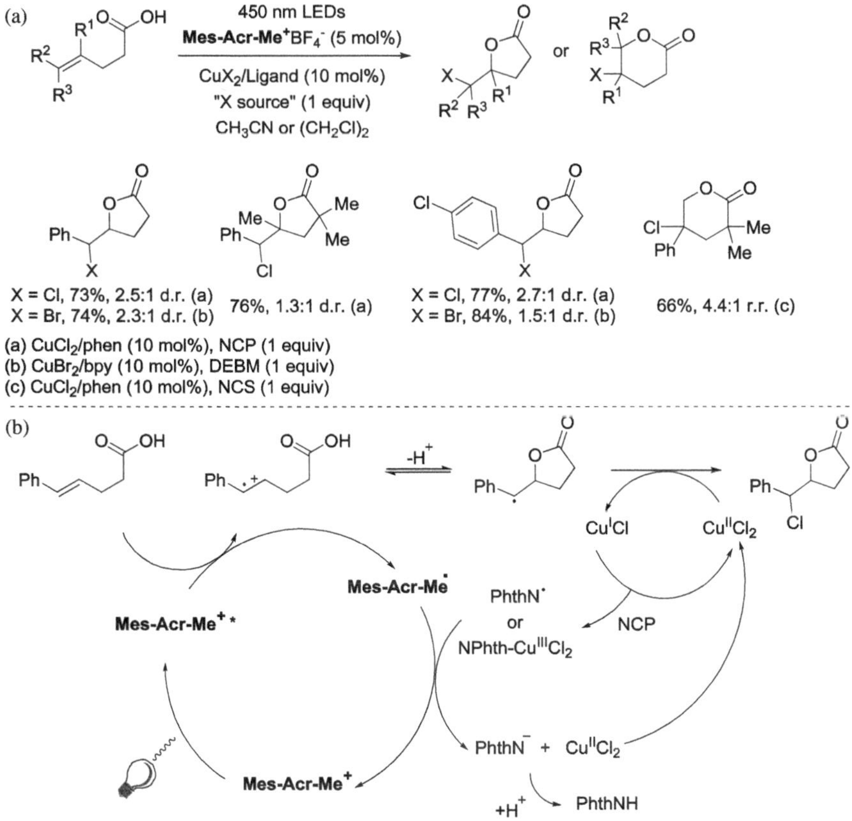

Figure 9.27. (a) Alkene halofunctionalization reaction; selected products. (b) Proposed mechanism.

alkene hydrofunctionalization reactions. The excited state of **Mes-Acr-Me**$^+$ oxidizes the alkene by a single electron oxidation to afford a cation radical, which undergoes reversible nucleophilic addition by the internal nucleophile, providing a lactone radical upon deprotonation. The halide is then transferred from copper to generate the desired product. The authors proposed that the so-formed Cu^{I}-species can be reoxidized by the halogenating agent to afford the active halogenating reagent. Turnover of the photocatalyst could be achieved oxidatively by either an imidyl radical (PhthN$^\bullet$) or by the Cu^{III} species (Figure 9.27(b)).

9.3.3 *Formation of C–N bonds*

Nicewicz and co-workers have extended their hydrofunctionalization method to include nitrogen nucleophiles, which has paved the way for anti-Markovnikov hydroamination protocols of alkenes (Figure 9.28) [57]. Given that primary amines are readily oxidized, the authors utilized amides as appropriate surrogates. The substrate scope parallels that of the hydroetherification and the products are obtained with complete regiocontrol. In addition, the method can be applied for intermolecular alkene hydroamination. In particular, reaction of

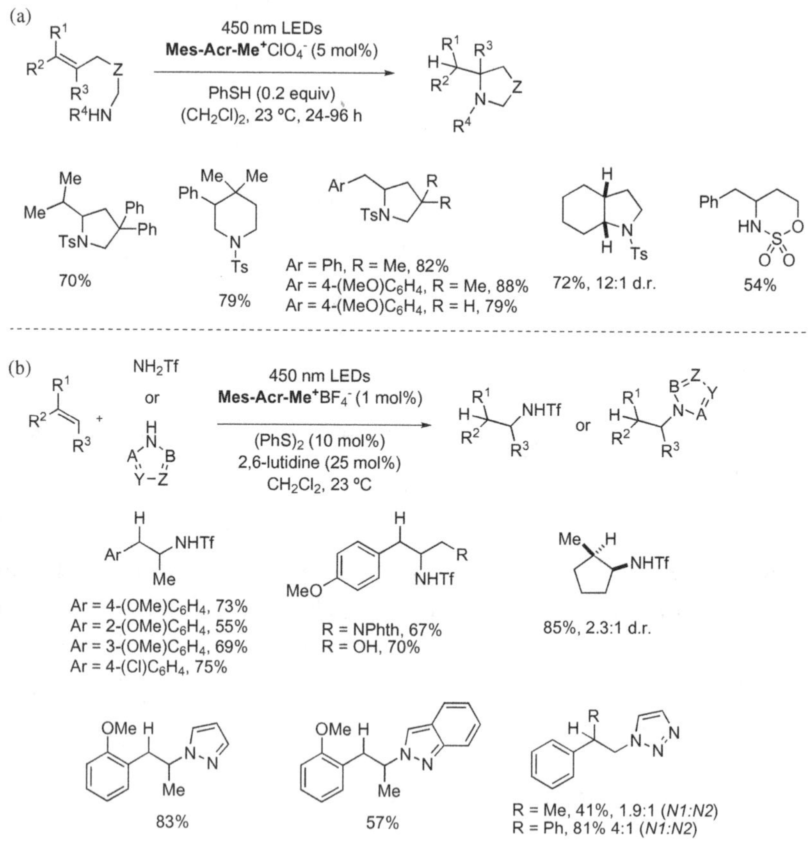

Figure 9.28. (a) Intramolecular and (b) intermolecular anti-Markovnikov hydroamination of alkenes; selected products.

styrenyl and alkenyl substrates with triflamide was feasible in the presence of catalytic 2,6-lutidine, which represents the first example of an organocatalytic intermolecular anti-Markovnikov alkene hydroamination. In a subsequent report, the authors developed an intermolecular addition with broader scope of nitrogen nucleophiles, providing access to phenethylamine derivatives, as well as allowing for the use of nitrogen heteroaromatic nucleophiles [58].

Nicewicz expanded the PRCC reaction manifold (Sec. 9.3.2) to include amides and amines as nucleophiles, which provides access to γ-lactams and pyrrolidines (Figure 9.29) [59]. A challenge associated with amide substrates was steering reactivity towards *N*-addition (lactam formation) over *O*-addition (imidate formation). Various amide-protecting groups were studied in a model system (β-methylstyrene and *N*-activated cinnamamide nucleophiles, **Mes-Acr-Me**$^+$ as catalyst and 4-methoxythiophenol as H-atom donor),

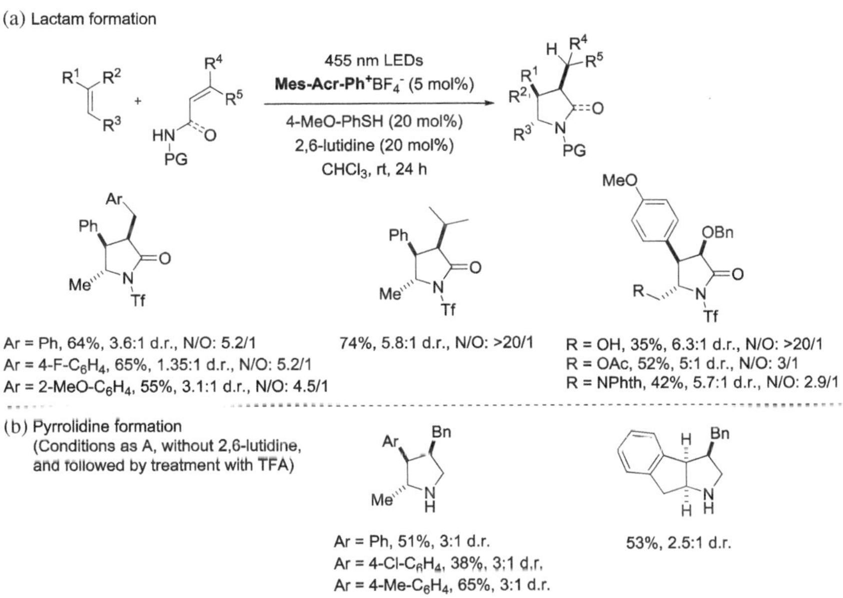

Figure 9.29. PRCC reaction between protected unsaturated amides or allyl amines and alkenes to form γ-lactams (a) or pyrrolidines (b), respectively; selected products.

and sulfonyl-protected amides (*N*-Tf) were found to produce clean reactions with synthetically useful *N*/*O* ratios. The reaction displayed good functional group compatibility (Figure 9.29(a)) and by employing Boc-protected allylic amines, pyrrolidine heterocycles could be obtained upon *in situ* deprotection. (Figure 9.29(b)).

Lei reported, together with the dehydrogenative cross-coupling of alkenes with alcohols (see Sec. 9.3.2), the synthesis of *N*-vinylazoles, by using azoles instead of alcohols as nucleophiles (Figure 9.30) [49]. The reaction is compatible with a variety of *N*-heterocyclic nucleophiles (e.g. pyrazole, pyrrole, benzotriazole, and indazole), but not with amines such as aniline, diphenylamine, or dibenzylamine.

9.3.4 *Formation of C–halogen bonds*

Nicewicz and co-workers have shown that the typical Markovnikov selectivity observed for the addition of mineral acids to styrenes can be completely inverted under photoredox conditions [43]. Initial studies involved the hydrochlorination of β-methylstyrene derivatives (Figure 9.31). HCl was generated *in situ* from pivaloyl chloride and 2,2,2-trifluoroethanol (**TFE**), and the desired anti-Markovnikov adducts could be formed using **Mes-Acr-Me**$^+$ as

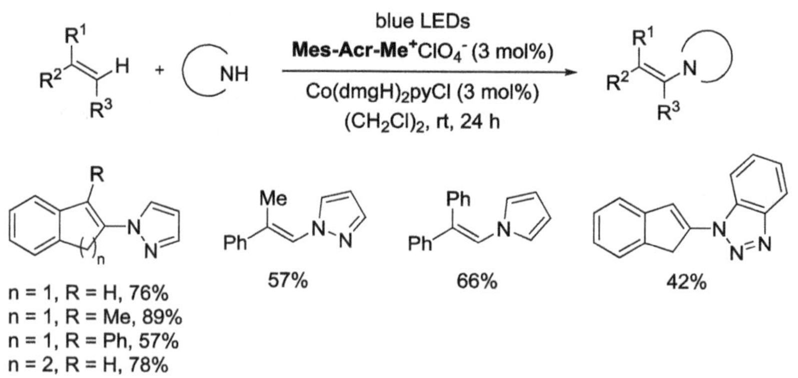

Figure 9.30. Coupling of alkenes with azoles for the synthesis of *N*-vinylazoles; selected products.

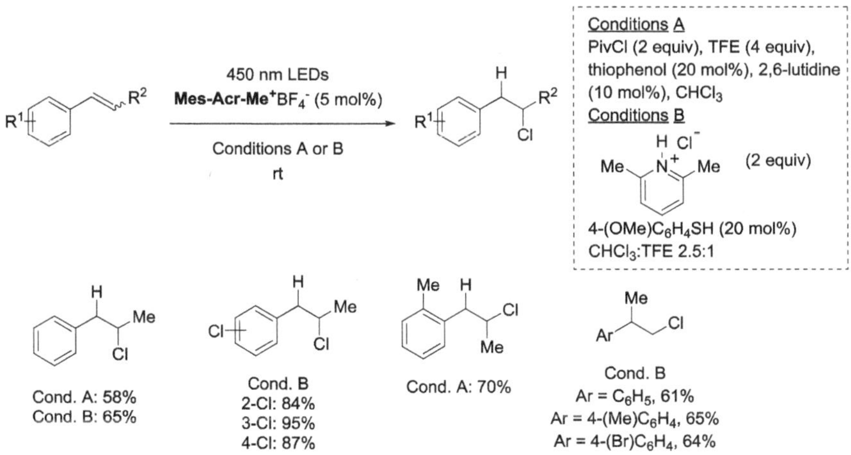

Figure 9.31. Anti-Markovnikov hydrochlorination of styrenes; selected products.

catalyst and thiophenol as H-donor. Depending on the electronic properties of the substituents on the alkene, small amounts of the undesired Markovnikov addition products were formed *via* the polar background electrophilic addition. In order to suppress this background reaction, and thus achieve complete regioselectivity, an amine base was added to the reaction that effectively disabled this competitive Brønsted acid-catalyzed reaction. The best results (highest yields with acceptable reaction times) were obtained with a combination of 2,6-lutidinium hydrochloride and 4-methoxythiophenol as the redox-active H-atom donor. This modification allowed for the use of electron-rich styrenes as substrates, giving the anti-Markovnikov hydrochlorinated products with good yields.

In the same report [43], the direct and regioselective anti-Markovnikov hydrofluorination of alkenes was also described. The challenge was to identify a suitable fluoride source, given that a nucleophilic fluoride ion was required for this photoredox system. Several common fluoride sources (e.g. quaternary ammonium salts, alkali-metal salts paired with crown ethers or phase-transfer reagents, triphenyldifluorosilicate or Olah's reagent) caused either decomposition of the catalyst or failed to yield the product.

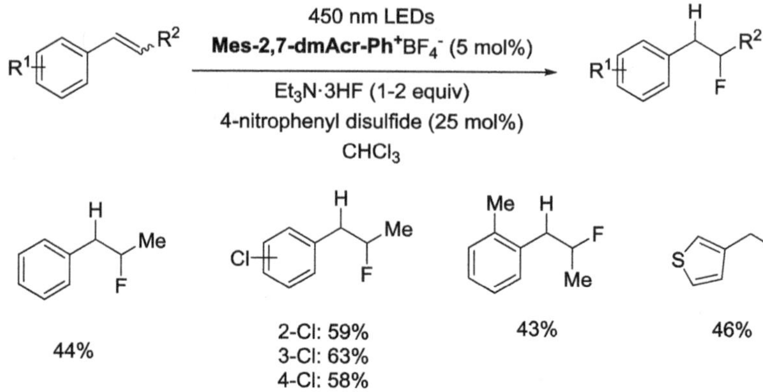

Figure 9.32. Anti-Markovnikov hydrofluorination of styrenes; selected products.

The only suitable fluoride source found was triethylamine trihydrofluoride. The optimal system involved the catalyst **Mes-2,7-dmAcr-Ph**$^+BF_4^-$, and the H-atom donor 4-nitrophenyl disulfide, which yielded the desired anti-Markovnikov addition products with complete regioselectivity (Figure 9.32).

9.4 Carboxylates, Boronates, and Sulfinates as Quenchers of the Photoexcited State

From a synthetic point of view, carboxylic acids, organoborates, and sulfinates share several advantageous characteristics: they are cheap, abundant, and stable. Carboxylic acids have the additional benefit that they are readily available from renewable feedstock, which makes them very favorable building blocks in environmentally sustainable synthetic endeavors. Mechanistically, the generation of reactive radicals from these classes of compounds using photoredox catalysis also shares some similarities. After single-electron oxidation, the functional group is oxidized and departs as boric acid derivatives, SO_2 or as CO_2. In all cases, the side products typically do not participate in radical reactions. Taken together, the common features of these functionalities make them very useful when constructing complex moleculars.

9.4.1 *Formation of C–C bonds*

Koike and co-workers reported in 2015, a method to generate radicals from organotrifluoroborates (Figure 9.33(a)) and carboxylic acids (Figure 9.33(b)), using the **Mes-Acr-Me**$^+$ catalyst [60]. This method allows for Giese-type reaction without the need for transition metals or co-oxidants. The proposed mechanism involves a visible-light-promoted oxidation of the substrates, which upon fragmentation gives C-centered free radicals. The resulting organic radicals reacts with alkenes resulting in a radical adduct, which undergoes

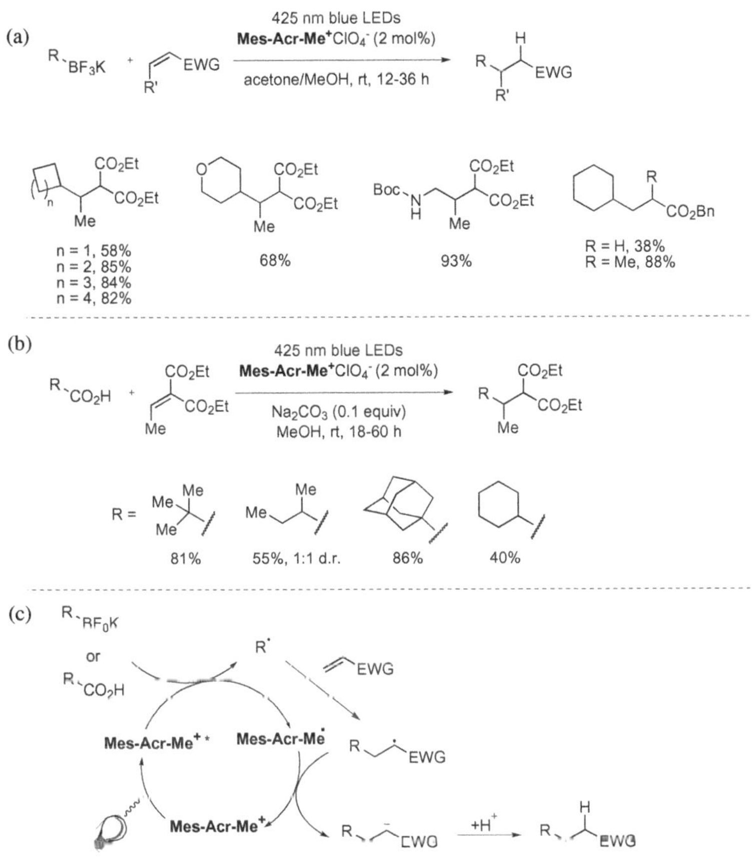

Figure 9.33. Radical addition of: (a) organotrifluoroborates and (b) carboxylic acids to alkenes; selected products. (c) Proposed mechanism.

single-electron reduction to give a carbanion intermediate that is protonated by the solvent (MeOH) furnishing the final product (Figure 9.33(c)). In a subsequent report, Gonzalez-Gomez and co-workers reexamined the reaction conditions to improve the applicability of this transformation to include a broader scope of carboxylic acids as substrates [61].

Even though the **Mes-Acr-Me**$^+$ catalyst is a powerful oxidant and useful in several types of transformations, it is still associated with a limited redox window and instability (bleaching). With the aim of developing an organic photoredox catalyst that is more robust, has a more useful redox window and is easily synthesized, DiRocco developed a series of 1,3,6,8-substituted acridinium salts [5]. The decarboxylative conjugate addition of Cbz-Proline to dimethyl maleate, also known to be catalyzed by [Ir(dF-CF$_3$-ppy)$_2$(dtbpy)](PF$_6$) [62], was used to evaluate this family of catalysts. It was shown that the new acridinium catalysts were viable alternatives to transition-metal-based photocatalysts (Figure 9.34), and that both the yield and stability of the **POC** as compared to the Fukuzumi catalyst could be improved.

Meggers and co-workers described an enantioselective version of the organotrifluoroborate-based Giese reaction. The Meggers system employed a chiral bis-cyclometalated rhodium catalyst [63] in combination with the **Mes-Acr-Me**$^+$ photoredox catalyst (Figure 9.35(a)) [64]. This protocol is limited to α,β-unsaturated

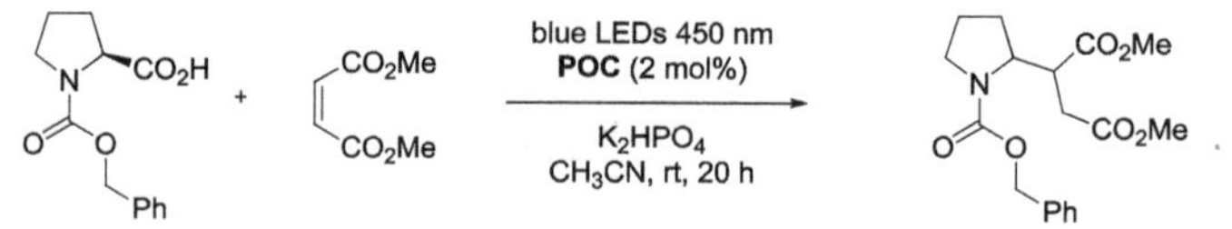

POC

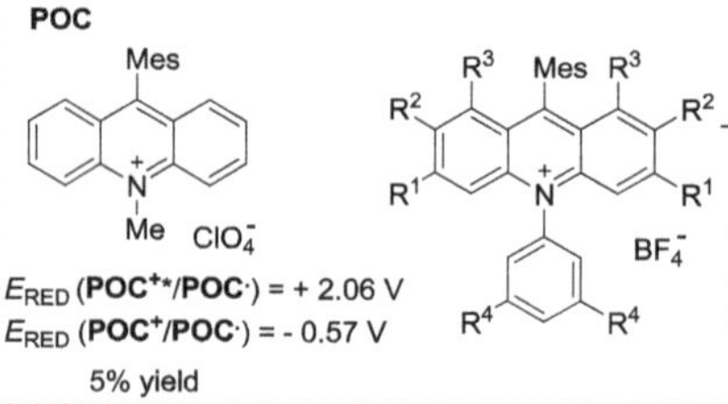

R^1	R^2	R^3	R^4	E_{RED} (POC^{+*}/POC·) (V) vs SCE	E_{RED} (POC$^+$/POC·) (V) vs SCE	yield (%)
*t*Bu	H	H	H	+2.08	-0.59	71
H	OMe	H	H	+1.90	-0.57	66
OMe	H	H	H	+2.01	-0.71	69
OMe	H	OMe	OMe	+1.65	-0.82	86
OMe	H	OMe	H	+1.62	-0.84	88

Figure 9.34. Decarboxylative conjugate addition of Cbz-proline to dimethyl maleate.

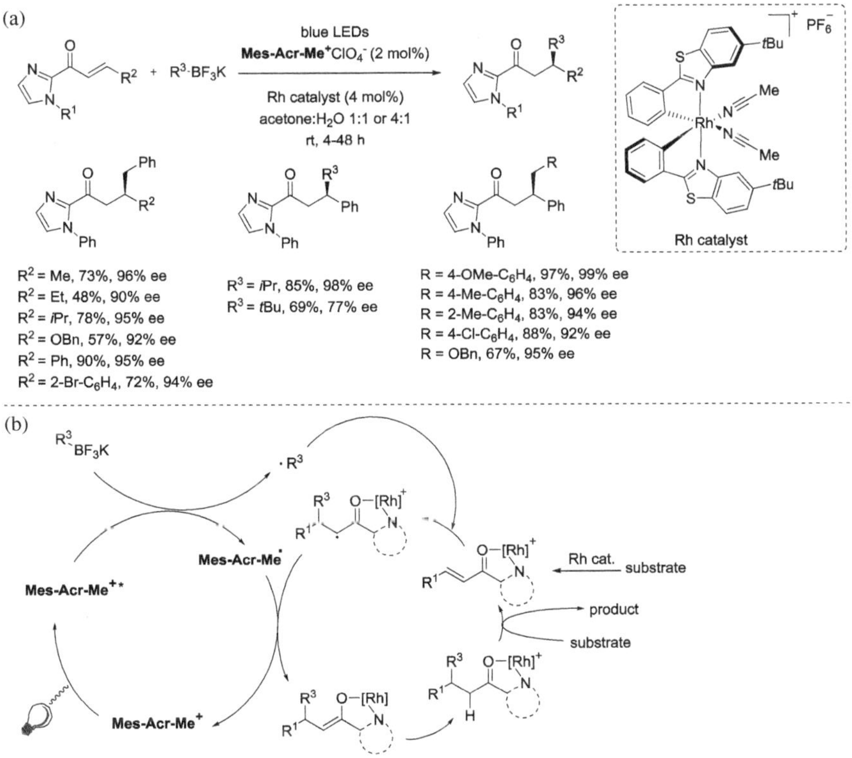

Figure 9.35. (a) Enantioselective addition of alkyl radicals to alkenes; selected products. (b) Proposed mechanism.

2-acyl imidazole and *N*-acyl pyrazole substrates, but typically gives the Giese products in high yields and with excellent enantiomeric excess. According to the proposed mechanism, a single-electron oxidation of the organotrifluoroborate by the excited state of **Mes-Acr-Me**$^+$ produces a C-centered radical, which adds to a *N,O*-rhodium-coordinated 2-acyl imidazole or *N*-acyl pyrazole substrate. This generates an adduct radical that is reduced by SET to a rhodium enolate, which upon protonation by water provides the rhodium-bound product (Figure 9.35(b)).

In addition to alkenes, Molander and co-workers have shown that C–H alkylation of heteroarenes can be achieved *via* a similar mechanism using alkyltrifluoroborates as radical precursors [65].

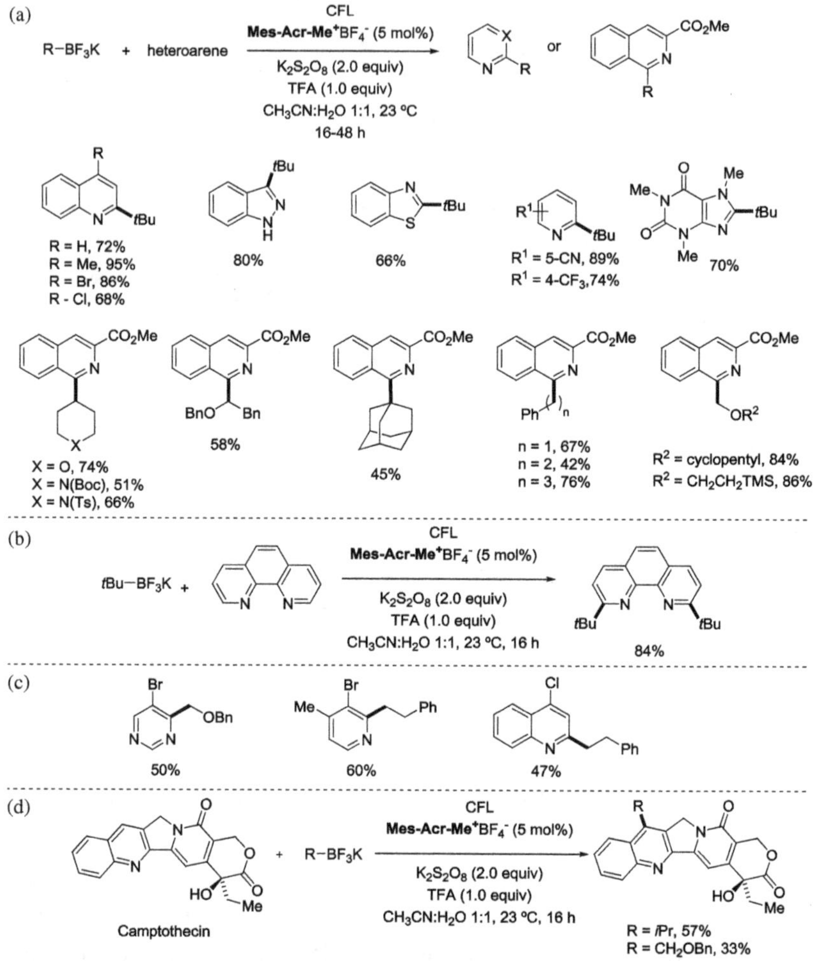

Figure 9.36. (a) C–H alkylation of heteroarenes with alkyltrofluoroborates; selected products. (b)–(d) Applications of the C–H alkylation of heteroarenes with alkyltrofluoroborates, with the formed bonds indicated in bold: (b) Ligand synthesis. (c) Utilizing heteroarenes with functional handles. (d) Late-stage functionalization of camptothecin.

A general method for the photoredox Minisci-type alkylation was developed for primary, secondary, and tertiary alkyltrifluoroborates, using the Fukuzumi's catalyst, the mild oxidant $K_2S_2O_8$, and trifluoroacetic acid (Figure 9.36). The mild conditions allow for the functionalization of complex heteroaromatic compounds, providing a valuable tool for late-stage derivatization. The same report describes several applications of the method: difunctionalization of bipyridine-based ligands, installation of alkyl subunits on brominated heteroarenes as a synthetic tool especially useful for medicinal chemists, and the late-stage functionalization of camptothecin (Figures 9.36(b)–9.36(d)). According to the proposed mechanism, the excited state of the acridinium catalyst oxidizes the alkyltrifluoroborate to release BF_3 and an alkyl radical. This alkyl radical can be intercepted by the protonated heteroarene to form a radical cation, which after H-atom abstraction by the sulfate radical anion leads to rearomatization and formation of the desired alkylated heteroarene. The sulfate radical anion is generated by reduction of persulfate by the reduced form of the acridinium salt, which regenerates the catalyst. This method was further extended to the heteroarylation of 2-trifluoroboratochromanones, which allowed the synthesis of a variety of heteroaryl-substituted chromanones [66].

Carboxylic acids have also been engaged in Minisci-type C–H functionalization of various heteroarenes in a similar fashion, but employing $[Ir(dF\text{-}CF_3\text{-}ppy)_2(dtbbpy)](PF_6)$ as the catalyst [67]. In terms of C–H functionalization of aromatic compounds using acridinium-based catalysts, Wang and co-workers have developed an acylation protocol of azo- and azoxybenzenes with α-keto acids by combining palladium and photoredox catalysis [68]. This transformation is achieved using the Fukuzumi's catalyst and $Pd(TFA)_2$, and the method tolerates a variety of disubstituted azo- and azoxybenzenes, as well as α-keto acids (Figure 9.37).

Similar to carboxylates and boronates, sulfinates are efficient quenchers of the excited state of acridinium salts. Liu and co-workers have developed chloro-, bromo-, and trifluoromethylthio-trifluoromethylations of unactivated alkenes using the Fukuzumi's catalyst, sodium triflinate (Langlois reagent), *N*-halophthalimide/*N*-trifluoromethylthiosaccharin as the halogen/SCF_3 source, and TFA or *p*-toluensulfonic acid (PTSA) as additive [69]. This method allows for the incorporation of a CF_3 group and Cl, Br or SCF_3 in one pot (Figure 9.38(a)). According to the proposed mechanism,

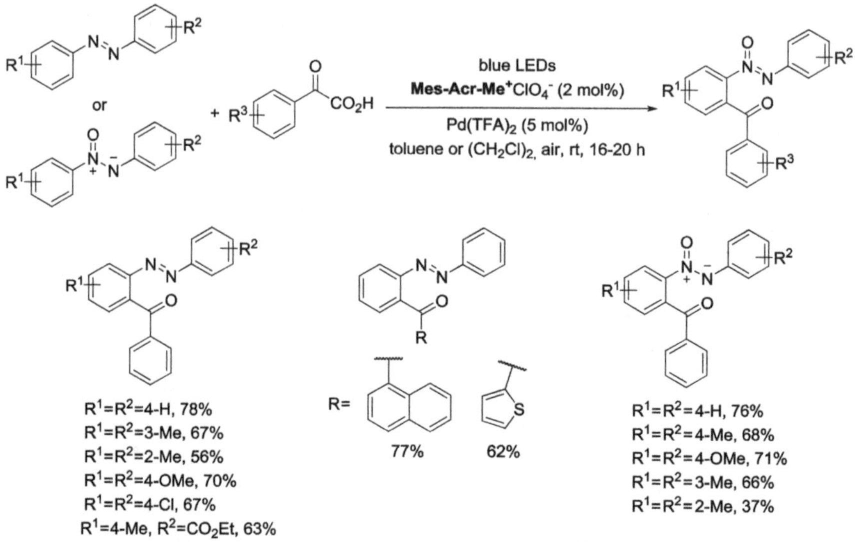

Figure 9.37. Acylation of azo- and azoxybenzenes with α-oxocarboxylic acids by combination of palladium and photoorganocatalysis; selected products.

single electron oxidation of the Langlois reagent by the excited state of the acridinium catalyst gives rise to a CF_3 radical and SO_2.

The CF_3 radical adds to the alkene leading to a β-trifluoromethyl C-centered radical. The *N*-halophthalimide transfers a halogen atom to that C-centered radical, furnishing the final product and an imidyl radical that could oxidize the acridine radical by a SET process in a turnover event of the catalytic cycle (Figure 9.38(b)).

Nicewicz described the **Mes-Acr-Me**$^+$-catalyzed hydrotrifluoromethylation of styrenes and unactivated aliphatic alkenes using the Langlois reagent CF_3SO_2Na (Figure 9.39) [70]. The reaction involves a coupled ternary catalytic system that provides the trifluoromethylated products with anti-Markovnikov selectivity. The method is rather functional group-tolerant and broadly applicable to mono, di, and trisubstituted aliphatic alkenes, as well as styrene substrates. A tentative mechanism was proposed (Figure 9.39(b)): oxidation of the Langlois reagent by photoexcited **Mes-Acr-Me**$^+$ generates the electrophilic trifluoromethyl radical, which adds to the alkene with high anti-Markovnikov selectivity producing a C-centered radical. Thiophenol

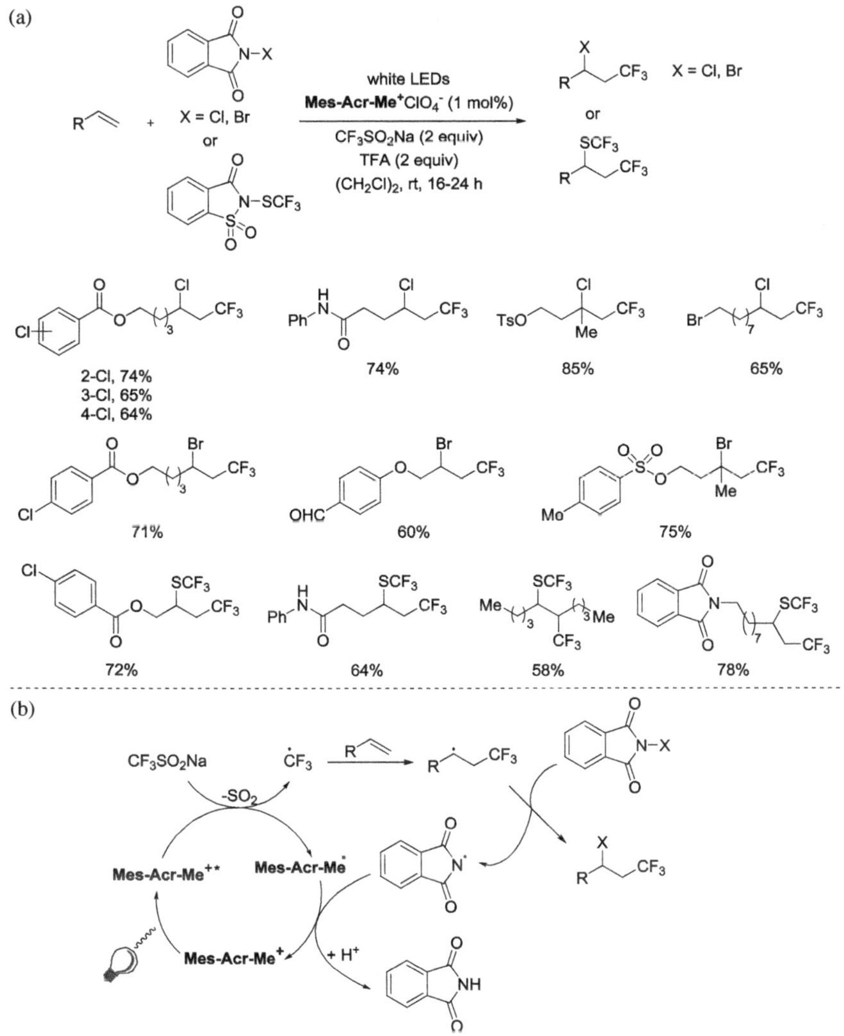

Figure 9.38. (a) Chloro-, bromo-, and trifluoromethylthio-trifluoromethylation of unactivated alkenes; selected products. (b) Proposed mechanism for the halotrifluoromethylation of alkenes.

or methyl thiosalicylate (**MTS**) together with **TFE** as a co-solvent act as a H atom shuttle, connecting the turnover event of the photoredox catalyst with product generation. Control experiments showed that alkyl-substituted alkenes provided the hydrotrifluoromethylated

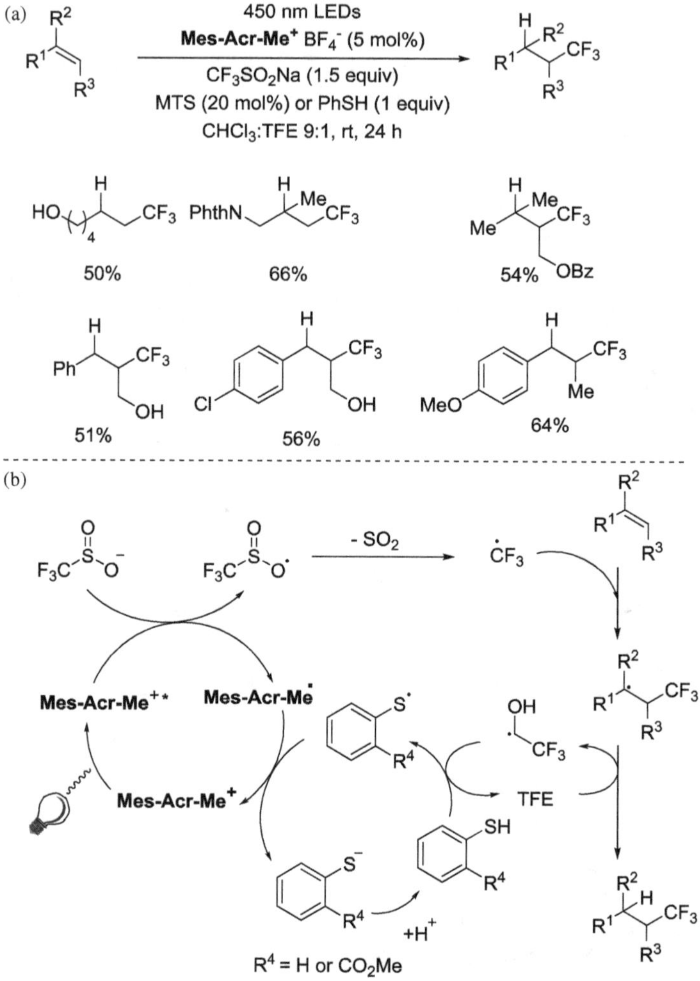

Figure 9.39. (a) Hydrotrifluoromethylation of aliphatic alkenes and styrenes; selected products. (b) Proposed mechanism.

products without the use of a thiol derivative as H-atom donor, but that **TFE** as a co-solvent was essential for the reactivity. Therefore, **TFE** is believed to act as the H-atom donor. However, the fact that **MTS** and thiophenol greatly improved the efficiency for most alkenes evaluated corroborated the proposed coupled ternary catalytic system being active under optimized conditions.

9.4.2 *Formation of C–O bonds*

The higher stability of benzoyloxy radicals compared to aliphatic equivalents allows for the intramolecular formation of C–O bonds, and thus ester formation. Based on this logic, Gonzalez-Gomez and co-workers developed a dehydrogenative lactonization of 2-arylbenzoic acids, using **Mes-Acr-Me**$^+$ in combination with $(NH_4)_2S_2O_8$, for the synthesis of benzo-3,4-coumarins (Figure 9.40(a)) [71]. A plausible mechanism involves the generation of a benzoyloxy radical intermediate *via* SET to the photoexcited acridinium catalyst, followed by a C–O bond forming cyclization event. After cyclization, the resulting delocalized radical is oxidized to the final product *via* a HAT or a SET/deprotonation

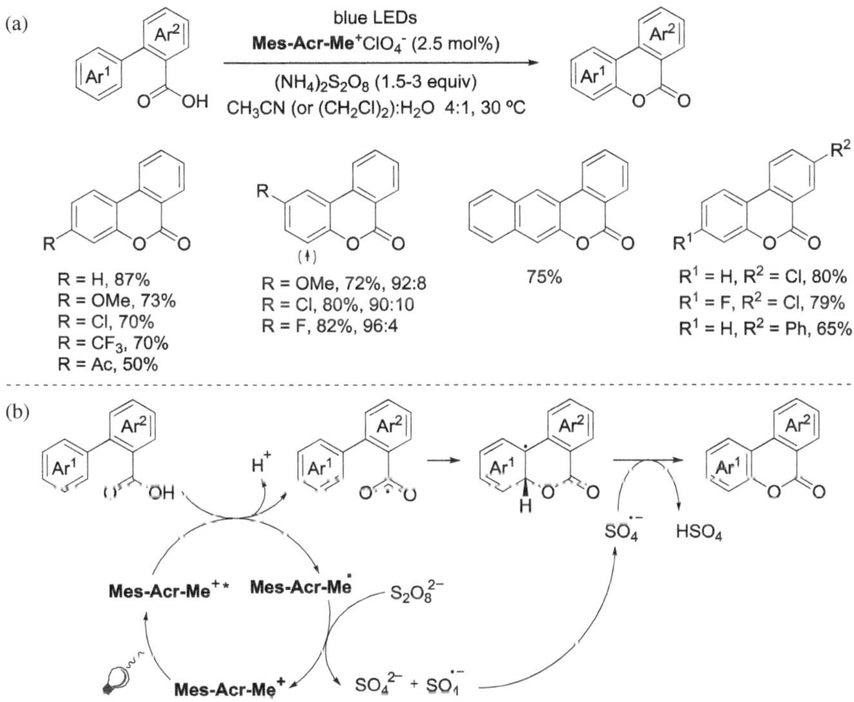

Figure 9.40. (a) Dehydrogenative lactonization of 2-arylbenzoic acids; selected products. (b) Proposed mechanism.

process. The acridine radical is oxidized by persulfate closing the catalytic cycle while the generated sulfate radical anion participates in the final aromatization step. (Figure 9.40(b)).

Lu and co-workers have developed a decarboxylative hydroxylation of carboxylic acids using the catalyst **Mes-Acr-Me**$^+$, a base, and molecular oxygen as oxidant [72]. This method enables the conversion of a variety of readily available carboxylic acids to alcohols in moderate to high yields (Figure 9.41(a)). According to the proposed mechanism, the carboxylate anion is oxidized by the excited state of the acridinium catalyst, forming a carboxyl radical that, after releasing CO_2, forms an alkyl radical. This radical is captured by O_2, which eventually leads to a mixture of alkylhydroperoxide and ketone. These are reduced during the work-up to yield the final alcohol. O_2 is also believed to function as an oxidant to regenerate the ground state of the photocatalyst (Figure 9.41(b)).

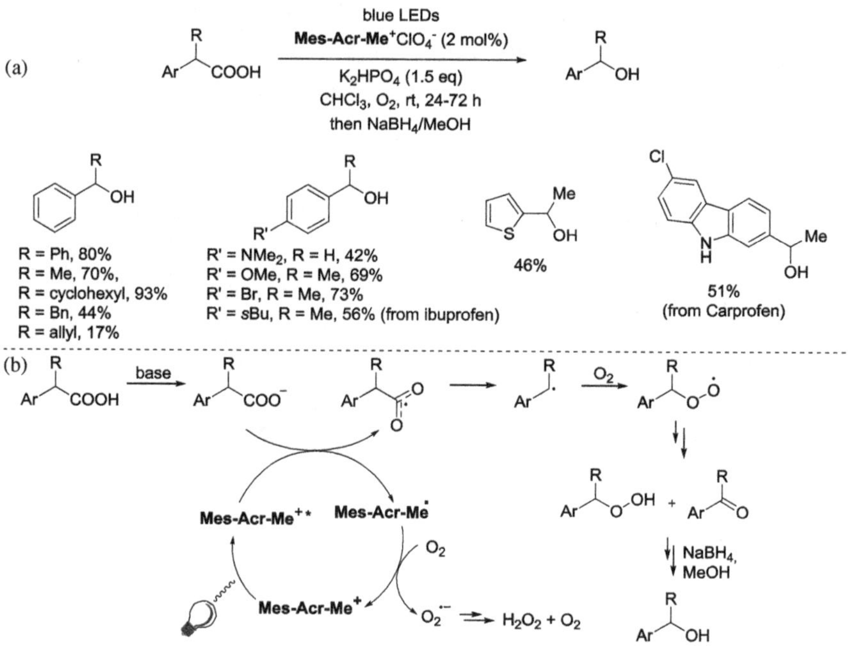

Figure 9.41. (a) Decarboxylative hydroxylation of carboxylic acids; selected products. (b) Proposed mechanism.

9.4.3 *Formation of C–N bonds*

Photoredox catalysis provides a useful alternative to the palladium-mediated formation of C–N bonds. As previously described in this chapter, both alkenes and electron-rich arenes can be engaged in the formation of C–N bonds. Tunge and co-workers have developed a protocol that employ unactivated alkanecarboxylic acids in an aminodecarboxylation reaction [73]. This method uses **Mes-Acr-Ph**$^+$ as catalyst, 1,8-diazabicyclo[5.4.0]undec-7-ene (DBU) as base, diisopropyl azodicarboxylate (DIAD), and readily available carboxylic acids to obtain aminated alkanes (Figure 9.42(a)). The proposed mechanism involves a SET from the carboxylate substrate to the excited **Mes-Acr-Ph**$^+$ catalyst, resulting in the formation of an alkyl radical. This radical is trapped by DIAD, which furnishes an *N*-centered radical that undergoes reduction by **Mes-Acr-Ph**$^•$ thereby turning over the catalyst. Protonation of the amide anion gives the

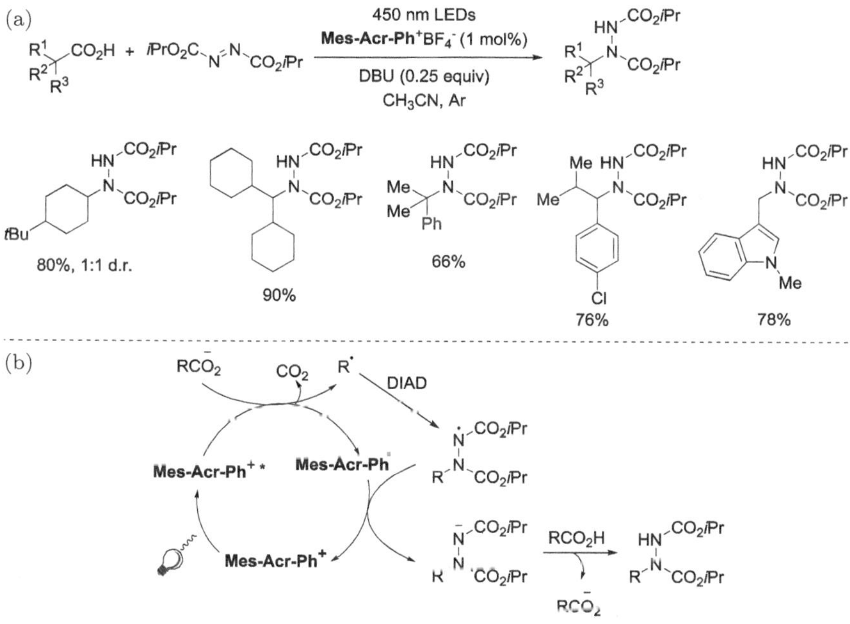

Figure 9.42. (a) Aminodecarboxylation of carboxylic acids; selected products. (b) Proposed mechanism.

final product (Figure 9.42(b)). The protocol is compatible with secondary, tertiary, and activated acids, providing the aminated products in moderate to high yields. Trapping C-centered radicals with DIAD, or equivalent, to form C–N bonds is a standard practice within the field of photoredox catalysis. It is of value to point out that this approach in amine synthesis has its limitations that stem from the necessity of subsequent synthetic steps to liberate the free amine.

9.4.4 *Formation of C–H bonds*

For a long time, Barton decarboxylation has been used as the standard method to reductively transform carboxylic acids to alkanes [74]. Wallentin and co-workers have developed a one-step photoredox catalytic version of this classical transformation [75]. This method utilizes **Mes-Acr-Me**$^+$ as the catalyst in combination with bis(4-chloro-phenyl)disulfide (DDDS) as H-atom donor and 2,6-lutidine. α-Amino-, α-hydroxy-, and phenylacetic acids react under these conditions, giving access to a variety of biologically-relevant compounds (Figures 9.43(a) and 9.43(b)). Under these conditions, the carboxylic acid is deprotonated, then the single electron oxidation by the excited state of **Mes-Acr-Me**$^+$ results in the formation of a C-centered radical after release of CO_2. Hydrogen-atom abstraction from thiophenol by that radical furnishes the final product. The thiyl radical then reoxidizes the reduced catalyst and is protonated to furnish the H-atom donor (Figure 9.43(c)). This catalytic system can thus be described as a binary hybrid or co-organocatalytic system.

The same transformation has been described by Nicewicz and co-workers, using **Mes-Acr-Ph**$^+$, rather than **Mes-Acr-Me**$^+$, as the catalyst [76]. Employing **TFE** as a solvent allowed for significant improvements in substrate compatibility, albeit at higher catalyst-loading and longer reaction times. Primary, secondary, and tertiary alkyl-substituted carboxylic acids, gave the desired products in good yields (Figure 9.44(a)). Hydrodecarboxylation of malonic acid derivatives is interesting from a synthetic point of view because it

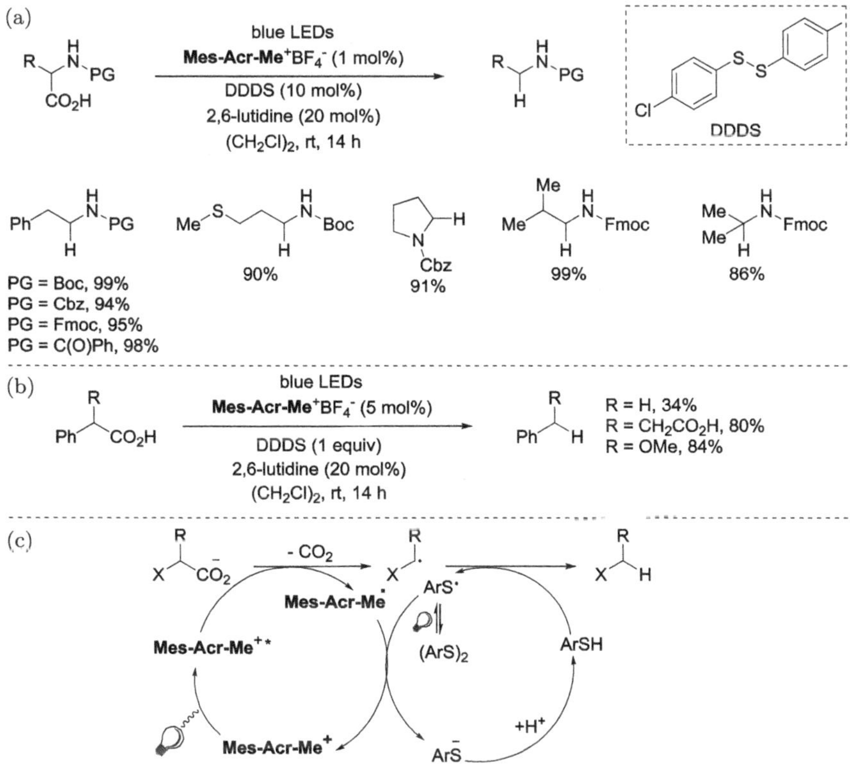

Figure 9.43. Decarboxylative reduction of carboxylic acids: (a) α-amino acids and (b) phenylacetic or α-oxy acids; selected products. (c) Proposed mechanism.

allows the use of malonate as a methylene synthon. In this case, a stronger base was necessary (KO*t*Bu instead of *i*Pr$_2$NEt) in order to obtain the desired doubly decarboxylated products (Figure 9.44(b)).

9.5 Other Quenchers of the Photoexcited State

As evident in this chapter, carboxylic acids, boronates, sulfonates, and electron-rich unsaturated systems are especially efficient quenchers of acridinium based photoredox catalysts. However, several other types of quenchers have been reported.

In a joint effort, the group of Lu and Xiao developed a formal [3+2] cycloaddition of 2H-azirines with electron-poor alkynes

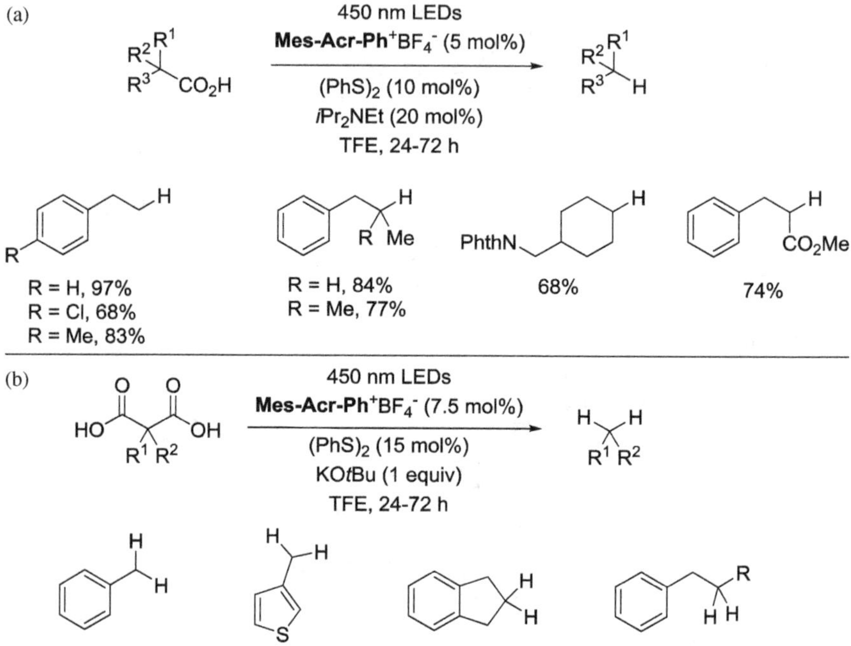

Figure 9.44. (a) Hydrodecarboxylation of carboxylic acids; selected products. (b) Hydrodecarboxylation of malonic acids; selected products.

(Figure 9.45(a)) [77]. 2H-azirines have a relatively high oxidation potential (E_{OX} = +1.65 V vs SCE); nonetheless, excited acridinium salts are suitable photoredox catalysts for achieving exergonic single electron oxidation of the 2H-azirines.

Tri- and tetra-substituted pyrroles were obtained in moderate to good yields. The authors also applied the developed methodology to the synthesis of an HMG-CoA reductase inhibitor. The mechanism (Figure 9.45(b)) begins with the single-electron oxidation of azirine. Subsequently, the oxidized 2H-azirine undergoes a ring-opening, resulting in a distonic cation radical that can react with an alkyne in a formal [3+2] cycloaddition. The turnover event of the acridinium cycle is the reduction of the vinylic radical, which subsequently closes the five-membered ring. In subsequent reports, the same group extended the methodology from alkynes to aldehydes,

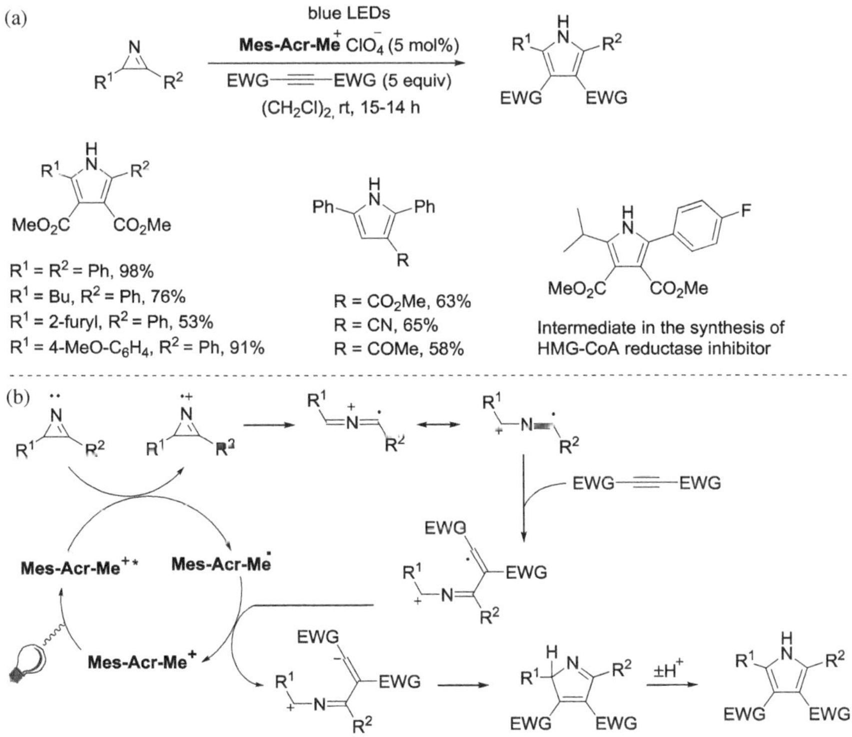

Figure 9.45. (a) Formal [3+2] cycloaddition of 2H-azirines with electron-poor alkynes; selected products. (b) Proposed mechanism.

providing a route to oxazoles [78]. In a similar methodology, alkynyl bromides were applied instead of aldehydes, allowing the authors to also construct isoxazoles [79].

The Xu group have expanded the redox manifold of the photoredox catalysts based on the acridine scaffold. By employing a reduced form of the typical acridinium salt, a dihydroacridine derivative, they have managed to address transformations that are initiated *via* oxidative quenching (Figure 9.46(a)). The Xu group applied diazonium salts as quenchers of the excited state, which allowed them to merge photoorganocatalysis with a Pd-catalyzed C–H activation of acetanilides and benzamides [80]. The reaction tolerates variously substituted diazonium salts and amides. In general, the

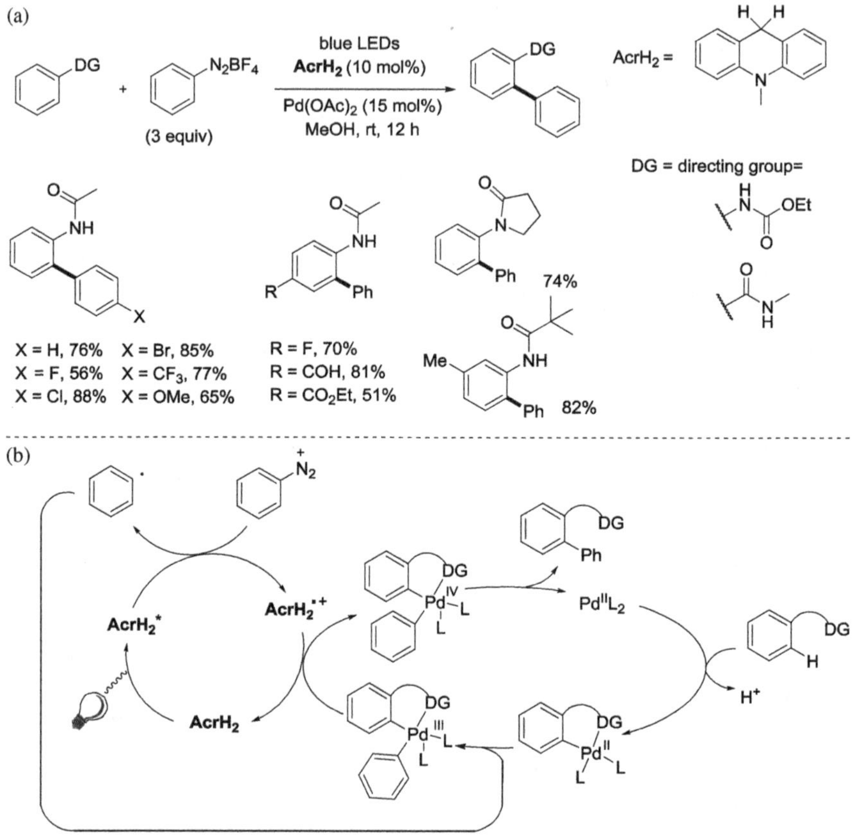

Figure 9.46. (a) Merger of photoorganoredox and Pd catalytic cycles; selected products. (b) Proposed mechanism.

products were obtained in yields ranging from 51% to 88%, but the method requires relatively high loadings in both the photoredox catalyst (10 mol%) and the Pd catalyst (15 mol%). Upon the initial SET (Figure 9.46(b)), the diazonium salts fragments into molecular nitrogen and reactive aryl radical that can further react with a cyclopalladated intermediate. The resulting Pd^{III} complex is then oxidized in a turnover event of the photoorganocatalytic cycle.

Further development of dual catalytic systems involving photoredox and transition metal catalysis can be exemplified with the work of Cornilleu and co-workers, who demonstrated that the Au-catalyzed coupling of diazonium salts with boronic acids benefits from the photoredox conditions [81]. When **Mes-Acr-Me**$^+$ was used, the known Au-catalyzed coupling was facilitated and provided the products at a rate roughly three times higher than the reaction without the **POC**. The conditions allowed for the synthesis of variously substituted biaryls in moderate to good yields (33%–92%). The authors also developed a second set of conditions based on $Ru(bpy)_3(PF_6)_2$. The efficiency of the catalysts with respect to the yields was complementary. The substrates that provided higher yields with the acridinium catalyst, provided lower yields with $Ru(bpy)_3(PF_6)_2$ and *viceversa.*

As illustrated above, transition metal complexes can provide a constructive redox communication with photoredox catalysts. However, and as previously pointed out in this chapter, photoredox catalysts can also work in concert with organic co-catalysts.

The Chen group has reported that hydrazones and oximes can afford the corresponding *N*- and *O*-centered radicals with acridinium-based catalysts in the presence of TEMPO as a quencher of the excited state (E_{OX} = (TEMPO$^+$/TEMPO$^\bullet$ = 0.62 V vs SCE)) and a coupled co-catalyst (Figure 9.47) [82]. The TEMPOnium generated upon the reductive quenching of the excited acridinium catalyst probably participates in a proton-coupled electron transfer with hydrazones or oximes leading to the corresponding *N*-centered or *O*-centered radicals. These are then trapped with an adjacent double bond and thereby cyclized to five- and six-membered heterocyclic intermediates. The O_2 present in the reaction mixture provides a turnover event for the reduced acridinium catalyst, and also traps the generated radical intermediates giving hydroaminated and alkoxylated products. The reaction tolerates halogen substitution on the adjacent arene as well as various electron-donating and electron-withdrawing substituents. Both aromatic and aliphatic hydrazones were suitable substrates. The ratio of 6-endo-cyclization and 5-exo cyclization is typically governed by the substitution pattern on the alkene.

While electron-rich heterocyclic arenes have relatively low oxidation potentials, they have been used as quenchers of the excited

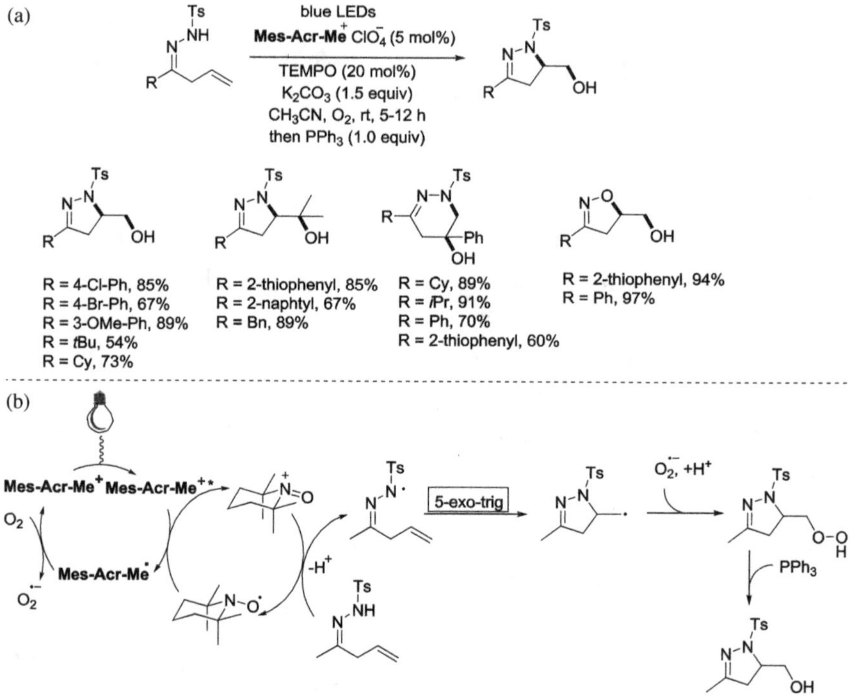

Figure 9.47. (a) Synthesis of dihydropyrazoles by a merger of photoorgano and organocatalysis; selected products. (b) Proposed mechanism.

state of acridinium-based catalysts [83]. The König group has developed a metal-free protocol for the sulfonamidation of pyrroles (E_{OX} (pyrrole) = +1.2 V vs SCE, based on an initial oxidation of the heteroaromatic substrate) (Figure 9.48) [84]. In the catalytic cycle, pyrrole is oxidized by the excited acridinium cation. The resulting electrophilic cation radical reacts with the deprotonated sulfonamide, producing a delocalized radical intermediate. The substituted pyrrole product is obtained upon loss of a hydrogen atom. Depending on the substitution pattern of the substrates, the desired sulfonamides were obtained in 10% to 99% yield. Interestingly, other electron-rich heteroarenes like furane or indole did not undergo any reaction.

A mechanistically different aminosulfonylation protocol was developed in a joint effort by the groups of Kuang and Wu [85].

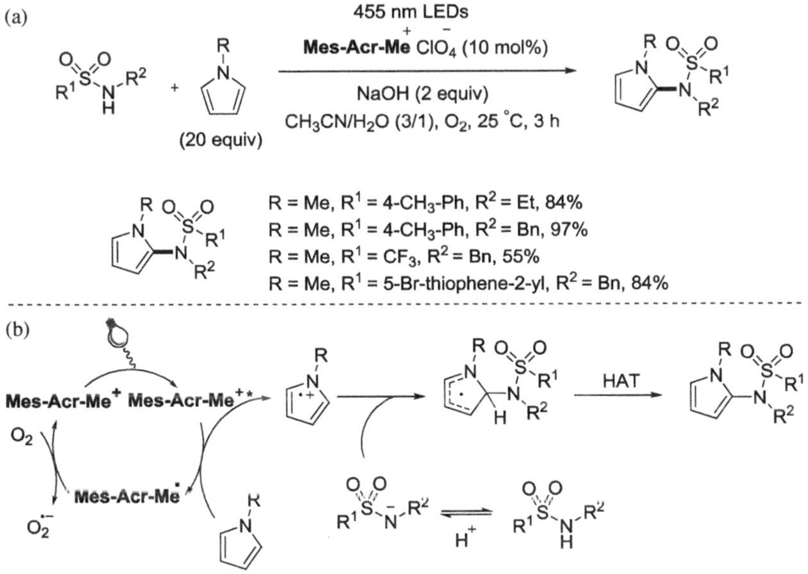

Figure 9.48. (a) Photoredox sulfonamidation of pyrroles; selected products. (b) Proposed mechanism.

In this example, the photoexcited state of acridinium is quenched by hydrazine (Figure 9.49). The turnover event of the catalyst involves a reduction of ethyl 2-bromo-2,2-difluoroacetate, which generates a difluoroacetyl radical. This electron-deficient radical adds to a third reaction partner, an alkyne, producing a new C(sp^2)-centered radical intermediate, which reacts further with SO_2 (the authors suggested a SO_2 complex with hydrazine) giving a *S*-centered radical, that recombines with the hydrazinyl radical affording the product. The reaction tolerates various electron-donating and electron-withdrawing substituents, halogens, and heteroaromatics as substituents on the alkyne. Furthermore, the protocol tolerates differently substituted hydrazines, giving products in 45% to 95% yield.

The group of Xu reported a Baeyer–Villiger oxidation of variously substituted cyclobutanones with **Acr-Me**$^+$ as the catalyst and hydrogen peroxide as the oxidant (Figure 9.50) [86].

Here, the strained ketone is the quencher of the excited state. The generated cation radical reacts with hydrogen peroxide to

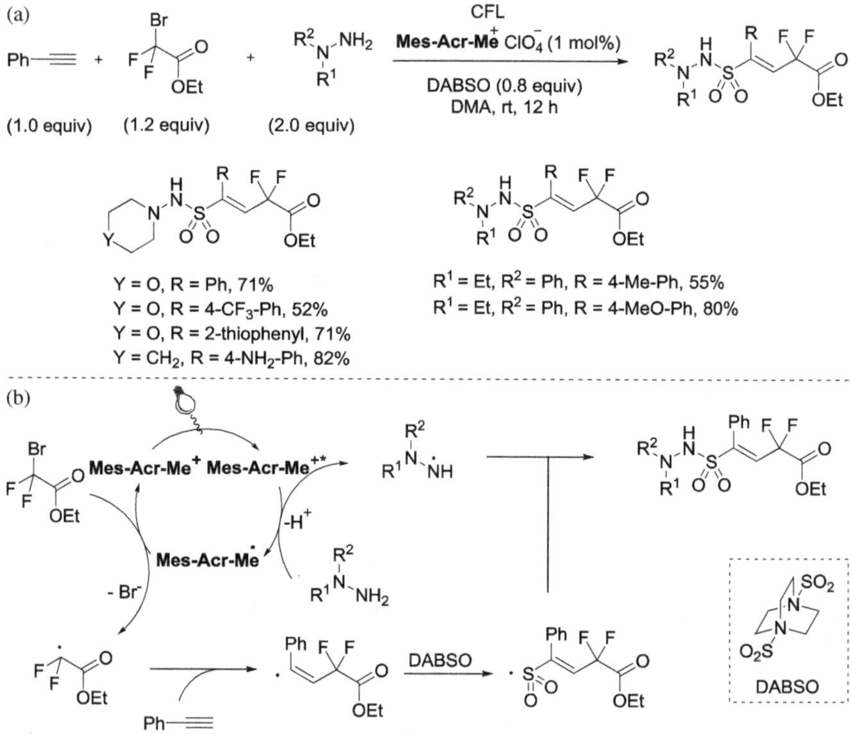

Figure 9.49. (a) Aminosulfonylation of alkynes; selected products. (b) Proposed mechanism.

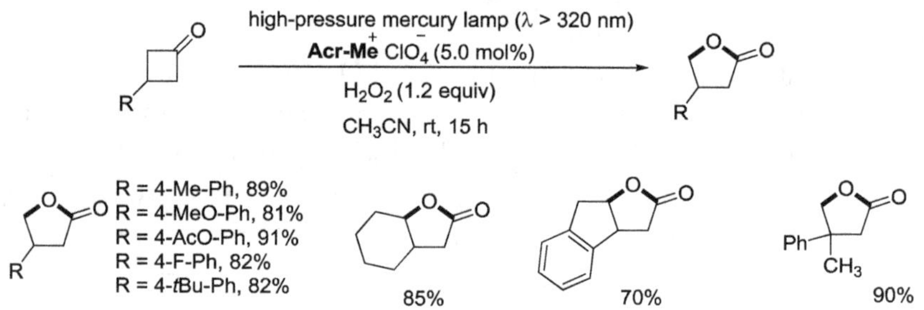

Figure 9.50. Photoredox catalyzed Baeyer–Villiger oxidation; selected products.

afford a radical, which participates in the turnover event of the catalyst and affords the Criegee intermediate that rearranges to the dihydrofuranones. The reaction tolerates a variety of functional groups, including halogens. While the majority of applications of

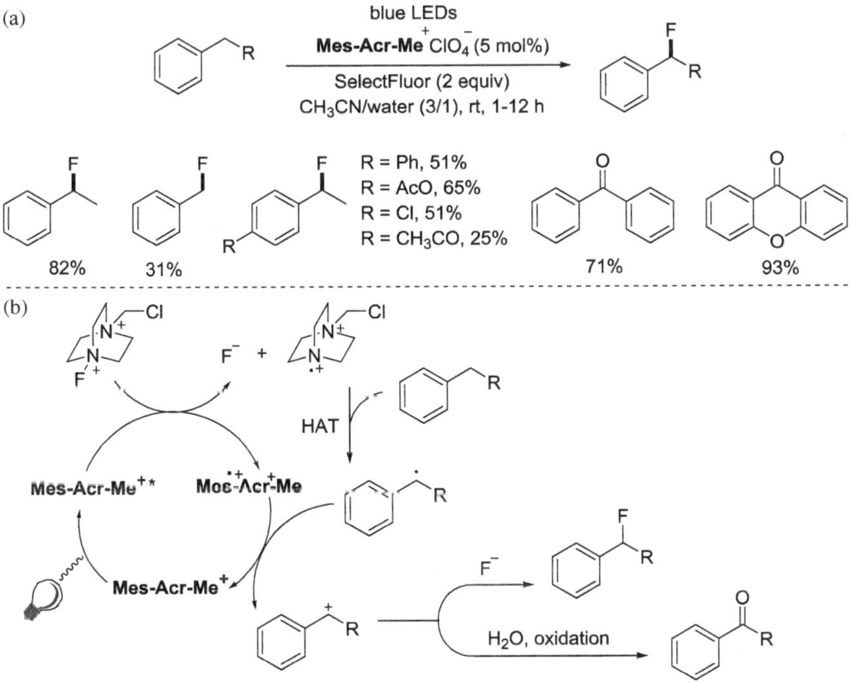

Figure 9.51. (a) Fluorination and oxidation of benzylic positions based on oxidative quenching of **Mes-Acr-Me**$^+$; selected products. (b) Proposed mechanism.

acridinium-based catalysts relies on reductive quenching, the group of Wu presented a transformation based on oxidative quenching (Figure 9.51) [87]. The photoexcited catalyst has an oxidation potential (E^*_{OX} (**POC*/POC$^•$**) = −0.57 V vs SCE), sufficiently low to render a SET reduction of SelectFluor (E_{RED} = −0.04 V vs SCE) exergonic. The generated *N*-centered radical cation is known to be a potent hydrogen-atom abstractor. After HAT from the benzylic position of the substrate (Figure 9.51), the benzylic radical that has been formed reduces the dicationic catalyst, and thereby closes the catalytic cycle. The benzylic carbocation can then react with the F^- liberated in the quenching event of the photoredox catalyst. Even though the mechanistic implications have a clear value, this specific protocol seems rather limited in terms of scope and provides the fluorinated products typically in low to modest yields. Interestingly, when increasing the proportion of water that was used as a co-solvent, diphenylketones were isolated as the major product, starting from diphenylmethanes.

9.6 Conclusions

Undoubtedly, acridinium-type catalysts belong to the class of highly popular **POCs**. They are cheap and non-toxic, but most importantly they open up for new synthetic opportunities that can be delineated by a set of simple principles. A hallmark of catalysis with acridinium ions is their ability to activate conjugated systems, converting them to potent electrophiles. Aromatic systems and alkenyl-bearing compounds can thus be engaged in C–H substitution, hydrofunctionalization, and difunctionalization reactions. The diversity of these catalytic systems can be utilized in late-stage functionalizations (e.g. aminations or cyanations) of arenes or anti-Markovnikov functionalizations of alkenes. When substrates like carboxylates, sulfinates, or boronates are used, they typically fragment to generate radicals that can be utilized in various radical reactions. The chemistry developed with acridinum catalysts has opened up for more challenging substrates that can be employed in an even more expanded manifold of chemical reactions. As a last remark, the history of acridinium-based catalysis is as short as it is rich, which bodes well for the future discovery of a plethora of novel transformations.

References

[1] Tsudaka, T.; Kotani, H.; Ohkubo, K.; Nakagawa, T.; Tkachenko, N. V.; Lemmetyinen, H.; Fukuzumi, S. *Chem. Eur. J.* **2017**, *23*, 1306–1317.

[2] Fukuzumi, S.; Kotani, H.; Ohkubo, K.; Ogo, S.; Tkachenko, N. V.; Lemmetyinen, H. *J. Am. Chem. Soc.* **2004**, *126*, 1600–1601.

[3] Kotani, H.; Ohkubo, K., Fukuzumi, S. *J. Am. Chem. Soc.* **2004**, *126*, 15999–16006.

[4] Benniston, A. C.; Harriman, V.; Li, P.; Rostron, J. P.; van Ramesdonk, H. J.; Groeneveld, M. M.; Zhang, H.; Verhoeven, J. W. *J. Am. Chem. Soc.* **2005**, *127*, 16054–16064.

[5] Joshi-Pangu, A.; Levesque, F.; Roth, H. G.; Oliver, S. F.; Campeau, L.-C.; Nicewicz, D.; DiRocco, D. A. *J. Org. Chem.* **2016**, *81*, 7244–7249.

[6] Fukuzumi, S.; Ohkubo, K. *Org. Biomol. Chem.* **2014**, *12*, 6059–6071.

[7] Ravelli, D.; Fagnoni, M.; Albini, A. *Chem. Soc. Rev.* **2013**, *42*, 97–113.

[8] Romero, N. A.; Nicewicz, D. A. *Chem. Rev.* **2016**, *116*, 10075–10166.

[9] Ohkubo, K.; Mizushima, K.; Iwata, R.; Fukuzumi, S. *Chem. Sci.* **2011**, *2*, 715–722.

[10] Romero, N. A.; Margrey, K. A.; Tay, N. E.; Nicewicz, D. A. *Science* **2015**, *349*, 1326–1331.

[11] Margrey, K. A.; McManus, J. B.; Bonazzi, S.; Zecri, F.; Nicewicz, D. A. *J. Am. Chem. Soc.* **2017**, *139*, 11288–11299.

[12] Niu, L.; Yi, H.; Wang, S.; Liu, T.; Liu, J.; Lei, A. *Nat. Commun.* **2017**, *8*: 14226, 1–7.

[13] Kahnt, A.; Peuntinger, K.; Dammann, C.; Drewello, T.; Hermann, R.; Naumov, S.; Abel, B.; Guldi, D. M. *J. Phys. Chem. A* **2014**, *118*, 4382–4391.

[14] McManus, J. B.; Nicewicz, D. A. *J. Am. Chem. Soc.* **2017**, *139*, 2880–2883.

[15] Bordwell, F. G.; Bausch, M. J.; Branca, J. C.; Harrelson, J. A. *J. Phys. Org. Chem.* **1988**, *1*, 225–241.

[16] Bordwell, F. G.; Cheng, J.-P.; Bausch, M. J.; Bares, J. E. *J. Phys. Org. Chem.* **1988**, *1*, 209–223.

[17] (a) Bordwell, F. G.; Cheng, J.-P. *J. Am. Chem. Soc.* **1989**, *111*, 1792–1795; (b) Xiang, M.; Meng, Q.-Y.; Li, J.-X.; Zheng, Y.-W.; Ye, C.; Li, Z.-J.; Chen, B.; Tung, C.-H.; Wu, L.-Z. *Chem. Eur. J.* **2015**, *21*, 18080–18084.

[18] Ohkubo, K.; Suga, K.; Fukuzumi, S. *Chem. Commun.* **2006**, 2018–2020.

[19] Xu, H. J.; Xu, X. L.; Fu, Y.; Feng, Y. S. *Chin. Chem. Lett.* **2007**, *18*, 1471–1475.

[20] Ohkubo, K.; Mizushima, K.; Iwata, R.; Souma, K.; Suzuki, N.; Fukuzumi, S. *Chem. Commun.* **2010**, *46*, 601–603.

[21] Yi, H.; Bian, C.; Hu, X.; Niu, L.; Lei, A. *Chem. Commun.* **2015**, *51*, 14046–14049.

[22] Zhang, L.; Yi, H.; Wang, J.; Lei, A. *Green Chem.* **2016**, *18*, 5122–5126.

[23] Yi, H.; Hu, X.; Bian, C.; Lei, A. *ChemSusChem* **2017**, *10*, 79–82.

[24] Kato, S.; Saga, Y.; Kojima, M.; Fuse, H.; Matsunaga, S.; Fukatsu, A.; Kondo, M., Masaoka, S.; Kanai, M. *J. Am. Chem. Soc.* **2017**, *139*, 2204–2207.

[25] Roberts, B. P. *Chem. Soc. Rev.* **1999**, *28*, 25–35.

[26] Barret, K. E. J.; Waters, W. A. *Discuss. Faraday Soc.* **1953**, *14*, 221–227.

[27] Nicolaou, K. C.; Snyder, S. A.; Montagnon, T.; Vassilikogiannakis, G. *Angew. Chem. Int. Ed.* **2002**, *41*, 1668–1698.

[28] Juhl, M.; Tanner, D. *Chem. Soc. Rev.* **2009**, *38*, 2983–2992.

[29] Lin, S.; Ischay, M. A.; Fry, C. G.; Yoon, T. P. *J. Am. Chem. Soc.* **2011**, *133*, 19350–19353.

[30] Zhang, X.; Yi, H.; Luo, Y.; Lei, A. *Chem. Asian J.* **2016**, *11*, 2117–2120.

[31] Wei, D.; Li, Y.; Liang, F. *Adv. Synth. Catal.* **2016**, *358*, 3887–3896.

[32] Zhou, R.; Liu, H.; Tao, H.; Yu, X.; Wu, J. *Chem. Sci.* **2017**, *8*, 4654–4659.

[33] Suga, K.; Ohkubo, K.; Fukuzumi, S. *J. Phys. Chem. A* **2003**, *107*, 4339–4346.

[34] Ohkubo, K.; Nanjo, T.; Fukuzumi, S. *Org. Lett.* **2005**, *7*, 4265–4268.

[35] Margrey, K. A.; Nicewicz, D. A. *Acc. Chem. Res.* **2016**, *49*, 1997–2006.

[36] Nicewicz, D. A.; Hamilton, D. S. *Synlett* **2014**, *25*, 1191–1196.

[37] Romero, N. A.; Nicewicz, D. A. *J. Am. Chem. Soc.* **2014**, *136*, 17024–17035.

[38] Hamilton, D. S.; Nicewicz, D. A. *J. Am. Chem. Soc.* **2012**, *134*, 18577–18580.

[39] Yang, Z.; Li, H.; Li, S.; Zhang, M.-T.; Luo, S. *Org. Chem. Front.* **2017**, *4*, 1037–1041.

[40] Morse, P. D.; Nicewicz, D. A. *Chem. Sci.* **2015**, *6*, 270–274.

[41] Schwekendiek, K.; Glorius, F. *Synthesis* **2006**, *18*, 2996–3002.

[42] Perkowski, A. J.; Nicewicz, D. A. *J. Am. Chem. Soc.* **2013**, *135*, 10334–10337.

[43] Wilger, D. J.; Grandjean, J.-M. M.; Lammert, T. R.; Nicewicz, D. A. *Nat. Chem.* **2014**, *6*, 720–726.

[44] Grandjean, J.-M. M.; Nicewicz, D. A. *Angew. Chem. Int. Ed.* **2013**, *52*, 3967–3971.

[45] Zeller, M. A.; Riener, M.; Nicewicz, D. A. *Org. Lett.* **2014**, *16*, 4810–4813.

[46] Cavanaugh, C. L.; Nicewicz, D. A. *Org. Lett.* **2015**, *17*, 6082–6085.
[47] Baiju, T. V.; Gravel, E.; Doris, E.; Namboothiri, I. N. N. *Tetrahedron Lett.* **2016**, *57*, 3993–4000.
[48] Zhang, G.; Hu, X.; Chiang, C.-W.; Yi, H.; Pei, P.; Singh, A. K.; Lei, A. *J. Am. Chem. Soc.* **2016**, *138*, 12037–12040.
[49] Yi, H.; Niu, L.; Song, C.; Li, Y.; Dou, B.; Singh, A. K., Lei, A. *Angew. Chem. Int. Ed.* **2017**, *56*, 1120–1124.
[50] Hu, X.; Zhang, G.; Bu, F.; Lei, A. *ACS Catal.* **2017**, *7*, 1432–1437.
[51] Yang, B.; Lu, Z. *Chem. Commun.* **2017**, *53*, 12634–12637.
[52] Zhang, Q.-B.; Ban, Y.-L.; Zhou, D.-G.; Zhou, P.-P.; Wu, L.-Z.; Liu, Q. *Org. Lett.* **2016**, *18*, 5256–5259.
[53] Griffin, J. D.; Cavanaugh, C. L.; Nicewicz, D. A. *Angew. Chem. Int. Ed.* **2017**, *56*, 2097–2100.
[54] Ledwith, A.; Russell, P. J. *J. Chem. Soc. Chem. Commun.* **1974**, 291–292.
[55] Mitani, M.; Kato, I.; Koyama, K. *J. Am. Chem. Soc.* **1983**, *105*, 6719–6721.
[56] Knorn, M.; Rawner, T.; Czerwieniec, R.; Reiser, O. *ACS Catal.* **2015**, *5*, 5186–5193.
[57] Nguyen, T. M.; Nicewicz, D. A. *J. Am. Chem. Soc.* **2013**, *135*, 9588–9591.
[58] Nguyen, T. M.; Manohar, N.; Nicewicz, D. A. *Angew. Chem. Int. Ed.* **2014**, *53*, 6198–6201.
[59] Gesmundo, N. J.; Grandjean, J.-M. M.; Nicewicz, D. A. *Org. Lett.* **2015**, *17*, 1316–1319.
[60] Chinzei, T.; Miyazawa, K.; Yasu, Y.; Koike, T.; Akita, M. *RSC Adv.* **2015**, *5*, 21297–21300.
[61] Ramirez, N. P.; Gonzalez-Gomez, J. C. *Eur. J. Org. Chem.* **2017**, *2017*, 2154–2163.
[62] Chu, L.; Ohta, C.; Zuo, Z.; MacMillan, D. W. C. *J. Am. Chem. Soc.* **2014**, *136*, 10886–10889.
[63] Ma, J.; Shen, X.; Harms, K.; Meggers, E. *Dalton Trans.* **2016**, *45*, 8320–8323.
[64] Huo, H.; Harms, K.; Meggers, E. *J. Am. Chem. Soc.* **2016**, *138*, 6936–6939.
[65] Matsui, J. K.; Primer, D. N.; Molander, G. A. *Chem. Sci.* **2017**, *8*, 3512–3522.

[66] Matsui, J. K.; Molander, G. A. *Org. Lett.* **2017**, *19*, 950–953.

[67] Garza-Sanchez, R. A.; Tlahuext-Aca, A.; Tavakoli, G.; Glorius, F. *ACS Catal.* **2017**, *7*, 4057–4061.

[68] Xu, N.; Li, P.; Xie, Z.; Wang, L. *Chem. Eur. J.* **2016**, *22*, 2236–2242.

[69] Fang, J.; Wang, Z.-K.; Wu, S.-W.; Shen, W.-G.; Ao, G.-Z.; Liu, F. *Chem. Commun.* **2017**, *53*, 7638–7641.

[70] Wilger, D. J.; Gesmundo, N. J.; Nicewicz, D. A. *Chem. Sci.* **2013**, *4*, 3160–3165.

[71] Ramirez, N. P.; Bosque, I.; Gonzalez-Gomez, J. C. *Org. Lett.* **2015**, *17*, 4550–4553.

[72] Song, H.-T.; Ding, W.; Zhou, Q.-Q., Liu, J.; Lu, L.-Q., Xiao, W.-J. *J. Org. Chem.* **2016**, *81*, 7250–7255.

[73] Lang, S. B.; Cartwright, K. C.; Welter, R. S.; Locascio, T. M.; Tunge, J. A. *Eur. J. Org. Chem.* **2016**, 3331–3334.

[74] Barton, D. H. R.; Crich, D.; Motherwell, W. B. *J. Chem. Soc. Chem. Commun.* **1983**, 939–941.

[75] Cassani, C.; Bergonzini, G.; Wallentin, C.-J. *Org. Lett.* **2014**, *16*, 4228–4231.

[76] Griffin, J. D.; Zeller, M. A.; Nicewicz, D. A. *J. Am. Chem. Soc.* **2015**, *137*, 11340–11348.

[77] Xuan, J.; Xia, X.-D.; Zeng, T.-T.; Feng, Z.-J.; Chen, J.-R.; Lu, L.-Q.; Xiao, W.-J. *Angew. Chem. Int. Ed.* **2014**, *53*, 5653–5656.

[78] Zeng, T.-T.; Xuan, J.; Ding, W.; Wang, K.; Lu, L.-Q., Xiao, W.-J. *Org. Lett.* **2015**, *17*, 4070–4073.

[79] Chen, L.; Li, H.; Li, P.; Wang, L. *Org. Lett.* **2016**, *18*, 3646–3649.

[80] Jiang, J.; Zhang, W.-M.; Dai, J.-J.; Xu, J.; Xu, H.-J. *J. Org. Chem.* **2017**, *82*, 3622–3630.

[81] Cornilleau, T.; Hermange, P.; Fouquet, E. *Chem. Commun.* **2016**, *52*, 10040–10043.

[82] Hu, X.-Q.; Chen, J.; Chen, J.-R.; Yan, D.-M.; Xiao, W.-J. *Chem. Eur. J.* **2016**, *22*, 14141–14146.

[83] Roth, H. G.; Romero, N. A.; Nicewicz, D. A. *Synlett* **2016**, *27*, 714–723.

[84] Meyer, A. U.; Berger, A. L.; König, B. *Chem. Commun.* **2016**, *52*, 10918–10921.

[85] Xiang, Y.; Li, Y.; Kuang, Y.; Wu, J. *Chem. Eur. J.* **2017**, *23*, 1032–1035.
[86] Xu, H.-J.; Zhu, F.-F.; Shen, Y.-Y.; Wan, X.; Feng, Y.-S. *Tetrahedron* **2012**, *68*, 4145–4151.
[87] Xiang, M.; Xin, Z.-K.; Chen, B.; Tung, C.-H.; Wu, L.-Z. *Org. Lett.* **2017**, *19*, 3009–3012.

Chapter 10

Other Nitrogen Heterocycles: Carbazoles, Imides and PDI, mpg-C_3N_4, Tetrazines, Riboflavin, and BODIPY

Pier Giorgio Cozzi*, Paola Ceroni†,
Andrea Gualandi, and Marianna Marchini

Department of Chemistry "Giacomo Ciamician",
University of Bologna, Bologna, Italy
**piergiorgio.cozzi@unibo.it*
†paola.ceroni@unibo.it

10.1 Introduction

The use of light to drive chemical reactions that are endergonic in the dark is common in nature: for example, photosynthesis, vision, and phototaxis. At the laboratory scale, until not many years ago, UV light was mainly used to excite one of the reactants and to promote organic photoreactions [1, 2]. Indeed, most organic substrates absorb only in this spectral region, thus requiring quartz apparatus and expensive irradiation sources. The recent explosion of photocatalysis in organic reactions [3–5] was brought about by the use of photosensitizers absorbing visible light, much more convenient from a practical view point. Conventional glassware or polymeric tubes in photochemical flow reactors are used, as well as less expensive light sources, particularly LEDs, or even sunlight. In most of the examples, the photocatalyst (**PC**) is involved in photoinduced electron-transfer processes. Indeed, the excited state is both a better reductant and a better oxidant than the corresponding ground state.

Metal complexes, particularly Ru(II) and Ir(III), constitute the first and most widely investigated class of photoredox catalysts because of the long lifetime of their luminescent excited state and their superior chemical stability even in the oxidized or reduced forms. Recently, organic **PCs** have gained an increased attention based on the following potential advantages: (i) simple and economical synthesis, (ii) possibility to synthesize libraries with a fine tuning of electronic and photophysical properties, which can affect the yield and the nature of the products of the organic reaction. Indeed, the large diversity obtained by modifying the well-known structures [6] offers promises for exploration of new reactions and optimization of synthetic methodologies.

In this chapter, we will illustrate the most significant and recent examples of photoredox catalysis based on nitrogen-containing heterocyclic systems. Before discussing each class of **PCs** and illustrating the main photophysical and electrochemical properties, we would like to remind some important photophysical and electrochemical principles that will be useful for the understanding of the mechanisms discussed in the following. For a thorough and general discussion, please refer to the introduction chapter.

10.2 Photoinduced Electron Transfer in Organic Photocatalysis

In the following paragraphs, we will discuss examples in which nitrogen-containing heterocyclic **PCs** are involved in photoinduced electron-transfer processes:

$$PC^* + A \rightarrow PC^- + A^+ \quad \textit{reductive quenching,} \qquad (10.1)$$

$$PC^* + A \rightarrow PC^+ + A^- \quad \textit{oxidative quenching.} \qquad (10.2)$$

Reactions (10.1) and (10.2) are potentially in competition, but usually only one is thermodynamically allowed. When both are thermodynamically allowed, kinetic factors establish which process wins the competition. A can be either the substrate of the organic reaction or a sacrificial reagent, which is irreversibly oxidized/reduced in reactions (10.1) and (10.2), respectively, so that it is consumed. On the other hand, the **PC** needs to be restored in order to have a cyclic process: this can be done by reaction of PC^- or PC^+ with a sacrificial reagent or a substrate of the organic reaction. For example, if the organic substrate A needs to be reduced to drive the chemical reaction, two mechanisms are possible, namely reaction (10.2) or (10.3):

$$PC^- + A \rightarrow PC + A^-. \tag{10.3}$$

In the case of electron transfer involving excited states, the corresponding redox potentials cannot be directly measured because of the short lifetime of the excited-state species, but it can be evaluated as follows:

$$E^0(PC^+/PC^*) = E^0(PC^+/PC) - E_{00}(PC^*/PC), \tag{10.4}$$

$$E^0(PC^*/PC^-) = E^0(PC/PC^-) + E_{00}(PC^*/PC), \tag{10.5}$$

where $E_{00}(PC^*/PC)$ is the one-electron potential corresponding to the zero–zero spectroscopic energy of the excited state. It is worth noting that the term $E_{00}(PC^*/PC)$ is related to the reactive excited state, usually S_1 or T_1 for organic **PCs**, and it does not necessarily coincide with the excited state populated by light absorption. In parallel, the excited-state redox potentials are not dependent on the excitation wavelength used. When irreversible processes are involved, the redox potential values are not experimentally available; although they can sometimes be estimated from other parameters, it is difficult to forecast whether the actual electron-transfer process will take place under the experimental condition used. On top of that, the reaction yield is not only affected by the efficiency of the electron-transfer step, but also by the stability of the generated radical species and by the kinetics of competing reactions.

Reaction mechanisms are usually hypothesized on the basis of thermodynamic considerations, but experimental evidence of the transient formation of the reduced/oxidized species would be desirable. Moreover, some of the reported examples are based on chain reaction mechanisms in which the photocatalytic step initiates a chain reaction. The presence of a chain mechanism is clearly evident when the reaction quantum yield is higher than 1, i.e. more than one product molecule is formed upon the absorption of one photon by the **PC**, or when the reaction quantum yield is higher than the quenching efficiency, thus demonstrating that quenching of the **PC** by electron transfer is not always necessary to drive the organic reaction, because the activated organic substrates can propagate the radical chain.

The study of the mechanism is a complex task, particularly in the reaction conditions, usually much different from the usual conditions of a photochemical study, but it is very useful for the design of new and improved photocatalytic systems.

Table 10.1 reports the most important photophysical and electrochemical data of the pristine nitrogen-containing heterocyclic chromophores (Figure 10.1) of each family discussed in the following. In particular, the data reported are: absorption maximum and molar absorption coefficient of the lowest energy band, fluorescence band maximum and quantum yield, lifetime of the lowest S_1 and T_1 excited states, quantum yield of population of the T_1 state (also called as the intersystem crossing quantum yield) and phosphorescence band maximum, as well as the half-wave potentials corresponding to the first reduction and/or oxidation processes. These data are important to locate the position of the relevant excited states, as displayed in the Jablonski diagram of each chromophore, and to evaluate the thermodynamics of photoinduced electron-transfer processes.

Table 10.1. Photophysical and electrochemical data of the pristine chromophores of each family (λ_{abs} is the maximum of the lowest energy absorption band and its corresponding molar absorption coefficient is ε; λ_F is the maximum of fluorescence emission band; Φ_F is the fluorescence quantum yield; τ_S is the lifetime of the lowest singlet excited state S_1; λ_p is the maximum of the phosphorescence emission band; Φ_T is the quantum yield for triplet formation; τ_T is the lifetime of the lowest triplet excited state T_1; $E_{1/2}$ half-wave potential).

	λ_{abs}/nm (ε/M^{-1} cm^{-1})	λ_F/nm	Φ_F	τ_S/ns	λ_p/nm	Φ_T	$\tau_S/\mu s$	$E_{1/2}$(**PC**$^+$/**PC**)/V vs SCE	$E_{1/2}$(**PC**/**PC**$^-$)/V vs SCE
1[a]	330 (Ref. [7]) (4,250)	336 (Ref. [7])	0.38 (Ref. [7])	16.1 (Ref. [7])	407 (Ref. [8])[b]	0.36 (Ref. [9])[c]	170 (Ref. [10])	+1.16 (Ref. [11])[d]	
2[e]	526 (Ref. [12]) (80,000)	537 (Ref. [13])	0.97 (Ref. [13])[c]	3.8 (Ref. [12])	1042 (Ref. [13])[f]	<10^{-3} (Ref. [13])[f]	140 (Ref. [13])[f]	+1.65 (Ref. [14])[g]	−0.53 (Ref. [14])[g]
3[a]	555 (Ref. [15]) (625)	538 (Ref. [16])[h]	1.5 × 10^{-4} (Ref. [17])	20 (Ref. [15])	820 (Ref. [15])		28.4 (Ref. [18])[c]		−0.91 (Ref. [19])[c]
4[i]	447 (Ref. [20]) (11,900)	523 (Ref. [20])	0.30 (Ref. [20])	5.65 (Ref. [21])	571 (Ref. [22])[j]	0.70 (Ref. [22])[j]	2.0 × 10^5 (Ref. [23])[k]		−0.71 (Ref. [24])[l]
5[m]	521 (Ref. [25]) (75,000) (Ref. [26])	538 (Ref. [25])	0.99 (Ref. [25])	6.2 (Ref. [27])				+0.95 (Ref. [25])	−1.37 (Ref. [25])
6[d]	529 (Ref. [28]) (89,000)	552 (Ref. [28])	0.03 (Ref. [28])	0.13 (Ref. [28])	760 (Ref. [29])[e]	0.97 (Ref. [28])[n]	220 (Ref. [29])[e]	+1.13 (Ref. [30])[o]	

Notes: [a] In cyclohexane. [b] In EPA (Et_2O:isopentane:EtOH, 5:5:2). [c] In benzene. [d] In acetonitrile. [e] In DMF. [f] Determined by energy transfer from a sensitizer. [g] In $CHCl_3$:CH_3CN (3:2 v/v). [h] In 3-methylpentane, 77 K. [i] In ethanol. [j] In EtOH:MeOH (9:1 v/v). [k] At 77 K. [l] In DMSO. [m] In DCM. [n] Calculated as $\Phi_T = 1 - \Phi_F$. [o] Reduction potential of 4,4-difluoro-2,6-diiodo-1,3,5,7-tetramethyl-4-bora-3a,4a-diaza-s-indacene in methanol.

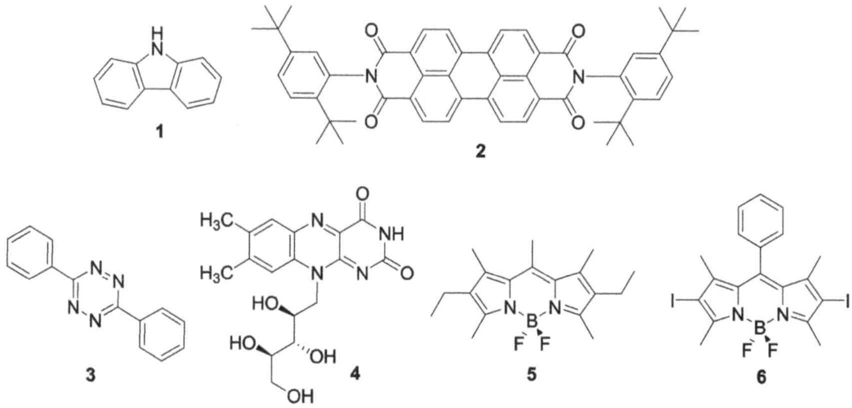

Figure 10.1. Chemical structure of the pristine chromophores of each family discussed in the text.

10.3 Carbazoles

Carbazole-based compounds are widely used in OLEDs as blue emitters [31–34]. The carbazole group can be readily modified in the 3- and 6-positions, thus giving the possibility to slightly tune the photophysical properties.

10.3.1 *Photophysical and electrochemical properties*

Carbazole **1** is not colored, so that its use as a **PC** is limited to UV irradiation (at *ca.* 330 nm, Table 10.1, and Figure 10.2). It displays a fluorescence band in the UV region ($\Phi_F = 0.38$) with maximum at 336 nm and a phosphorescence peaked at 400 nm; the phosphorescent excited state T_1 is populated with good quantum yield ($\Phi_T = 0.36$, Table 10.1).

The lowest energy electronic transition has been assigned to a π,π^* transition on the basis of computational studies and in agreement with the experimental observation that the fluorescence band of carbazole does not show strong a solvent and temperature dependence [35, 36].

The absorption spectra of the C- and *N*-substituted carbazoles in cyclohexane solution exhibit quite a small shift of the lowest energy band: for example, *N*-methylcarbazole and *N*-phenylcarbazole display a bathochromic shift of 12 nm and 6 nm, respectively;

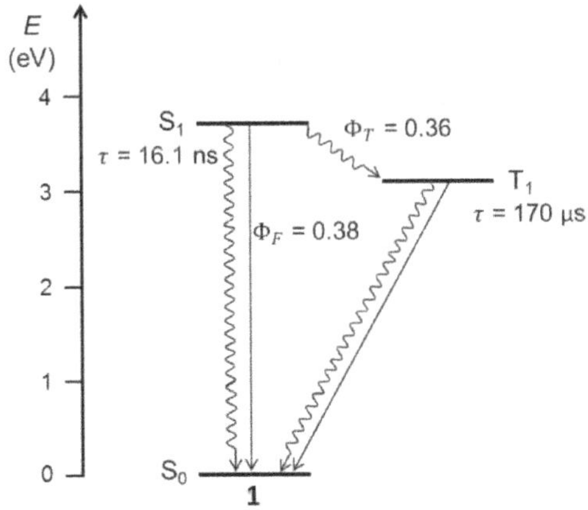

Figure 10.2. Schematic Jablonski diagram for molecule **1** in solution at 298 K [7–11].

N-acetylcarbazole shows a 20 nm hypsochromic shift compared to pristine carbazole [35].

From the electrochemical point of view, carbazole can be easily oxidized and the electron-transfer process is coupled to a chemical reaction, i.e. a radical coupling at the 3-position, leading to the formation of a bicarbazole for pristine carbazole and *N*-alkyl or *N*-aryl derivatives with extremely high coupling rates [37]. If substituents are inserted in the 3, 6, and 9 positions, the reversibility of the electrochemical oxidation is increased, indicating that their radical cations are more stable and longer-lived.

In search for visible-light **PCs** based on carbazoles, systems containing multiple carbazole units linked by π-conjugated groups were extensively studied for heterogeneous catalysis upon blue-light irradiation (see below) [38–40].

10.3.2 *Reactions and use in photoredox catalysis*

In an early example of photoredox-mediated reactions, Saito and co-workers reported a practical photoredox deoxygenation of esters, that avoided the use of toxic tin reagents [41]. Photoexcitation of *N*-methyl carbazole was followed by electron transfer to an aromatic ester substituted with the electron-withdrawing group. The so-formed ketyl radical of the ester was readily protonated and, if the ester of a secondary or tertiary alcohol was used, the oxygen-protonated radical underwent *β*-scission to produce stabilized secondary or tertiary alkyl radical and the corresponding aryl carboxylic acid. These alkyl radicals were able to abstract hydrogen from the hydrogen-donating source to furnish the desired deoxygenated product. The reaction required the use of stochiometric amount of *N*-methylcarbazole and was applied to the deoxygenation of 2′,3′,5′-tri-O-[*m*-(trifluoromethyl) benzoyl]-adenosine, 2′-O-[m-(trifluoromethyl) benzoyl]-3′,5′-O-TPDS-uridin and tri-O-benzoyluridine. In all cases, a nitrogen-purged *i*PrOH:H_2O solution was used and the irradiation was performed with a 100-W high-pressure mercury lamp equipped with a Pyrex filter. The proposed mechanism was proven by the near-diffusion-controlled rate ($k = 4.5 \times 10^9\ M^{-1}\ s^{-1}$ in THF:H_2O) for quenching of *N*-methylcarbazole fluorescence by aryl esters.

As was pointed out in the electrochemical properties, by blocking the 3- and 6-positions of carbazole, an improved stability of the radical cation is observed, and the better stability of the radical gave the possibility to use the photoredox mediator in catalytic amount. In fact, Rizzo and co-workers [42] improved the deoxygenation of benzoyl derivatives by blocking the 3- and 6-positions of carbazole with methyl groups (Scheme 10.1). The methyl was selected so that the energy of the excited states and the redox potentials are not

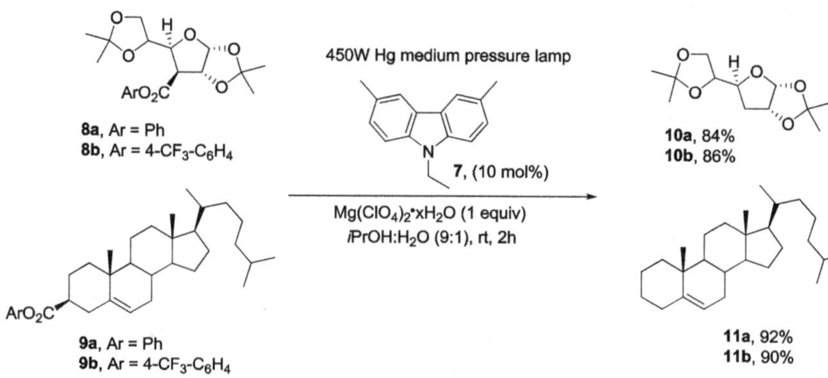

Scheme 10.1. Photoredox deoxygenation of benzoates and *p*-(trifluoromethyl) benzoates.

significantly affected in comparison to pristine carbazole. Deoxygenations were carried out with a 450-W medium-pressure Hg lamp in *i*PrOH:H_2O. Various sugar and cholesterol esters were deoxygenated by employing 10 mol% of the carbazole derivative **7** and fluorescence quenching was reported in all cases.

One of the limitations associated with the use of carbazole derivatives was the requirement of UV irradiation. To circumvent this problem, different strategies were reported. Zhang [39] incorporated the carbazole unit into a 3D π-conjugated framework, thereby preparing a heterogeneous photocatalytic system. The carbazole-based porous organic framework, **Cz-POF 12** (Scheme 10.2) is a highly efficient photoredox catalyst and was used in three different organic reactions: dehalogenation of phenacyl bromide and its derivatives, hydroxylation of arylboronic acids, and α-alkylation of aldehydes. **Cz-POF 12** was prepared using $FeCl_3$-induced, oxidative polymerization [43] of carbazole-based monomer 3,3′,5,5′-tetra(9*H*-carbazol-9-yl)-1,1′-biphenyl (**Cz-1**, **13**), giving a porous material (72% of the pore volume is contributed by mesopores) adapted for large-sized substrates or products. Interestingly, the monomer **Cz-1 13** has minimal visible-light absorption, while the polymer has a broad absorption band that extends to the visible region. **Cz-POF 12** shows an intense, long-lived (2.74 μs at 77 K) luminescence band at 464 nm. Cyclic voltammetry (CV) characterization of **Cz-POF 12** reveals an oxidation process ($E_{1/2}$) at +1.38 V (vs SCE). The $E_{1/2}$ (**Cz-POF**$^{\bullet+}$/**Cz-POF***) can be calculated at −1.53 V (vs SCE), more negative than that of $[Ru(bpy)_3]^{3+}/[Ru(bpy)_3]^{2+*}$ (−0.81 V) commonly used in photoredox catalysis. The carbazole catalyst was employed in the dehalogenation of phenacyl bromides **15** using diethyl-1,4-dihydro-2,6-dimethyl-3,5-pyridinedicarboxylate (Hantzsch ester) as the stoichiometric reductant (Scheme 10.2(a)), under irradiation with a compact fluorescent lamp (CFL). A Stern–Volmer study demonstrated that phenacyl bromide quenches the fluorescence of **Cz-POF 12** which suggests that the reaction proceeds *via* oxidative quenching. Furthermore, the catalyst was tested in the photomediated oxidative hydroxylation of arylboronic acids **17** (Scheme 10.2(b)) to the corresponding phenols **18** using a

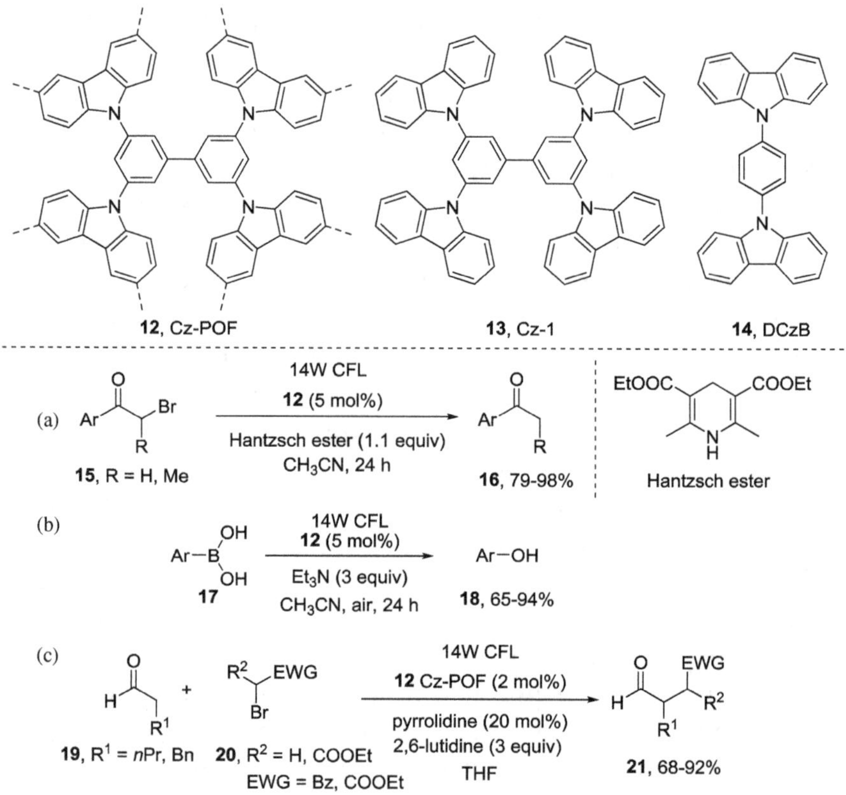

Scheme 10.2. Different photoredox reactions catalyzed by the carbazoles **12–14**.

sacrificial reductant (Et_3N) and molecular oxygen in air. Different from the above-reported example, the proposed mechanism is that the photoexcited state of **Cz-POF 12** is able to reduce oxygen to superoxide anion ($O_2^{\bullet-}$) that is reacting with boronic acid. By successive hydrogen abstraction, oxidation, and rearrangement, the reaction yielded the corresponding phenol. Finally, the photocatalytic α-alkylation of aldehydes **19** in the presence of pyrrolidine, acyl bromide, and bromo malonate **20** was reported. Also, in this case, the direct reduction of bromo derivative is occurring, and the radical formed is added onto the intermediate enamine formed *in situ* by the reaction of pyrrolidine and aldehyde used in excess (Scheme 10.2(c)).

Carbazolic POFs (**CzCPs**) were prepared by oxidative statistical copolymerization of carbazole-based compounds, namely **14** (1,4-di(9H-carbazol-9-yl)-benzene) and 1,2,3,5-tetrakis(carbazol-9-yl)-4,6-dicyanobenzene (**4CzIPN**) at different ratios using $FeCl_3$ [40]. Compound **12** is an excellent electron donor while the introduction of two cyano groups in **4CzIPN** gave the molecule strong electron-accepting capability. A polymer formed by **14** will be a strong reductant, while the polymer formed by **4CzIPN** will be a strong oxidant. By varying the ratio of the two components (**14** and **4CzIPN**), it is possible to modulate the redox potentials of the polymer obtained. With this modular capability, the material was applied as catalyst in the lignin degradation *via* stepwise oxidation and reduction approach. The heterogeneous catalyst was applied for degradation of the lignin *β*-O-4 models, *via* aerobic oxidation of *β*-O-4 alcohols and reductive cleavage of *β*-O-4 ketones [40].

In a similar application, Loh and co-workers [44] have prepared a carbazole-conjugated microporous polymer (**22**, **C-CMP**) through a carbazole-based oxidative coupling polymerization. The material **C-CMP** exhibits high porosity, and a good specific surface area, and it is thermally stable. It was used as a heterogeneous **POC** for a wide range of reactions, including the oxidative coupling of primary benzylic amines **23** to imines **24** (Scheme 10.3(a)), aerobic

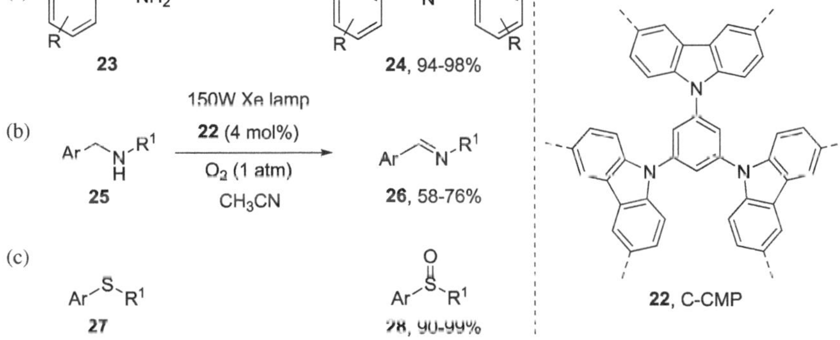

Scheme 10.3. Oxidations catalyzed by the carbazole derivative **22** of: (a) primary benzylic amines; (b) secondary benzylic amines; (c) sulfides.

dehydrogenation of dibenzylamine, 1,2,3,4-tetrahydroisoquinoline, and tetrahydroquinazolines (Scheme 10.3(b)). In addition, the catalyst was employed in the synthesis of substituted aryl-alkyl sulfoxides **28** (Scheme 10.3(c)) from the corresponding sulfides **27**. The estimated reduction potential of **C-CMP**$^{•+}$/**C-CMP*** is −1.42 V vs SCE. The radical $O_2^{•-}$ was formed by PET from **C-CMP*** to oxygen. The **C-CMP**$^{•+}$ obtained is able to oxidize the amines and sulfides to their radical cations, that undergo subsequent reactions.

10.4 Imides and PDI

3,4,9,10-perylene diimides (**PDIs**), a class of organic dyes with excellent chemical, thermal, and photostability, among the best *n*-type organic semiconductors [45], have found many applications in opto-electronic devices, such as light-emitting diodes [46], photovoltaics [47,48], electrochemiluminescent devices [14], and organic field-effect transistors [49–51].

10.4.1 *Photophysical and electrochemical properties*

PDI compounds are intensely colored and they have been extensively exploited as industrial dyes and pigments. The rigidity of the aromatic planar structure reduces the rate of non-radiative decays, such that the **PDIs** are highly fluorescent molecules. Substituents on the nitrogen imide are not strongly influencing the photophysical as well as electrochemical properties, while, bay substitution can greatly affect the energy of the $S_0 \rightarrow S_1$ transition, as well as the redox properties [47]. In particular, bay π-donor substituents destabilize the HOMO, while bay π-acceptors stabilize the LUMO, leading to bathochromic shifts of the spectra in either case.

Molecule **2** (Figure 10.1) presents four bulky tert-butyl groups, which favors solubility in organic solvents. The lowest energy transition presents a band with a maximum at 526 nm, characterized by vibrational peaks (separated by *ca.* 1,400 cm^{-1}) and a very high molar absorption coefficient ($\varepsilon_{max} = 8 \times 10^4$ M^{-1} cm^{-1}, Table 10.1). The main deactivation pathway of S_1 is fluorescence, characterized by a fluorescent band symmetrical to the last absorption band

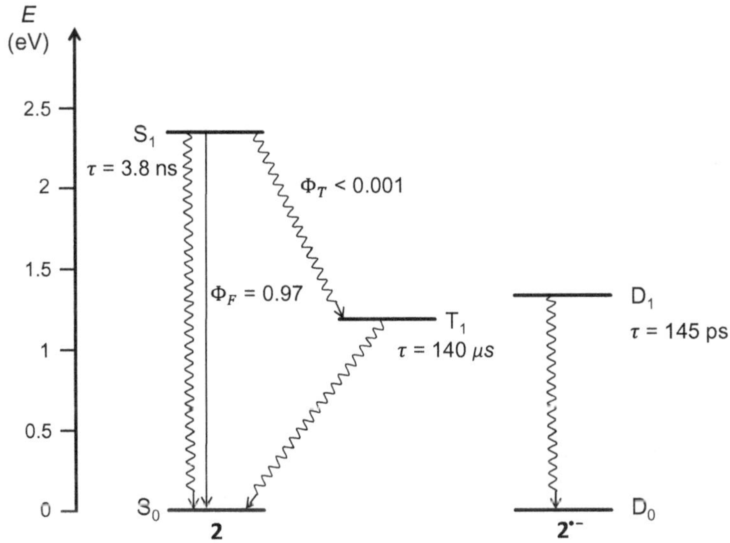

Figure 10.3. Schematic Jablonski diagram for molecule **2** and its radical anion in solution at 298 K [12–14].

($\Phi_F = 0.97$, $\tau = 3.8$ ns, Table 10.1 and Figure 10.3). Population of the T_1 state by intersystem crossing from S_1 cannot be experimentally detected ($\Phi_T < 0.001$, Table 10.1) and the position of T_1 has been determined by energy transfer from suitable sensitizers [13].

A red-shifted fluorescence is observed in the solid state [13] or in polar solvents [52], and it has been attributed to the formation of aggregates.

PDI dyes and, in particular, molecule **2** at scan rates above 0.2 V/s, show chemically reversible waves for the first reduction and oxidation processes [14]. At scan rates below 0.2 V/s, the oxidation wave of **2** becomes less reversible, suggesting that the radical cation undergoes slow decomposition. On the other side, the radical anion is quite stable and can also be produced by photoirradiation of **2** in the presence of triethylamine as the sacrificial electron donor [53], according to the following processes:

$$\mathbf{2} + h\nu \rightarrow \mathbf{2}^*, \tag{10.6}$$

$$\mathbf{2}^* + Et_3N \rightarrow \mathbf{2}^{\bullet-} + Et_3N^{\bullet+}. \tag{10.7}$$

The radical anion $\mathbf{2}^{\bullet-}$ shows intense absorption bands in the near-infrared (NIR) spectral region: it has a ground electronic configuration with an unpaired electron, so that the ground state and the first electronic excited states are doublets and the lowest energy transition has been assigned to $D_0 \rightarrow D_1$ transition [12]. Internal conversion from higher-lying D_n states occurs rapidly, without intersystem crossing to a long-lived quartet state. The subnanosecond excited-state lifetime of the diimide radical anion contributes to its exceptional photostability, despite its strong reducing power.

10.4.2 *Reactions and use in photoredox catalysis*

One of the first examples of the employment of imides, in particular, naphthalene monoimide (**NI, 29**) in photocatalytic reaction was described by Melone and co-workers (Figure 10.4) [54] who reported the aerobic photoactivation of *N*-hydroxyphthlimide (NHPI) **30** to the phthalimido-*N*-oxyl (PINO) **31** radical for the selective oxidation of isopropylbenzene and cyclohexene derivatives. UV irradiation was necessary to induce the formation of ROOH peroxides. The photoredox catalyst was able to increase the concentration of the PINO radical by generation of the superoxide radical $O_2^{\bullet-}$ which was

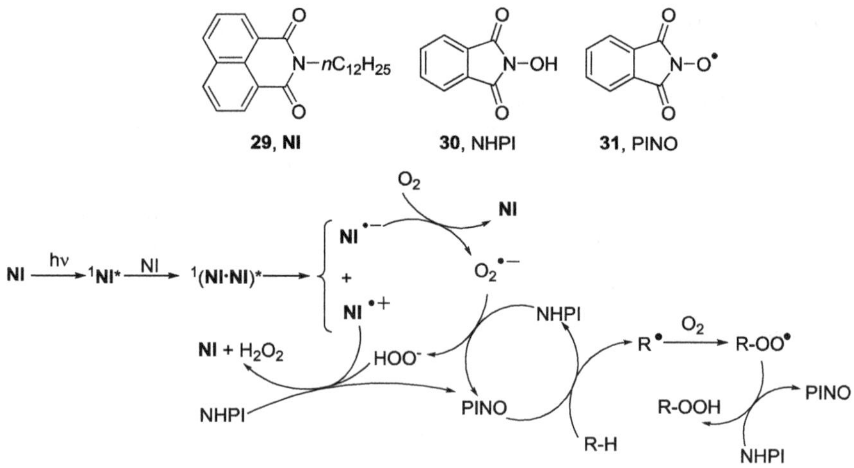

Figure 10.4. Aerobic photoactivation of NHPI.

capable of abstracting a hydrogen atom from NHPI forming the PINO radical. The superoxide was obtained by reduction of O_2 *via* a photoactivated electron transfer under aerobic conditions. The generation of $NI^{\bullet-}$ responsible for the reduction of O_2 was ascribed to the formation of excimer ^{1}(**NI**·**NI**)* [55] according to the equation ^{1}(**NI**·**NI**)*→ $\mathbf{NI}^{\bullet-}$ + $\mathbf{NI}^{\bullet+}$, that was favored by the high concentration in which the reaction is performed. It also worth mentioning that the specie $\mathbf{NI}^{\bullet+}$ was able to act as an electron scavenger, favoring the formation of H_2O_2 and closing the catalytic cycle.

One of the major problems associated with the use of **NI** was the need of UV light for inducing the PET reaction. On this perspective, another class of imides, the **PDIs**, allowed the use of visible light for photoexcitation. An intriguing application of **PDI** was reported by König [56].

As discussed above, photoexcitation of **PDI** in the presence of a sacrificial amine yields the formation of the radical anion $\mathbf{PDI}^{\bullet-}$ (see, for instance, reactions (7) and (8)), which is a potential reductant. However, organic species like aryl chlorides have very negative reduction potentials, so that $\mathbf{PDI}^{\bullet-}$ cannot perform their reduction (Scheme 10.4). König proposed a Z-scheme [56] reminiscent of the natural photosynthetic process in which two photosystems absorb light to drive energy-demanding electron-transfer reactions.

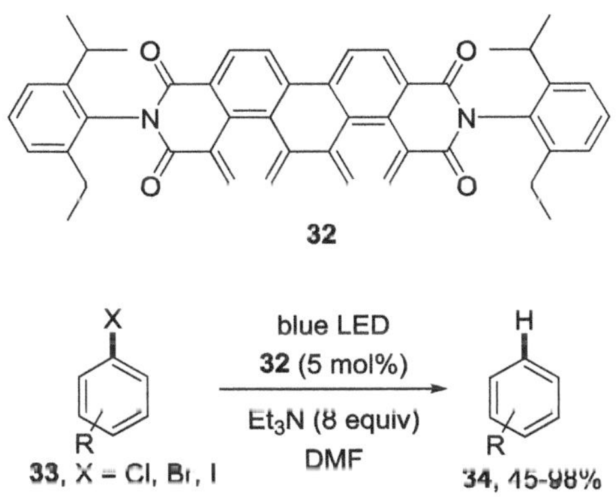

Scheme 10.4. Photoredox dehalogenation of aryl compounds catalyzed by **PDI 32**.

Perylene diimide **32** was reduced by visible light PET in the presence of triethylamine (Et_3N) to its stable and colored radical anion **32**$^{\bullet-}$. Visible-light excitation of the so-formed radical anion **32**$^{\bullet-}$ promoted the reduction of stable aryl chlorides giving aryl radicals, trapped by hydrogen atom donors. The reaction was performed in the presence of **PDI 32** (5 mol%) and Et_3N with blue light (455 nm) in DMF, in the presence of different arylchlorides. Aryl chlorides with quite high reduction potentials (e.g. −2.09 V vs SCE for chlorobenzene) cannot be reduced by this approach. Although the mechanism is intriguing and open many possibilities for the design of new **POCs** based on the Z-scheme, the proposed mechanism is questionable [57]. In particular, the lowest-energy excited state of **PDI** radical anion is short-lived ($\tau = 145$ ps, as reported by Wasielevski [12]), and dynamic quenching by the organic substrate is expected to be very inefficient on the basis of the Stern–Volmer equation. The same approach was exploited by other authors. In fact, Duan and co-workers [58] incorporated **PDI** into a metal-based polymer to form a heterogeneous **PC** Zn(II)-**PDI**, applied to the same aryl halides described in the previous publication [56]. In addition, the polymer was found to be a good **PC** for oxidation of benzylic substrates performed in the presence of oxygen [58].

Cao and Li have developed a highly efficient retro-Smile rearrangement promoted by **PDI 32** under visible-light irradiation. The process was mediated by a reductive quenching of the **PDI** excited state by a carboxylate anion followed by intramolecular aryl migration [59].

10.5 mpg-C_3N_4

Polymeric graphitic carbon nitride materials (g-C_3N_4) have attracted considerable attention in recent years as analogs of graphene [60, 61]. The material is a semiconductor, composed by C, N, and some minor H content: the 2D-planes are arranged as tri-s-triazine units connected by planar amino groups. The easy synthesis and the possibility to perform chemical modifications allows a range of applications, particularly as an oxidation catalyst [60].

10.5.1 *Photophysical and electrochemical properties*

Graphitic carbon nitride (g-C_3N_4) is a stable semiconductor material with a band gap of 2.7 eV (corresponding to 460 nm), employed as a visible-light **POC**. The conduction band (CB) of g-C_3N_4 is located at −1.39 V (vs SCE) and the valence band (VB) is located at +1.22 V (vs SCE) at pH = 6.6 [62].

The photophysical and electrochemical properties are strongly affected by the presence of surface defects or impurities in the core, so the electronic properties, which show a marked dependence on the synthetic procedures cannot be discussed in detail [63] and are not completely reproducible from one batch to the other, as it is always the case for nanomaterials compared to molecules.

10.5.2 *Reactions and use in photoredox catalysis*

mpg-C_3N_4 has an appropriate bandgap for water splitting, i.e. production of H_2 and O_2 by photolysis of water, which resulted in the conversion of solar energy into chemical energy stored in fuels (H_2 in the present case). Visible-light photoexcitation of mpg-C_3N_4 produces an electron–hole pair (Figure 10.5), called exciton. In principle, the hole in the VB is expected to drive water oxidation, while the electron in the CB is expected to drive water reduction. However, the production of one molecule of O_2 requires the

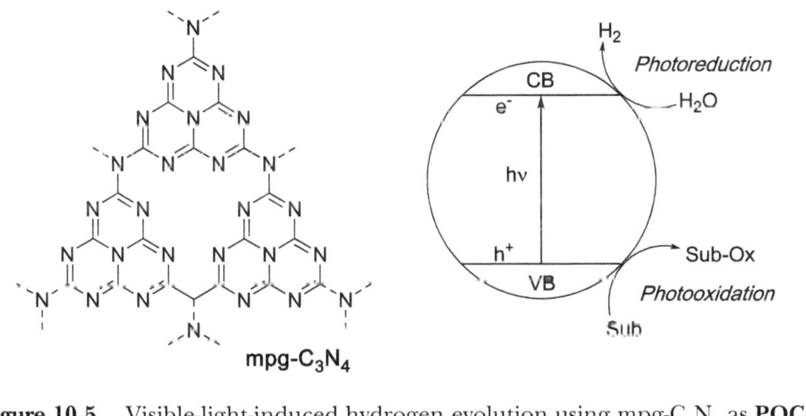

Figure 10.5. Visible-light-induced hydrogen evolution using mpg-C_3N_4 as **POC**.

exchange of four electrons, while two electrons are involved for H_2 production. Therefore, absorption of at least four photons is required to drive water splitting, and other catalysts, Pt complexes and RuO_2, are used in the following to store electrons and holes, respectively.

Antonietti and co-workers have described the use of this polymeric material in the generation of hydrogen from water, although in low efficiency (0.1% by irradiation at 420–460 nm) [64]. The efficiency of hydrogen evolution was dependent on the porosity and doping (i.e. fluorine) of the material. A recent modification with sulfur exhibited an improved catalytic activity [65a, 65b]. In order to promote the generation of oxygen, a RuO_2 catalyst was introduced [66], but the activity for O_2 evolution was an order of magnitude lower compared to that for H_2 evolution. To favor the generation of hydrogen, homogenous platinum complexes were used [61]. In order to extend visible-light absorption of C_3N_4, dyes, e.g. magnesium phthalocyanines [67], were deposited on the **POC**: upon excitation of the dye, photoinduced electron transfer with the semiconductor C_3N_4 is observed and H_2 evolution was reported in the presence of triethanolamine (**TEOA**) sacrificial reagent. The photophysical properties, stability, and recyclability of g-C_3N_4 are interesting for effective metal-free photocatalytic oxidation reactions that use molecular oxygen or hydrogen peroxide as the primary oxidant. C_3N_4 can be used in the oxidation of benzene to phenol [68], cyclohexane to cyclohexanone [69a, 69b], benzyl compounds to corresponding ketones (using oxygen as primary oxidant) [70], and oxidation of alcohols to aldehydes or ketones [71]. The system was able to promote the selective oxidation of benzylic alcohols with visible light, and up to 100% conversion for *p*-MeO benzylic alcohol. The oxidation of alkyl substrates, like 2-hexanol, was also obtained with a lower conversion (35%). The mechanism of the reaction involving the formation of $O_2^{\bullet-}$ by PET with g-C_3N_4 was confirmed by ESR and spin-trap experiments. The $O_2^{\bullet-}$ deprotonated the alcohols forming $HO_2^{\bullet}$, that abstracted hydrogen radical from alkoxide to give the corresponding ketyl radical. The latter gave the carbonyl *via* SET with g-C_3N_4. In an alternative mechanism, the deprotonated

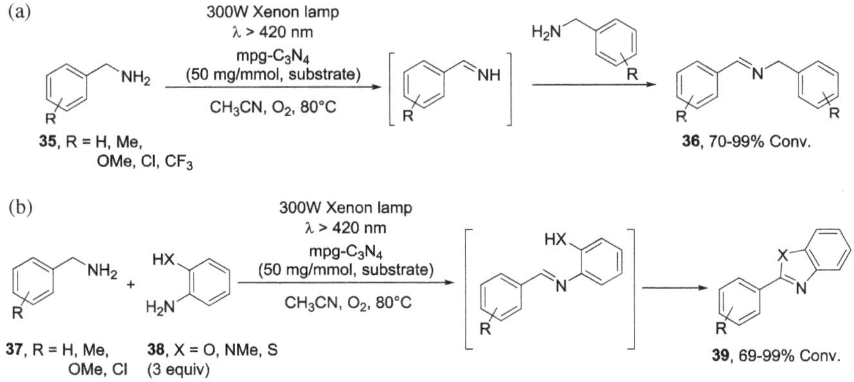

Scheme 10.5. Photooxidation of benzylic amines to imines (a) and to benzoxazoles, benzimidazoles, and benzothiazoles (b) promoted mpg-C_3N_4.

alcohol was oxidized by the **POC** to oxygen center radical, that was oxidized to carbonyl by $HO_2^{\bullet}$. The oxidation of benzyl substrates can be extended to benzylic amines to give benzylic imines (Scheme 10.5(a)). The intermediate imine is trapped by the benzylic amine forming a benzylic imine [72]. The methodology was also employed in the synthesis of various heterocycles, such as benzoxazoles, benzimidazoles, and benzothiazoles (Scheme 10.5(b), up to 98% yield).

The catalytic photooxidation of *N*-phenyltetrahydroisoquinoline derivatives (**40**) in MeOH with mpg-C_3N_4 in the presence of oxygen gave the corresponding iminium derivative, that reacted *in situ* with allylstannanes and allenyl stannanes (**41**), yielding the desired products **42** in quite good yields (Scheme 10.6) [73]. It is also important to mention that weak nucleophiles, such as allylsilanes and allylboranes could be employed. Unfortunately, the use of chiral allylborane (**43**) gave almost racemic products.

The polymeric mpg-C_3N_4 material was employed in the presence of NHPI (**30**) and oxygen for allylic oxidations through formation of PINO (**31**) radical [74]. Among the substrates investigated, cyclohexene was oxygenated to 2-cyclohexen-1-one in 93% conversion and 77% selectivity. Phenylethane, α-isophorone, and α-pinene were also oxidized with a good selectivity for ketones (66%–84%),

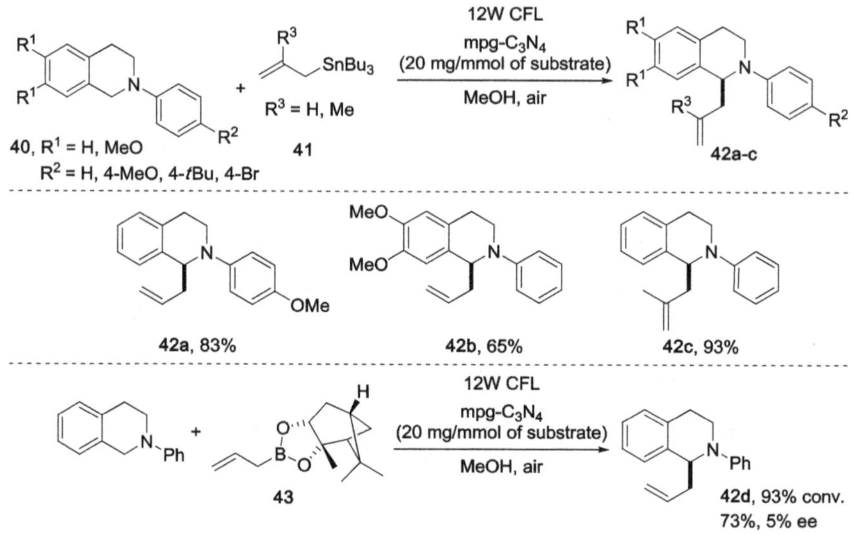

Scheme 10.6. Allylation of *N*-phenyltetrahydroisoquinoline derivatives through photooxidation with mpg-C_3N_4.

but in moderate conversions (26%–71%). Oxidation is also possible with compounds containing heteroatoms. As an example, Li and co-workers employed mpg-C_3N_4 in the selective oxidation of sulfides to sulfoxides with O_2 at room temperature, in the presence of isobutyraldehyde. Different aromatic and, in one case, benzylic sulfides were oxidized to sulfoxides in high yields (92%–98%). The plausible reaction mechanism involves the photoinduced electron transfer from mpg-C_3N_4 to molecular oxygen, forming $O_2^{\bullet-}$ that abstracts a hydrogen from aldehyde giving HO_2^-. The produced acyl radical formed a peroxide radical or a peracid, able to oxidize sulfides to the corresponding sulfoxides [75].

Blechert described an interesting application of mpg-C_3N_4 in C–C bond forming photoredox reactions, showing the broad applicability of the material in the context of organic synthesis. In the first example, the radical cyclization of alkenyl and alkynyl 2-bromo-1,3-dicarbonyl compounds to afford the corresponding five-membered derivatives was described. In the reaction, mpg-C_3N_4 (2.5 wt.%) was employed in diluted THF solution and irradiation was performed by a blue LED. The desired products were obtained from moderate to good yields (58%–85%) after chromatography [76]. It is worth mentioning that THF was used as the reaction solvent because it is able to donate a hydrogen atom to the radical intermediate obtained upon cyclization. In the second example, Blechert reported trifluoromethylations and perfluoroalkylations of heteroaromatic compounds promoted by mpg-C_3N_4 employing fluorinated sulfonylchlorides. Reduction of the sulfonyl chlorides occurs upon photoexcitation of mpg-C_3N_4. Successive decomposition with loss of SO_2 gives the corresponding perfluorinated carbon radicals. The reaction was performed in CH_3CN employing K_2HPO_4 as a base, to trap the acid liberated from the rearomatization of the heterocycles employed. It was possible to trifluoromethylate different heteroaromatic substrates, namely pyrroles, oxazoles, furans, thiophenes, indoles, and pyrazines in moderate to excellent yields (51%–99%). The reaction was also tolerant with respect to aldehyde, ester, halide, and amide functional groups present on the heterocyclic ring [77].

10.6 Tetrazine and Triazines

1,2,4,5-Tetrazine consists of an electron-poor aromatic ring with four nitrogens: they are strong electron acceptors because of the four electronegative nitrogen atoms in the aromatic ring. Therefore, the excited states of tetrazines are good oxidants. The synthesis of s-tetrazines was reported by Pinner many years ago [78].

10.6.1 *Photophysical and electrochemical properties*

Tetrazines have a low-energy absorption band in the visible region, which is weakly influenced by the nature of the substituents [17, 79a–79d]. In many derivatives, the S_1 excited state has a very long decay time ($\tau = 10$–150 ns), involving a forbidden $n \rightarrow \pi^*$ transition from the non-bonding orbitals of the nitrogens to the π^* orbitals of the tetrazine ring [80]. Many tetrazines are known to be fluorescent, although the parent H,H-tetrazine and 3,6-dimethyltetrazine are not photostable [81]. The increase of the size of the substituents improves their photostability, especially when going to aromatics, but it strongly reduces their fluorescence.

The absorption spectrum of molecule **3** (Figure 10.1) in cyclohexane displays a maximum at 558 nm in the visible spectral region, with a molar absorption coefficient of ca. 625 M^{-1} cm^{-1} (Table 10.1) [17]. In acetonitrile, the absorption maximum is blue-shifted to 543 nm. Because of the extremely low molar absorption coefficient in the visible region, **3** is not a good candidate as a visible-light **POC**.

Upon visible-light excitation, very weak fluorescence from S_1 is observed in the region 600–750 nm (Φ_F ca. 1.5×10^{-4} in cyclohexane, Table 10.1 and Figure 10.6). Unexpectedly, fluorescence from higher-energy S_n π–π* state is observed, violating the Kasha rule [82]. The triplet excited state T_1 can be populated by excitation at 450 nm of a proper sensitizer, e.g. $[Ru(bpy)_3]^{2+}$ [19], but it is not

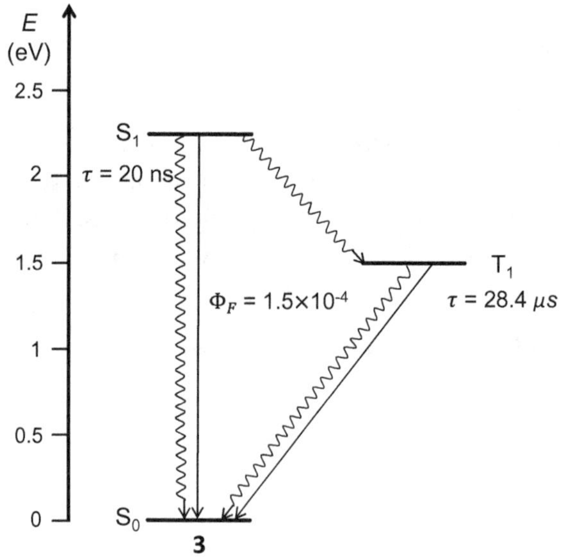

3

Figure 10.6. Schematic Jablonski diagram for molecule **3** in solution at 298 K [15–19].

detected by direct excitation of **3** at 355 nm, suggesting that the triplet population quantum yield is extremely low.

From the electrochemical point of view, molecule **3** is easily reduced to the corresponding radical anion at −0.91 V (vs SCE, Table 10.1).

10.6.2 *Reactions and use in photoredox catalysis*

The photoredox catalyst **PYTZ 44** was used in the presence of oxygen to promote the reaction between aldehydes, *o*-phenylenediamine, or *o*-aminothiophenol, to give the corresponding 2-substituted benzimidazoles and benzothiazoles (**47**), under visible-light irradiation (Scheme 10.7) [83].

The reaction showed high yields (60%–95%) for aromatic aldehydes containing both electron-donating and electron-withdrawing substituents. The reaction occurred through an imine intermediate formed by condensation of the aldehyde with the amino group.

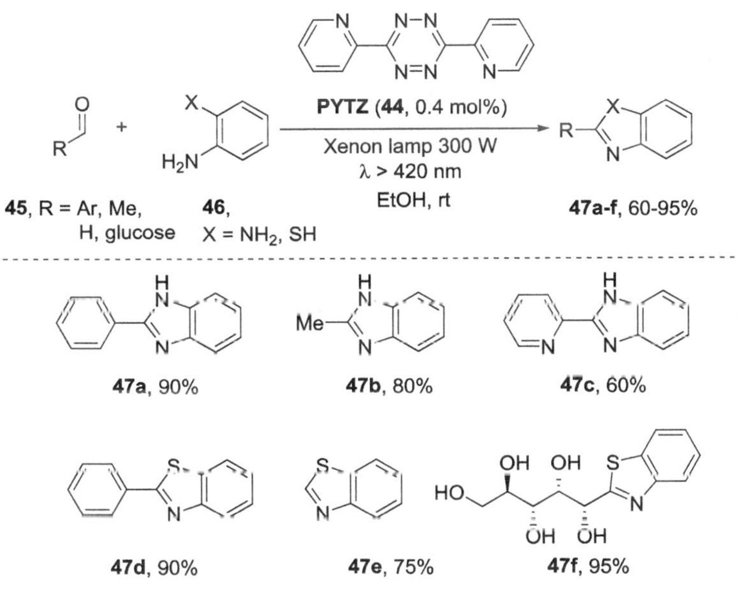

Scheme 10.7. Photocatalytic preparation of benzimidazoles and benzothiazoles.

PYTZ 44, excited by visible light to **PYTZ***, was reductively quenched by the (mono)imine intermediate to produce **PYTZ**$^{\bullet-}$ radical anion and nitrogen- or sulfur-centered radical cations. The radical cation, upon deprotonation, gave an intramolecular nucleophilic attack on the imine forming a nitrogen-centered radical. Reduction of this radical by **PYTZ**$^{\bullet-}$ regenerated the **PYTZ** catalyst and formed the corresponding nitrogen anion. Protonation of the cyclized aza-anion and subsequent oxidation mediated by oxygen gave the observed products.

Triazine-containing molecules are displaying analogous photoredox properties. Zhang recently described the use of a thiophene-based covalent triazine framework (CTF) as purely organic and visible-light-absorbing **POC** for the selective oxidation of alcohols at room temperature, in the presence of oxygen [84]. The reaction gave high conversion and selectivity for variously substituted benzylic alcohols. In addition, benzhydryl and cinnamyl alcohols were also compelling substrates giving 93% and 78% of conversion, respectively.

10.7 Riboflavin

Flavins are a huge family of compounds, mediating biological processes like photosynthesis and phototropism. As a coenzyme factor of flavoenzymes, they take part in redox reactions in the respiratory chain. They are also components of food products [85]. Riboflavin (vitamin B2) is one of the best-known members of the flavin family. Riboflavin, as an endogenous photosensitizer, can inactivate a broad range of dangerous viruses and bacteria in blood components, preserving their activity and functionality. On the other hand, riboflavin is present in food; therefore, its light-induced reactions are often the reason of food degradation: food products can lose their flavor, color, texture, and nutritive value. Indeed, riboflavin chromophores exhibit strong absorption in the visible region and act as sensitizers in the photodegradation of a wide range of substrates by free radical or singlet oxygen-mediated mechanisms: these processes may cause damage to the biological system [86]. Photochemistry of riboflavin was described years ago [87].

10.7.1 *Photophysical and electrochemical properties*

The electronic absorption spectrum of **4** in aqueous solution exhibits a maximum at 447 nm in the visible region with a high molar absorption coefficient, indicative of π–π^* transition [86]. The positions of the absorption maxima and the values of ε depend on the environment. Molecule **4** exhibits an intense yellow–green fluorescence at 523 nm ($\Phi_F = 0.30$) in aqueous solution. Only the neutral forms of riboflavins are fluorescent and their anions or cations are non-fluorescent. Population of the T_1 excited state is efficient ($\Phi_T = 0.70$, Table 10.1 and Figure 10.7), and many photoreactions take place from this long-lived T_1 excited state [86]. Photodegradation of riboflavin involves photoreduction, photodealkylation, and photoaddition. Riboflavin can be easily and reversibly reduced at −0.71 V vs SCE [24].

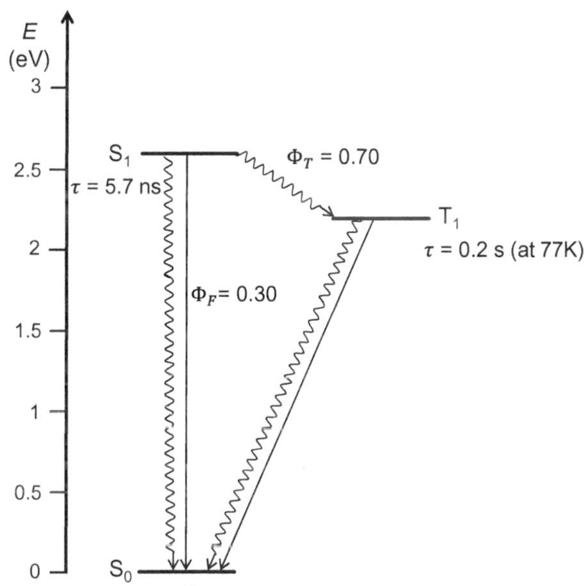

4

Figure 10.7. Schematic Jablonski diagram for molecule **4** in solution at 298 K [20–24].

10.7.2 *Reactions and use in photoredox catalysis*

The good photocatalytic properties of Riboflavin (**RF**, **4**, Figure 10.8) and Riboflavin tetraacetate (**RFTA**, **48**) in acid solution were noted by Fukuzumi [88]. The reaction reported was the photocatalytic aerobic oxidation of substituted benzylic alcohols. The $RFTAH^+$ is able to oxidize the $ArCH_2OH$ to ArCHO forming $RFTAH_2^{\bullet+}$, that is then reduced by O_2. However, the paper does not report the reaction yield and the reaction was not considered by the synthetic community.

After the renewed interest in photochemistry catalyzed by pioneering results reported by MacMillan [89], the possibility to use RF in photoredox catalysis was reinvestigated.

König reported [90] that modified flavin-containing thiourea groups were able to oxidize 4-methoxybenzyl alcohol with conversion up to 99% in air. The thiourea group acts as an electron shuttle. When the flavin derivatives were prepared, by introducing fluorinated hydrophobic aliphatic chains, an immobilization on solid support was possible (fluorous silica gel, reversed phase silica gel, or in polyethylene by entrapping) [91]. By using this supported catalyst, oxidation of benzylic alcohols was possible with up to three recycles. The oxidation of more challenging substrates, such as methylbenzenes, styrenes, and phenylacetic acids to arylaldehydes was also possible (yields 24%–69%) [92]. Oxidation of benzylamines [93] was reported with **RFTA 48** (Scheme 10.8(a)). Interestingly, the procedure was applied to a mild deprotection of

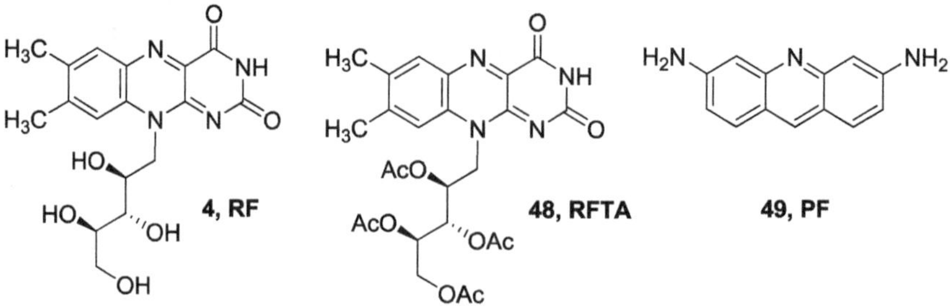

Figure 10.8. Different flavin derivatives.

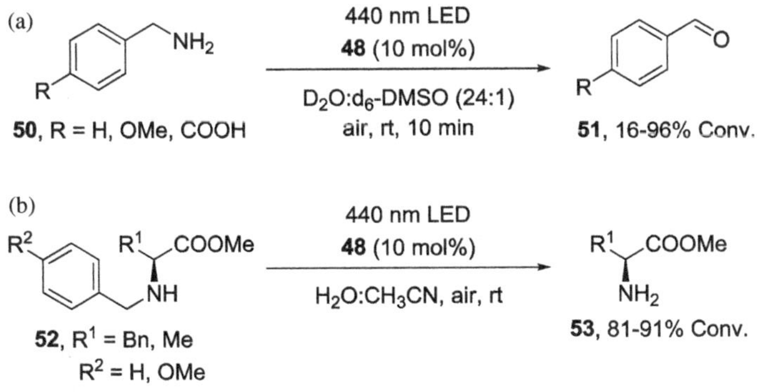

Scheme 10.8. Photooxidation of benzylic amines.

the 4-methoxybenzyl group in primary amines, showing a good conversion in short reaction time. The reaction was tolerant to esters, alkenes, and BOC groups, and it was applied in the deprotection of aminoacid esters (Scheme 10.8(b)). In addition, methoxybenzyl-protected carboxylic acids and carboxybenzoyl (**Cbz**)-protected amines were found inert under the photodeprotection conditions. It also noteworthy that benzylic alcohols can be oxidized with **RFTA 48** under solvent-free conditions [94].

The basis for the aforementioned oxidation was photoinduced electron transfer involving the S_1 excited state of **RFTA 48**. However, the redox potential *E* (RFTA*/**RFTA**$^{\bullet-}$) is 1.67 V vs SCE, and it is not sufficient to drive the oxidation of electron-poor benzylic substrates. Coordination with the Lewis acid Sc^{3+} allowed to move the redox potential *E* (**RFTA***/**RFTA**$^{\bullet-}$) to more positive values [95]. Alkylarenes were oxidized in the presence of **RFTA** (10 mol%) and $Sc(OSO_2CF_3)_3$ (20 mol%) in CH_3CN, in the presence of HCl (37%), using blue light for irradiation (440 nm, 3 W). For example, *p*-tert-butylbenzaldehyde and *p*-chlorobenzaldehyde were formed in 68% and 84% yields, respectively. The conditions were also suitable for the oxidation of primary and secondary benzylic alcohols. The reaction was performed in air, as oxygen is able to regenerate the **RFTA**-$2Sc^{3+}$ complex by the oxidation of $RFTAH_2^{\bullet+}$-$2Sc^{3+}$ forming H_2O_2. **RFTA** was also employed in the photooxidation of sulfides to sulfoxides employing CH_3CN:water (85:15 v/v) as reaction solvent under visible-light irradiation (450 nm) [96]. The reaction was described for electron-rich and electron-poor sulfides, as both singlet oxygen generation and SET were operative in the reaction mechanism. The reaction was working with quite low catalyst loading (0.1 mol%) and was giving quite good yields in a short reaction time, with no traces of over-oxygenated products. In addition, a practical synthesis of the drug Modafinil (90% yield) was reported. Oxidation of non-benzylic alcohols can be carried out by coupling the riboflavin photooxidation and regeneration of the enzyme co-factor NADH, using a blue LED light source [97]. A quite remarkable application of **RF** was described by Gilmour [98a, 98b] that reported a highly selective photocatalytic method for the isomerization of activated olefins (up to 99:1 Z/E, Scheme 10.9 (a)).

The reverse isomerization was possible by introducing a substituent in the alkene moiety, increasing the 1,3-allylic strain. A series of enones were isomerized in quite good yields, using CH_3CN as reaction solvent and 5 mol% of the **RF** catalyst under 24 h light irradiation (402 nm). Mechanistically, it revealed a photostationary state with constant Z/E ratios. E isomer was able to quench the photoexcited state of **RF** with a better efficiency (1 order of

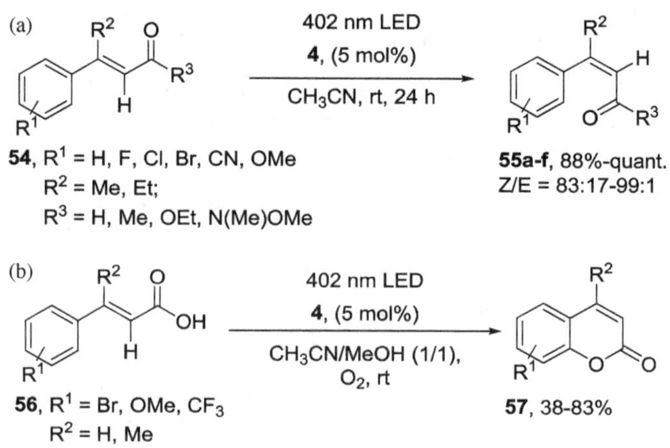

Scheme 10.9. Photoisomerization of cinnamyl derivatives (a) followed by cyclisation to coumarin (b).

magnitude) than the corresponding Z isomer. The quenching mechanism was an energy transfer to populate the T_1 excited state of the two isomers. Gilmour was able to extend this approach to a simple and effective synthesis of various substituted coumarins (Scheme 10.9(b)) under air [99]. The isomerization was occurring *via* energy transfer, while the successive cyclization of the isomerized acids was favored by an ET pathway, again promoted by the photo-excited state of **RF**. Recently, Gilmour has also described the catalytic E → Z isomerization of α,β-unsaturated nitriles [100].

RFTA POC was recently employed in interesting C–C bond forming reactions. For example, the [2+2] cycloaddition reaction was reported by visible photoexcitation [101]. Cibulka studied an isomer of **RF**, namely, 1-butyl-7,8-dimethoxy-3-methylalloxazine **58** which displays a higher energy S_1 state, corresponding to absorption near 400 nm. Molecule **58** is supposed to have a higher energy T_1 state compared to RF **4**, able to populate the T_1 excited state of styrene derivatives to promote [2+2] cycloadditions, as reported in Scheme 10.10.

The reaction proceeded with quantitative conversions for electron-rich styrenes. In the reaction, the exo-diastereoisomers

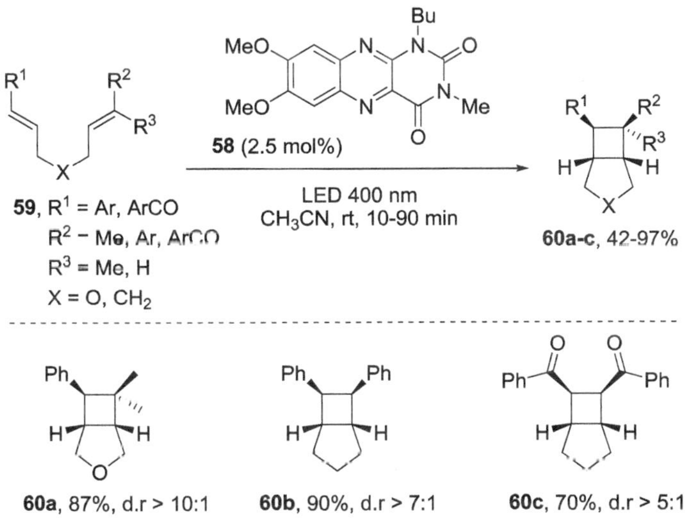

Scheme 10.10. [2+2] cycloadditions promoted by methylalloxazine **58**.

were formed as major products. Protonated RFTAH$^+$ or 1,3-dimethyl-8-trifluoromethylalloxazinium perchlorate were also promoting the reversed reactions, the [2+2] cycloreversion of cyclobutane ring with high oxidation potential in the presence of sensitive groups (e.g. furan) [102]. Another interesting application of **RF** in C–C bond forming reactions was the Mitsunobu reaction [103a–103c]. Mitsunobu reaction is quite important in organic chemistry, but, unfortunately is producing a significant amount of by-products. Normally, it is carried out with stoichiometric amounts of Ph_3P and azodicarboxylate derivatives (**DIAD**). The azodicarboxylate is reduced to hydrazine dicarboxylate and Ph_3P is oxidized to $Ph_3P{=}O$ in the reaction with alcohols or in esterification reactions. An oxidation would be necessary for recycling hydrazine formed during the Mitsunobu reaction. The usefulness of **RF**-based aerial photooxidation allowed the recycling/regeneration from the corresponding dialkyl hydrazine dicarboxylate of the dialkyl azodicarboxylate (Scheme 10.11). In addition, the photoinduced electron transfer from **RF** to Ph_3P, giving the corresponding $Ph_3P^{\bullet+}$, allowed the use of hydrazine dicarboxylate in catalytic amount (10 mol%), without the need of azodicarboxylate for the activation of Ph_3P. Enantiopure phenyl ethanol and ethyl lactate were successfully employed in esterification with no racemization [104].

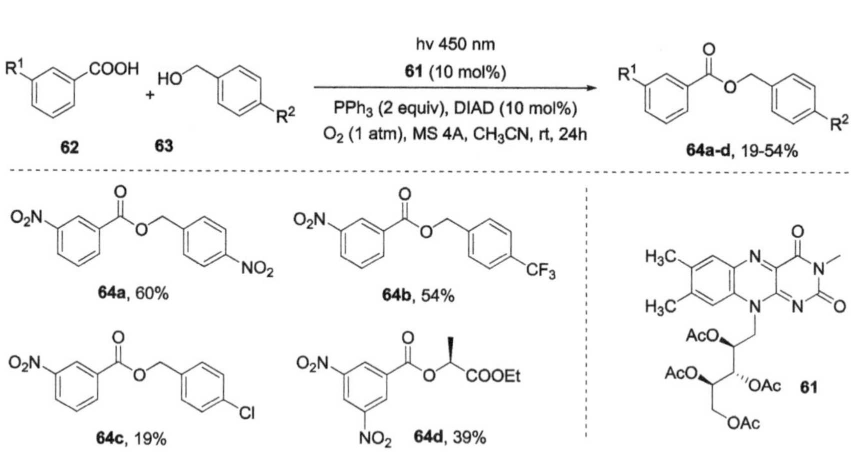

Scheme 10.11. Photocatalytic Mitsunobu reaction.

RF POC was also used for promoting the formation of C–halogen bonds. Aromatic chlorination was carried out in Friedel–Crafts conditions using chlorine gas or electropositive chlorine reagents [105]. In nature, the electrophilic chlorination is realized by enzymatically catalyzed oxidation of abundant chloride ions, *via* FAD-dependent halogenases, where the O_2 acts as the oxidant. **RFTA** was described as an efficient photooxidant for chlorination of arene substrates [106]. The reaction was performed in acidic conditions using CH_3CN as the reaction solvent with 10 mol% of **RFTA** as the **POC**. A broad scope was revealed, as substituted arenes (*N,N*-dialkyl aniline, anilide, alkyl, or aryl phenol ethers) were successfully chlorinated. If the substrates are too activated, an oxidation of the substrates was reducing the yields. On the other hand, electron-poor aromatic rings were not reacting. Enolizable substrates in acidic conditions (such as acetophenone) were giving the corresponding chlorination in the α-position of the carbonyl group. Photoredox catalysis with organic dyes in the presence of metals or metal complexes has been defined as synergistic catalysis [107]. The **POC** is able to behave either as an oxidant or as a reductant. König reported photoreduction of aldehydes to alcohols in the presence of ketones using **TEOA** as sacrificial electron donor, proflavine (**PF**) (**49**, Figure 10.8) as the **POC** (10 mol%), and [Cp*Rh(III)(bpy)Cl]Cl (**64**) as the synergic catalyst (Scheme 10.12) [108]. The rhodium complex was able to coordinate water which, in the reaction conditions, was transformed in the active hydride [Cp*Rh(III)(bpy)H]Cl (Rh(III)–H). The reaction was conducted in a mixture of DMF:water and was giving a good yield of the desired aromatic aldehydes, while the reaction with aliphatic aldehydes gave only moderate yields. The reaction was also conducted in flow conditions. The T_1 excited state of PF was reductively quenched by TEAO to give the corresponding radical anion $PF^{\bullet-}$. Two successive reductions by $PF^{\bullet-}$ occurred, forming Rh(II) and then Rh(I) complex. The Rh(I) complex oxidatively reacted with proton giving a Rh(III)–H complex that was the effective reductant. In another example of synergistic catalysis, RFTA and a non-heme iron complex $[Fe(TPA)\text{-}(CH_3CN)_2](ClO_4)_2$ (TPA=tris(2-pyridylmethyl)amine)

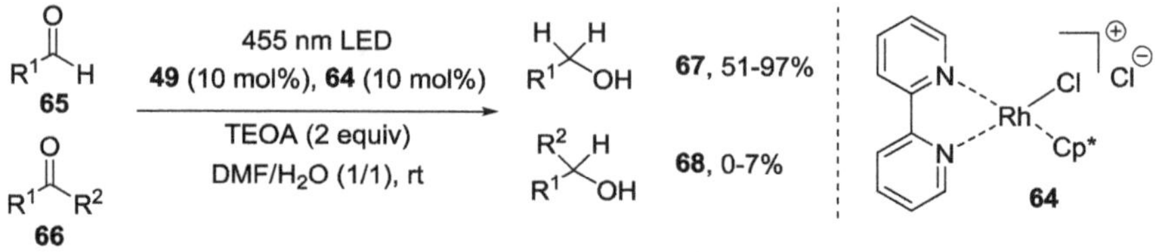

Scheme 10.12. Synergistic photoreduction of aldehydes to alcohols in the presence of ketones.

were the effective catalytic combination for promoting the visible-light-driven aerobic oxidation of alkyl benzenes to ketones and carboxylic acids [109]. Benzyl ethers were transformed in the catalytic conditions (**RFTA** (10 mol%), $[Fe(TPA)(CH_3CN)_2](ClO_4)_2$, 2 mol% in CH_3CN/H_2O, by irradiation with blue light at 440 nm) to the corresponding carboxylic acids in very good yields. Xanthene and thioxanthene gave the corresponding ketones. Finally, the procedure could be applied to isochroman, indane, and tetrahydronaphthalene, all transformed to the corresponding ketones. From the investigations carried out, it seemed that H_2O_2 was generated and iron complexes were acting as disproportionation catalyst for H_2O_2, forming an active Fe-oxo species that was oxygenating the substrates.

10.8 BODIPY

The first BODIPY compounds were synthesized in 1968 [110], but the current interest began mostly in the 1990s from their applications as laser dyes [111]. Currently, BODIPY dyes are commercially available: their efficient absorption of visible light combined with their stability and simple synthesis have attracted considerable interest in recent years for biological labeling experiments of DNA, proteins, and lipids [112–114].

10.8.1 *Photophysical and electrochemical properties*

BODIPY dyes are chemically and photochemically stable, display intense visible-light absorption (ε in the range between 4 to 11 × 10^4 M^{-1} cm^{-1}), high fluorescence quantum yields (normally between

60% and 90% if heavy atoms are not present), narrow absorption and emission bands (half-widths typically being around 25–35 nm), good solubility and thus no aggregation in solution, and versatile redox properties [115–117]. Moreover, their photophysical and electrochemical properties can be finely tuned by proper substituents on the difluoroboron dipyrromethene core.

Molecule **5** displays absorption maximum at 521 nm with vibrational fine structure typical of the molecular C=C framework of the core. The fluorescence band has a mirror-shape with a maximum at 538 nm and a very narrow Stokes shift and high fluorescence quantum yield (Φ_F *ca.* 1, Table 10.1 and Figure 10.9). Based on the unitary efficient fluorescence, population of T_1 is not possible by direct excitation of molecule **5**.

The effect of substituents on the difluoroboron dipyrromethene core has been widely investigated and allows a fine tuning of the energy of the lowest energy transition, as well as the fluorescence quantum yield. For example, alkyl substituents in positions 2 and 6 cause a bathochromic shift of the absorbance and fluorescence spectra, due to the stabilization of the excited state through a positive inductive effect. On the contrary, alkyl or phenyl substituents at the *meso* position does not significantly affect the energy of the absorption and emission bands. However, the fluorescence quantum yield of the *meso*-phenyl compound is strongly enhanced by the presence of 1,7-alkyl substituents, which prevent free rotation of the phenyl group reducing loss of energy from the excited states *via* non-radiative decays. Electron-withdrawing substituents at the *meso* position preferentially stabilize the LUMO, resulting in a decreased HOMO–LUMO gap.

It is worth mentioning the effect of heavy-atom substitution on the photophysical properties: the 2,6-diiodo-substituted derivative **6** is weakly fluorescent ($\Phi_F = 0.027$, Table 10.1 and Figure 10.9) and displays a very efficient population of the T_1 excited state ($\Phi_T = 0.97$, estimated assuming only fluorescence and intersystem crossing decays for the S_1 excited state), in contrast with the previously discussed chromophore **5**. The presence of the heavy iodine atom in the π-conjugated core strongly increases the spin–orbit coupling, thus leading to a very fast intersystem crossing S_1–T_1, which

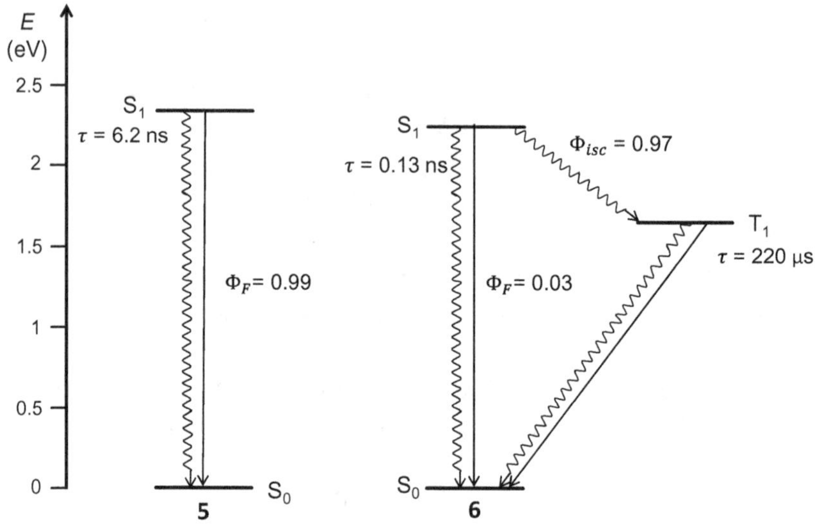

Figure 10.9. Schematic Jablonski diagram for molecules **5** and **6** in solution at 298 K [25–30].

effectively competes with fluorescence decay of the S_1 excited state. The resulting T_1 excited state is long lived ($\tau = 220\ \mu s$) and can be exploited in photocatalytic processes (see below).

In agreement with the photophysical properties, the electrochemical properties of BODIPY dyes strongly depend on the nature of the substitution in the *meso* position, α (3- and 5-positions), and β positions (1-,2-, 6-, and 7-positions) [25]. The completely substituted dye **5** shows reversible one-electron oxidation and reduction, with the formation of stable radical ions. Absence of substitution in positions 2-, 3-, 5-, or 6-causes instability of the radical cation produced on oxidation, and lack of the substitution in the *meso* position causes the radical anion produced on reduction to be more reactive.

10.8.2 *Reactions and use in photoredox catalysis*

Due to the photophysical properties, iodo-substituted BODIPY dyes such as compound **6** (Figure 10.9), were used as organic catalysts for three different photoredox catalytic organic reactions by Zhao (Scheme 10.13) [118].

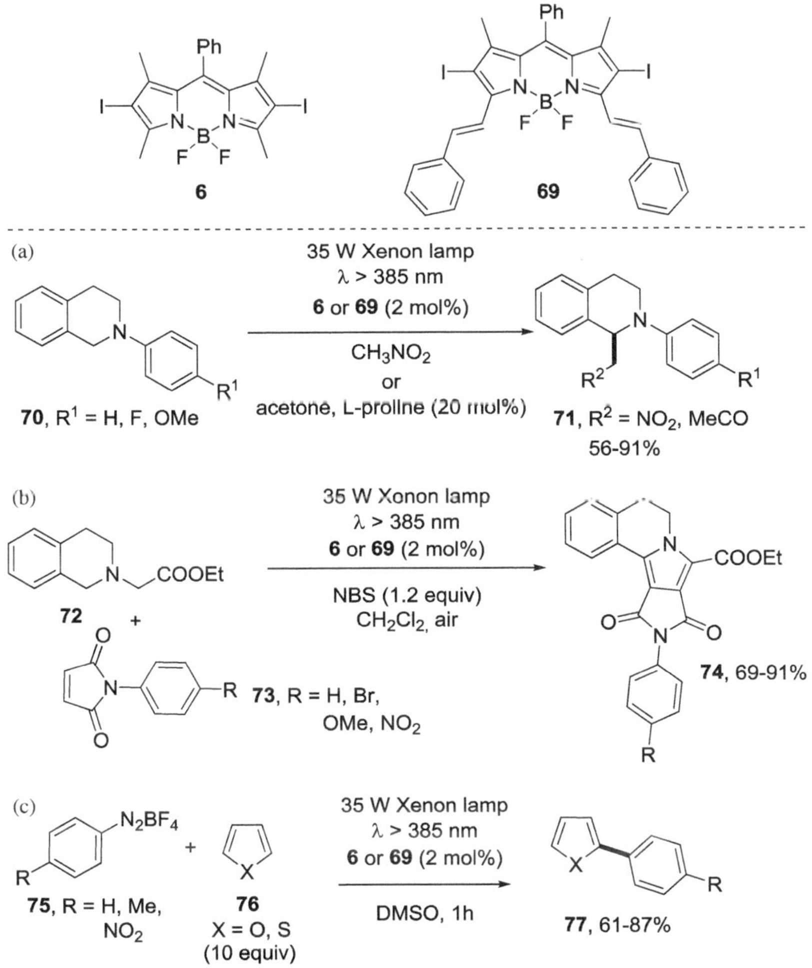

Scheme 10.13. Different photocatalytic reactions catalyzed by BODIPY derivatives: (a) alkylation of N-phenyltetrahydroisoquinolines with nitromethane and acetone; (b) oxidation/[3 + 2] cycloaddition/aromatization tandem reaction with tetrahydroisoquinoline derivatives; and (c) C–H arylation of furan and thiophene with diazonium salts.

POCs **6** and **69** are able to act either as electron acceptors (reductive quenching) or electron donors (oxidative quenching) in the reactions discussed in the following. Zhao studied the aza-Henry reaction of tetrahydroisoquinoline **70** (Scheme 10.13(a)), the oxidation/[3+2] cycloaddition/oxidative aromatization tandem reaction between tetrahydroisoquinolines **72** and maleimides **73** (Scheme 10.13 (b)), and C–H arylation of heteroarenes **76** with diazonium salts **75** (Scheme 10.13 (c)). In the case of the Henry reaction, *N*-aryl tetrahydroisoquinolines (aryl = phenyl, 4-F-C_6H_4, 4-MeO-C_6H_4) were employed and the nucleophiles $MeNO_2$ or $EtNO_2$ were used as reaction solvent. The reaction took place *via* reductive quenching and led to the oxidation of tetrahydroisoquinoline to the corresponding radical cation. The Iodo-BODIPY$^{\bullet-}$ formed was oxidized to Iodo-BODIPY by oxygen, that was necessary to carry out the reaction. Yields reported were 55%–91%, using 2 mol% of the **POC**. The cycloaddition was obtained by changing the protecting group of the tetrahydroisoquinoline from aryl to *N*-CH_2COOR. In the reaction conditions, this group was forming an ylide that underwent cycloaddition. The tetrahydroisoquinoline derivative quenched the **POC** in a way similar to the aza-Henry reaction. The reaction was carried out in the presence of NBS, as this oxidant is able to oxidize the pyrrolidine obtained in the cycloaddition to an aromatic pyrrole ring. Finally, in the reaction with diazonium salts with aromatic substrates (electron-rich furan and thiophene), Iodo-BODIPY was reducing the diazonium salt to the corresponding aryl radical with loss of N_2.

Zhao has also introduced different BODIPY derivatives such as 2,6-diiodo-aza-BODIPY **78** and the BODIPY-2,6-diiodo-aza-BODIPY triad **79** (Figure 10.10).

The two catalysts were employed for cycloaddition reactions with maleimide, and for promoting an alkynylation of tetrahydroisoquinoline derivatives by a synergistic catalysis in the presence of a Cu(I) salt. Triad **79** was designed to populate the excited state of the central Iodo-BODIPY chromophore (acceptor) by energy transfer from the two terminal BODIPY dyes (donors). Molecules **78** and **79** are strong visible absorbers: ε ca. 73,000 M^{-1} cm^{-1} at 683 nm for **78**, and ca. 165,000 M^{-1} cm^{-1} at 504 nm, and 71,000 M^{-1} cm^{-1} at 683 nm for **79** [119]. The triad was simply prepared through Cu(I)-promoted "click" reaction [120] from suitable precursors. In the cycloaddition reaction, promoted by the use of 1.5 mol% of the **POC** in DCE, and carried out in the presence of NBS as oxidant, system

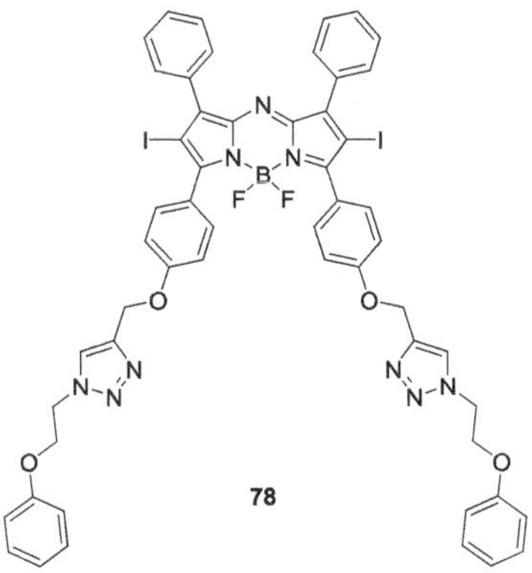

78

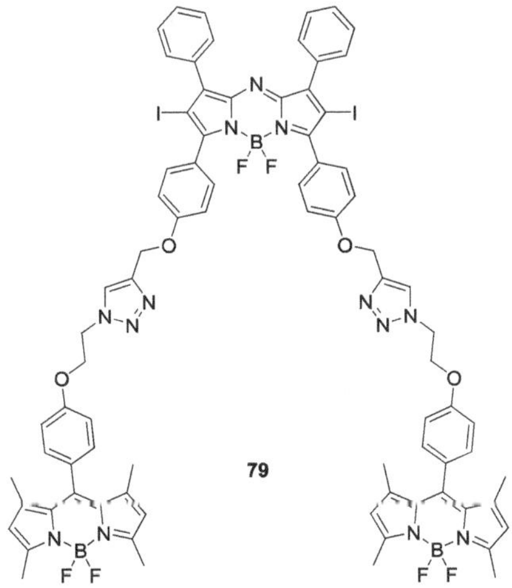

79

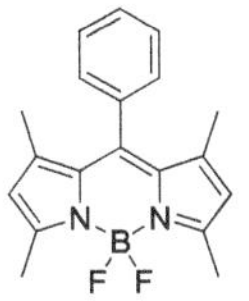

80

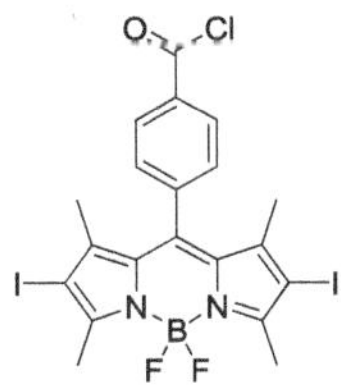

81

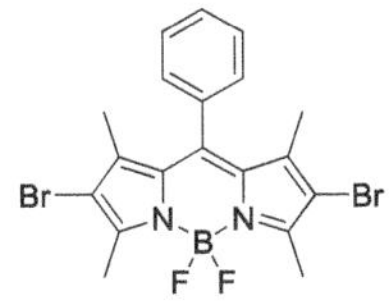

82

Figure 10.10. Effective BODIPY **POC** described in the literature.

79 gave better yields (78%–86%). Also, in the addition of phenylacetylene to *N*-phenyl-tetrahydroisoquinoline in CH_3CN, in the presence of 10 mol% of $(CH_3CN)_4CuPF_6$ as a co-catalyst, **79** furnished up to 80% of the desired product. Meng and Wu [121] investigated oxidation of *N*-aryl-tetrahydroisoquinolines for Henry reactions and addition of phosphonates. They reported that the S_1 excited state of BODIPY **80** (Figure 10.10) was able to transfer an electron from *N*-phenyltetrahydroisoquinoline, generating **80**$^{\bullet-}$, as demonstrated by time-resolved spectroscopy and electron spin resonance (ESR) studies. The oxidation potential of *N*-phenyltetrahydroisoquinoline was experimentally determined at +0.82 V vs SCE, while the E_0 reduction potential of **80** is −1.20 V vs SCE. As the BODIPY is absorbing at 505 nm, the excited state energy E_{00} is 2.46 eV. The calculated delta energy for the SET from tetrahydroisoquinoline to **80** is

–0.44 eV, giving a thermodynamically favored process. The yields for the Henry reaction were similar to the previously cited examples (46%–80% with CH_3NO_2, $CH_3CH_2NO_2$, $CH_3CH_2CH_2NO_2$). The reaction with *N*-Aryl tetrahydroisoquinolines and $HPOP(OEt)_2$ in the presence of **80** (1 mol%) in CH_3CN gave the desired product in 65%–85% yields. Iodo-BODIPY **80** was conjugated with a Zr-MOF, producing a heterogeneous **POC** [122].

Xue and co-workers have developed a synthesis of amides from aromatic aldehydes, employing the photo oxidative properties of BODIPY. A novel BODIPY **82** (Figure 10.10) was used in 2 mol% in the presence of 2 equiv of 3,5-di-tert-butyl-4-hydroxytoluene (**BHT**) under air (Scheme 10.14) [123]. Normally, BHT blocks radical processes, but, in this case, the formation of BHTOOH was observed in the reaction mixture. The BHTOOH species is more active than H_2O_2 and was formed by reaction with singlet oxygen, generated by action of **POC** and light. Secondary amines such as pyrrolidine, piperidine, or morpholine were used as reaction partners. Piperidines substituted in 3- or 4- positions were also tolerated, as well as

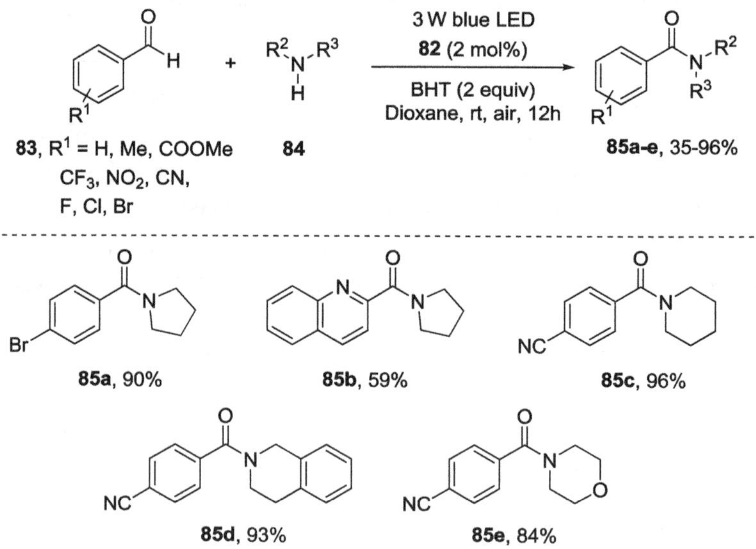

Scheme 10.14. Photo oxidative synthesis of amides.

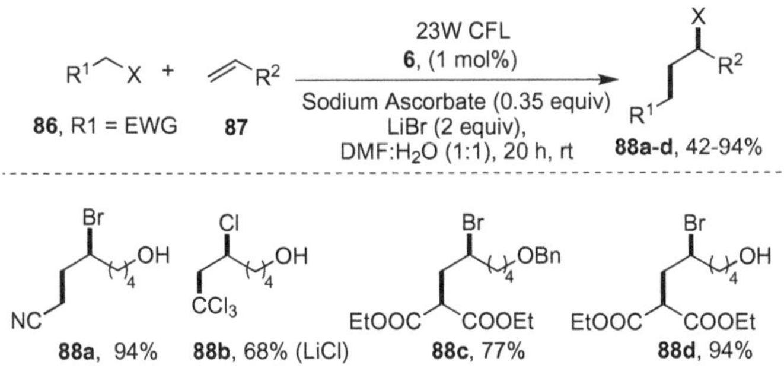

Scheme 10.15. ATRA reactions promoted by the iodo-BODIPY **6**.

N-methyl piperazine, and azepine. From the point of view of aldehydes, aromatic electron-rich aldehydes (pyrenecarboxyaldehyde, thiophenecarboxyaldehyde) or electron-poor (*p*-NO_2^-, *p*-CN^-, *p*-COOMe) benzaldehydes were tolerated. The reaction was possible also with 2- or 4-pyridine carboxyaldehyde. Yields from 35% to 96% were observed.

The employment of Iodo-BODIPY **6** in atom-transfer radical addition (ATRA) reaction was recently reported [29]. In the ATRA reaction [124a, 124b], different bromo- and chloro-derivatives were employed in addition to alkenes to afford the corresponding halogenated derivatives (Scheme 10.15) in the presence of sodium ascorbate and LiBr or LiCl as Lewis acid additive.

In this reaction, the Iodo-BODIPY **6** was an effective catalyst, while the same molecule without iodine atoms was not able to promote the reaction. Sodium ascorbate was necessary as sacrificial reagent, while other reductants, such as aliphatic tertiary amines, were not suitable. Different substituted organic halides were employed in the reaction, that was carried out in a solvent mixture DMF:water in the presence of 1 mol% of **6**. Quite interestingly, the reaction conditions allowed the employment of rather challenging substrates, like bromoacetate and haloalkyl pyridine salts to generate the corresponding products in good yields. The reaction was also possible with branched and cyclic alkenes, like cyclohexene and cyclooctene. Natural products like Limonene or (−)-Pinene could be employed in this reaction giving a simple access to functionalized compounds. The mechanism of the reaction was carefully investigated: sodium ascorbate was the main quencher of the lowest triplet excited state of **6**. Based on this finding and other Stern–Volmer experiments, the proposed catalytic mechanism was the formation of $\mathbf{6}^{\bullet-}$ from the ascorbate that was acting as the electron donor. The $\mathbf{6}^{\bullet-}$ species was a strong reductant able to reduce the organic halides employed. The so-formed organic radical reacted with the alkene, giving the corresponding radical. The observed products could be formed by two alternative pathways. In the first case, a radical chain mechanism could be involved, in which the alkyl radical, a strong reductant, was able to abstract a halogen atom from the starting halides. In the second case, oxidation of the radical intermediate to the corresponding carbenium ion happened, and the carbenium ions were intercepted by the halides. Further experiments conducted in the presence of different halide anions suggest that the radical chain mechanism was favored in these conditions.

10.9 Conclusions

Organic molecules containing nitrogen display interesting properties for photoinduced electron transfer and were introduced in photoredox catalysis. In this chapter, we have examined strong oxidants as well as reductant and their applications in organic reactions. The field is expanding quite rapidly, and it is predicted that new classes of N-containing dyes will be described in organic reactions, replacing the expensive **POCs** based on rare iridium and ruthenium metal ions. Although other cheap metals like copper [125], iron [126], or cobalt have been described recently in interesting photochemical reactions, organic molecules gave great advantage for the smooth preparation and the possibility to tailor the photophysical and electrochemical properties. These perspectives are exciting, particularly for the use of new concepts and reactions to transform complex natural feedstocks and materials. The future will be "illuminated" by photocatalysis.

References

[1] Turro, N. J.; Ramamurthy, V.; Scaiano, J. C. *Modern Molecular Photochemistry of Organic Molecules,* University Science Books, Sausalito, 2010.

[2] Klán, P.; Wirz, J. *Photochemistry of Organic Compounds: From Concepts to Practice,* Wiley, Chichester, 2009.

[3] Stephenson, C.; Yoon, T. (Eds.)., *Acc. Chem. Res.* **2016**, *49*, Issue 10, 1429–2315.

[4] Ravelli, D.; Protti, S.; Fagnoni M. *Application of Visible and Solar Light in Organic Synthesis* in *Applied Photochemistry when Light Meets Molecules,* Eds. Bergamini, G. and Silvi, S., Springer-Verlag, Berlin, 2016.

[5] Beeler, A. (Eds.). *Chem. Rev.* **2016**, *116*, Issue 17, 9629–10342.

[6] Pitre, S. P.; McTiernan, C. D.; Scaiano, J. C. *ACS Omega* **2016**, *1*, 66–76.

[7] Berlman, I. B. *Handbook of Fluorescence Spectra of Aromatic Molecules,* 2nd edn., Academic Press, New York and London, 1971.

[8] Heckman, R. C. *J. Mol. Spectrosc.* **1985**, *2*, 27–41.

[9] Lamola, A. A.; Hammond, G. S. *J. Chem. Phys.* **1965**, *43*, 2129–2135.

[10] Yamamoto, S.; Kikuchi, K.; Kokubun, H. *Chem. Lett.* **1977**, *6*, 1173–1176.

[11] Mann, C. K.; Barnes, K. K. *Electrochemical Reactions in Nonaqueous Systems,* Marcel Dekker Inc., New York, USA, 1970.

[12] Gosztola, D.; Niemczyk, M. P.; Svec, W.; Lukas, A. S.; Wasielewski, M. R. *J. Phys. Chem. A* **2000**, *104*, 6545–6551.

[13] Ford, W. E.; Kamat, P. V. *J. Phys. Chem.* **1987**, *91*, 6373–6380.

[14] Lee, S. K.; Zu, Y.; Herrmann, A.; Geerts, Y.; Müllem, K.; Bard, A. J. *J. Am. Chem. Soc.* **1999**, *121*, 3513–3520.

[15] Ghosh, S.; Chowdhury, M. *Chem. Phys. Lett.* **1982**, *85*, 233–238.

[16] Chowdhury, M; Goodman, L. *J. Chem. Phys.* **1962**, *36*, 548–549.

[17] Plugge, M.; Alain-Rizzo, V.; Audebert, P.; Brouwer, A. M. *J. Photochem. Photobiol. A Chem.* **2012**, *234*, 12–20.

[18] Yuasa, J.; Fukuzumi, S. *Chem. Commun.* **2006**, *0*, 561–563.

[19] Yuasa, J.; Fukuzumi, S. *J. Am. Chem. Soc.* **2006**, *128*, 14281–14292.

[20] Koziol, K. *Photochem. Photobiol.* **1966**, *5*, 41–54.

[21] Fugate, R. D.; Song P.-S. *Photochem. Photobiol.* **1976**, *24*, 479–481.

[22] Chambers, R. W.; Kearns, D. R. *Photochem. Photobiol.* **1969**, *10*, 215–219.

[23] Moore, W. M.; McDaniels, J. C.; Hen, J. A. *Photochem. Photobiol.* **1977**, *25*, 505–512.

[24] Tatwawadi, S. V.; Santhanam, K. S. V.; Bard, A. J. *J. Electroanal. Chem.* **1968**, *17*, 411–420.

[25] Nepomnyashchii, A. B.; Bard, A. J. *Acc. Chem. Res.* **2012**, *45*, 1844–1853.

[26] Nepomnyashchii, A. B.; Cho, S.; Rossky, P. J.; Bard, A. J. *J. Am. Chem. Soc.* **2010**, *132*, 17550–17559.

[27] Ziessel, R.; Goze, C.; Ulrich, G. *Synthesis* **2007**, *6*, 936–949.

[28] Wu, W.; Guo, H.; Wu, W.; Ji, S.; Zhao, J. *J. Org. Chem.* **2011**, *76*, 7056–7064.

[29] Magagnano, G.; Gualandi, A.; Marchini, M.; Mengozzi, L.; Ceroni, P.; Cozzi, P. G. *Chem. Commun.* **2017**, 53, 1591–1594.

[30] Yogo, T.; Urano, Y.; Ishitsuka, Y.; Maniwa, F.; Nagano, T. *J. Am. Chem. Soc.* **2005**, *127*, 12162–12163.

[31] Almeida, K. D.; Bernede, J. C.; Marsillac, S.; Godoy, A.; Diaz, F. R. *Synth. Met.* **2001**, *122*, 127–129.

[32] Lee, J.; Woo, H.; Kim, T.; Park, W. *Opt. Mater.* **2003**, *21*, 225–229.

[33] Morin, J. F.; Boudreault, P. L.; Leclerc, M. *Macromol. Rapid Commun.* **2002**, *23*, 1032–1036.

[34] Grigalevicius, S.; Ma, L.; Xie, Z.-Y.; Scheri, U. *J. Polym. Sci., Part A: Polym. Chem.* **2006**, *44*, 5987–5994.

[35] Bonesi, S. M.; Erra-Balsells, R. *J. Lumin.* **2001**, *93*, 51–74.

[36] Pinkham, C. A.; Wait Jr., S. C. *J. Mol. Spect.* **1968**, *27*, 326–342.

[37] Ambrose, J. F.; Nelson, R. F. *J. Electrochem. Soc.* **1968**, *115*, 1159–1164.

[38] Prudhomme, D. R.; Wang, Z.; Rizzo, C. J. *J. Org. Chem.* **1977**, *62*, 8257–8260.

[39] Luo, J.; Zhang, X.; Zhang, J. *ACS Catal.* **2015**, *5*, 2250–2254.

[40] Luo, J.; Zhang, X.; Lu, J.; Zhang, J. *ACS Catal.* **2017**, *7*, 5062–5070.

[41] Saito, I.; Ikehira, H.; Kasatani, R.; Watanabe, M.; Matsuura, T. *J. Am. Chem. Soc.* **1986**, *108*, 3115–3117.

[42] Prudhomme, D. R.; Wang, Z.; Rizzo, C. J. *J. Org. Chem.* **1997**, *62*, 8257–8260.

[43] Zhang, X.; Lu, J.; Zhang, J. *Chem. Mater.* **2014**, *26*, 4023–4029.

[44] Su, C.; Tandiana, R.; Tian, B.; Sengupta, A.; Tang, W.; Su, J.; Loh, K. P. *ACS Catal.* **2016**, *6*, 3594–3599.

[45] Newman, C. R.; Frisbie, C. D.; da Silva Filho, D. A.; Bredas, J. L.; Ewbank, P. C.; Mann, K. R. *Chem. Mater.* **2004**, *16*, 4436–4451.

[46] Saikia, D.; Sarma, R. *J. Elec. Mat.* **2018**, *47*, 737–743.

[47] Huang, C.; Barlow, S.; Marder, S. R. *J. Org. Chem.* **2011**, *76*, 2386–2407.

[48] Li, C.; Wonneberger, H. *Adv. Mater.* **2012**, *24*, 613–636.

[49] Seo, H.-S.; Kim, D.-K.; Oh, J.-D.; Shin, E.-S.; Choi, J.-H. *J. Phys. Chem. C* **2013**, *117*, 4764–4770.

[50] Chesterfield, R. J.; McKeen, J. C.; Newman, C. R.; Ewbank, P. C.; da Silva Filho, D. A.; Brédas, J.-L.; Miller, L. L.; Mann, K. R.; Frisbie, C. D. *J. Phys. Chem. B* **2004**, *108*, 19281–19292.

[51] Delgado, M. C. R.; Kim, E.-G.; da Silva Filho, D. A.; Bredas, J.-L. *J. Am. Chem. Soc.* **2010**, *132*, 3375–3387.

[52] Balakrishnan, K.; Datar, A.; Naddo, T.; Huang, J.; Oitker, R.; Max Yen, M.; Zhao, J.; Zang, L. *J. Am. Chem. Soc.* **2006**, *128*, 7390–7398.

[53] Tauber, M. J.; Kelley, R. F.; Giaimo, J. M.; Rybtchinski, B.; Wasielewski, M. R. *J. Am. Chem. Soc.* **2006**, *128*, 1782–1783.

[54] Melone, L.; Franchi, P.; Lucarini, M.; Punta, C. *Adv. Synth. Catal.* **2013**, *355*, 3210–3220.

[55] Reszka, K. J.; Takayama, M.; Sik, R. H.; Chignell, C. F.; Saito, I. *Photochem. Photobiol.* **2005**, *81*, 573–580.

[56] Ghosh, I.; Ghosh, T.; Bardagi, J. I.; König, B. *Science* **2014**, *346*, 725–728.

[57] Marchini, M; Gualandi, A.; Mengozzi, L.; Franchi, P.; Lucarini, M.; Cozzi, P. G.; Balzani, V.; Ceroni, P. *Phys. Chem. Chem. Phys.* **2018**, *20*, 8071–8076.

[58] Zeng, L.; Liu, T.; He, C.; Shi, D.; Zhang, F.; Duan, C. *J. Am. Chem. Soc.* **2016**, *138*, 3958–3961.

[59] Wang, S. F.; Cao, X.-P.; Li, Y. *Angew. Chem. Int. Ed.* **2017**, *56*, 13809–13813.

[60] Wang, Y.; Wang, X; Antonietti, M. *Angew. Chem. Int. Ed.* **2012**, *51*, 68–89.

[61] Liu, J.; Wang, H.; Antonietti, M. *Chem. Soc. Rev.* **2016**, *45*, 2308–2326.

[62] Zhang, J. S.; Chen, X. F.; Takanabe, K.; Maeda, K.; Domen, K.; Epping, J. D.; Fu, X. Z.; Antonietti, M.; Wang, X. C. *Angew. Chem. Int. Ed.* **2010**, *49*, 441–444.

[63] Thomas, A.; Fischer, A.; Goettmann, F.; Antonietti, M.; Müller, J.-O.; Schlögl, R.; Carlsson, J. M. *J. Mater. Chem.* **2008**, *18*, 4893–4908.

[64] Wang, X.; Maeda, K.; Thomas, A.; Takanabe, K.; Xin, G.; Carlsson, J. M.; Domen, K.; Antonietti, M. *Nat. Mater.* **2009**, *8*, 76–80.

[65] (a) Liu, G.; Niu, P.; Sun, C. H.; Smith, S. C.; Chen, Z. G.; Lu, G. Q.; Cheng, H. M. *J. Am. Chem. Soc.* **2010**, *132*, 11642–11648; (b) Zhang, J.; Sun, J.; Maeda, K.; Domen, K.; Liu, P.; Antonietti, M.; Fu, X.; Wang, X. *Energy Environ. Sci.* **2011**, *4*, 675–678.

[66] Maeda, K.; Wang, X.; Nishihara, Y.; Lu, D.; Antonietti, M.; Domen, K. *J. Phys. Chem. C* **2009**, *113*, 4940–4947.

[67] Takanabe, K.; Kamata, K.; Wang, X.; Antonietti, M.; Kubota, J.; Domen, K. *Phys. Chem. Chem. Phys.* **2010**, *12*, 13020–13025.

[68] Chen, X.; Zhang, J.; Fu, X.; Antonietti, M.; Wang, X. *J. Am. Chem. Soc.* **2009**, *131*, 11658–11659.

[69] (a) Silva, T. F. S.; Mishra, G. S.; da Silva, M. F. G.; Wanke, R.; Martins, L. M. D. R. S.; Pombeiro, A. J. L. *Dalton Trans.* **2009**, *0*, 9207–9215; (b) Wang, Y.; Zhang, J.; Wang, X.; Antonietti, M.; Li, H. *Angew. Chem. Int. Ed.* **2010**, *49*, 3356–3359.

[70] Wang, Y.; Li, H.; Yao, J.; Wang, X.; Antonietti, M. *Chem. Sci.* **2011**, *2*, 446–450.

[71] Su, F.; Mathew, S. C.; Lipner, G.; Fu, X.; Antonietti, M.; Blechert, S.; Wang, X. *J. Am. Chem. Soc.* **2010**, *132*, 16299–16301.

[72] Su, F.; Mathew, S. C.; Möhlmann, L.; Antonietti, M.; Wang, X.; Blechert, S. *Angew. Chem. Int. Ed.* **2011**, *50*, 57–660.

[73] Möhlmann, L.; Blechert, S. *Adv. Synth. Catal.* **2014**, *356*, 2825–2829.

[74] Zhang, P.; Wang, Y.; Yao, J.; Wang, C.; Yan, C.; Antonietti, M.; Lia, H. *Adv. Synth. Catal.* **2011**, *353*, 1447–1451.

[75] Zhang, P.; Wang, Y.; Li, H.; Antonietti, M. *Green Chem.* **2012**, *14*, 1904–1908.

[76] Woźnica, M.; Chaoui, N.; Taabache, S.; Blechert, S. *Chem. Eur. J.* **2014**, *20*, 14624–14628.

[77] Baar, M.; Blechert, S. *Chem. Eur. J.* **2015**, *21*, 526–530.

[78] Pinner, A. *Chem. Ber.* **1893**, *26*, 2126–2135.

[79] (a) Audebert, P.; Miomandre, F.; Clavier, G.; Vernières, M. C.; Badré, S.; Méallet-Renault, R. *Chem. Eur. J.* **2005**, *11*, 5667–5673; (b) Gong, Y. H.; Miomandre, F.; Meallet-Renault, R.; Badré, S.; Galmiche, L.; Tang, J.; Audebert, P.; Clavier, G. *Eur. J. Org. Chem.* **2009**, *35*, 6121–6128; (c) Qing, Z.; Audebert, P.; Clavier, G.; Miomandre, F.; Tang, J.; Vu, T. T.; Méallet-Renault, R. *J. Electroanal. Chem.* **2009**, *632*, 39–44; (d) Gückel, F.; Maki, A. H.; Neugebauer, F. A.; Schweitzer, D.; Vogler, H. *Chem. Phys.* **1992**, *164*, 217–227.

[80] Clavier, G.; Audebert, P. *Chem. Rev.* **2010**, *110*, 3299–3314.

[81] Hochstrasser, R. M.; King, D. S.; Smith III, A. B. *J. Am. Chem. Soc.* **1977**, *99*, 3923–3933.

[82] Balzani, V.; Ceroni, P.; Juris, A. *Photochemistry and Photophysics: Concepts, Research, Applications*, Wiley-VCH, Weinheim, 2014.

[83] Samanta, S.; Das, S.; Biswas, P. *J. Org. Chem.* **2013**, *78*, 11184–11193.

[84] Huang, W.; Ma, B. C.; Lu, H.; Li, R.; Wang, L.; Landfester, K.; Zhang, K. A. I. *ACS Catal.* **2017**, *7*, 5438–5442.

[85] Insińska-Rak, M.; Golczak, A.; Sikorski, M. *J. Phys. Chem. A* **2012**, *116*, 1199–1207.

[86] Silva, E.; Edwards A. M. *Flavins: Photochemistry and Photobiology*, RCS Publishing, Cambridge, 2007.

[87] Halwer, M. *J. Am. Chem. Soc.* **1951**, *73*, 4870–4874.

[88] Fukuzumi, S.; Tanii, K.; Tanaka, T. *J. Chem. Soc. Chem. Commun.* **1989**, *13*, 816–818.

[89] Nicewicz, D. A.; MacMillan, D. W. C. *Science* **2008**, *322*, 77–80.

[90] Svoboda, J.; Schmaderer, H.; Köning, B. *Chem. Eur. J.* **2008**, *14*, 1854–1865.

[91] Schmaderer, H.; Hilgers, P.; Lechner, R.; König, B. *Adv. Synth. Catal.* **2009**, *351*, 163–174.

[92] Lechner, R.; Kümmel, S.; König, B. *Photochem. Photobiol. Sci.* **2010**, *9*, 1367–1377.

[93] Lechner, R.; König, B. *Synthesis* **2010**, *10*, 1712–1718.

[94] Obst, M.; König, B. *J. Org. Chem.* **2016**, 12, 2358–2363.

[95] Mühldorf, B.; Wolf, R. *Chem. Commun.* **2015**, *51*, 8425–8428.

[96] Neveselý, T.; Svobodov, E.; Chudoba, J.; Sikorski, M.; Cibulka, R. *Adv. Synth. Catal.* **2016**, 358, 1654–1663.

[97] Rauch, M.; Schmidt, S.; Arends, I. W. C. E.; Oppelt, K.; Karac, S.; Hollmann, F. *Green Chem.* **2017**, *19*, 376–379.

[98] (a) Metternich, J. B.; Gilmour, R. *J. Am. Chem. Soc.* **2015**, *137*, 11254–11257; (b) Metternich, J. B.; Gilmour, R. *Syn.* **2016**, *27*, 2541–2552.

[99] Metternich, J. B.; Gilmour, R. *J. Am. Chem. Soc.* **2016**, *138*, 1040–1045.

[100] Metternich, J. B.; Artiukhin, D. G.; Holland, M. C.; von Bremen-Kühne, M.; Neugebauer, J., Gilmour, R. *J. Org. Chem.* **2017**, *82*, 9955–9977.

[101] Albini, A.; Fagnoni, M. *Handbook of Synthetic Photochemistry*, Wiley-VCH Verlag GmbH & Co. KGaA, Weinheim, 2010.

[102] Hartman, T.; Cibulka, R. *Org. Lett.* **2016**, *18*, 3710–3713.

[103] (a) Fletcher, S. *Org. Chem. Front.* **2015**, *2*, 739–752; (b) Swamy, K. C. K.; Kumar, N. N. B.; Balaraman, E.; Kumar, K. V. P. P. *Chem. Rev.* **2009**, *109*, 2551–2651; (c) But, T. Y. S.; Toy, P. H. *Chem. Asian J.* **2007**, *2*, 1340–1355.

[104] März, M.; Chudoba, J.; Kohouta, M.; Cibulka, R. *Org. Biomol. Chem.* **2017**, *15*, 1970–1975.

[105] Littke, A. F.; Fu, G. C. *Angew. Chem. Int. Ed.* **2002**, *41*, 4176–4211.

[106] Hering, T.; Mîhldorf, B.; Wolf, R.; König, B. *Angew. Chem. Int. Ed.* **2016**, *55*, 5342–5345.

[107] Allen A. E.; MacMillan D. W. C. *Chem. Sci.* **2012**, *3*, 633–658.

[108] Ghosh, T.; Slanina, T.; König, B. *Chem. Sci.* **2015**, *6*, 2027–2034.

[109] Mühldorf, B.; Wolf, R. *Angew. Chem. Int. Ed.* **2016**, *55*, 427–430.

[110] Treibs, A.; Kreuzer, F. H. *Liebigs Ann. Chem.* **1968**, *718*, 208–223.

[111] Sunahara, H.; Urano, Y.; Kojima, H.; Nagano, T. *J. Am. Chem. Soc.* **2007**, *129*, 5597–5604.

[112] Yee, M.-C.; Fas, S. C.; Stohlmeyer, M. M.; Wandless, T. J.; Cimprich, K. A. *J. Biol. Chem.* **2005**, *280*, 29053–29059.

[113] Wang, F.; Guo, Z.; Li, X.; Li, X.; Zhao, C. *Chem. Eur. J.* **2014**, *20*, 11471–11478.

[114] Wang, X.; Cao, J.; Zhao, C. *Org. Biomol. Chem.* **2012**, *10*, 4689–4691.

[115] Boens, N.; Leen, L.; Dehaen, W. *Chem. Soc. Rev.* **2012**, *41*, 1130–1172.

[116] Ulrich, G.; Ziessel, R.; Harriman, A. *Angew. Chem. Int. Ed.* **2008**, *47*, 1184–1201.

[117] Loudet, A.; Burgess, K. *Chem. Rev.* **2007**, *107*, 4891–4932.

[118] Huang, L.; Zhao, J. *RSC Adv.* **2013**, *3*, 23377–23388.

[119] Guo, S.; Tao, R.; Zhao, J. *RSC Adv.* **2014**, *4*, 36131–36139.

[120] Kolb, H. C.; Finn, M. G.; Sharpless, K. B. *Angew. Chem. Int. Ed.* **2011**, *40*, 2004–2021.

[121] Wang, X.-Z.; Meng, Q.-Y.; Zhong, J.-J.; Gao, X.-W.; Lei, T.; Zhao, L.-M.; Li, Z.-J.; Chen, B.; Tung, C.-H.; Wu, L.-Z. *Chem. Commun.* **2015**, *51*, 11256–11259.

[122] Quan, Y.; Li, Q.-Y.; Zhang, Q.; Zhang, W.-Q.; Lu, H.; Yu, J.-H.; Chen, J.; Zhao, X.; Wang, X.-J. *RSC Adv.* **2016**, *6*, 23995–23999.

[123] Wang, X.-F.; Yu, S.-S.; Wang, C.; Xue, D.; Xiao, J. *Org. Biomol. Chem.* **2016**, *14*, 7028–7037.

[124] (a) Curran, D. P. *Synthesis* **1988**, *7*, 489–513; (b) Kamigaito, M.; Ando, T.; Sawamoto, M. *Chem. Rev.* **2011**, *101*, 3689–3746.

[125] Reiser, O. *Acc. Chem. Res.* **2016**, *49*, 1990–1996.

[126] Gualandi, A.; Marchini, M.; Mengozzi, L.; Natali, M.; Lucarini, M.; Ceroni, P; Cozzi, P. G. *ACS Catal.* **2015**, *5*, 5927–5931.

Chapter 11

Photoorganocatalysts for Polymerization

Didier Gigmes*, Frédéric Dumur, Nicolas Zivic, Ségolène Villotte and Jason C. Morris

Aix-Marseille University, CNRS, Institut de Chimie Radicalaire ICR, Marseille, France
**didier.gigmes@univ-amu.fr*

Jacques Lalevée[†], and Aude Héloïse Bonardi

Institut de Science des Matériaux de Mulhouse IS2M, UMR CNRS 7361, University of Haute-Alsace, 15, rue Jean Starcky, 68057 Mulhouse Cedex, France
[†]jacques.lalevee@uha.fr

11.1 Introduction

The range of polymer applications is a rapidly changing and expanding field that now exceeds that of any other class of materials. Photopolymerization offers several advantages over classical thermal polymerizations. The photopolymerization process can be extremely rapid and completed within a few seconds. Another advantage of photopolymerization is that the formulation is solvent-free, reducing the emission of volatile organic compounds. Faced with growing environmental concerns, photopolymerization has become an increasingly interesting tool for the production of polymers with reduced environmental impact. Since the initiation of photopolymerization is dependent on the presence of light, the polymerization can be performed within a well-defined area (irradiation zone) and immediately stopped when the light source is switched off. This feature provides a good spatial and temporal control, rendering this polymerization process extremely attractive for 2D and 3D applications (e.g. this property is very attractive in 3D printing using the stereolithography approach) [1]. Within this context, photoinitiator (PA) and/or PA systems exhibiting remarkable efficacy under soft irradiation conditions are highly needed. In this way, the concentration of the PAs in the resin formulation decreases. Recently, photoredox catalysis, originally applied to organic chemistry, has merged with polymer chemistry. This merger has led to the implementation of new photochemical reactions to produce polymers with increased efficiency and precision, whilst allowing for a drastic reduction in the amount of initiator required. Furthermore, due to the low operational costs and the future prospects of utilizing the Sun as a source of energy, light is undoubtedly the most attractive source of energy for initiating chemical reactions.

11.2 Mechanism of Photoinitiation

A PA, or photoinitiating system, can generate a variety of reactive species in the form of radicals, acids, cations, or anions. However, only PAs and photoinitiating systems, which generate radicals, acids, and cations will be discussed herein, due to their high reactivity and broader applicability to photopolymerization systems. Indeed, PAs for anionic polymerizations are extremely rare in industry due to the high reactivity of the propagating species and their sensitivity to nucleophilic attack, requiring the use of extremely pure compounds [1]. In this section, we will discuss the PA performance optimization by means of photosensitization or the addition of a third component. Photosensitization is based on the addition of a chromophore to facilitate the excited-state energy transfer to the PA. This process is used to extend the range and efficiency of absorption, as well as to increase the population of PA-excited states responsible for the initiation step. The use of catalytic cycles through the photoredox catalysis approach for high performance photoinitiating systems will be also presented.

11.2.1 *Initiation of radical polymerization*

PAs capable of initiating radical polymerizations are undoubtedly the most widely studied PAs. These compounds can be classified into two categories, namely type I and type II PAs [2–4]. The two PA mechanisms (type I and type II) and the main families of molecules exploiting these initiation processes are described herein.

11.2.2 *Mechanism of type I photoinitiation*

The unimolecular formation of reactive radical species *via* type I processes are typically encountered in aromatic carbonyls derived from the acetophenone scaffold, which absorb radiation in the UV region (320–330 nm). Consequently, activation of these PAs is typically performed with UV light sources. The most representative families of PAs utilizing type I processes are α-hydroxyalkylacetophenones (**HAAP**), derivatives of benzoine (**B**), acylphosphine oxide (**OAP**), and α-aminoketones (**AC**). An example of each family is presented below (Figure 11.1).

Radicals are formed through excited-state homolytic cleavage of the bond in the α-position to the carbonyl function [5]. The homolytic cleavage efficiency is dependent on several parameters. Firstly, the excited-state energy should exceed that of the bond dissociation energy (BDE) in order to achieve bond homolysis. Secondly, the BDE should be sufficiently high so as to avoid unwanted thermal decomposition of the PA at room temperature (BDEs are typically in the 45–65 kcal mol^{-1} range). An example of photochemical dissociation of an aromatic carbonyl is presented below (Scheme 11.1).

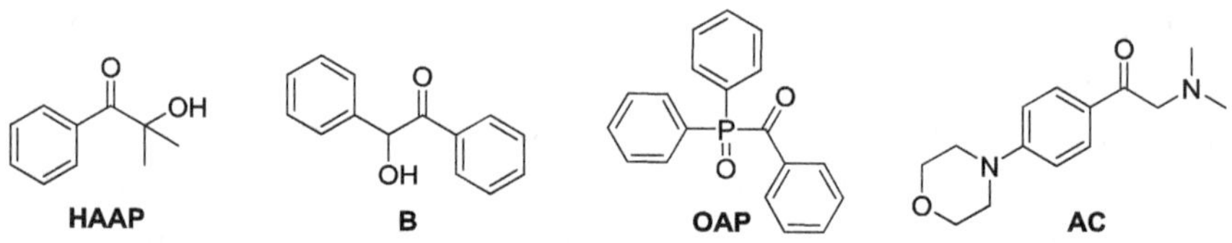

Figure 11.1. A selection of type I PAs: 2-hydroxy-2-methyl-1-phenylpropan-1-one (**HAAP**), 2-hydroxy-1,2-diphenylethanone (**B**), (diphenylphosphoryl)(phenyl) methanone (**OAP**), 2-(dimethylamino)-1-(4-morpholinophenyl)ethanone (**AC**).

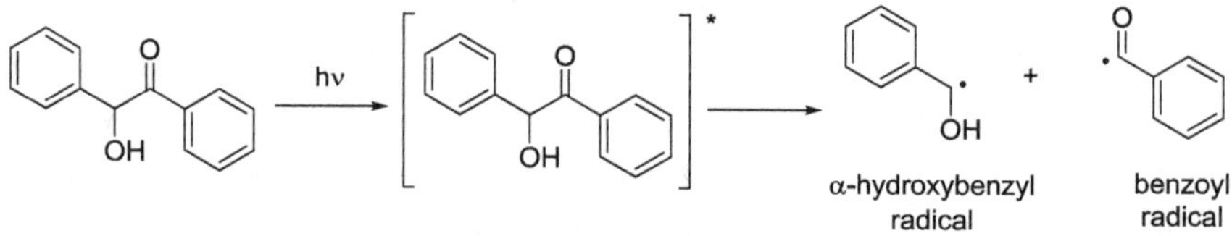

Scheme 11.1. Mechanism of homolytic cleavage for type I PAs.

The generated benzoyl radical (Scheme 11.1) is characterized by a high reactivity with vinylic and acrylic monomers. Furthermore, the other radical fragment produced (i.e. the α-hydroxybenzyl radical) is also capable of initiating polymerization. Benzoine derivatives are the most widely used PAs for radical polymerization, notably for styrenic monomers, as these radicals exhibit a high reactivity and can be formed in a high yield [6, 7].

11.2.3 *Mechanism of type II photoinitiation*

Contrary to type I photoinitiation, type II photoinitiation occurs through bimolecular processes. PAs utilizing this mechanism are typically aromatic carbonyls (Figure 11.2). However, for aromatic carbonyls employing a type II mechanism, the excited-state energy is typically lower than the BDE of the bond in the α-position to the carbonyl function preventing homolytic cleavage. Similar to type I PAs, the absorption bands of type II PAs are strongly centered in the UV region. However, appropriate functionalization of the aromatic system is capable of shifting the absorbance into the visible range.

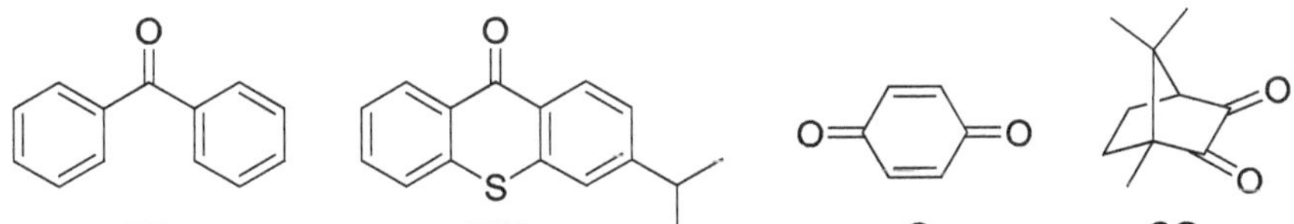

Figure 11.2. Selection of type II PAs: Benzophenone (**BP**), **ITX**, Quinone (**Q**), **CQ**.

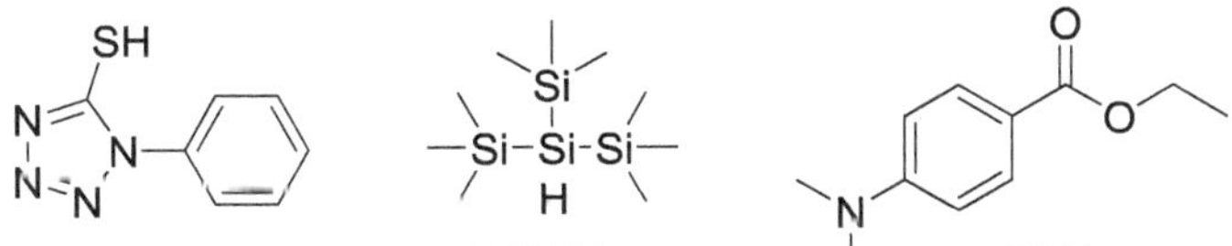

Figure 11.3. Selection of co-initiators: Phenyltetrazolethiol (**PTZT**), tris(trimethylsilyl) silane (**TTMSS**), ethyldimethylaminobenzoate (**EDB**).

Indeed, aromatic carbonyls such as isopropylthioxanthone (**ITX**) or camphorquinone (**CQ**), commonly used in dentistry, absorb beyond 400 nm.

Through type II processes, radicals are formed by proton, or hydrogen-atom, transfer from a hydrogen-atom donor to the carbonyl moiety of the excited-state PA. The proton (or a hydrogen atom) donor, which ideally does not absorb light under the applied conditions, is designated as the co-initiator (Figure 11.3). Type II processes may occur *via* one of two mechanisms. Firstly, hydrogen-atom transfer can occur directly from the hydrogen-atom donor to the excited-state PA. Secondly, single-electron transfer from the hydrogen-atom donor is followed by proton transfer.

11.2.4 *Mechanism involving a hydrogen abstraction*

Excited-state hydrogen-atom transfer occurs from a hydrogen donor (DH) (i.e. an ether, an alcohol, or a thiol) to the excited-state carbonyl to generate a ketyl radical and a second radical derived from the co-initiator, as illustrated in the scheme below (Scheme 11.2).

Polymerization is initiated by the co-initiator derived $D^{\bullet}$ radical. Conversely, the ketyl radical, stabilized by electronic delocalization

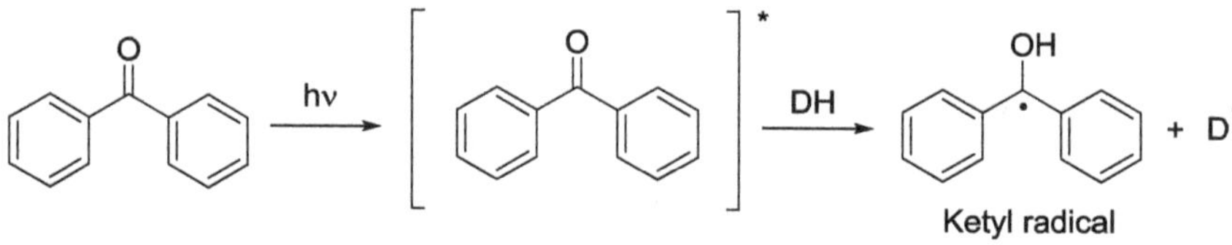

Scheme 11.2. Mechanism of type II initiation by hydrogen-atom abstraction.

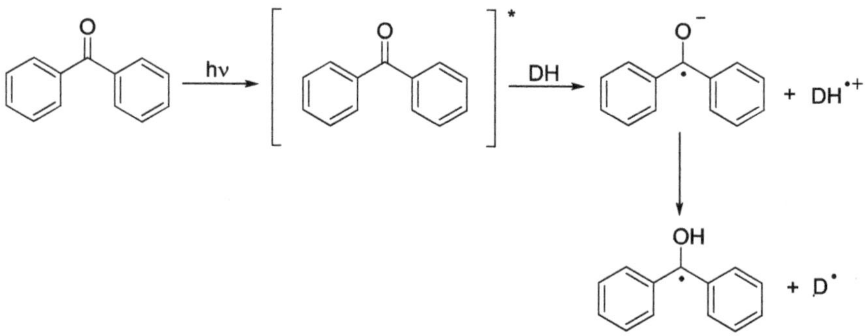

Scheme 11.3. Mechanism of type II photoinitiation by electron/proton transfer.

and steric hindrance, is generally incapable of initiating polymerization. However, the ketyl radical typically undergoes self-recombination or reaction with a growing polymer chain [8].

11.2.5 *Mechanism by electron/proton transfer*

Hydrogen transfer by electron/proton transfer can occur from either the excited triplet or singlet states. The initial transfer of an electron occurs from the co-initiator to the PA to form a radical ion pair. Once formed, deactivation of the radical ion pair proceeds *via* proton transfer to form a ketyl radical and a co-initiator-derived $D^{\bullet}$ radical (Scheme 11.3). The hydrogen donor (co-initiator) is generally an electron-donating molecule, such as a tertiary amine, bearing a labile hydrogen in the α-position to the nitrogen atom.

The polymerization is initiated in a similar way to that previously mentioned for the mechanism by hydrogen abstraction. The advantage of this mechanism is to form α-aminoalkyl radicals that exhibit high reactivity toward vinylic monomers. The efficiency of this process can be controlled by the appropriate selection of the aromatic carbonyl/amine combination. Indeed, the formation of radical species can be adversely affected by: an unfavorable back-electron transfer in competition with proton transfer, a low reactivity of the α-aminoalkyl radicals, competitive quenching of the excited singlet or triplet states of the PA as well as side reactions of the reactive species [9, 10].

11.2.6 *Mechanism of type II monocomponent initiation*

One of the main problems associated with type II PAs is the potential migration of malodorous molecules of low molecular weight (e.g. amines, thiols, **BP**) within the polymer matrix, as well as the yellowing usually obtained with amines (poor color stability of the polymer). This migration can adversely affect the properties of the materials (e.g. a lower adhesion of the material). Additionally, depending on the desired application of the prepared polymers, the potential toxicity may render these materials unsuitable (e.g. food storage). To address this issue, the first strategy investigated consisted of introducing a polymerizable functionality within the PA structure to limit its extractability. Hence, the PA is inserted in the polymer backbone during the polymerization process. Similarly, the second strategy involved the synthesis of type II monocomponent PAs to minimize migration inside the material [11, 12]. Type II monocomponent PAs combine the chromophore and the hydrogen donor within the same molecule. The combination of both functionalities within a single molecule permits improved PA solubility. Furthermore, the PA efficiency is improved by the intramolecular interaction which does not require the collision of the reaction partners. This approach also benefits from the reduced migration of low molecular species within the polymer matrix. Indeed, both functionalities are covalently connected so that, following the initiation of polymerization, both are covalently attached to the growing polymeric chains. Additionally, this technique benefits from increased molecular weights of the monocomponent systems, reducing their volatility.

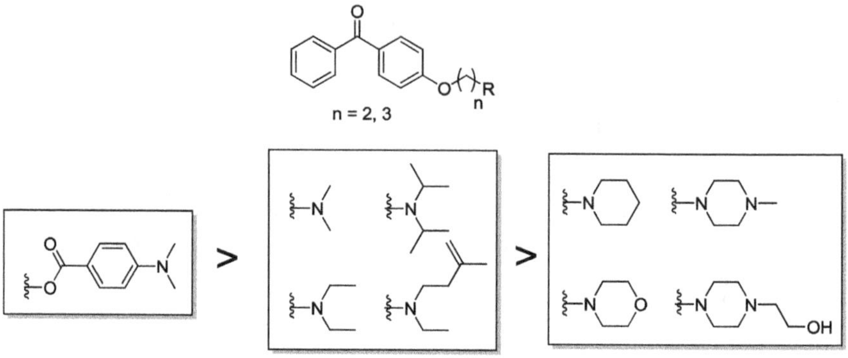

Figure 11.4. Derivatives of **BP** used to form type II monocomponent initiators proposed by the group of Allen [13] and classified by increasing reactivity.

The first type II monocomponent PAs were synthesized by the group of Allen (Figure 11.4) [13]. Through the covalent combination of the **BP** and tertiary amine components within a single molecule, this research group demonstrated the increased efficiency of the monocomponent system over the bicomponent system. The efficiency is largely dependent on the nature of the tertiary amine (i.e. its ionization potential) and the size of the aliphatic chain used to combine the two components. Indeed, when the **BP** motif is substituted by an aromatic amine, a higher efficiency is observed when compared to those substituted with aliphatic or cyclic amines. **BPs** substituted with cyclic amines are the least efficient monocomponent PAs. Moreover, **BPs** substituted with a tertiary amine, connected *via* aliphatic linkages composed of three carbons, exhibited a higher reactivity than aliphatic linkages composed of two carbons. These results support hydrogen transfer occurring through an intramolecular process, with the three-carbon linkage providing greater accessibility between the excited-state carbonyl and the tertiary amine.

Investigation of these new derivatives by laser flash photolysis and examination of the photolysis products by mass spectrometry were used to distinguish between the inter and intramolecular hydrogen-transfer processes [14]. The length of the spacer, the concentration of PA, and the viscosity of the resin were examined. Indeed, when the size of the aliphatic chain was appropriately selected, the intramolecular interaction between the ketone and the amine was favored. Additionally, at low concentrations of PA and/or low resin viscosities, the intermolecular interaction was reduced. Whilst at high concentrations of PA and/or high resin viscosities, the intermolecular interaction was enhanced.

Following these pioneering works, the group of Liska has observed a similar behavior with **CQs** substituted with tertiary aromatic amines (Figure 11.5) [15]. By photo-differential scanning calorimetry (Photo-DSC) on an acrylic resin, this group demonstrated that the monocomponent system could exhibit a similar or greater reactivity than that of the equivalent bimolecular system. The influence of the spacer length between the **CQ** and the tertiary amine was also examined. Derivatives of **CQ** connected to the tertiary aromatic amine by one carbon bond furnished better results than their analogs connected by two carbon bonds. The influence of the substitution pattern was also examined. Derivatives of **CQ** connected to a tertiary aromatic amine substituted with two methyl groups were found to be the most reactive. This increased reactivity could be ascribed to reduced steric interactions, allowing for better interaction between the **CQ** and amine components. Furthermore, derivatives of **CQ** connected to the tertiary aromatic amine by

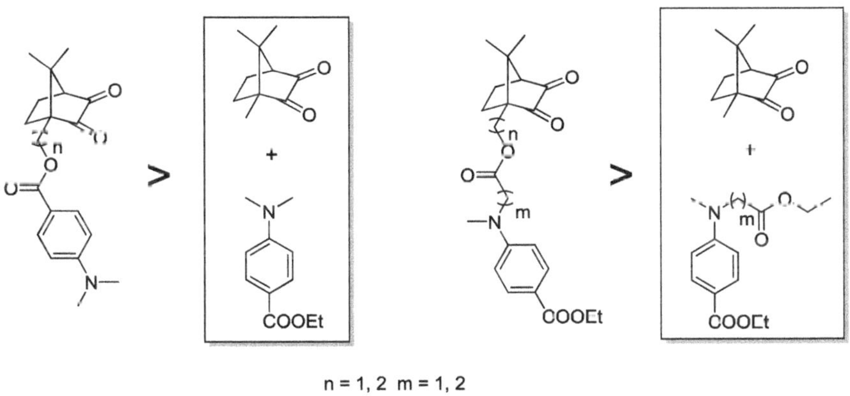

Figure 11.5. Derivatives of **CQ** for type II monocomponent initiators proposed by the group of Liska [15] and their equivalent systems in a bimolecular version.

means of a one- or two-carbon aliphatic spacer exhibited the best reactivity. The increased size of the aliphatic chain between the ester and the tertiary amine was reported as favoring the intramolecular hydrogen transfer (see Figure 11.5).

To prove the formation of radicals, the double bond conversion (DBC) of the monomer, as a function of the concentration of PA, was examined. This study demonstrated that the formation of radicals results from intramolecular hydrogen transfer (Figure 11.6). Indeed,

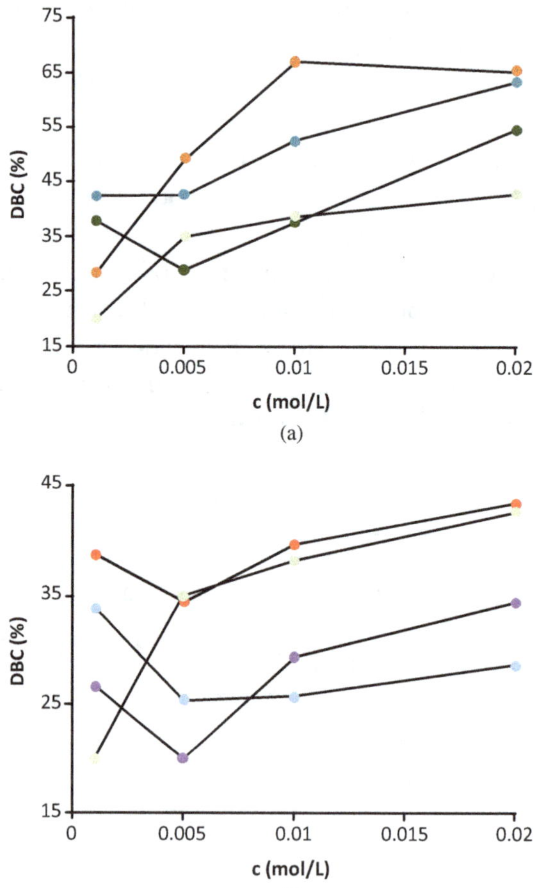

Figure 11.6. DBC of the monomer (lauryl acrylate) as a function of the concentration of PA. (a) (•) $n = 1$ (•) $n = 1$, $m = 1$ (•) $n = 1$, $m = 2$ ( ) **CQ/EDB.** (b) (•) $n = 2$ (•) $n = 2$, $m = 1$ (•) $n = 2$, $m = 2$ ( ) **CQ/EDB.**

the final conversion of the double bond is only slightly affected by a decrease in the concentration of PA when the amine is connected to CQ. Whilst for bicomponent systems, a dramatic decrease of the DBC is observed with decreasing concentration, consistent with a reduced probability for the two partners to react at a low concentration.

Paralleling this work, the group of Yagci extended this class of monocomponent PAs to thioxanthones, functionalized by thiol or carboxylic groups (Figure 11.7) [16, 17]. Thus, the formation of a thiyl radical or an alkyl radical in the α-position to the sulfur atom is possible by means of a type II monocomponent mechanism. The thiyl radical is highly reactive with monomers such as styrenes [18].

2-mercaptothioxanthone (**TX-SH**) and 2-thioxanthone-thioacetic acid (**TX-SCH$_2$COOH**) in monocomponent systems exhibit a higher reactivity than bimolecular systems (thioxanthone/*N*-methyldiethanolamine in acrylic resins). In order to prove the mechanism of radical formation, the excited triplet-state lifetime was measured by LFP. At low concentrations, the excited-state lifetime of **TX-SH** is similar to that of thioxanthone. This similarity indicates that the **TX-SH** PA can form radicals by mean of an intermolecular electron and proton-transfer mechanism (Scheme 11.4). Indeed, the rigidity of the fused aromatic structure impedes interaction between the thiol functionality and the ketone within same molecule.

However, at both high and low concentrations, the excited-state lifetime of **TX-SCH$_2$COOH** were identical demonstrating that, in this specific configuration, an intramolecular interaction could occur [19]. Two mechanisms controlled by the concentration of PA in the resin are proposed. For concentrations higher than 5 mM, radicals are formed though intermolecular processes, whereas for concentrations lower than 5 mM radicals are formed through

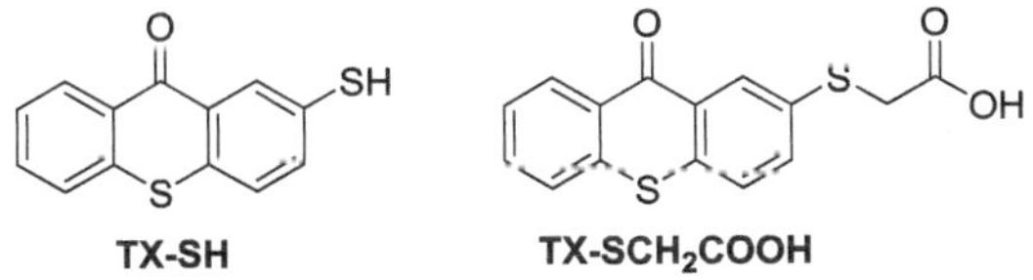

Figure 11.7. Derivatives of thioxanthone for type II monocomponent systems proposed by Yagci.

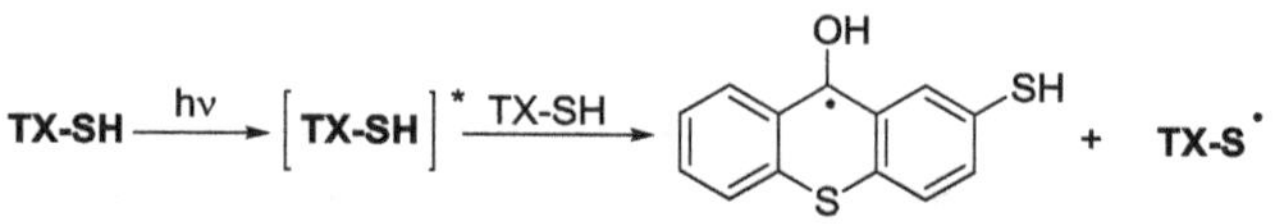

Scheme 11.4. Mechanism of formation of radical for **TX-SH**.

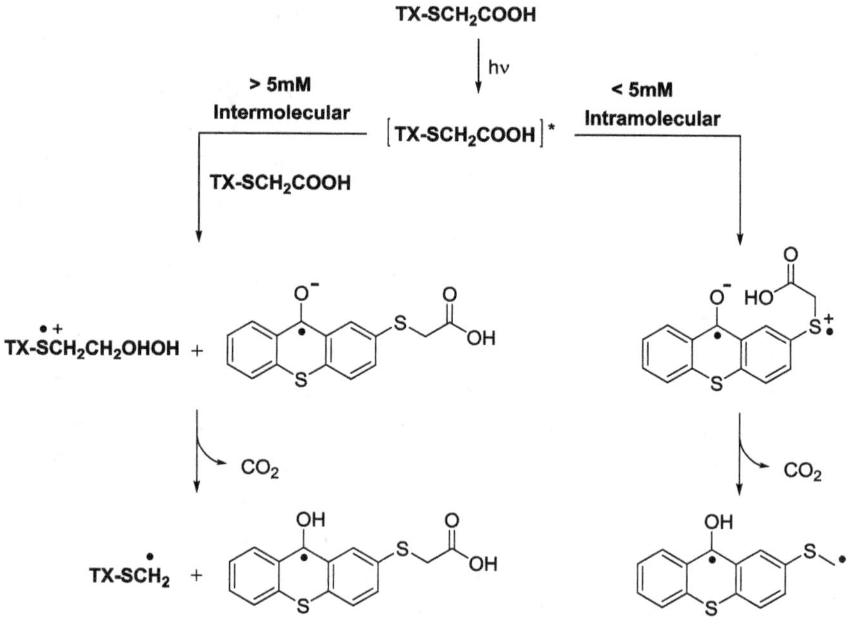

Scheme 11.5. Mechanisms of formation of radicals for **TX-SCH$_2$COOH**.

intramolecular interactions. The release of CO_2 occurs in both cases (Scheme 11.5).

11.3 Photoorganocatalysis with Three-Component Systems

11.3.1 *Photosensitization*

Photosensitization involves the indirect formation of the PA-excited state by means of an energy transfer or an electron-transfer process from the chromophore (photosensitizer) to the PA. This approach provides a pathway for overcoming the poor absorbance properties of the PA (Scheme 11.6) [20, 21]. The photosensitizer, when excited

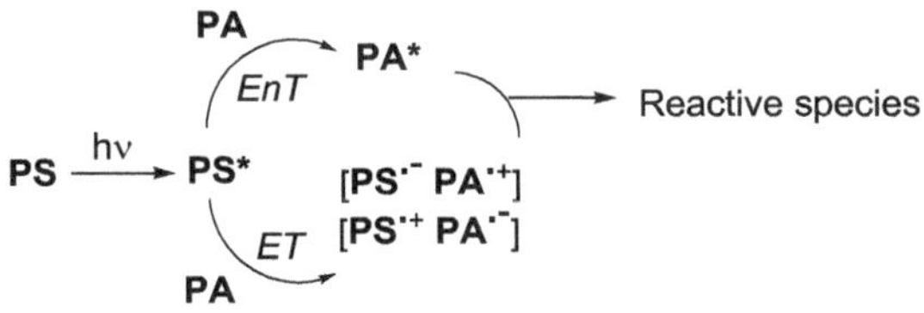

EnT = Energy transfer; *ET* = Electron transfer

Scheme 11.6. Mechanisms of photosensitization.

to its singlet or triplet state, can transfer energy to the PA in its ground state upon physical interaction. The PA is promoted to its excited singlet or triplet state (depending on the photosensitizer), whilst the return of the photosensitizer to the ground state is jointly observed [9]. Once in the appropriate excited state, the PA can form reactive species to initiate polymerization, through the same processes that operate under direct irradiation of the PA. For an efficient photosensitization, it is compulsory that the excited-state energy of the PA is comparable, or lower, to the excited-state energy of the photosensitizer.

By activation *via* electron transfer (ET), an electron is transferred from the photosensitizer to the PA (or *vice versa*) so that the ion pairs [PS$^{\bullet\ \mathrm{ion}}$ PA$^{\bullet\ \mathrm{ion}}$] or [PS$^{\bullet\ \mathrm{ion}}$ PA$^{\bullet}$] ([PS$^{\bullet}$ PA$^{\bullet\ \mathrm{ion}}$]) are formed. The ET is favored when the Gibbs free energy variation (ΔG) is negative. This latter can be calculated using the oxidation potential of the electron donor (E^0_{OX}), the reduction potential of the electron acceptor (E^0_{RED}), the excited-state energy of the photosensitizer (E_T or E_S, for excited-triplet and singlet states, respectively), and the Coulomb stabilization energy associated with the intermediate radical ions pair (C). These values, when introduced into the Rehm–Weller equation, allow for the determination of the driving force of the reaction, and the likelihood for the reaction to occur (Eq. (11.1)) [22]. The Coulomb stabilization energy is generally negligible for polar solvents.

$$\Delta G = E^0_{\mathrm{OX}} - E^0_{\mathrm{RED}} - E_T\ (\text{or } E_S) + C. \qquad (11.1)$$

To date, various families of structures have been developed (e.g. (PS/PA), (PS/Co-initiator), or (PS/PA/Co-initiator)) to initiate

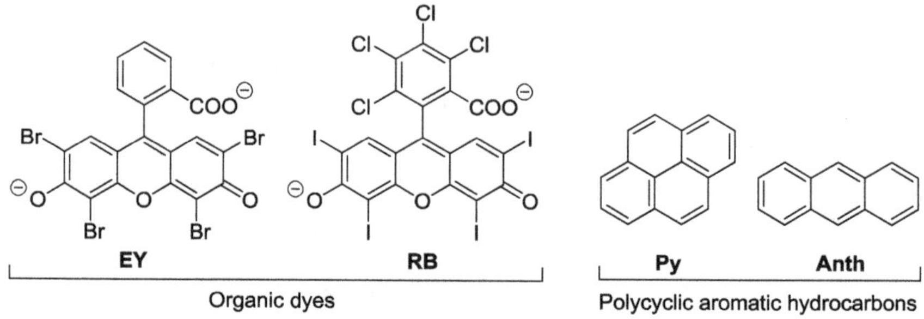

Figure 11.8. Examples of **POCs** used as photosensitizers. Organic dyes: Eosin Y (**EY**) and Rose Bengal (**RB**), **PAHs:** Pyrene (**Py**) and Anthracene (**Anth**).

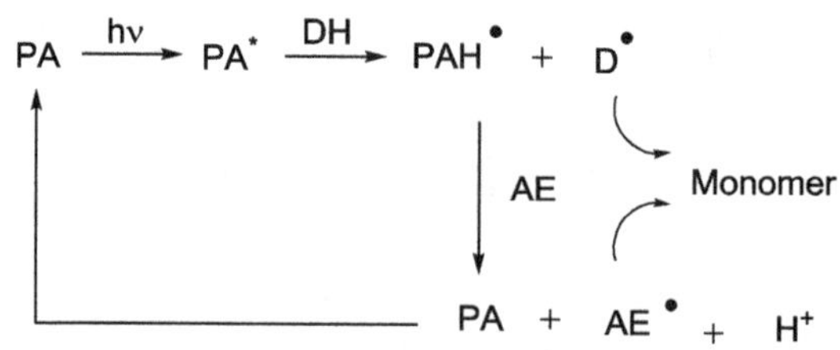

Scheme 11.7. Three-component systems including a type II PA, a co-initiator DH, and an electron acceptor AE.

polymerization [39, 40]. Amongst these families, photoorganocatalysts (**POCs**) such as organic dyes and polycyclic aromatic hydrocarbons (**PAHs**) are common (Figure 11.8) [23–28].

11.3.2 *Three-component systems*

The use of three-component PA systems is constantly expanding as it constitutes an elegant method to increase the efficacy of photoinitiating systems. The introduction of an electron acceptor (AE), typically an iodonium salt or halogenated derivative (e.g. CBr_4), in type II photoinitiating systems (e.g. PA/DH) allows for the consumption of radicals ($PAH^•$) incapable of initiating polymerization, to generate more reactive species ($AE^•$), whilst simultaneously regenerating the PA. More precisely, the electron acceptor limits the probability for the irreversible termination of growing polymer chains by the unreactive $PAH^•$ radical (Scheme 11.7) [29–31].

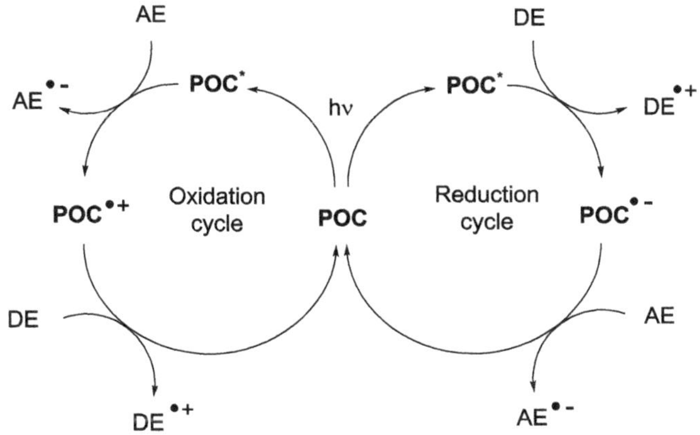

Scheme 11.8. Three-component systems including a **POC**, a DE, and an AE.

Following the first generation of three-component photoinitiating system, new systems comprising a photosensitizer, an electron donor (DE) (typically an amine or a silane derivative), and an electron acceptor proved to be highly efficient under visible-light irradiation [32, 33]. These systems can also be catalytic, enabling the photosensitizer to act as a **POC** (Scheme 11.8) [34]. The latter is characterized by a high redox potential in its excited state, which can react with an oxidant or reductant to form various reactive species. Indeed, radical and cationic species can be generated by means of oxidative or reductive processes, enabling the polymerization of various acrylates, vinylic derivatives, or epoxides.

11.4 From UV to the Visible Range by Structural Modification

The search for new photoinitiating systems to polymerize under visible light with irradiation sources such as LEDs, halogen lamps, or laser diodes is a highly active area of research, both from academic and industrial perspectives. These setups are more ecological and consume less energy compared to conventional mercury lamps. Visible light is also safer for the user than higher energy UV irradiation. In this section, the different strategies developed to adapt conventional UV PAs to visible-light irradiation are discussed.

11.4.1 *Type I and II PAs*

The modification of conventional UV PAs to confer absorbance in the visible range has been the focus of numerous studies. Typically, the most straightforward strategy consists of the introduction of electron-donating auxochromes (e.g. methoxy, methylthio, dimethylamino groups) within the PA framework [35–37]. Through this approach, a bathochromic shift in the absorption maximum (λ_{max}) can be observed. Nevertheless, this strategy is restricted to moderate bathochromic shifts and the development of visible-light PAs requires the investigation of new structures. The covalent linkage of an additional chromophore and/or the expansion of the conjugated system to lower the bandgap and shift the absorption to longer wavelengths have been widely investigated. These two strategies will be detailed in this part.

Among the PAs that readily form radical species, type II PAs such as **BP** and **ITX**, and type I PAs such as 2,2-dimethoxy-2-phenylacetophenone (**DMPA**) have been substituted with **Py** or covalently linked to the truxene scaffold (**Tr**) [38–40]. In the case of **BP** or **DMPA**, these modifications drastically increased the molar extinction coefficients as well as shifted the absorbance maximum by approximately 10 nm. Interestingly, the modification of the **ITX** core did not cause a bathochromic shift, but drastically increased the molar extinction coefficient (Table 11.1).

Although bathochromic shift is not observed for each molecule, a significant increase in the molar extinction coefficient is demonstrated, as illustrated in the figure below (Figure 11.9).

Theoretical calculations have been used to examine the influence of substitution on the BDE of PAs. Notably, the excited-state energies and BDE for homolytic cleavage of the C–C bonds of **DMPA** derivatives were examined and are provided below (Table 11.2).

A decrease in the excited triplet energy levels is observed with increasing conjugation. Where **DMPA** is substituted with **Py**, homolysis of the C–C bond is unfavorable as the BDE exceeds the excited triplet energy level of the PA. Cleavage of the C–C bond can only occur from the higher-energy excited singlet state (71.5 kcal mol^{-1}). Interestingly, incorporation of the **Py** and truxene chromophores almost had no impact on the PA BDE. Indeed, due to the

Table 11.1. Optical properties of PAs before/after modification.

	Type II PA		Type I PA
	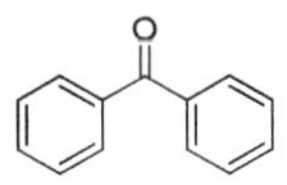 **BP**	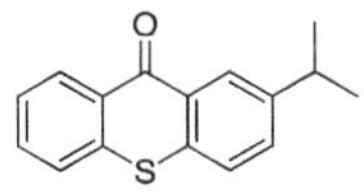 **ITX**	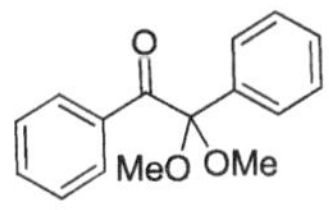 **DMPA**
λ_{max} (nm)	335	383	332
ε_{max} (M^{-1} cm^{-1})	150	4,700	400
R = 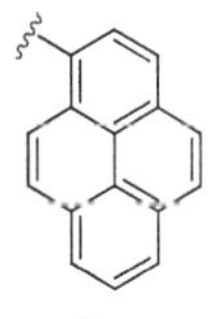 Py	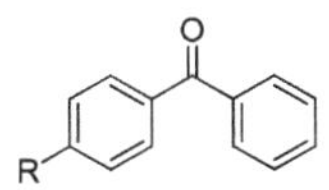 **Py-BP**	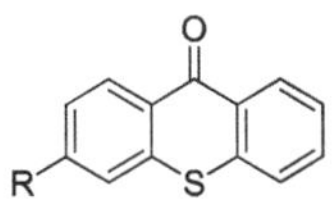 **Py-TX**	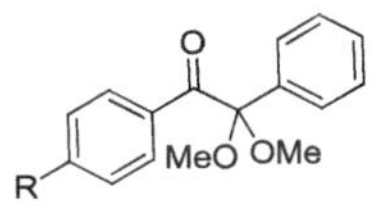 **Py-DMPA**
λ_{max} (nm)	348	357	345
ε_{max} (M^{-1} cm^{-1})	26,000	32,000	35,000
R = 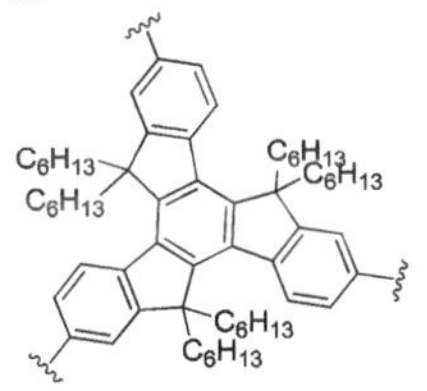 Tr	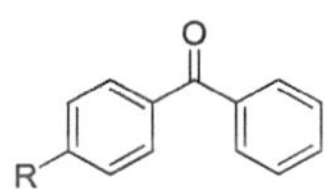 **Tr-BP**	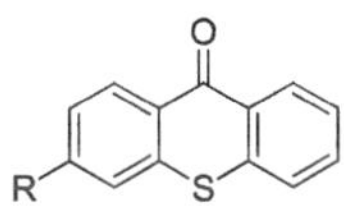 **Tr-TX**	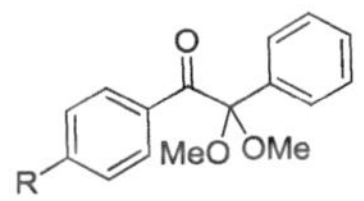 **Tr-DMPA**
λ_{max} (nm)	345	353	338
ε_{max} (M^{-1} cm^{-1})	105,000	123,000	63,000

orthogonality between the **Py** and truxene chromophores and central **DMPA**, no electronic delocalization can occur to lower the BDE.

11.4.2 *PAs derived from the iodonium salt*

To date, only two iodonium salts substituted with chromophores have been reported in the literature, one being a 9-fluorenone

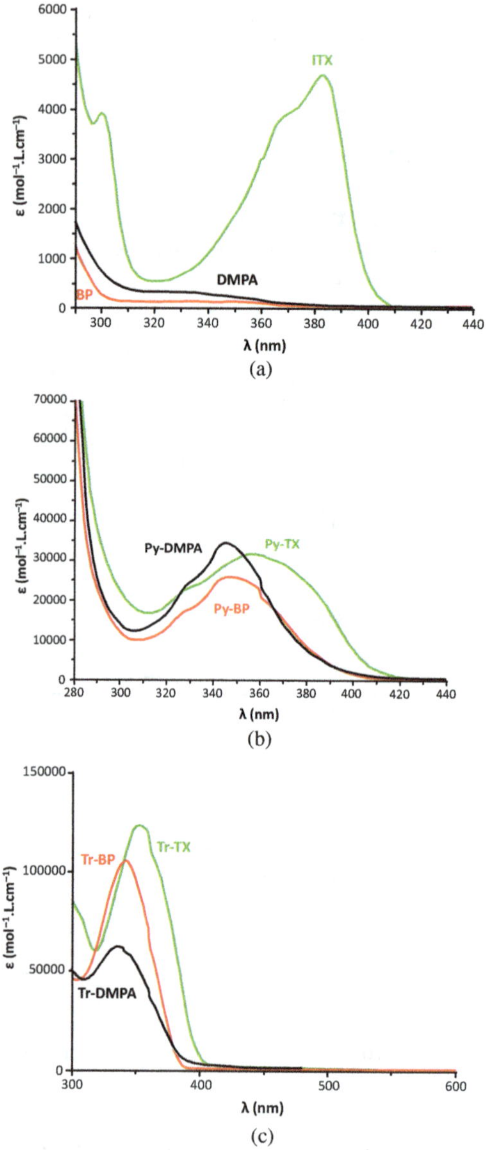

Figure 11.9. (a) Absorption spectra of **BP**, **ITX**, and **DMPA**. (b) Absorption spectra of **Py-BP**, **Py-TX**, and **Py-DMPA.** (c) Absorption spectra of **Tr-BP**, **Tr-TX**, and **Tr-DMPA**.

Table 11.2. BDE and energy level of the triplet excited state (i.e. E_T) determined by theoretical calculations.

	DMPA	Py-DMPA	Tr-DMPA
BDE (kcal mol^{-1})	56	53.1	54.5
E_T (kcal mol^{-1})	68.2	46.3	57.5

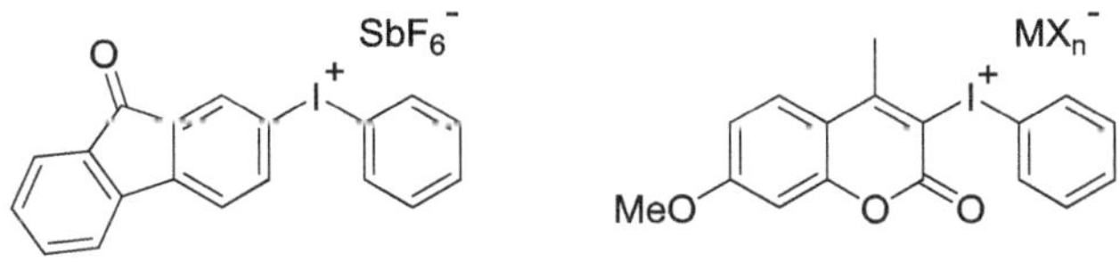

Figure 11.10. Derivatives of iodonium salts comprising as chromophores, a fluorenone (i.e. **Fl-Sb**) or a coumarin (i.e. **P3C-P** and **P3C-Sb**).

derivative (**Fl**) and the second being a derivative of coumarin (**P3C**) (Figure 11.10) [41, 42].

Compared to the parent iodonium salt, the substitution with fluorenone and coumarin chromophores caused a clear bathochromic shift of the absorbance maxima (Figure 11.11). The replacement of a phenyl ring, from the parent iodonium salt, with the fluorenone moiety led to a two-fold increase in the molar extinction coefficients at 267, 298, 380, and 394 nm. This effect was further enhanced for the coumarin derivative. The influence of the counterion was also examined and minor differences were detected in the UV–visible absorbance spectra. Thus, better absorption properties were evidenced for the hexafluoroantimonate (SbF_6^-) salt compared to the hexafluorophosphate (PF_6^-) salt. Due to the lower nucleophilicity, the hexafluoroantimonate counterion less efficiently stabilizes the positive charge on the iodine atom, improving the reactivity for photopolymerization [43]. The introduction of substituents on the parent iodonium salt also influenced the BDE of C–I bond. Thus,

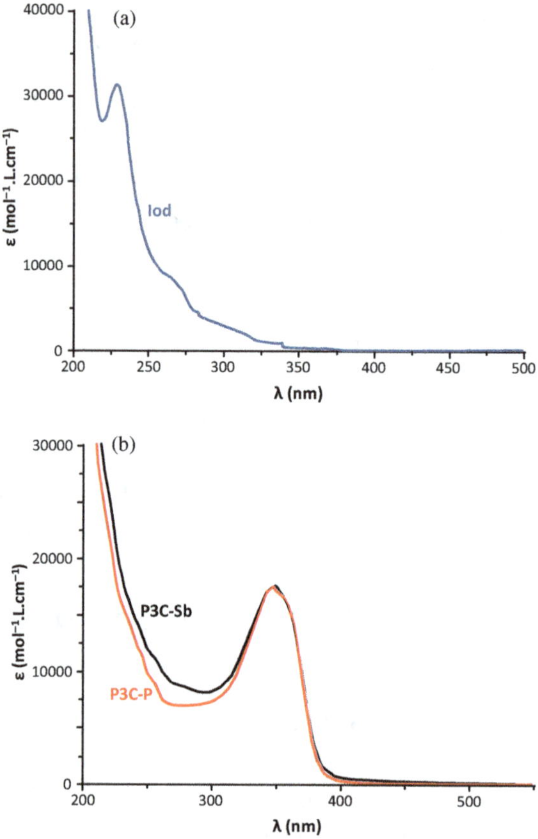

λ_{max} (nm)	ε_{max} (M^{-1} cm^{-1})
Fl-Sb	**Fl-Sb**
267	42,000
298	8500
380	430
394	400

Figure 11.11. (a) Absorption spectrum of the parent iodonium salt (Iod). (b) Absorption spectra of **P3C-P** and **P3C-Sb**. Table: ε_{max} and λ_{max} for **Fl-Sb.**

for the coumarin-substituted iodonium salt, a BDE of 45.66 kcal mol^{-1} for the Ph–I bond and 61.37 kcal mol^{-1} for the Coumarin–I bond were determined by theoretical calculations, indicating that cleavage of the bond is favored on the benzene ring side.

11.4.3 *Development of PS/PC*

The development of photosensitizers and photoredox catalysts (PCs), as an alternative to improve the absorption properties of conventional PAs, proved to be a more fruitful strategy to develop PAs with strong absorbances in the visible light region. Indeed, chemical modification of conventional PAs can be synthetically demanding. In this context, a great deal of interest has been devoted to the development of push–pull chromophores, these compounds being well-known to exhibit a strong absorption in the visible range. Amongst visible-light PAs, some proved to be useful PCs. This aspect is highly desired in order to decrease the PA content within photocurable resins, and also to drastically improve the photoinitiating ability.

11.4.4 *Modification of PAHs*

Examples of PAHs in the literature are numerous (e.g. **Py**), as these structures can easily initiate a photopolymerization [44–46]. Through standard functional group modification of their scaffolds, highly conjugated compounds displaying remarkable molar extinction coefficients in the visible range are accessible. As an example, the symmetrically substituted linear polyacenes reported in Figure 11.12

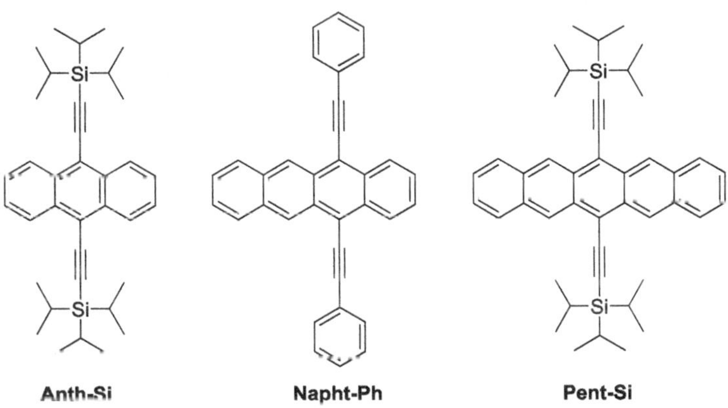

Anth-Si **Napht-Ph** **Pent-Si**

Figure 11.12. Derivatives of PAHs: 9,10-bis[(triisopropylsilyl)ethynyl]anthracene (**Anth-Si**), 5,12-bis(phenylethynyl)naphthacene (**Napht-Ph**), and 6,13-bis(triisopropylsilyl ethynyl)pentacene (**Pent-Si**).

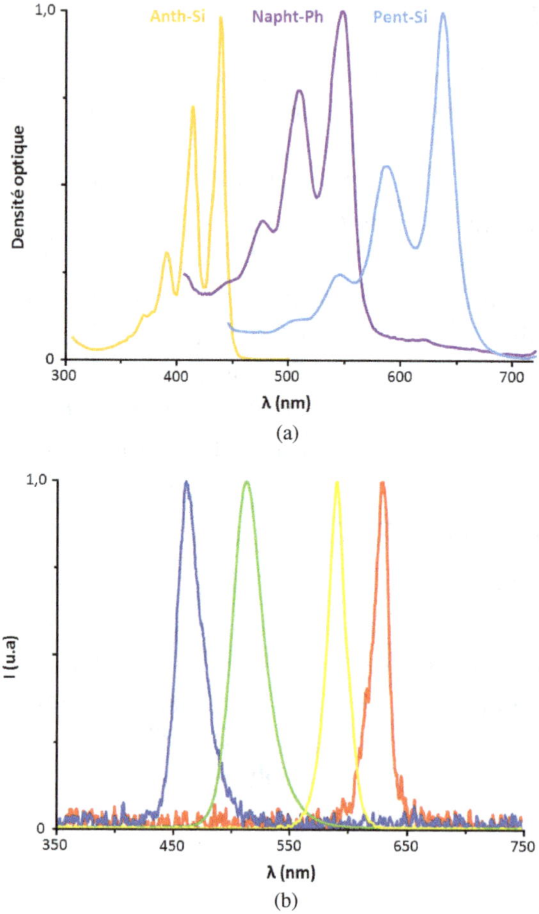

Figure 11.13. (a) UV–visible absorption spectra of **Anth-Si**, **Napht-Ph**, and **Pent-Si**. (b) Emission spectra of blue, green, yellow, and red-emitting LEDs.

efficiently absorb in the 400–700 nm range [47, 48]. Extensively used in organic electronics, these derivatives of the well-known acene chromophore, with enhanced solubilization properties, displayed interesting photoinitiating properties [49–51].

In particular, a bathochromic shift in the absorbance maxima is readily observed with increasing conjugation of the central core. As a result, candidates suitable for an irradiation with visible light LEDs were obtained (Figure 11.13).

These acene derivatives proved to be efficient in three-component systems, both in oxidative of reductive catalytic cycles. For example, in an oxidative cycle in the presence of an iodonium salt and a silane derivative R_3SiH, the formation of highly reactive R_3Si^+ cations, responsible for the polymerization of epoxides, could be evidenced. Pentacenes can also be used as **POC** in the presence of a tertiary amine and alkyl halides in a reductive catalytic cycle, i.e. the generation of alkyl radicals capable of initiating the free radical polymerization of acrylates were observed.

11.4.5 *"Push–pull" compounds*

"Push–pull" chromophores are compounds comprising of an electron-donating substituent connected to an electron-accepting substituent *via* a π-conjugated spacer. Due to this specific arrangement, compounds exhibiting a strong absorption in the visible range can be prepared. For all "push–pull" chromophores, a characteristic absorption band can be detected which corresponds to the transfer of an electron from the HOMO to the LUMO level. This band, termed the intramolecular charge transfer (ICT) band, is thus a characteristic of the chromophore and, from the position of the ICT band, the HOMO–LUMO gap of the chromophore can be determined [52]. The use of "push–pull" chromophores as photosensitizers has been developed by the group of Lalevée and Gigmes who have described "push–pull" derivatives of barbituric acid (Figure 11.14) [53].

The absorption maxima are observed at around 440 nm for the barbituric derivative and at around 470 nm for the thiobarbituric derivative. Due to the improved electron-accepting ability of the thiobarbituric acid derivative, relative to that of the barbituric derivative, a clear increase in the absorbance maxima could be detected. Importantly, the extension of the absorbance maxima for the thiobarbituric derivative facilitated the use of a laser diode emitting at 473 nm. In the presence of an iodonium salt (as an AE) and *N*-vinylcarbazole (i.e. **NVK**), as the electron donor, the formation of $\mathbf{NVK}^+$ cations responsible for the polymerization of epoxides could be observed. Furthermore, in the presence of a tertiary amine (as a

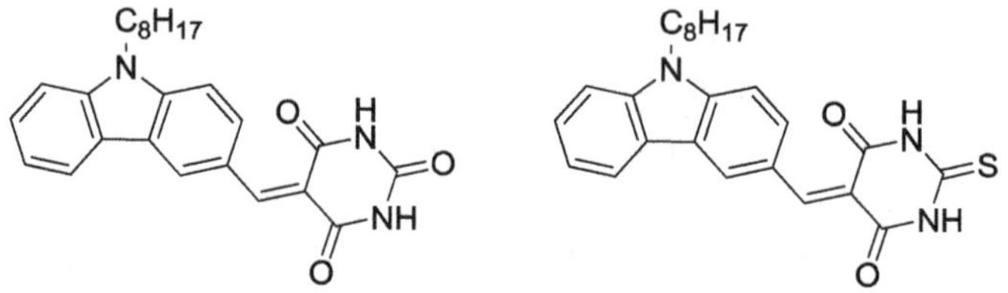

Figure 11.14. First "push–pull" compounds used as photosensitizers by the group of Lalevée and Gigmes [53].

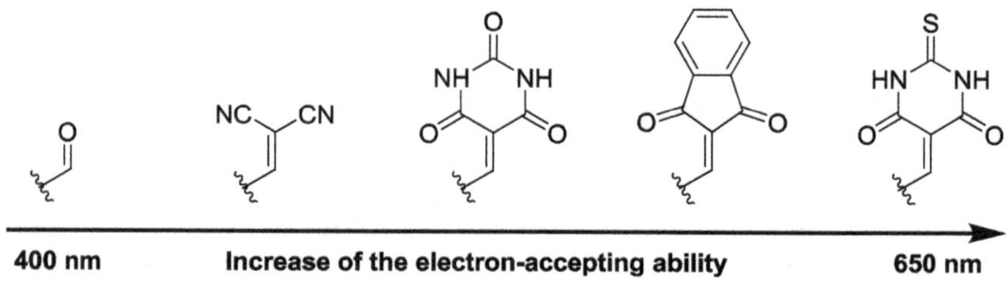

Figure 11.15. "Push–pull" compounds derived from Michler's ketone.

DE) and an alkyl halide (AE), the alkyl radicals formed could initiate the polymerization of acrylates. Conversion yields of 50% were obtained with the two systems. The role of **POCs** was also evidenced as the photosensitizer was regenerated.

Following this initial study, numerous "push–pull" chromophores derived from Michler's ketone were investigated [54]. Within the Michler's ketone scaffold, aldehyde, dicyanovinyl, barbituric, indanedione, and thiobarbituric groups were introduced as the electron acceptors (Figure 11.15). Through modification of the electron-accepting ability, dyes absorbing between 400 nm and 650 nm were obtained.

11.5 The Golden Rules of Photoredox Catalysis for the Development of New PCs

In an attempt to find highly sensitive PAs, a strategy is followed based on using a catalytic cycle, and the PA is a PC. Thus, PC can initiate radical and cationic polymerizations and is capable of

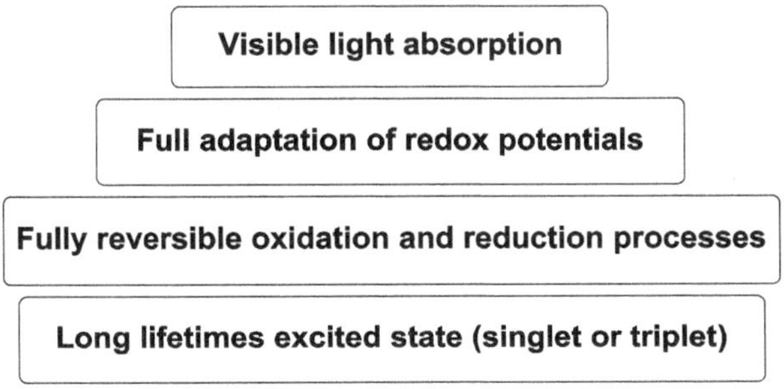

Figure 11.16. Criteria to respect to obtain an efficient PC.

being regenerated during polymerization reactions. Thus, low concentrations of PC are sufficient to initiate the polymerization. This, in turn, reduces the toxicity of the formulation and maintains an excellent efficiency over the irradiation period due to the regeneration of PC.

In general, a group of golden rules should be obeyed to obtain an efficient PC (Figure 11.16). Initially, the PC should absorb visible light for its absorption spectra to overlap with the emission spectra of the utilized visible LEDs. The PC redox potentials must be adapted for efficient interaction with the additives in the oxidative or reductive cycles. Furthermore, the PC oxidation or reduction processes must be fully reversible to ensure a precise catalytic cycle. Additionally, the PC should possess long excited-state lifetimes (1st singlet S_1 or 1st triplet T_1) to enhance the efficiency of interaction with the additives. In this context, highly fluorescent materials are good candidates as these chromophores typically possess long excited-state lifetimes. Moreover, recent developments in the area of fluorescent materials have revealed the photophysical concept of thermally activated delayed fluorescence (TADF) [55]. A short S_1–T_1 energy gap marks TADF compounds such that the S_1 state can be thermally repopulated (at room temperature) from the triplet state (T_1) by means of a reverse intersystem crossing (RISC). In concept, this behavior is promising for PAs and PCs due to the extended excited-state lifetimes accessible, allowing for more efficient reactions from the excited singlet state.

11.6 POCs Used in Photopolymerization

Photocatalysts based on transition-metal complexes are appealing candidates for photopolymerization due to the high conversion rate they can achieve. As drawbacks, these compounds can be highly toxic and rather expensive, so that low-cost alternatives have been actively researched [56, 57]. In this field, a great deal of interest has been focused on **POCs** due to their ease of synthesis, their low cost, and low toxicity. At present, **POCs** are mostly organic dyes or polyaromatic compounds. Organic dyes have been extensively studied in three-component polymerization systems [30], but photocurable resins prepared with these compounds suffer from a low stability under storage [58], and a severe degradation is also observed during polymerization [59] so that their catalytic role can be clearly discussed. In this field, the methylene blue (**MB**$^+$)/amine/iodonium salt combination is efficient to initiate the generation of phenyl radicals for the free radical polymerization of (meth) acrylates by reaction of methylene blue (**MB**) in the excited state with an amine (Figure 11.17) [60, 61]. The polymerization of

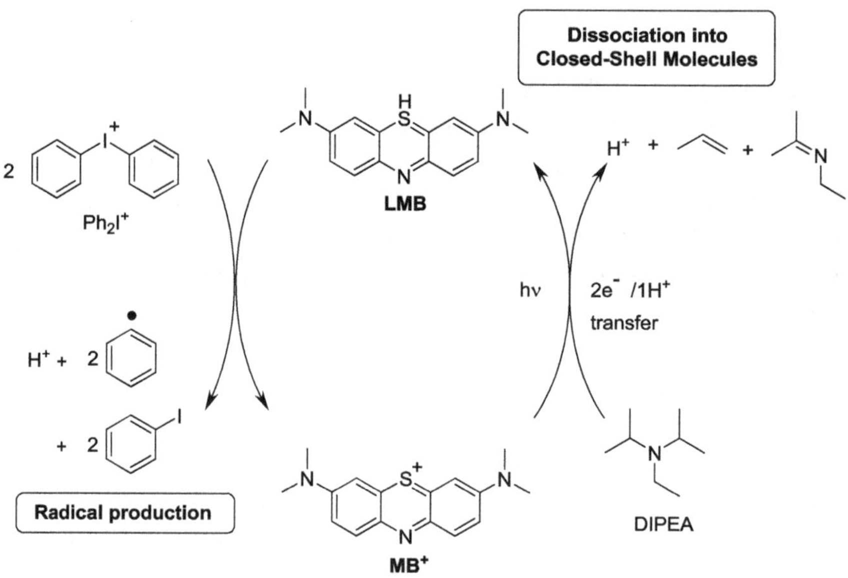

Figure 11.17. Reaction mechanism for the **MB**$^+$/DIPEA/Ph_2I^+ system.

(meth)acrylates could be carried out through a photoredox catalytic mechanism with this three-component system [62].

Precisely, **MB** could react with a sacrificial amine, i.e. *N,N*-diisopropylethylamine (DIPEA) and the iodonium salt, enabling to create a visible-light three-component system that could function *via* a two electron/one proton $2e^{-}/H^{+}$ transfer mechanism. In fact, this is the first report mentioning such a mechanism. Interestingly, the polymerization could continue even in the absence of light, and this specificity is of crucial interest to allow the polymerization in areas that are not exposed to light. More recently, polyaromatic compounds, such as **Py** [45], carbazoles and phenothiazines [63], naphthacene [47], bis[(triisopropyl)silyl]anthracene [48], bis-[(triisopropyl)silyl]pentacene [48], truxene derivatives [44, 64], and triazine derivatives [46] have been examined as **POCs** for photopolymerization. As evidenced in the Figures 11.18–11.21, a wide range of structures have been examined over the years [65–68].

With these molecules, the catalysis occurred by means of an oxidative cycle involving a three-component system **POCs**/iodonium salt/silane (R_3SiH). In this cycle, R_3Si^+ cations are formed that are capable to initiate the ring-opening polymerization of epoxides [48]. Parallel to this, by replacing the silane by an alkyl halide, the resulting three-component system could operate *via* a reductive cycle, so that radicals capable of initiating the free radical polymerization of acrylates were formed [60]. The absorption maxima of polyaromatic structures can be easily controlled by the length of the conjugation, enabling to cover the visible range [69]. Polyaromatics can also operate *via* an oxidative cycle, and representative examples of this are truxene–acridinediones **18, 19** (Figure 11.16) [69].

11.7 Controlled Radical Photopolymerization

Since polymer research is constantly evolving, both in importance and complexity, synthetic procedures that allow for a more efficient access to precise molecular weights and architectures are actively researched. In recent years, numerous controlled photopolymerization techniques have been developed [69]. These techniques that are derived from their thermal analogs provide unprecedented

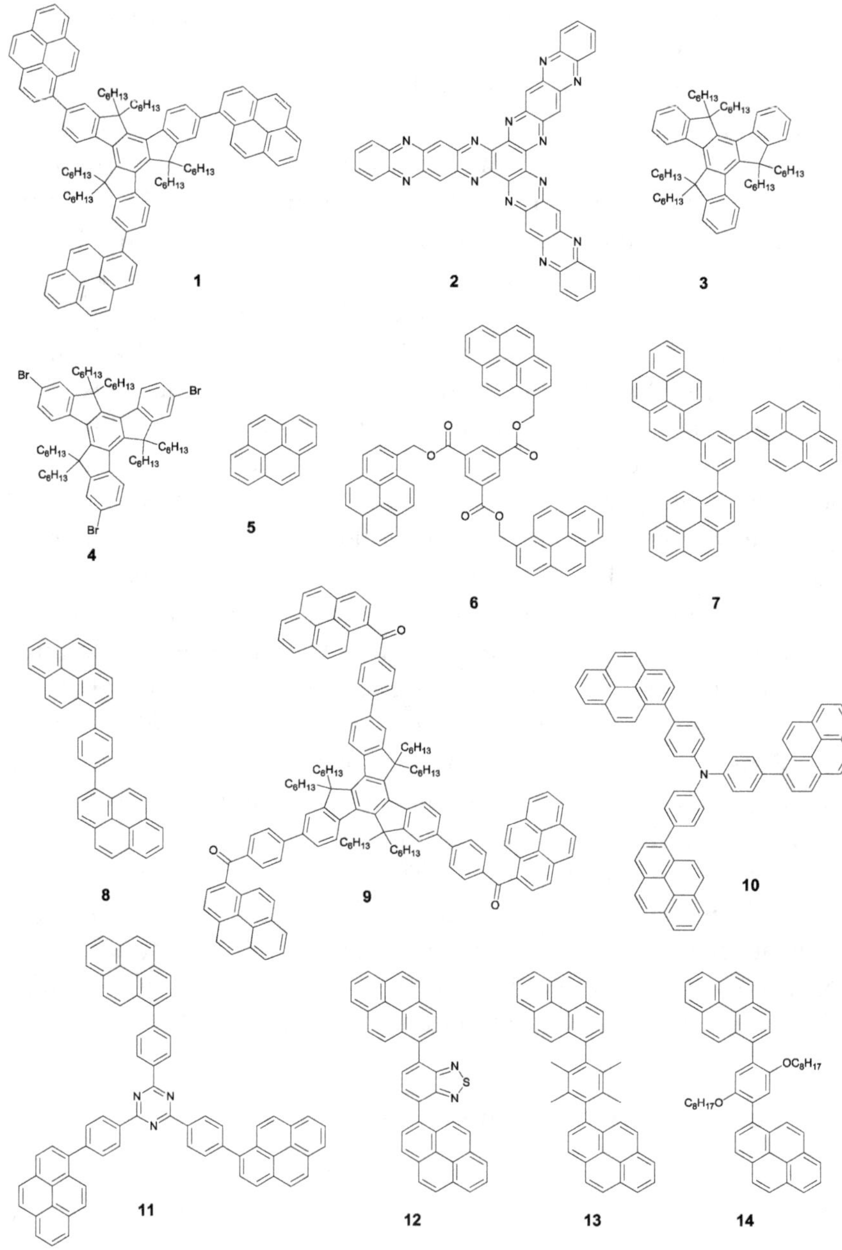

Figure 11.18. POCs 1–14 used in photopolymerization reactions.

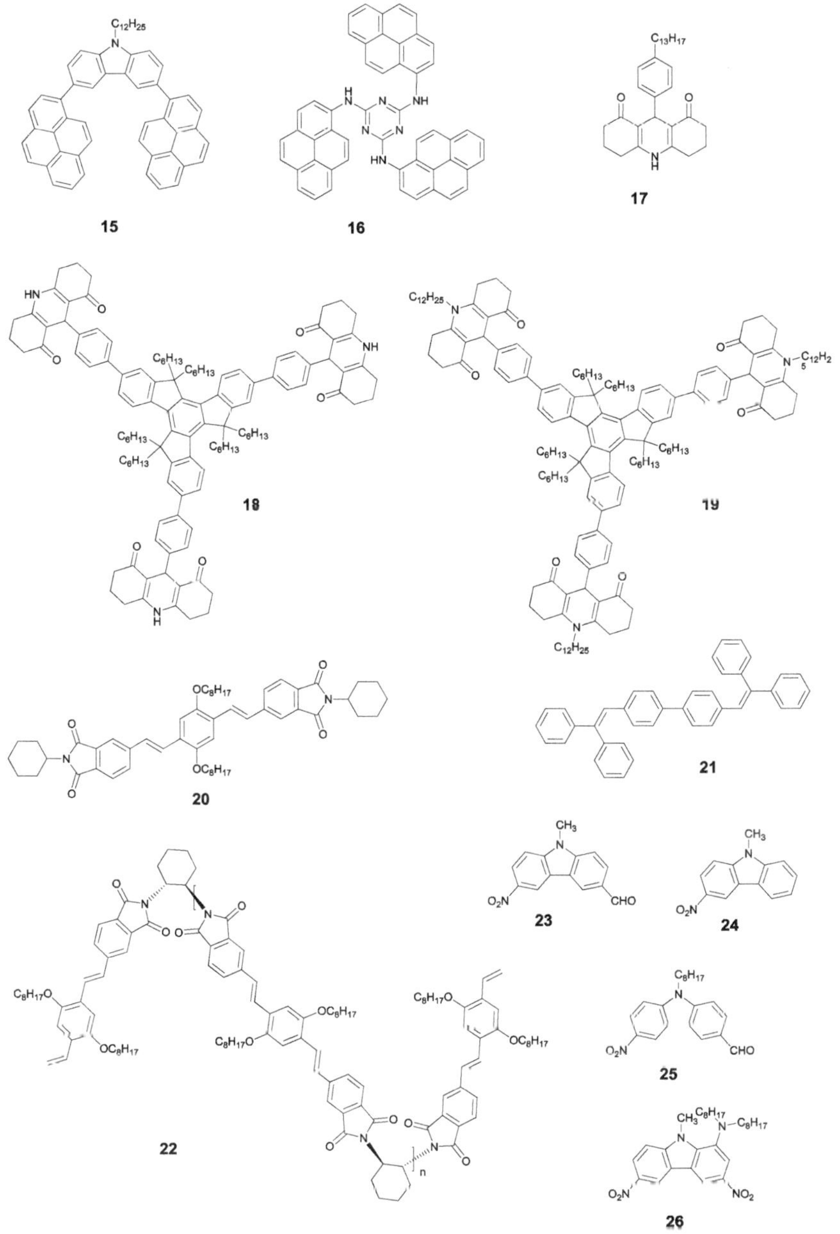

Figure 11.19. POCs 15–26 used in photopolymerization reactions.

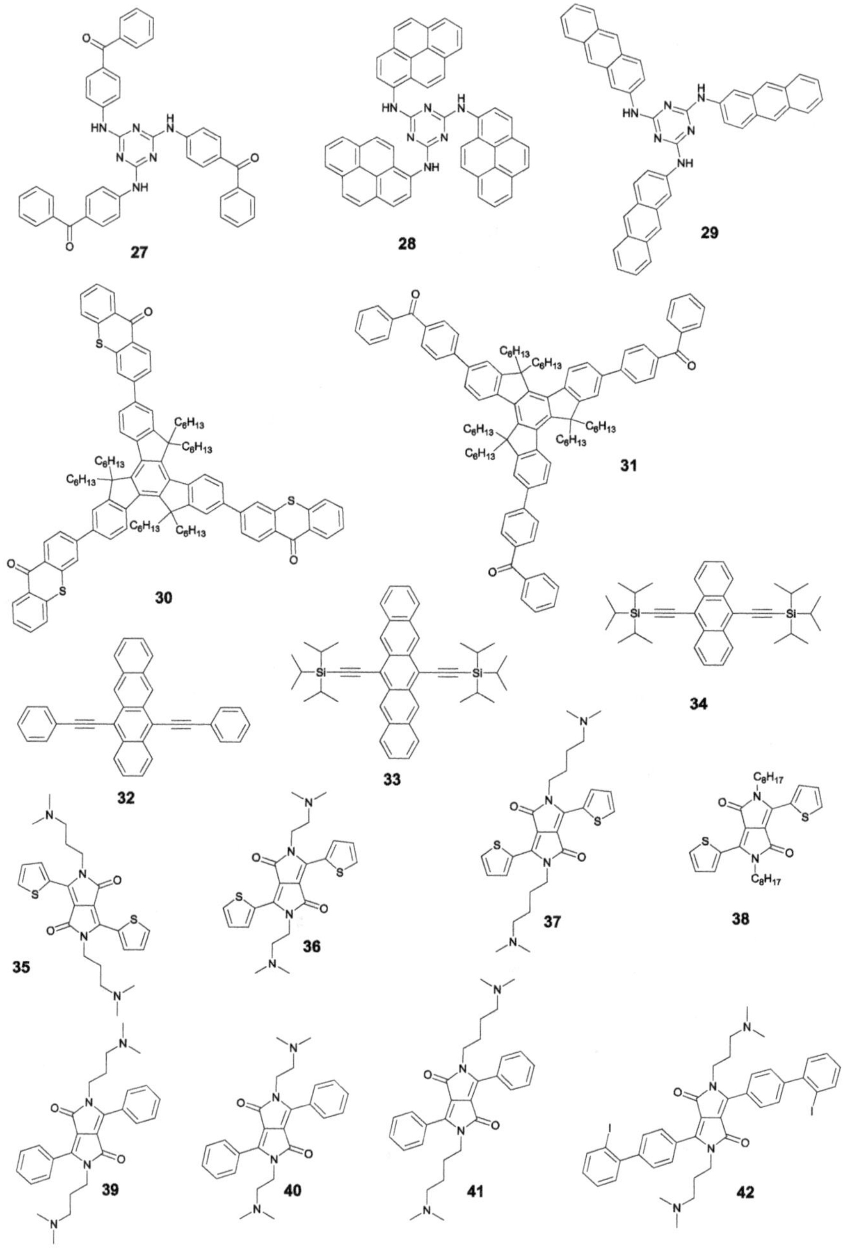

Figure 11.20. POCs 27–42 used in polymerization reactions.

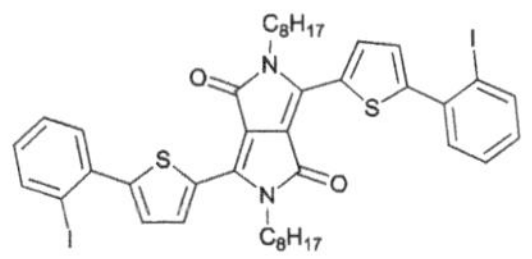

43

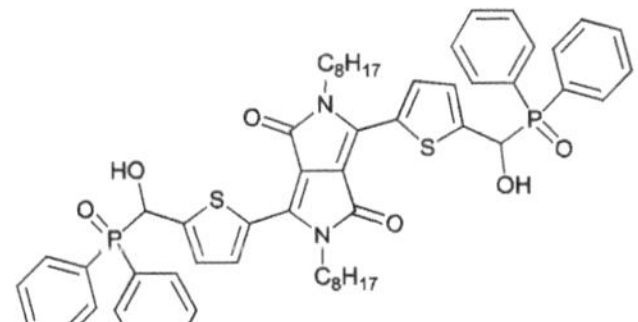

44

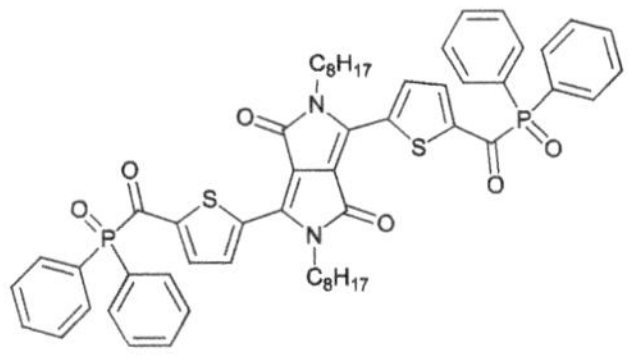

45

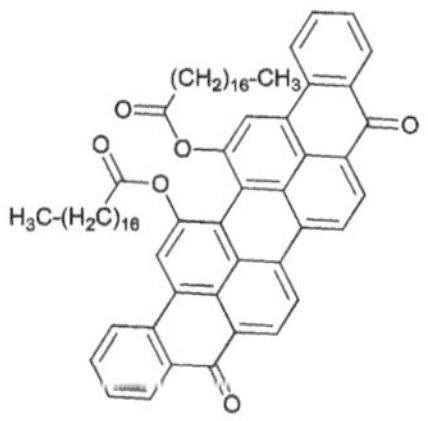

46

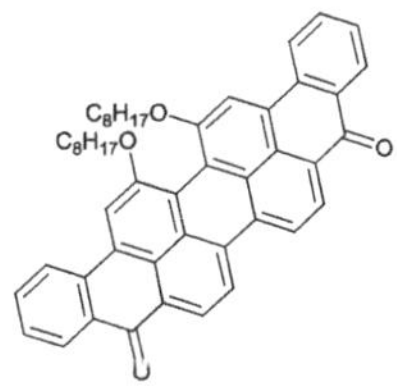

47

Figure 11.21. POCs 43–47 used in polymerization reactions.

opportunities to construct polymers with precise properties. Typically, the use of light to mediate radical photopolymerizations can occur through two different processes: intramolecular photochemical processes and/or photoredox processes. While different photocontrolled radical polymerization techniques have been proposed, this section describes the main developments obtained during these last years in photoorganocatalyzed-controlled radical polymerization i.e. photoreversible addition fragmentation chain transfer (photo-RAFT) and photoatom transfer radical polymerization (photo-ATRP).

11.7.1 *RAFT*

RAFT polymerizations can utilize thiocarbonylthio compounds as chain-transfer agents to mediate the polymerization of conventional radical polymerization systems [70, 71]. Polymerization is initiated by a free radical source which, either by thermolysis or photolysis, generates two radical species. The propagation of these radical species leads to growing polymeric chains. Control over this process is obtained by the reversible deactivation of the polymeric chains through interaction with the RAFT agent, allowing for the

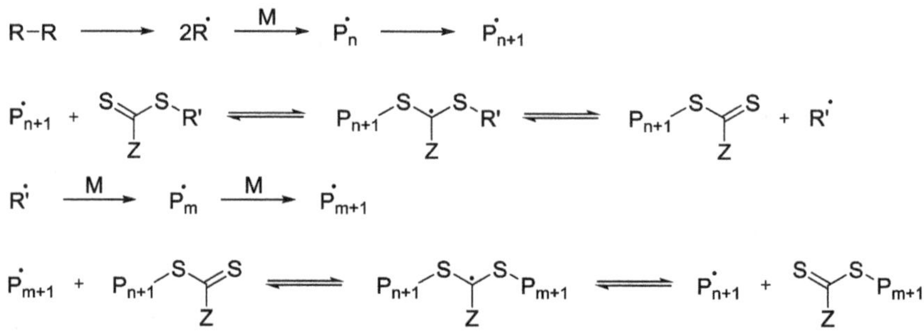

Scheme 11.9. General RAFT polymerization mechanism.

controlled addition of monomers to a growing polymeric chain (Scheme 11.9) [70, 71].

RAFT photopolymerization has been achieved through the combination of RAFT agents with type I and type II PAs. Photochemical generation of radical species derived from type I or type II processes initiates polymerization while the RAFT agent independently mediates the polymerization [72–80]. Whilst this process allows for a controlled polymerization, the external radical source leads to enhanced chain termination processes.

More recently, a powerful addition to RAFT photopolymerization has been presented [81]. Photoinduced electron transfer (PET) has been employed to facilitate the efficient photopolymerization of monomers over a broad range of molecular weights (1,000 to 2,000,000 g mol^{-1}) with narrow molecular weight distributions ($Mw/Mn < 1.3$) [86]. This approach negates the need of external radical sources as PET [82] and facilitates cleavage of the RAFT agent to generate the initiating radical species [86]. PET was shown to be remarkably efficient, requiring ppm levels of photosensitizer relative to the monomer, under low-energy visible-light irradiation [86]. Additionally, whilst radical polymerizations are sensitive to the presence of molecular oxygen, necessitating the exclusion of molecular oxygen from a polymerization medium, PET processes were found to consume molecular oxygen, thereby avoiding the need to exclude molecular oxygen from the polymerization medium (Scheme 11.10) [86].

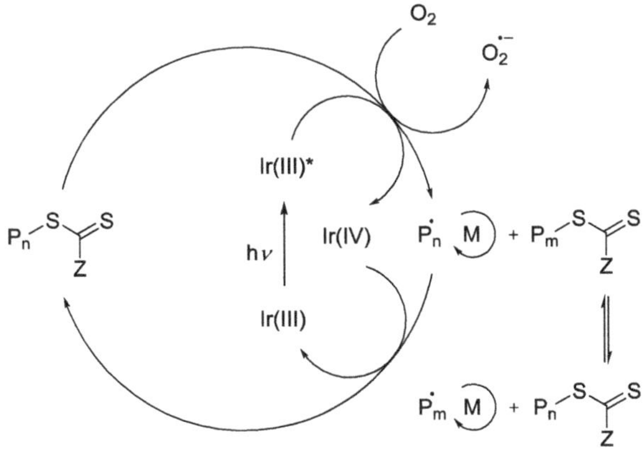

Scheme 11.10. Photosensitized RAFT polymerization mechanism.

The versatility of this approach has been demonstrated by the range of PCs employed. Organic dyes such as fluorescein, rhodamine, nile red, and **MB** have been employed as **POCs** for PET-RAFT [83] in place of the more expensive iridium complexes. Furthermore, when **EY** was employed, a similar insensitivity to molecular oxygen was observed in the presence of triethylamine as a sacrificial electron donor.

More recently, porphyrins such as chlorophyll a have been likewise employed [84]. The use of chlorophyll facilitated the polymerization of a range of monomers under low-energy excitation wavelength (635 nm). The excitation energy was further reduced with the bio-photocatalyst bacteriochlorophyll a (**Bachl a**). **Bachl a** realized the first controlled radical polymerization under far-red (780 nm) and near-IR (850 nm) irradiation [85]. Photopolymerizations performed under near-IR irradiation are particularly advantageous as wavelengths in this range avoid internal filter effects, allowing for deep penetration within the polymerization medium.

The PET-RAFT polymerization of methacrylates under UV irradiation has been also performed by means of a previously prepared solid-state photocatalyst. The photocatalyst was obtained by grafting polymer chains containing an amine i.e. poly-2-(dimethylamino)-ethyl methacrylate using surface-initiated ATRP technique [86].

11.7.2 *ATRP*

ATRP utilizes transition-metal-based catalysts to temporarily activate/deactivate growing polymeric chains, thereby affording control over the polymerization (Scheme 11.11) [87].

For a controlled polymerization, there are some restrictions concerning the transition metal catalyst. This catalyst must possess two accessible oxidation states separated by one electron, an affinity for halogens, and an expandable coordination sphere when oxidized, so as to accommodate the halogen [87]. Furthermore, for the polymerization of a desired monomer, consideration must be given to the choice of catalyst, initiator, solvent, and temperature. Each variable can have a large impact on the properties of the resulting polymers [87].

With consideration to the above features, the polymerization of a range of monomers can be obtained with high control over the molecular weight distributions. Furthermore, as the polymerization proceeds through reversible activation/deactivation processes, a range of complex polymer architectures can be obtained [87].

The effect of light irradiation on ATRP processes has been examined. Two distinct mechanisms of photo-ATRP have emerged: photoinitiated and photoredox photo-ATRP [88]. While both mechanisms maintain the fundamental ATRP control mechanism, the photochemical processes show different mechanisms (Scheme 11.12).

Photoinitiated photo-ATRP proceeds through photochemical decomposition of a PA to generate two radical species through either type I or type II processes [88]. The generated radical species reduce Cu(II) to the Cu(I) polymerization activator. Equilibrium between active and dormant species proceeds through continued regeneration of the Cu(I) activator by the photochemically derived radical species. In this case, the PA acts as an initiator for continuous activator regeneration (ICAR-ATRP) [88].

Scheme 11.11. General ATRP polymerization mechanism.

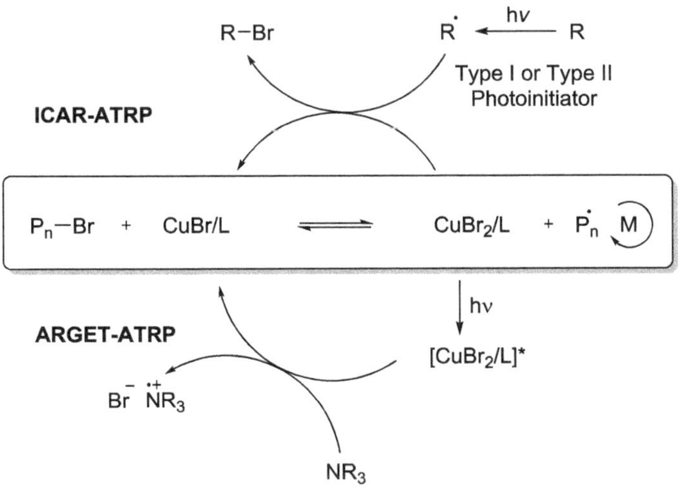

Scheme 11.12. ATRP photopolymerization mechanism.

Unlike photoinitiated photo-ATRP, photoredox photo-ATRP proceeds through PET processes (Scheme 11.12) [88]. PET facilitates reduction of Cu(II) to generate the Cu(I) activator species through an activator regenerated by an ET process (ARGET-ATRP). Irradiation of the Cu(II) deactivator species leads to an excited state, which in the presence of an electron donor facilitates the photochemical reduction of Cu(II) to generate the Cu(I) polymerization activator [88]. Early investigations into this process found that upon UV irradiation polymerization rates were increased, which was assigned to enhance oxidative processes between the redox states of the metal center [89–101]. **POCs** can also be employed to facilitate photochemical reduction of Cu(II) to generate the Cu(I) polymerization activator species [102].

Yagci *et al.* compared both photoinitiated and photoredox photo-ATRP with a type I PA and PCs under similar experimental conditions [102]. Polymerizations performed with **EY** and erythrosine B PCs evidenced a linear increase in molecular weights with conversion. However, the molecular weights obtained experimentally were considerably higher than the theoretical values. Furthermore, the molecular weight distributions were relatively high (Mw/Mn = 1.28–1.60), demonstrating only a moderate control over the polymerization. In contrast, bis(2,4,6-trimethylbenzoyl) phenylphosphine oxide (**BAPO**)-photoinitiated polymerizations afforded polymers with molecular weights much closer to the theoretical values. Furthermore, the **BAPO** system evidenced a good control over the polymerization with narrow molecular weight distributions obtained (Mw/Mn = 1.11–1.18) [102].

More recently, ATRP has been achieved without the use of a transition metal activator/deactivator species. The **POC** 10-phenylphenothiazine (**PTH**) was found to be an efficient ATRP agent under UV irradiation [103]. Photopolymerizations of methyl methacrylate mediated by **PTH** demonstrated a good control over polymer molecular weights, low polymer dispersities (Mw/Mn = 1.18–1.32), as well retention of polymer chain end group functionality. Moreover, as typical ATRP initiators were employed, the **PTH** system could be combined with conventional ATRP and iridinium-based systems, providing high synthetic versatility. Importantly, the exclusion of a transition-metal catalyst in the **PTH** system avoids the potential for toxic metal contamination of the generated polymers, which is a limiting factor of conventional metal-based ATRP processes.

Recently Miyake *et al.* demonstrated that diaryl dihydrophenazines (Figure 11.22) were remarkable compounds to perform highly efficient photoorganocatalyzed ATRPs under visible light [104]. Using this approach, various well-defined poly(methyl methacrylate)-based co-polymers were obtained.

Recently, through the direct observation of transient intermediates by ultrafast absorption spectroscopy, Orr-Ewing *et al.* [105]

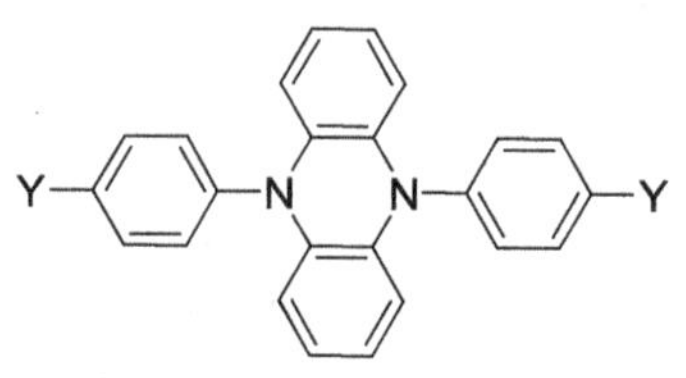

With Y = H, OMe, CF_3, CN

Figure 11.22. Structures of the diphenyl dihydrophenazine used by Miyake [104].

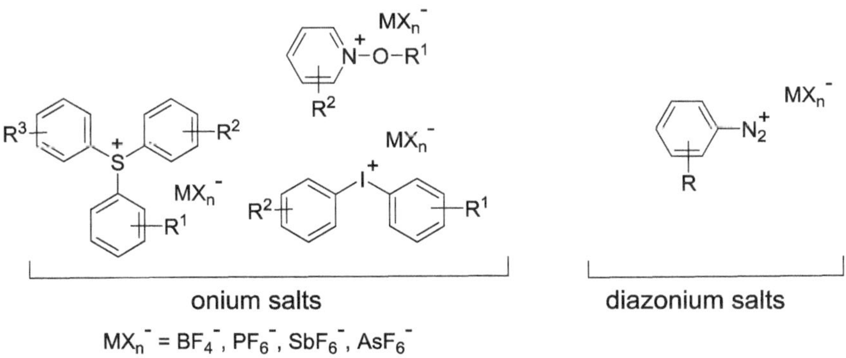

Figure 11.23. Selection of PAs comprising of onium and diazonium salts.

suggested that in the case of *N,N*-diaryl-5,10-dihydrophenazines used in photoorganocatalyzed atom-transfer radical polymerization, PET from short-lived excited singlet states govern the control of both the polymer molecular weight and dispersity.

11.8 Cationic Photopolymerization

Regarding polymerization by means of ionic species, cationic polymerizations are the most widely studied. Through this technique, epoxides and vinylic resins can be polymerized. Monomers used in cationic polymerization are less volatile, less toxic and exhibit improved mechanical properties to their acrylic counterparts [106–108]. Contrary to radical polymerizations, the reactive species are insensitive to oxygen and, consequently, higher monomer conversions can be achieved. Indeed, generally speaking, polymerization stops in the presence of water, by intermolecular reactions, or in the presence of a counterion. PAs involved in cationic polymerization are typically onium salts and diazonium salts (Figure 11.23) [109–112].

Onium salts, such as triarylsulfonium or diaryliodonium salts, are amongst the most widely used PAs for cationic photopolymerization, since their first use by the group of Crivello [113]. These salts are comprised of positively charged iodine or sulfur atoms that bridge two or three aromatic rings, respectively, and predominantly

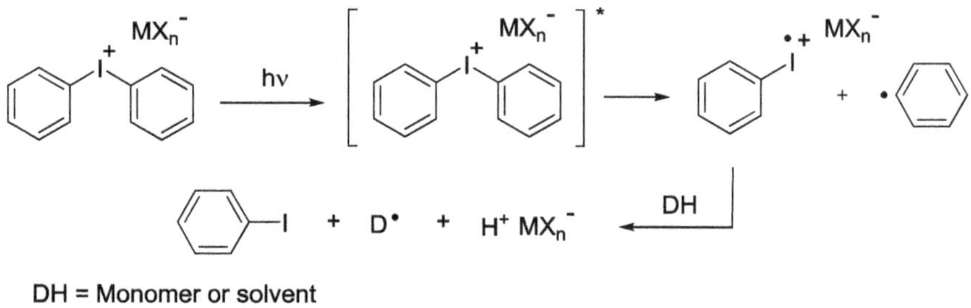

Scheme 11.13. Mechanism of acid photorelease from a diaryliodonium salt.

absorb in the 230–300 nm region [114]. Under UV irradiation, the iodonium salt decomposes, giving rise to a radical cation and a phenyl radical by homolytic cleavage of the C–I bond. The reaction of a hydrogen donor with the radical cation forms a Brønsted acid (Scheme 11.13) [115].

However, the low molar extinction coefficient of onium salts at wavelengths greater than 300 nm drastically limits their application. At wavelengths greater than 300 nm, most UV light sources are unable to deliver a sufficient intensity to efficiently decompose the onium salts. Additionally, the photolysis of these PAs can form secondary aromatic compounds that absorb at similar wavelengths. This effect is known as "internal filter" and can affect the absorption by the PA and reduce the formation of reactive species. Moreover, the diaryliodonium salt is consumed during the photoinitiation process, and therefore it cannot be used in catalytic quantities.

Interestingly, using a combination of a pyrylium derivative as **POC** and a chain-transfer agent, Fors *et al.* developed a strategy allowing the preparation of well-defined poly(vinyl ether) under visible irradiation, also following a cationic mechanism [116]. The proposed mechanism (Scheme 11.14) for such an approach [117] is supposed to involve a single electron transfer from the polymeric chain-transfer agent to the excited photocatalyst, generating both the corresponding radical cation and radical species (step I). Then, a mesolytic decomposition of the radical cation affords the cationic propagating polymer chain and the chain-transfer agent under its radical form CTA$^•$ (step II). Step III consists of the

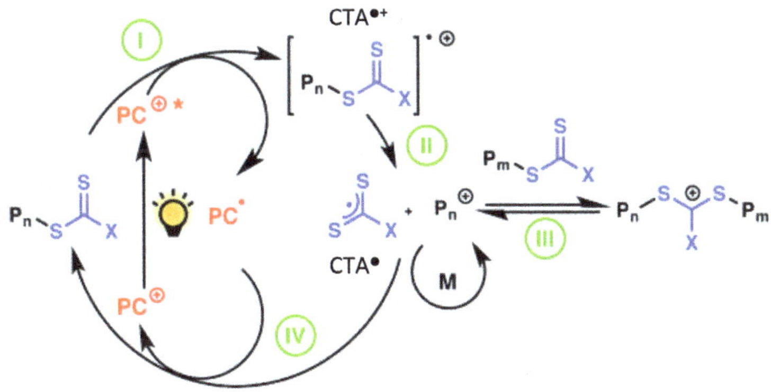

Scheme 11.14. Proposed mechanism for the cationic photopolymerization of vinyl ethers in the presence of a PC and CTA (reproduced with permission from Ref. [117]. © The American Chemical Society).

polymerization process through a cationic RAFT mechanism (step III). Finally, in step IV, the photocatalyst is regenerated after having reduced the CTA•.

11.9 Conclusions and Perspectives

In the present chapter, an overview of PAs/photoinitiating systems for polymer synthesis is described. As illustrated, the selection or the development of these photoinitiating systems is decisive for the performance of the photosensitive resins (i.e. with a strong effect on the polymerization rates and final conversions) as well as the final properties of the obtained polymers. These photoinitiating systems must also be carefully selected or designed for the selected experimental conditions (excitation wavelengths, light intensity, polymerization of a large range of monomers, etc.).

During these last years, significant contributions have been already achieved on organophotoinitors designed for photopolymerization [118]; however, important challenges still remain, for example, through the development of (i) high-performance systems characterized by ever higher polymerization rates (e.g. for access to larger objects in 3D printing or a higher productivity in coatings, paints, inks, composites, etc.), (ii) less or non-toxic photoinitiating systems (for the REACH registration) with potential use in (bio) medical applications but also for the safety of the users, (iii) photoinitiating systems for the access to composites with improved mechanical properties (the light penetration is currently strongly limited with UV light — few hundred micrometers for the most favorable cases; access to mm or cm ranges is a huge challenge, but important improvements are already under way), (iv) photoinitiating systems for safer irradiation devices with access to red or even NIR lights, (v) photoinitiating systems for access to hydrogels and water-based formulations,etc., and (vi) photoinitiating systems for new photopolymerization processes (emulsion photopolymerization under visible light, controlled photopolymerization, hybrid polymerization, dual curing combining, polymerization of bio-sourced resins, etc.).

Abbreviations

AC	2-(dimethylamino)-1-(4-morpholinophenyl) ethanone
AE	electron acceptor
Anth	Anthracene
Anth-Si	9,10-bis[(triisopropylsilyl)ethynyl]anthracene
ATRP	atom transfer radical polymerization
B	2-hydroxy-1,2-diphenylethanone
BAPO	bis(2,4,6-trimethylbenzoyl)phenylphosphine oxide
BDE	bond dissociation energy
BP	benzophenone
CQ	camphorquinone
DBC	double bond conversion
DH	hydrogen donor
DIPEA	*N,N*-diisopropylethylamine
DMPA	2,2-dimethoxy-2-phenylacetophenone

E^0_{OX}	oxidation potential
E^0_{RED}	reduction potential
EDB	ethyldimethylaminobenzoate
EY	Eosin Y
HAAP	2-hydroxy-2-methyl-1-phenylpropan-1-one(HAAP), 2-hydroxy-1,2-diphenylethanone
HOMO	highest occupied molecular orbital
ICT	intramolecular charge transfer
ITX	isopropylthioxanthone
LED	light-emitting diode
LUMO	lowest unoccupied molecular orbital
Napht-Ph	5,12-bis(phenylethynyl)naphthacene
NVK	*N*-vinylcarbazole
OAP	(diphenylphosphoryl)(phenyl)methanone
PA	photoinitiator
PC	photoredox catalyst
Pent-Si	6,13-bis(triisopropylsilylethynyl)pentacene
PET	photoinduced electron transfer
photo-ATRP	photoatom transfer radical polymerization
Photo-DSC	photodifferential scanning calorimetry
photo-RAFT	photoreversible addition-fragmentation chain-transfer
PTH	10-phenylphenothiazine
PTZT	phenyltetrazolethiol
Py	pyrene
Q	quinone
RAFT	reversible addition fragmentation chain transfer
RB	rose Bengal
REACH	registration, evaluation, authorization, and restriction of chemicals
RISC	reverse intersystem crossing

TX-SCH2COOH	2-thioxanthone-thioacetic acid
TX-SH	2-mercaptothioxanthone
TADF	thermally activated delayed fluorescence
TTMSS	tris(trimethylsilyl)silane
UV	ultraviolet

References

[1] Scranton, A. B.; Bowman, C. N.; Peiffer, R. W. (Eds.), *Photopolymerization: Fundamentals and Applications*, ACS Symposium Series (Book 673), American Chemical Society, Washington, 1997.

[2] Monroe, B. M.; Weed, G. C. *Chem. Rev.* **1993**, *93*, 435–448.

[3] Gruber, H. F. *Prog. Polym. Sci.* **1992**, *17*, 953—1044.

[4] Hageman, H. J. *Prog. Org. Coat.* **1985**, *13*, 123–150.

[5] Bamford, C. H.; Norrish, R. G. W. *J. Chem. Soc.* **1935**, 1504–1511.

[6] Yagci, Y.; Jockusch, S.; Turro, N. J. *Macromolecules* **2010**, *43*, 6245–6260.

[7] Shrestha, N. K.; Yagi, E. J.; Takatori, Y.; Kawai, A.; Kajii, Y.; Shibuya, K.; Obi, K. *J. Photochem. Photobiol. A Chem.* **1998**, *116*, 179–185.

[8] Hutchison, J.; Lambert, M. C.; Ledwith, A. *Polymer* **1973**, *14*, 250–254.

[9] Kim, D.; Stansbury, J. W. *J. Polym. Sci. A Polym. Chem.* **2009**, *47*, 887–898.

[10] Kavarnos, G. J.; Turro, N. J. *Chem. Rev.* **1986**, *86*, 401–449.

[11] Wang, Y.; Jiang, X.; Yin, J. *Eur. Polym. J.* **2009**, *45*, 437–447.

[12] Wen, Y.; Jiang, X.; Yin, J. *Polym. Eng. Sci.* **2009**, *49*, 1608–1615.

[13] Allen, N. S.; Lam, E.; Howells, E. M.; Green, P. N.; Green, A.; Catalina, F.; Penado, C. *Eur. Polym. J.* **1990**, *26*, 1345–1353.

[14] Allen, N. S.; Lam, E.; Kotecha, J. L.; Green, W. A.; Timms, A.; Navaratnam, S.; Parsons B. J. *J. Photochem. Photobiol. A Chem.* **1990**, *54*, 367–388.

[15] Ullrich, G.; Herzog, D.; Liska, R.; Burtscher, P.; Moszner, N. *J. Polym. Sci. A Polym. Chem.* **2004**, *42*, 4948–4963.

[16] Cokbaglan, L.; Arsu, N.; Yagci, Y.; Jockusch, S.; Turro, N. J. *Macromolecules* **2003**, *36*, 2649–2653.

[17] Aydin, M.; Arsu, N.; Yagci, Y. *Macromol. Rapid Commun.* **2003**, *24*, 718–723.

[18] Karasu, F.; Arsu, N.; Yagci, Y. *J. Appl. Polym. Sci.* **2006**, *103*, 3766–3770.

[19] Aydin, M.; Arsu, N.; Yagci, Y.; Jockusch, S.; Turro, N. J. *Macromolecules* **2005**, *38*, 4133–4138.

[20] Eaton, D. F. *Adv. Photochem.* **1986**, *13*, 427–487.

[21] Crivello, J. V.; Lam, J. H. W. *J. Polym. Sci. Polym. Chem. Ed.* **1978**, *16*, 2441–2451.

[22] Rehm, D.; Weller, A. *Isr. J. Chem.* **1970**, *8*, 259–271.

[23] Bi, Y.; Neckers, D. C. *J. Photochem. Photobiol. A Chem.* **1993**, *74*, 221–230.

[24] Mallavia, R.; Amat-Guerri, F.; Fimia, A.; Sastre, R. *Macromolecules* **1994**, *27*, 2643—2646.

[25] Encinas, M. V.; Majmud, E.; Lissi, E. A.; Scaiano, J. C. *Macromolecules* **1991**, *24*, 2111—2112.

[26] Oster, G. *Nature* **1954**, *173*, 300—301.

[27] Nelson, E. W.; Carter, T. P.; Scranton, A. B. *Macromolecules* **1994**, *27*, 1013—1019.

[28] Crivello, J. V.; Jiang, F. *Chem. Mater.* **2002**, *14*, 4858–4866.

[29] Fouassier, J. P.; Erddalane, A.; Morlet-Savary, F.; Sumiyoshi, I.; Harada, M.; Kawabata, M. *Macromolecules* **1994**, *27*, 3349–3356.

[30] Fouassier, J.-P.; Lalevée, J. *RSC Adv.* **2012**, *2*, 2621–2629.

[31] Fouassier, J.-P.; Ruhlmann, D.; Takimoto, Y.; Harada, M.; Kawabata, M. *J. Polym. Sci. A Polym. Chem.* **1993**, *31*, 2245–2248.

[32] Fouassier, J. P.; Chesneau, E. *Makromol. Chemie* **1991**, *192*, 1307–1315.

[33] Kim, D.; Scranton, A. *J. Polym. Sci. A Polym. Chem.* **2004**, *42*, 5863–5871.

[34] Zivic, N.; Bouzrati-Zerelli, M.; Kermagoret, A.; Dumur, F.; Fouassier, J.-P.; Gigmes, D.; Lalevée, J. *ChemCatChem* **2016**, *8*, 1617–1631.

[35] Ruhlmann, D.; Wieder, F.; Fouassier, J. P. *Eur. Polym. J.* **1992**, *28*, 591–599.

[36] Ruhlmann, D.; Fouassier, J. P.; Schnabel, W. *Eur. Polym. J.* **1992**, *28*, 287—292.

[37] Esen, D. S.; Arsu, N.; Da Silva, J. P.; Jockusch, S.; Turro, N. J. *J. Polym. Sci. A Polym. Chem.* **2013**, *51*, 1865–1871.

[38] Tehfe, M.-A.; Dumur, F.; Graff, B.; Morlet-Savary, F.; Gigmes, D.; Fouassier, J.-P.; Lalevée, J. *Polym. Chem.* **2013**, *4*, 2313–2324.

[39] Lalevée, J.; Tehfe, M.-A.; Dumur, F.; Gigmes, D.; Graff, B.; Morlet-Savary, F.; Fouassier, J.-P. *Macromol. Rapid Commun.* **2013**, *34*, 239–245.

[40] Tehfe, M.-A.; Dumur, F.; Graff, B.; Clément, J. L.; Gigmes, D.; Morlet-Savary, F.; Fouassier, J.-P.; Lalevée, J. *Macromolecules* **2013**, *46*, 736–746.

[41] Hartwig, A.; Harder, A.; Lühring, A.; Schröder, H. *Eur. Polym. J.* **2001**, *37*, 1449–1455.

[42] Mokbel, H.; Toufaily, J.; Hamieh, T.; Dumur, F.; Campolo, D.; Gigmes, D.; Fouassier, J. P.; Ortyl, J.; Lalevée, J. *J. Appl. Polym. Sci.* **2015**, *132*, 42759.

[43] Gu, H.; Ren, K.; Grinevich, O.; Malpert, J. H.; Neckers, D. C. *J. Org. Chem.* **2001**, *66*, 4161–4161.

[44] Tehfe, M.; Contal, E.; Gigmes, D.; Bertin, D.; Nechab, M.; Morlet-Savary, F.; Fouassier, J. P. *Macromolecules* **2012**, *45*, 4454–4460.

[45] Telitel, S.; Dumur, F.; Faury, T.; Graff, B.; Tehfe, M.-A.; Gigmes, D.; Fouassier, J.-P. ; Lalevée, J. *Beilstein J. Org. Chem.* **2013**, *9*, 877–882.

[46] Tehfe, M.-A.; Dumur, F.; Graff, B.; Morlet-Savary, F.; Fouassier, J.-P.; Gigmes, D.; Lalevée, J. *Macromolecules* **2012**, *45*, 8639–8647.

[47] Tehfe, M.-A.; Lalevée, J.; Morlet-Savary, F.; Graff, B.; Blanchard, N.; Fouassier, J.-P. *Macromolecules* **2012**, *45*, 1746–1752.

[48] Tehfe, M.; Lalevée, J.; Morlet-Savary, F.; Graff, B.; Blanchard, N.; Fouassier, J.-P. *ACS Macro Lett.* **2012**, *1*, 198–203.

[49] Anthony, J. E.; Eaton, D. L.; Parkin, S. R. *Org. Lett.* **2002**, *4*, 15–18.

[50] Chung, D. S.; Park, J. W.; Park, J.; Moon, D.; Kim, G. H.; Lee, H.; Lee, D. H.; Shim, H.; Kwon, S.; Park, C. E. *J. Mater. Chem.* **2010**, *20*, 524–530.

[51] Griffith, O. L.; Jones, A. G.; Anthony, J. E.; Lichtenberger, D. L. *J. Phys. Chem. C* **2010**, *114*, 13838–13845.

[52] Bureš, F. *RSC Adv.* **2014**, *4*, 58826–58851.

[53] Tehfe, M.-A.; Dumur, F.; Graff, B.; Morlet-Savary, F.; Gigmes, D.; Fouassier, J.-P.; Lalevée, J. *Polym. Chem.* **2013**, *4*, 3866–3875.

[54] Tehfe, M.-A.; Dumur, F.; Graff, B.; Morlet-Savary, F.; Fouassier, J.-P.; Gigmes, D.; Lalevée, J. *Macromolecules* **2013**, *46*, 3761–3770.

[55] Rodrigues, M. R.; Neumann, M. G. *Macromol. Chem. Phys.* **2001**, *202*, 2776–2782.

[56] Tehfe, M.-A.; Lepeltier, M.; Dumur, F.; Gigmes, D.; Fouassier, J.-P.; Lalevée, J. *Macromol. Chem. Phys.* **2017**, *218*, 1700192.

[57] Fukuzumi, S.; Ohkubo, K. *Org. Biomol. Chem.* **2014**, *12*, 6059–6071.

[58] Monroe B. M.; Weed, G. C. *Chem. Rev.* **1993**, *93*, 435–448.

[59] Avens H. J.; Bowman, C. N. *J. Polym. Sci. A* **2009**, *47*, 6083–6094.

[60] Padon K. S.; Scranton, A. B. *J. Polym. Sci. A* **2000**, *38*, 2057–2066.

[61] Kim, D.; Scranton A. B.; Stansbury, J. W. *J. Appl. Polym. Sci.* **2009**, *114*, 1535–1542.

[62] Aguirre-Soto, A.; Lim, C.-H.; Hwang, A. T.; Musgrave C. B.; Stansbury, J. W. *J. Am. Chem. Soc.* **2014**, *136*, 7418–7427.

[63] Pan, X.; Fang, C.; Fantin, M.; Malhotra, N.; So, W. Y.; Peteanu, L. A.; Isse, A. A.; Gennaro, A.; Liu P.; Matyjaszewski, K. *J. Am. Chem. Soc.* **2016**, *138*, 2411–2425.

[64] Tehfe, M.-A.; Dumur, F.; Contal, E.; Graff, B.; Gigmes, D.; Fouassier; J.-P.; Lalevée, J. *Macromol. Chem. Phys.* **2013**, *214*, 2189–2201.

[65] Ay, E.; Raad, Z.; Dautel, O.; Dumur, F.; Wantz, G.; Gigmes, D.; Fouassier, J.-P. Lalevée, J. *Macromolecules* **2016**, *49*, 2124–2134.

[66] Al Mousawi, A.; Dumur, F; Garra, P.; Toufaily, J.; Hamieh, T.; Graff, B.; Gigmes, D.; Fouassier J.-P.; Lalevée, J. *Macromolecules* **2017**, *50*, 2747—2758.

[67] Tehfe, M.-A.; Lalevée, J.; Morlet-Savary, F.; Graff B.; Fouassier, J.-P. *Macromolecules*, **2011**, *44*, 8374—8379.

[68] Bouzrati-Zerelli, M.; Zivic, N.; Dumur, F.; Gigmes, D.; Graff, B.; Fouassier J. P.; Lalevée, J. *Polym. Chem.* **2017**, *8*, 2028–2040.

[69] Chen, M.; Zhong, M.; Johnson, J. A. *Chem. Rev.* **2014**, *136*, 6513–6533.

[70] Keddie, D. J.; Moad, G.; Rizzardo, E.; Thang, S. H. *Macromolecules* **2012**, *45*, 5321–5342.

[71] Moad, G.; Rizzardo, E.; Thang, S. H. *Aust. J. Chem.* **2005**, *58*, 379–410.

[72] Omrane, K.; Partch, R. E.; Shipp, D. A. *PMSE Prepr.* **2005**, *93*, 566–567.

[73] Buback, M.; Junkers, T.; Vana, P. *Polym. Prepr.* **2005**, *46*, 362—363.

[74] Muthukrishnan, S.; Pan, E. H.; Stenzel, M. H.; Barner-Kowollik, C.; Davis, T. P.; Lewis, D.; Barner, L. *Macromolecules* **2007**, *40*, 2978—2980.

[75] Ran, R.; Yu, Y.; Wan, T. *J. Appl. Polym. Sci.* **2007**, *105*, 398–404.

[76] Lu, L.; Zhang, H.; Yang, N.; Cai, Y. *Macromolecules* **2006**, *39*, 3770–3776.

[77] Kitchin, A. D.; Velate, S.; Chen, M.; Ghiggino, K. P.; Smith, T. A.; Steer, R. P. *Photochem. Photobiol. Sci.* **2007**, *6*, 853–856.

[78] Bouchekif, H.; Narain, R. *Polymer Preprints* **2007**, *48*, 403–404.

[79] Zhang, H.; Deng, J.; Lu, L.; Cai, Y. *Macromolecules* **2007**, *40*, 9252–9261.

[80] Luo, Q.; Zheng, H.; Peng, Y.; Gao, H.; Lu, L.; Cai, Y. *J. Polym. Sci. A Polym. Chem.* **2009**, *47*, 6668–6676.

[81] Xu, J.; Jung, K.; Atme, A.; Shanmugam, S.; Boyer, C. *J. Am. Chem. Soc.* **2014**, *136*, 5508–5519.

[82] Dadashi-Silab, S.; Doran, S.; Yagci, Y. *Chem. Rev.* **2016**, *116*, 10212–10275.

[83] Xu, J.; Shanmugam, S.; Duong, H. T.; Boyer, C. *Polym. Chem.* **2015**, *6*, 5615–5619.

[84] Shanmugam, S.; Xu, J.; Boyer, C. *Chem. Sci.* **2015**, *6*, 1341–1349.

[85] Shanmugam, S.; Xu, J.; Boyer, C. *Angew. Chem. Int. Ed.* **2016**, *55*, 1036–1040.

[86] Fu, Q.; McKenzie, T. G.; Ren, J. M.; Tan, S.; Nam, E.; Qiao G. G. *Sci. Rep.* **2016**, *6*, 20779.

[87] Matyjaszewski, K. *Macromolecules* **2012**, *45*, 4015–4039.

[88] Dadashi-Silab, A.; Tasdelen, M. A.; Yagci, Y. *J. Polym. Sci. A Polym. Chem.* **2014**, *52*, 2878–2888.

[89] Muftuoglu, A. E.; Cianga, I.; Colak, D.; Yagci, Y. *Des. Monomers Polym.* **2004**, *7*, 563–582.

[90] Roof, A. C.; Tillman, E. S.; Malik, R. E.; Roland, A. M.; Miller, D. J.; Sarry, L. R. *Polymer* **2006**, *47*, 3325–3335.

[91] Johnson, J. A.; Finn, M. G.; Koberstein, J. T.; Turro, N. J. *Macromolecules* **2007**, *40*, 3589–3598.

[92] Cummins, D.; Wyman, P.; Duxbury, C. J.; Thies, J.; Koning, C. E.; Heise, A. *Chem. Mater.* **2007**, *19*, 5285–5292.

[93] Ishizu, K.; Kobayakawa, N.; Takano, S.; Tokuno, Y.; Ozawa, M. *J. Polym. Sci. A Polym. Chem.* **2007**, *45*, 1771–1777.

[94] Ishizu, K.; Murakami, T.; Takano, S. *J. Colloid Interface Sci.* **2008**, *322*, 59–64.

[95] Durmaz, Y. Y.; Karagoz, B.; Bicak, N.; Yagci, Y. *Polym. Intern.* **2008**, *57*, 1182–1187.

[96] Cohen, N. A.; Tillman, E. S.; Thakur, S.; Smith, J. R.; Eckenhoff, W. T.; Pintauer, T. *Macromol. Chem. Phys.* **2009**, *10*, 263–268.

[97] Durmaz, Y. Y.; Kumbaraci, V.; Demirel, L. A.; Talinli, N.; Yagci, Y. *Macromolecules* **2009**, *42*, 2899–2905.

[98] Acik, G.; Kahveci, M. U.; Yagci, Y. *Macromolecules* **2010**, *43*, 9198–9201.

[99] Tasdelen, M. A.; Uygun, M.; Yagci, Y. *Macromol. Chem. Phys.* **2010**, *211*, 2271–2275.

[100] Tasdelen, M. A.; Uygun, M.; Yagci, Y. *Macromol. Rapid Comm.* **2011**, *32*, 58–62.

[101] Kwak, Y.; Matyjaszewski, K. *Macromolecules* **2010**, *43*, 5180–5183.

[102] Tasdelen, M. A.; Ciftci, M.; Yagci, Y. *Macromol. Chem. Phys.* **2012**, *213*, 1391–1396.

[103] Treat, N. J.; Sprafke, H.; Kramer, J. W.; Clark, P. G.; Barton, B. E.; Alaniz, J. R.; Fors, B. P.; Hawker, C. J. *J. Am. Chem. Soc.* **2014**, *136*, 16096–16101.

[104] Theriot, J. C.; Lim, C.-H.; Yang, H.; Ryan, M. D.; Musgrave, C. B.; Miyake, G. M. *Science* **2016**, *352*, 1082–1086.

[105] Koyama, D.; Dale, H. J. A.; Orr-Ewing A. J. *J. Am. Chem. Soc.* **2018**, *140*, 1285–1293.

[106] Yagci, Y. *Macromol. Symp.* **2006**, *240*, 93–101.

[107] Cho, J.-D.; Hong, J.-W. *J. Appl. Polym. Sci.* **2005**, *97*, 1345–1351.

[108] Sangermano, M.; Malucelli, G.; Bongiovanni, R.; Priola, A.; Annby, U.; Rehnberg, N. *Eur. Polym. J.* **2002**, *38*, 655–659.

[109] Schlesinger, S. I. *Polym. Eng. Sci.* **1974**, *14*, 513–515.

[110] Marek, M.; Toman, L. *J. Polym. Sci. Polym. Symp.* **1973**, *42*, 339–343.

[111] Toman, L.; Marek, M.; Jokl, J. *J. Polym. Sci. Polym. Chem. Ed.* **1974**, *12*, 1897–1903.

[112] Klingert, B.; Riediker, M.; Roloff, A. *Comments Inorg. Chem.* **1988**, *7*, 109–138.

[113] Crivellzo, J. V.; Lam, J. H. W. *Macromolecules* **1977**, *10*, 1307–1315.

[114] Gomurashvili, Z.; Crivello, J. V. *J. Polym. Sci. A Polym. Chem.* **2001**, *39*, 1187–1197.

[115] Yagci, Y.; Schnabel, W. *Makromol. Chem. Macromol. Symp.* **1988**, *13/14*, 161–174.

[116] Kottisch, V.; Michaudel, Q.; Fors B. J. *J. Am. Chem. Soc.* **2016**, *138*, 15535–15538.

[117] Michaudel, Q.; Chauviré, T.; Kottisch, V.; Supej, M. J.; Stawiasz, K. J. Shen, L.; Zipfel, W. R. Abruña, H. D.; Freed, J. H.; Fors P. B. *J. Am. Chem. Soc.* **2017**, *139*, 15530–15538.

[118] Corrigan, N.; Shanmugam, N.; Xu, J.; Boyer, C. *Chem. Soc. Rev.* **2016**, *45*, 6165–6212.

Chapter 12

Photoorganocatalysis in Flow

Manuel Anselmo, Andrea Basso*, and Lisa Moni

Dipartimento di Chimica e Chimica Industriale,
Università degli Studi di Genova,
Via Dodecaneso 31, 16146 Genova, Italy
**andrea.basso@unige.it*

12.1 Introduction

Continuous-flow methods have gained great notoriety in photochemistry as they represent one of the most important tools to circumvent the old and unsolved problems associated with photocatalyzed reactions, such as scale-up, reaction efficiency, and safety issues. While it is generally possible to increase the scale of a classical thermal reaction simply by raising the dimensions, the same does not apply to photochemical reactions. This is connected with an intrinsic problem related to batch photochemistry, i.e. light penetration. Photon transport through a solution decreases logarithmically with the path length, according to the Lambert–Beer law. Moreover, a larger reactor and a longer reaction time can lead to over-irradiation of certain zones of the reaction mixture, forming side-products, lowering yields, and hampering purification processes. In addition, many photochemical reactions involve reactive gases (i.e. O_2); therefore, they are considered as multiphase systems. Three different entities must be added and mixed in a proper manner in order to achieve high selectivities and conversion rates, and these are as follows: (1) the liquid phase, containing organic substrates, the photoactive component (catalyst or sensitizer), the solvent, and possible additives; (2) the gas phase, containing the gaseous reagents; and (3) the light. When the interfacial area between the phases is low and poorly defined, as it happens in batch conditions, the efficiency of the reaction decreases. Also, the explosive nature of oxygenated organic liquids is well known and can lead to significant health and safety issues.

For all these reasons, flow photochemistry is extremely appealing both for laboratory and industrial scale, as most of the problems listed above are suppressed or dramatically reduced. In addition, the combination of sequential chemical transformations in a streamline flow system suppresses the isolation and purification of intermediates, thus enabling flow chemistry to mimic the efficiency of cell-based multistep biosynthetic pathways. Indeed, photoorganocatalysis has also benefited from flow chemistry; many valuable examples have been reported during the last decades, and the field is in dramatic expansion. This chapter, however, is not simply aimed at showing those examples, but is actually designed to be a manual with general principles and instructions for those wishing to put into practice flow photochemistry on their own. Therefore, the first part is dedicated to the description of the components required to set up a flow photochemical reactor. In the second part, different flow systems employed in photoorganocatalysis are illustrated and critically analyzed; finally, the last part is dedicated to examples, taken from the relevant literature, where the application of a flow approach has resulted in an improvement of reaction efficiency, compared to the existing methodologies.

12.2 Components of Flow Photochemical Reactors

The reactors performing flow photochemistry typically comprise a number of components, which include several types of tubings and channels, pumps for delivering reactants, flow meters/controllers, feedback-controlled heating/cooling units, separators for chemical work-up between reaction stages, various kinds of valves, back-pressure regulators, chemical analysis equipment, and, last but not least, light sources. All these elements will be shortly discussed in the following sections.

12.2.1 *Pumps, flow meters/controller, heating/cooling units, backpressure regulators*

The most employed pumping systems are syringe pumps, HPLC or peristaltic pumps, and gas cylinders (which often act as true pumps). Syringe pumps are certainly the cheaper available option for research groups, especially for smaller ones. Their main disadvantage though, is the limited volume which they can pump and their need to be interrupted in order to refill the syringe. In addition, the residence time of the reactant solution in the syringe should be taken into account because of possible side reactions or substrate decomposition. More expensive are the HPLC and peristaltic pumps, but, differently from syringe pumps, they allow to continuously suck the solution from a container in which the reactant solution can be continuously replenished. In the case of biphasic gas–liquid reactions, gas cylinders (O_2, CO_2, and compressed air, among the most common) often also constitute the pumping system. Both cylinders and gas lines can be used. Flow meters and flow controllers instead are quite advanced systems which allow for a more precise quantification of the bulk of liquid or gas/liquid in movement; they become especially important when the volumes in play get bigger and scalability becomes an issue. Heating/cooling units are not exactly distinguishing units for flow photochemical reactors: often, this chemistry is performed at room temperature, without any need for heat. Reactions in play are rarely exothermic and light sources are less and less problematic in terms of overheating; nonetheless, cooling jackets are often employed to hamper the heat derived from the light source. But every time it is needed, a temperature controlling unit becomes fundamentally important. Cooling is often obtained by running water jackets or by immersion well systems. Heating units play a role in just a handful of situations instead and they typically work with electrical power. Backpressure regulators are fundamental components of flow reactors, especially in the case of pressurized biphasic gas–liquid systems. Often, the gas–liquid interaction generates homogeneous bubbles and, consequently, film layers, which require to be regularized in order for the irradiation to be uniformly delivered.

12.2.2 *Tubings*

Nowadays, photoreactors are frequently constructed from polymer-based materials, such as poly(methyl methacrylate) (PMMA), polydimethylsiloxane (PDMS), perfluoroalkoxyalkane (PFA) or polytetrafluoroethylene (PTFE), and many more. These polymer-based materials have excellent light transmission, are easy to handle, and are commercially available at very affordable prices. Unfortunately, they are not stable to heating, contrary to glass and quartz reactors, which dominated the scene in the past years, but they are commonly employed in room temperature (or lower) processes. However, PMMA and PDMS are susceptible to swelling when subjected to organic solvents for a prolonged time, rendering them impractical for some photochemical transformations. In contrast, fluorinated polymers (i.e. PTFE, PFA, and fluorinated ethylene propylene (FEP)) provide a high chemical stability and good optical properties toward both UV and especially visible light, while maintaining excellent physical properties. These materials are commercially available as capillary tubings in a wide variety of different lengths and diameters. Given the ease of handling of these capillaries and their low cost, such materials have become increasingly popular in the photochemical community. In addition, high-purity polymers are commonly used to avoid leaching of the plasticizers from the polymer (i.e. PFA), which would have a negative impact on the photocatalytic process.

12.2.3 *Light sources*

Normally, the selection of a proper light source for photochemical applications depends mainly on the overlap between the emission spectrum of the radiation source and the absorption characteristics of the photoactive reactants or catalysts. In the case of current visible light flow photochemistry, the panorama of the most used photocatalysts is quite standardized in terms of wavelengths, while greater efforts would be spent when designing novel **POCs**. All the previously cited and most commonly used materials are perfectly visible light transparent, and the market offers countless light sources irradiating at a precise wavelength. Other important selection criteria are cost, energy efficiency, lifetime of the light source, and a proper match between the light source dimensions and the photoreactor (which can go down to the millimeter scale). Among the most common light sources used in the past years are compact fluorescent light (CFL) bulbs, which are nowadays finding next to complete replacement by light emitting diodes (LEDs). These devices are certainly the best option. They are cheap and available in thousands of dimensions and colors; colored LED bulbs can be bought in most electronical device stores for as low as a few eurocents each. LEDs are cooler than incandescent lights, sturdier (they are more resistant to breakage due to robust epoxy lenses), and consume less electricity while maintaining a bright and powerful luminescence.

12.3 Reactors for Flow Photochemistry: Concepts

A wide range of flow photochemical devices has been investigated in recent years, and there are a variety of aspects that can be discussed in this regard when different models are taken into consideration. These devices can be bought from suppliers ready or, much more commonly, they can be built *ad hoc* by researchers in order to shoot a particular target they aim at. The following paragraphs will propose a discussion on the main categories and the relative concepts of such devices, and then a selection of appositely selected examples will be described in the subsequent paragraphs.

12.3.1 *Micro/mesoflow reactors*

A flow device can be categorized as microfluidic or mesofluidic depending on the channel or tube size [1]. A microfluidic reactor

Table 12.1. Microflow vs mesoflow reactors.

Microflow reactors: PROs	Efficient mixing capabilities, ideal for reaction conditions optimization, lower production of wastes, higher surface
Microflow reactors: CONs	Susceptible to clogging, difficult to clean
Mesoflow reactors: PROs	Improved flow capabilities, low pressure drops, compatibility with packed bed reactors, no clogging, easy to clean and dismantle
Mesoflow reactors: CONs	Low heat transfer capability, large quantities of solvents required, high production of wastes

ranges from 10 to 500 μm size, while a mesofluidic reactor typically goes from 500 μm to few millimeters. For microreactors, surface parameters are commonly between 5000 and 50,000 m^2/m^3 while for mesoreactors the common values are between 100 and 10,000 m^2/m^3. Thus, high heat transfer capacities are achieved with small diameter channels favoring micro- over mesoflow reactors. A list of advantages and disadvantages of the two categories is listed in Table 12.1.

12.3.2 *Monophase/multiphase reactors*

The standard flow device is typically denoted as the single-channel monophase reactor. This device is composed of a single tubing with an irradiation area (typically a coil or a 2D-serpentine) inside which a liquid solution flows, as depicted in Figure 12.1 [2].

Multiple channels can be connected by means of T-mixers in order to add different reactants at appropriate times, allowing to chain several reactions in a row. Normally, the observed flow pattern of monophase single channels is laminar flow. Under these conditions, the fluid is flowing in parallel layers and its mixing is governed by diffusion across parallel *lamellae.*

On the other hand, many flow photochemical reactions require the interaction between different phases. Among the most common cases, there are reactions exploiting singlet oxygen chemistry, which

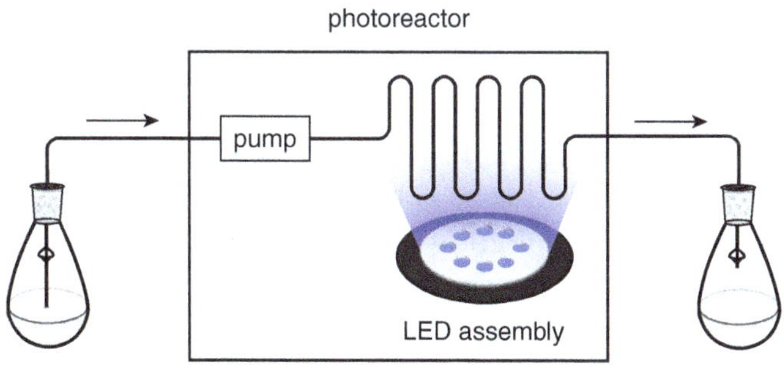

Figure 12.1. Simple standard single-channel monophase reactor scheme (reprinted with permission from Ref. [2]. © Elsevier).

require the efficient mixing between the liquid phase and molecular oxygen. In these cases, appropriate interactions among the liquid and gaseous sources should be allowed for and, depending on these, a different result will be obtained in terms of heterogeneous gas–liquid flow. There are three main categories of flow devices allowing optimal gas–liquid interaction: single-channel, falling-film, and membrane reactors. Single-channel reactors are characterized by the so-called Taylor flow [3], which will be described shortly. In falling-film devices, the liquid flows down a microstructured reaction plate under the action of gravity and spreads to form an expanded thin film in each microchannel. The gas flows over the liquid film contained within the microchannels. This design allows for specific gas–liquid interfacial areas of up to 20,000 m^2/m^3 to be achieved, which is noticeably higher than the other gas–liquid-contacting reactors. Membrane reactors base their operation on a semipermeable membrane separating the two phases that maintain contact with each other by diffusing through the membrane [4].

12.3.3 *Taylor flow/annular flow reactors*

In microscale multiphase photoreactions conducted in single-channel microreactors, the observed flow may be of two types: Taylor flow (also known as segmented flow or slug flow) and annular flow (or pipe flow). Taylor flow is a type of gas–liquid flow pattern that consists of elongated bubbles possessing an equivalent diameter, separated by liquid slugs. In this case, toroidal vortices are established which cause a powerful mixing effect and intensify the mass transfer from one phase to the other. A Taylor flow can be achieved by two main strategies: colliding or perpendicular shearing [5]. The formation of the Taylor flow occurs in a T-mixer connecting: (a) the gas phase coming from the cylinder, (b) the liquid phase coming from the pump, and (c) the flow distribution leading to the irradiation zone. The colliding shearing, as it can be deduced from its name, implies an impact between the two phases in a straight line of the T-mixer. The perpendicular shearing instead happens *via* a 90 degrees meeting of gas and liquid. On the other hand, annular flow is obtained when one phase flows at a higher velocity in the center of the capillary, while the wetting phase forms a thin layer at the channel wall. The flow core can contain entrained liquid droplets. In this case, the region is often referred to as annular-dispersed flow, where the entrained fraction may vary from zero (a pure annular flow) to a value close to unity (a dispersed flow). The establishment of a determined flow pattern depends on the volume fluxes, the way they collide with each other, and the fluid properties, such as viscosity, density, and surface tension (Figure 12.2).

12.3.4 *Internal/external irradiation*

Optimal irradiation is obviously crucial in this area of chemistry. The major parameter that is currently assessed with respect to light irradiation is the amount of illuminated surface per unit of reaction liquid volume inside the reactor (m^2 irradiation/m^3 reactor). This value has been denoted as kay, κ, by Ray and Beenackers [6]. They calculated κ for different photoreactor configurations [7]. It should not be forgotten that in many photoreactor designs the light has to travel through the fluid- or gas-containing reagents and other species, and also through a transparent wall. Part of the originally emitted light can (or, to better say it, it certainly will) be absorbed by these components before it reaches the catalyst. A further

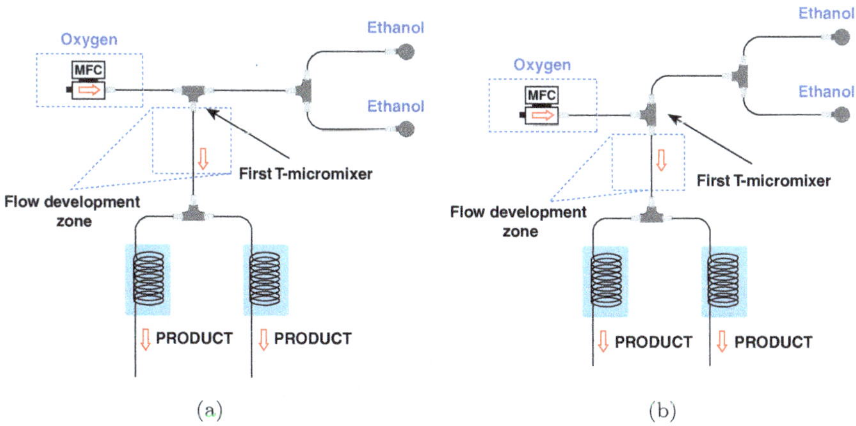

Figure 12.2. Taylor flow and its formation (two types of shearing: (a) colliding shearing, (b) perpendicular shearing) (reprinted with permission from Ref. [5]. © The Royal Society of Chemistry).

classification regards the way the solution is irradiated. There are two possible configurations: internal [8] or external [9] irradiation. The most typical configuration of internal irradiation comprises the classic device with the tubings wrapped around the light source. External irradiation is achieved instead every time the light source is externally directed toward the irradiation area. This happens every time that the sun is exploited as a natural light source. The distinction might look trivial: the direction from which the light arrives should not matter as long as it efficiently irradiates the selected area with enough photons. Instead, it makes a difference, especially in terms of design. Harvesting the maximum number of photons is economically relevant; even though this is unimportant at a laboratory scale, it becomes an issue when scaled up at an industrial level. For example, the industrial production of artemisinin [10] requires a huge reactor which exploits internal irradiation to better harvest the light produced by the energetically demanding light source (several square meters of LED bulbs). The literature reports the efforts that have been put into insulating this reactor and the economic advantage that has been obtained after optimization. While an internal irradiation ensures the highest possible light harvesting, it also generates problems connected with the possibility of dismantling, cleaning, and substituting the light sources. On the other hand, external irradiation has a lower photon efficiency, although it is a mandatory choice for devices unable to contain the source inside, such as those based on rotating surfaces. In addition, an external source of irradiation can be designed to reduce overheating problems, exploiting the surrounding space to dissipate heat. For this reason, a cooling unit is almost never required when the light source is external, despite its necessity in the case of internal irradiation. Studies addressing the performance improvement by optimizing photon transfer have been performed and could likely bring improvements in the future years; the introduction of a multitude of micro- or even nanoscale light emitting sources close to the catalyst particles is a very attractive perspective.

12.3.5 *Homogeneous/heterogeneous catalysis*

The principles discussed above can be applied to any photochemical process; however, in dealing with organocatalysis, the choice of the promoter is also obviously important. Heterogeneous catalysis and catalyst support have not yet become common in flow photochemistry yet, but it is presumably one of the future achievements of these techniques. The combination of such enabling technologies with continuous flow started to become popular in recent years, and it is expected to also have a place in the flow photochemistry territory [11]. It should not be forgotten that the idea of photocatalyst support *per se* is not new; in fact, porphyrins can naturally generate singlet oxygen [12, 13]. Using heterogeneous photocatalysts shows two obvious advantages: the first one is the simplified purification procedure since there is no need to remove the catalyst from the product (considering that the downstream separation of the catalyst is often one of the major expenses of a process running cost). The second advantage is an economically convenient scalability, since the same catalytic support can be reused several times. Evidently, this is not an easy issue because of the difficulty in irradiating a supported catalyst (much more difficult than irradiating a solution of it) and the eventual photocatalyst leaching that has to be avoided. A number of heterogenization approaches have been suggested in the literature [14]. These include the deposition of photocatalysts on tubular reactor walls, membranes, or other supports such as monoliths, structural packing, or directly onto light sources such as side scattering optical fibers or packaged LEDs. Heterogeneous catalysts can be employed as slurries, but this system suffers from clogging problems and it generates difficulties in the recovery and separation of the catalyst. An alternative approach is the immobilization of this catalyst on the reactor walls or in a packed bed. A packed bed reactor suffers from the problem of light penetration through the packed catalyst. Packing the walls with a thin film of catalyst instead is a way more performing strategy.

12.4 Reactors for Flow Photochemistry: Examples

Lately, researchers are dedicating considerable attention to visible light flow photochemistry, coming up with countless appealing ideas. In order to accomplish their needs, they rarely can allow themselves to employ standard devices, while designing *ad hoc* reactors to shoot a particular target they are aiming at. The mainstream principles that lead the researchers to these particular setups are mainly a scalability improvement, the search for a higher efficiency, the promotion of multiphase interaction, and the fulfillment of the principles of green chemistry. Some case studies will be illustrated in the following sections.

12.4.1 *Microflow reactors*

The typical microflow reactor resembles the one shown in Figure 12.3, consisting of a borosilicate glass chip having dimensions $45.3 \times 15.3 \times 2.2$ mm with a maximum channel width of 600 μm, a channel depth of 500 μm, and a maximum volume of 100 μL.

The pumping is achieved by means of a syringe pump. As is always the case for microflow reactors, the irradiation source comes

Figure 12.3. Classic microflow device (reprinted with permission from Ref. [4]. © The Royal Society of Chemistry).

from one side of the chip (which is usually located in an appropriate housing). An example of application is brought by the work of Wooton [15] which showed the use of this glass chip for the [4+2]-cycloaddition of singlet oxygen to α-terpinene with the formation of ascaridole, catalyzed by Rose Bengal. For this gas–liquid reaction, the authors utilized a one-channel microreactor (5 × 2 cm) with a channel length of 50 μm, a depth of 50 μm, and a width of 150 μm. The flow rate for the liquid phase was only 1 μL/min, and it was 15 μL/min for the gaseous phase.

A very elegant application of a microflow device for the optimization of reaction conditions has been set up by Jensen and co-workers [16].

They presented an automated microfluidic platform for in-flow studies for both liquid and gas–liquid reactions. A computer-controlled liquid handler prepares and injects the reaction mixture (a 15 μL droplet) in the system that is then moved inside a gas-filled fluoropolymer tubing to a horseshoe-shaped reactor illuminated from the top by means of a high-intensity LED. Gas/liquid mass transfer is promoted by a constant oscillatory motion of the droplet within the reactor. After a given time, a sample of the reaction is automatically injected into an LC/MS unit for analysis. The computer-aided automation makes this set-up ideal for a fast and easy optimization of the process (including photocatalyst and flow rate screening).

A case study of the oxidative hydroxylation of phenylboronic acids and dimerization of thiophenols has been performed *via* the automated microfluidic platform; Rose Bengal and Eosin Y gave the best results as photocatalysts. The key aspect of this system is the oscillatory flow: the constant oscillatory motion of the droplet carried by the gas ensures that the reaction mixture remains well mixed over the course of the reaction.

12.4.2 *Mesoflow reactors*

Coming to mesoflow devices instead, a very common reactor design is the Booker-Milburn apparatus, named after its designer [17]. Zeitler [18] employed it for the dehalogenation of activated halogenides and for Aza-Henry reactions occurring *via* the alpha-amino C–H functionalizations catalyzed by Eosin Y. Its structure consists of an HPLC pump, pushing the liquid through a FEP tubing (O.D. 1.6 mm and I.D. 0.8 mm) wrapped around a traditional water cooled immersion well (diameter = 60 mm, the tube length corresponded to an internal volume of 10 mL), bearing in its cavity a UV lamp. Such internal irradiation and cooling solution are the footprints of this kind of reactor. A receiver flask collects the product solution (Figure 12.4).

Some photochemical reactions need a long time to be completed, and thus they require a long irradiation time (having, as a consequence, high energetic consumption). In these situations, recycling the solution might be advantageous. In fact, Rueping [19] optimized the Rose Bengal-catalyzed photoredox oxidative Ugi multicomponent reaction conditions in a mesoflow reactor similar to the one described above, with a recycling flow of 3 mL/min for 20 hrs. An HPLC pump sucked the solution from the collection flask and pumped it in the reaction coil again, and then recollected it inside the previous flask, thus recycling it.

While the two previously reported examples of mesoflow reactors were characterized by internal irradiation, the following one

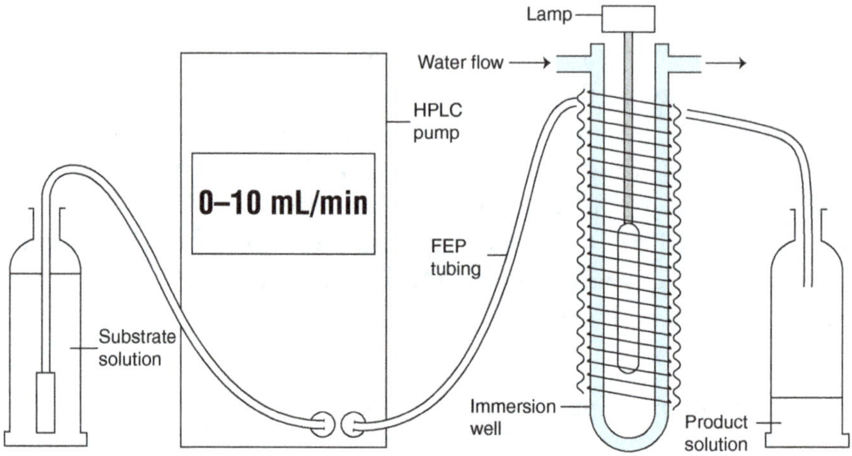

Figure 12.4. Booker-Milburn reactor (reprinted with permission from Ref. [17]. © American Chemical Society).

definitely belongs to the external irradiation class, as can be easily deduced by its name: the sunflow reactor [20]. This reactor exploits the sun as a light source, thus avoiding energetic consumption in maximum accordance with the principles of green chemistry. While this system is cheap, safe, and sustainable, it is not free from defects. First of all, the solar radiation is a complex and rich one, and it might induce side reactions and product decomposition. Secondly, the sun is not equally intense during the days and during the year; the utilization of the reactor requires a carefully scheduled planning. In addition, the rotation of the system during the day, in order to focus the sun radiation, is preferable. Last but not least, the weather conditions should be taken into account to avoid the equipment from being ruined. Opatz exposed the sunflow reactor on the roof from 9:30 to 16 CEST in Mainz, Germany, between May and June, on sunny cloudless days. He used it to perform the Rose Bengal- and Eosin Y-catalyzed oxidative α-cyanation of tertiary amines to give the corresponding α-aminonitriles [20]. They wrapped the FEP tubings on an aviary fence to protect it from the wind.

12.4.3 *Gas–liquid devices*

An example of single-channel reactor is that by Park [21] who tested the photooxygenation of beta-pinene promoted by several organic sensitizers such as Rose Bengal, Eosin Y, and methylene blue. In this device, two T junctions were inserted in order to dose, firstly, two different reactant solutions and, secondly, the so-formed solution and the oxygen gas. The resulting slug was allowed to react in an irradiated wrapped coil of 0.5/1 mm I.D. FEP tubings. The irradiating LED source was internal (Figure 12.5).

Jahnisch [22] employed a falling-film reactor to perform the photooxydation of olefins. This reactor consisted of a bottom housing section with integrated heat exchanger, a contact-zone mask, a top housing section with open space, and a reaction plate. The latter was constituted of 32 parallel microchannels of 66 mm length, 600 μm width, and 300 μm depth designed by the Institut für Mikrotechnik (IMM, Mainz). The liquid flows *via* a horizontal slot into the upper part of the bottom housing section and *via* another slot to the upper site of the reaction plate. Here, it flows downwards due to gravity. The reaction takes place at the gas–liquid interface on the reaction plate (Figure 12.6).

Another interesting device is the tube-in-tube reactor (TIT-R). The outer layer of this device consists of a transparent and gas-nonpermeable polyetheretherketone (PEEK) tube, while the inner

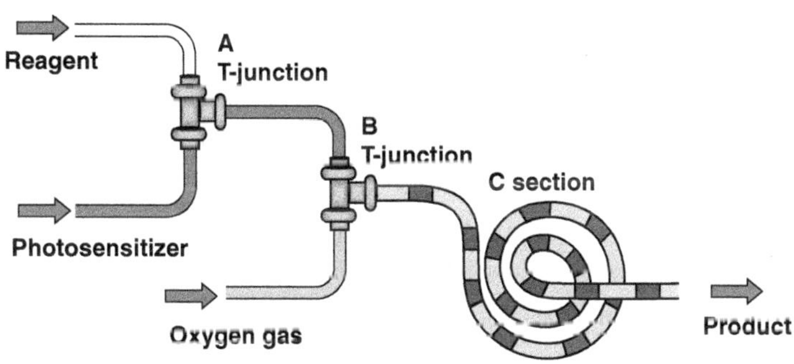

Figure 12.5. Single-channel reactor developed by Park (reprinted with permission from Ref. [21]. © The Royal Society of Chemistry).

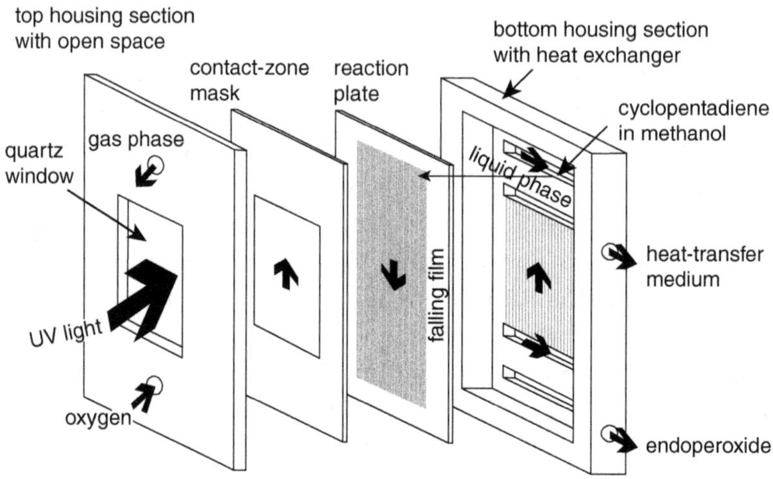

Figure 12.6. Falling-film reactor developed by Jahnisch (reprinted with permission from Ref. [22]. © Wiley-VCH Verlag GmbH & Co. KGaA).

layer is made of a chemically resistant and gas-permeable Teflon AF-2400 tube. Park [21] employed it for the photooxygenation of monoterpenes. Permeation of the gas through the membrane allows contact with the liquid, and thus the occurrence of the reaction. The advantage of this system is the possibility of larger scale ups when compared to single-channel and falling-film microreactors, due to a considerably larger (up to 300 times) working volume.

Another intriguing use of permeable membranes concerns the so-called "through wall mass transport". de Mello [23] exploited microcapillary films (MCFs) for the synthesis of ascaridole starting from α-terpinene with Rose Bengal catalysis. MCFs are comprised of multiple parallel channels embedded in a plastic film. They are made in FEP, with an internal diameter of around 100 μm and an average wall thickness of 60 μm. The refractive index is close to that of water and they possess a high trasmittance in the UV/Vis region, providing an excellent light penetration in their assembly. No gaseous oxygen is supplied directly to the reaction, but rather mass transport occurs exclusively through the reactor walls. The author confined the microcapillaries inside an LED-irradiated glass reactor placed inside a miniclave under controlled oxygen pressure. The reaction solution was injected *via* a syringe pump and FEP tubing connections. MCFs provide a cheap alternative as an enabling technology, especially for scale-up applications in singlet oxygen chemistry.

In order to promote efficient mixing of the liquid and gas phases, Poliakoff [24] developed a particular sapphire tubular reactor for the photocatalytic hydroxylation of arylboronic acids to phenols with Rose Bengal. The organic feed is delivered by means of an HPLC pump while air is dosed *via* a Rheodyne valve. Two different systems are used to promote mixing; at first, a static micromixer is located at the junction between the HPLC pump and the gas cylinder (or air). Secondly, inside the reactor, glass beads are used to further promote an efficient mixing.

12.4.4 *Rotating surface devices*

Rotating surfaces have recently become attractive, as demonstrated by Poliakoff [25] who designed a semicontinuous reactor for photooxidation reactions such as the photodesymmetrization of benzene-1,4-diboronic acids catalyzed by Rose Bengal. This reactor, named *Photovap*, is based on a standard rotary evaporator. The rapid rotation of the reactor at a tilt angle above the horizontal position generates a dynamic thin film on the walls of the tube, where an intense micromixing and efficient absorption of the externally delivered light takes place. After the desired reaction time, the reactor stops, the product solution is pumped out, and the procedure is repeated. All operations are carried out under computer control (Figure 12.7).

Another very elegant rotating surface device, named microfluidic vortex device, has been designed by Raston and Stubbs [26]. The formation of C–C and C–P bonds *via* Rose Bengal-catalyzed aza-Henry reactions have been performed exploiting intense micromixing due to high-speed rotation (on average 4000 rpm) inside a borosilicate glass NMR tube. The device can be operated in two different modes: a confined mode for a finite volume of liquid and

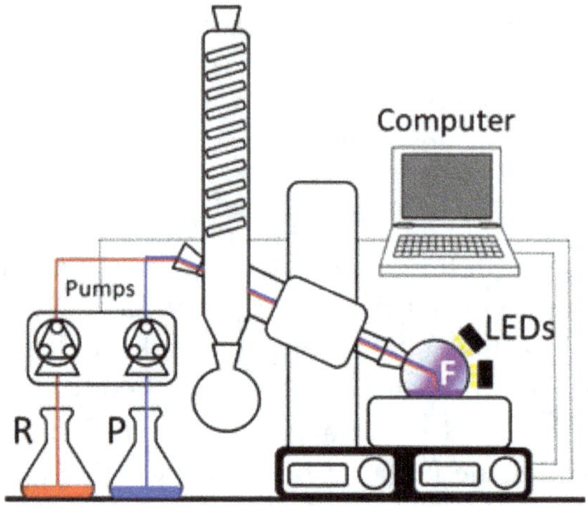

Figure 12.7. The *Photovap* developed by Poliakoff [25] (reprinted with permission from Ref. [25]. © American Chemical Society).

a continuous-flow mode. In the latter, the solution is directed to the bottom part of the tube through a jet feed inlet; when the liquid, thanks to the high-speed rotation, reaches the upper part of the tube, it is diverted to a drain for the collection of the product. The thickness of the film (which ranges around 200 μm) can be modulated according to the rotating speed (2000 to 10,000 rpm) and to the jet feed, governed by a peristaltic pump (Figure 12.8).

12.4.5 *Nebulization system devices*

Of a similar mindset, Vassilikogiannakis [28] came up with an innovative idea, inspired by nature where a great deal of oxygen photochemistry happens in the atmosphere. He built a pneumatic nebulizer-based photooxygenation system, named *NebPhotOx*, which he used for the photooxidation of β-citronellol catalyzed by Rose Bengal. The reactor operates by nebulizing the reaction mixture inside a cylindrical chamber (thus generating a "reaction cloud") surrounded by a low-energy LED strip. As a consequence, the interfacial area between the two phases is maximized, leading to

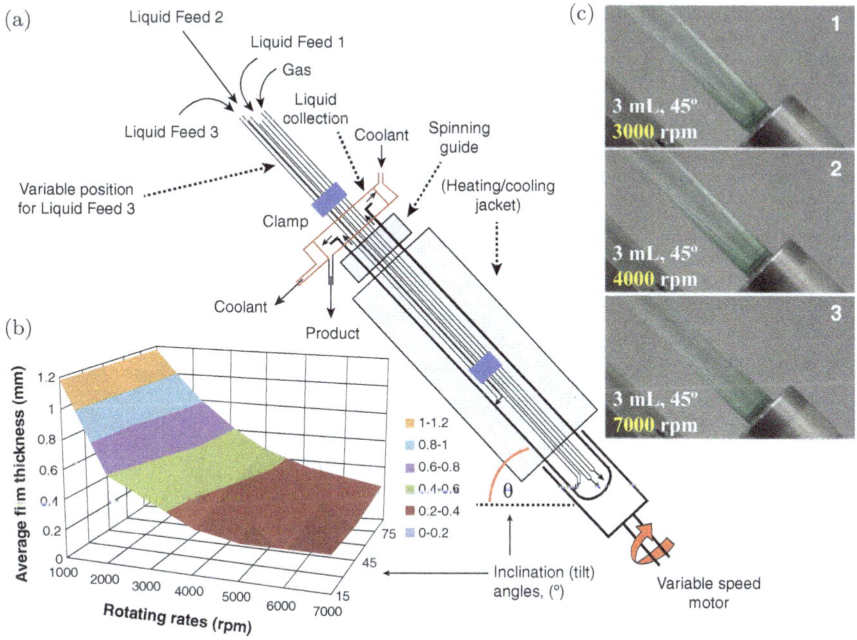

Figure 12.8. The rotating surface device developed by Raston and Stubbs reprinted with permission from Ref. [27]. © Nature Research Journals.

greatly improved reaction rates and optimized light penetration. The reactor uses pneumatically generated aerosols to maximize the surface area to volume ratio for the reaction solution, thus effectively eliminating the traditional problems associated with light penetration and oxygen solubility (Figure 12.9). A great advantage of this system regards the possibility to scale up singlet oxygen reactions (reaching flow rates of nearly 1 mL/min) and, due to its high productivity, it might in future become the model for industrial applications. At the moment, however, an industrial scale-up seems arduous, in view of the several measures that were taken to eliminate all possible ignition sources from the fume hood area (sparks or flames, e.g. the electricity transformer for the LEDs was kept outside of the fume hood) in which the *NebPhotOx* system was operated. Moreover, it was essential to maintain the fume hood always

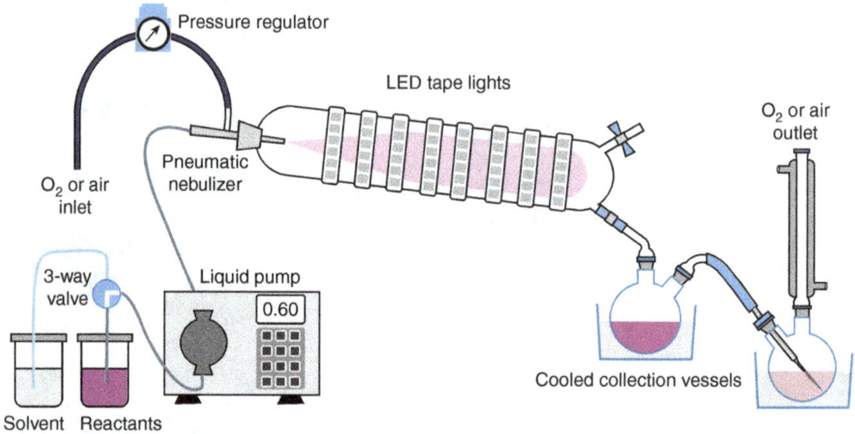

Figure 12.9. The *NebPhotOx* developed by Vassilikogiannakis (reprinted with permission from Ref. [28]. © Wiley-VCH Verlag GmbH & Co. KGaA).

adequately ventilated with a high air flow, in order to avoid oxygen stagnation in the system, and to equip the operators with safety glasses with side shields and flame-resistant safety clothing.

12.5 Reactions in Flow: Examples of Improved Processes

In this section, the photoorganocatalyzed reactions in which flow methods have produced an improvement of reactivity have been reviewed. The examples have been classified on the basis of the type of reactions involved and molecular changes occurring. This part is focused mainly on the scalability and efficiency issues, improved with flow methodologies. More details on the different reactor devices used can be found in the previous sections.

12.5.1 *Singlet oxygen-mediated oxidations*

Singlet oxygen ($^1O_2$) is a traceless reagent that can be generated from air using visible-spectrum light (sunlight or low-energy LED technologies) and natural photosensitizers. Furthermore, the reactions of singlet oxygen can be undertaken at high concentrations and in "preferred green" solvents, such as MeOH or water.

Furthermore, $^1O_2$ has the added benefit of being a very selective oxidant. The reactions of $^1O_2$ with a broad range of functional groups are too fast to allow good selectivity. In addition, $^1O_2$ does not react with alcohols and can be used in the presence of unprotected amines, thus not requiring the use of protecting groups, quite essential for other oxidants. Finally, the number of methodologies using singlet oxygen is currently multiplying very rapidly, because $^1O_2$ can be used to initiate high-yielding cascade reaction sequences which involve rapid and dramatic increase in molecular complexity.

In these kinds of reactions, the utility of continuous-flow methods is essentially correlated to the scalability of the process and safety. The use of microreactors enables one to obtain a large surface area to volume ratio and uniform light distribution across the reaction solution, allowing better performances and decreasing the danger due to the use of large amounts of oxygen. Here, the prominent examples present in the literature are reported, classified according to the type of oxidation.

12.5.1.1 *Oxygen addition to carbon–carbon double bonds: Generation of oxygenated hydrocarbons*

The reaction of α-terpinene **1** with photogenerated singlet oxygen ($^1O_2$) to form ascaridole **2** (Scheme 12.1) was first reported in the 1940s [29]. The batch synthesis suffers from well-known problems, such as the formation of *p*-cymene **3**, the most common by-product. Karlberg and co-workers proposed that this aromatic compound is formed as the result of a single electron-transfer reaction with triplet ground-state oxygen [30]. Moreover, ascaridole **2** is one of the most stable endoperoxide compounds, and probably for these reasons the continuous-flow photooxidation of terpinene **1** is one of the most investigated.

In Table 12.2, all the experiments reported in the literature in the last few years are collected. Since it is quite difficult to rationalize the results obtained using different flow systems, because the data are not homogeneous or missing, we decided to use the daily output

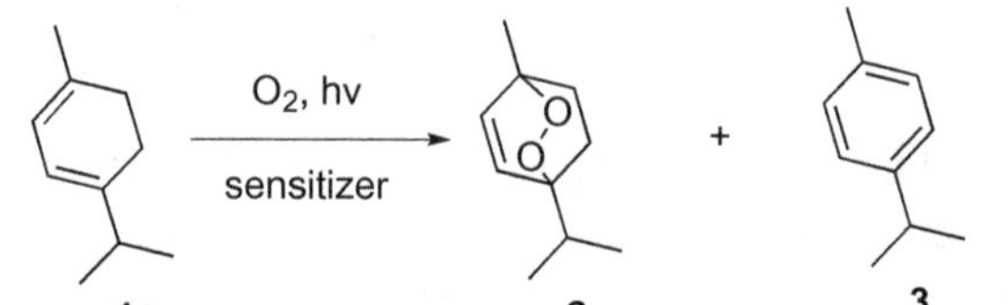

Scheme 12.1. Synthesis of ascaridole from α-terpinene.

Table 12.2. Photocatalytic conversion of α-terpinene **1** to **2** with different flow system devices.

Entry	Reactor	Sensitizer	Yield of **2** (Conversion)	DOP (mmol day^{-1})
1 [15]	Glass microchip	Rose Bengal	– (85%)	n.d.[a]
2 [15]	Batch	Rose Bengal	67% (–)	98
3 [23]	MCFs	Rose Bengal	– (90%)	3793
4 [21]	Batch	Methylene blue	–92% (–)	18
5 [21]	Monochannel microreactor (MC-MR)	Methylene blue	90% (–)	2
6 [21]	TIT-R	Methylene blue	87% (–)	623
7 [28]	*Neb-PhotOx*	Rose Bengal	93% (99%)	107,136

Notes: All reactions are carried out in MeOH, O_2, LED.
[a]No sufficient data were provided.

(DOP) as the parameter to compare the efficiency of the different reactors (when not given by the authors, DOP has been calculated using the following formula: reagent concentration (mmol mL^{-1})* reactor volume (mL)* yield (%)* 1440 (min)/reaction time (min). de Mello [15] was the first to demonstrate the feasibility of the reaction in a microflow reactor and its efficiency with respect to the batch conditions (Table 12.2, entries 1 and 2), but this preliminary experiment was performed on a small scale without the isolation of the product (isolated yield was not reported) and with a partial conversion. Recently, the same group proposed a new, more sophisticated system [23] based on MCFs (more details in Sec. 12.4.3), reaching excellent conversion and DOP value (entry 3). They suggested that this system could be particularly interesting for the large-scale production of **2**, though they did not perform the experiment. In 2015, Park published a paper on the photooxidation of monoterpenes [21], where they compared bulk conditions and two different flow systems (entry 4–6). While batch reaction produced various side-products, the yields obtained when using a MC-MR and a TIT-R were excellent, with significantly shortened reaction times. More importantly, the TIT-R has greater productivity compared with that of the MC-MR and the batch system. Thanks to its excellent productivity and greater working volume (7.92 mL against 0.026 mL of MC-MR), TIT-R can be considered a potential device for a feasible scale-up process in industrial photooxygenation processes.

Recently, Vassilikogiannakis and co-workers reported the same reaction using a conceptually new photoreactor: the "nebulizer-based photooxygenation system (*Neb-PhotOx*)" [28]. At first, this peculiar device (see Sec. 12.4.5) appears absolutely the most efficient with exceptionally high productivities (entry 7). The reason lies in the fact that the surface area to volume ratio is maximized, as previously discussed. Anyway, the scalability of this method could be affected by safety problems, as previously discussed.

Another well-studied reaction is the photooxidation of citronellol **4** using $^1O_2$. Moreover, it represents one of the few photochemical reactions used at an industrial scale to produce rose oxide **5**, an important fragrance (Scheme 12.2) [31].

For what concerns the photoorganocatalyzed reactions performed with flow methods, the principal results that appeared in the

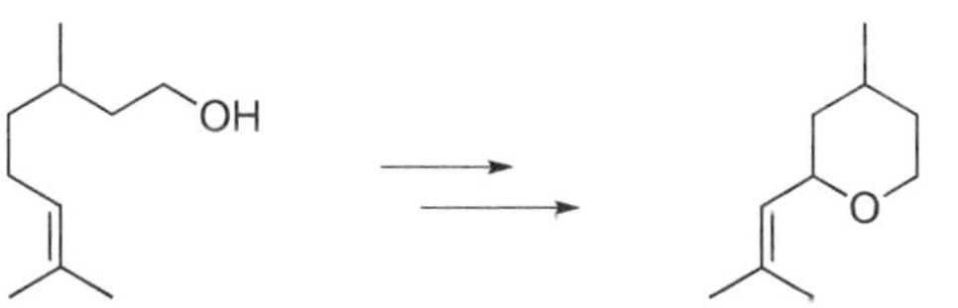

4
citronellol

5
Rose oxide

Scheme 12.2. Preparation of rose oxide from citronellol.

Table 12.3. Photocatalytic conversion of citronellol **4** into diols **7** performed with different flow-system devices.

Entry	Reactor	Sensitizer	Yield of **7** (Conversion)	DOP (mmol day^{-1})
1 [32]	Triple-channel microreactor	Methylene blue	– (99%)	70
2 [28]	*NebPhotOx*	Rose Bengal	87% (–)	100,224
3 [25]	Semicontinuous mode rotatory evaporator	Rose Bengal EtOH	– (77%)	165,988
4 [21]	Batch	Methylene blue	92% (–)	25.8
5 [21]	MC-MR	Methylene blue	99% (–)	6.5
6 [21]	TIT-R	Methylene blue	88% (–)	1752

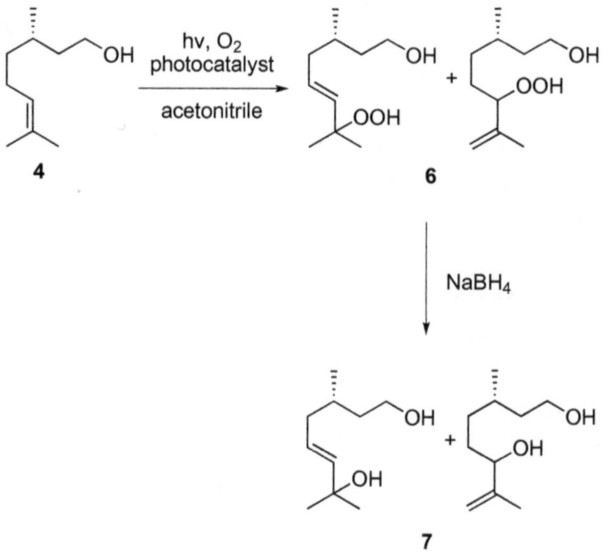

Scheme 12.3. Photocatalytic conversion of citronellol **4** in **6** and subsequent reduction to diols **7**.

literature have been summarized in Table 12.3. Due to the instability of the hydroxyperoxides **6** obtained, the yields or the conversions are generally calculated on the corresponding alcohols **7**, isolated after reduction of the reaction mixture (Scheme 12.3). In all cases, a mixture of isomers is obtained in an almost equimolar ratio. Impressive differences in efficiency have been achieved using various reactors. Again, the nebulizer-based system proposed by Vassilikogiannakis [28] showed excellent virtual productivity, otherwise all the limitations regarding the safety issue seen before still remain (entry 2). The same drawbacks are valid for the *PhotoVap* reactor built by Poliakoff (entry 3), which warns: "Any experiment involving flammable organic solvents and air or pure oxygen is potentially hazardous, especially when partially contained, as is the case of the flask of our reactor" [25].

Good results have been obtained by Kim employing a triple-channel microreactor (TCM, entry 1) [32]. Thanks to the higher gas–liquid interfacial area/volume (80 cm^{-1} for TCM against 0.76 cm^{-1} for batch reactor), the substrate was completely converted within a few minutes, whereas it took several hours to go to reach a complete conversion of the substrate in the classical batch conditions.

Again, the use of Park's TIT-R showed optimal performances, producing a 67.9-fold greater DOP than the batch system and 270.0-fold greater DOP than the MC-MR (entries 4–6) [21].

The photooxygenation of commercially available 1,5-dihydroxynaphthalene **8** to give 5-hydroxy-1,4-naphthoquinone (Juglone, **9**) in microflow systems has been studied independently by Oelgemöller [33] and de Oliveira [34] (Scheme 12.4).

The first group used a falling-film microreactor, Rose Bengal as the photosensitizer, and a green mixture of solvent (*i*PrOH/water), obtaining conversion rates of up to 31% after just 160 s of irradiation, while a conversion of 14% was achieved after 10 min under conventional batch conditions. More recently, de Oliveira published a similar reaction mediated by porphyrin derivatives realized in an in-house engineered photoreactor [34]. Interestingly, they demonstrated the scalability of their reaction, performing two experiments on a large scale using the optimized conditions. Running the reactions for 24 hrs, they obtained Juglone (**9**), respectively, in 72% (1.35 g) and 74% (1.40 g) yield, proving this protocol to be reproducible and in agreement with the predicted

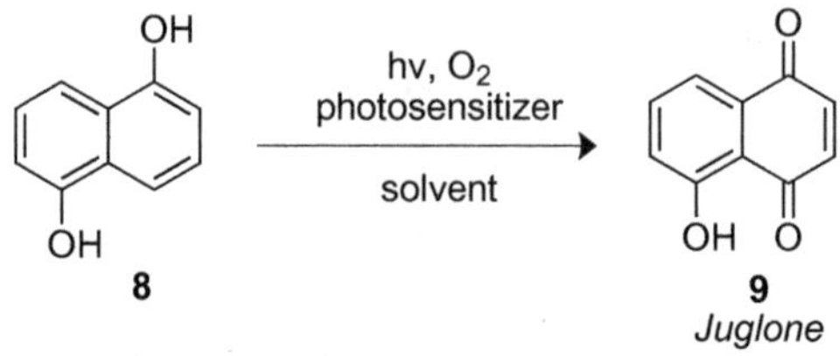

Scheme 12.4. Photocatalyzed synthesis of Juglone **9**.

space–time yield (STY, 1.54 g). This is a significant example of the benefits gained by flow technology combining a simple device, with mild and cost-competitive conditions.

Photoorganocatalyzed oxidations in flow have been also used as a green method to prepare natural products. Seeberger reported a convenient synthesis of Artemisinin (**10**), a very popular antimalarial drug, starting from dihydroartemisinic acid, using tetraphenylporphyrin (**TPP**) as photosensitizer and CH_2Cl_2 as solvent (Scheme 12.5) [35].

The assembled reactor results are quite complex because all the five steps needed for the conversion of Artemisinic acid (**11**) to **10** are conducted in flow. The sequence involves: (a) a double bond reduction, (b) a singlet oxygen reaction to form hydroperoxide **13,** (c) an acid-mediated Hock cleavage to give **14,** (d) an oxidation, and (e) a final condensation. The crucial steps were individually optimized and ultimately integrated in a fully continuous process, obtaining the desired product in 39% isolated yield. The authors declared that their setup could produce 200 g of artemisinin per day. This process was further scaled for industrial application in a follow-up paper [36]. In particular, **TPP** was substituted with 9,10-dicyanoanthracene (**DCA**) as a singlet oxygen sensitizer, as it provided a higher quantum efficiency and acid stability. In addition, higher selectivity towards the formation of the desired product was found to be performing the ene reaction at –20°C instead of room temperature. The industrial suitability of this process was demonstrated in 2014 when Sanofi started to exploit this novel semi-synthetic route to artemisinin [37].

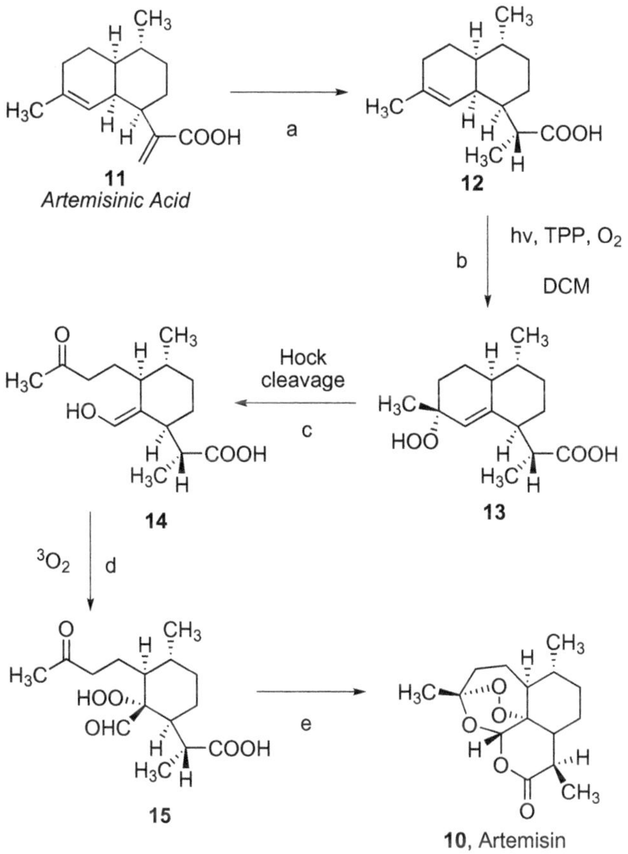

Scheme 12.5. Five-step synthesis of Artemisin **10** from Artemisinic acid, using photoorganocatalysis in flow conditions.

In 2015, Kappe studied the feasibility of the photooxidation of 5-hydroxymethylfurfural **16**, a very popular bio-based building block (Scheme 12.6) [38]. The reaction in batch conditions was already known, but suffered from several drawbacks, such as difficult scalability, low yields, long reaction times, and safety issues [39]. Using the optimized conditions (0.2 M of **16**, 1 mM Rose Bengal, *i*PrOH/H_2O 1:1, flow 0.5 mL/min), product **17** was obtained with an excellent conversion (98% determined by NMR analyses) and isolated yield after column chromatography (93%). The throughput under

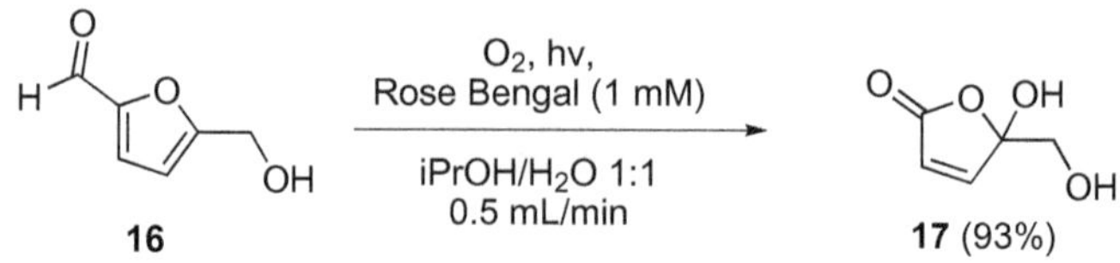

Scheme 12.6. Selective photooxidation of 5-hydroxymethylfurfural **16** in continuous flow.

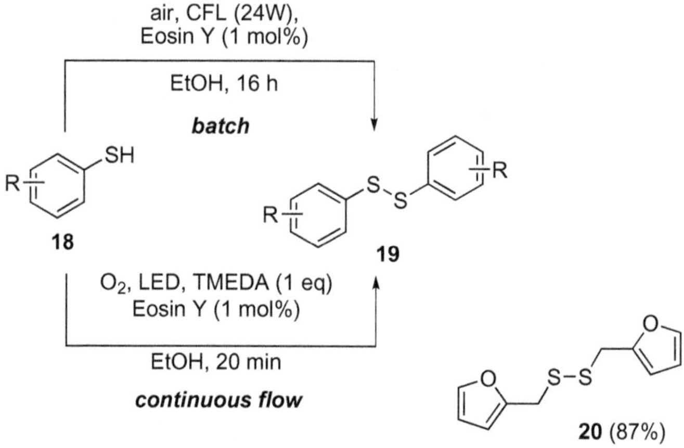

Scheme 12.7. Comparison of photocatalyzed oxidation of thiols in batch and continuous flow.

these conditions was approximately 0.75 g/h, and the reactor could be run for 5 h without loss of conversion or selectivity.

12.5.1.2 *Oxidation of thiols to disulfides*

Disulfides represent significant molecular and structural features in various biologically active compounds and fine chemicals. Therefore, the development of mild, efficient, and sustainable methods to access disulfides is of great importance. In 2015, Nöel's group reported the first example of metal-free photocatalytic aerobic oxidation of thiols to disulfides (Scheme 12.7) [40]. They optimized the reaction under classical batch conditions with Eosin Y as a photosensitizer, obtaining the product **19** with excellent yield in 16 hrs with no evidence of over-oxidation. Anyway, they observed that the rate of the reaction in batch was not completely reproducible, as it was strongly influenced by the mixing efficiency. That pushed them to explore a continuous-flow system.

The corresponding disulfides **19** were obtained in excellent yield and high purity, with a substantial acceleration compared to the corresponding batch reactions (all runs could be completed within a residence time of 20 min). The authors demonstrated the advantages of the flow reactor synthesizing difurfuryl disulfide **20**, an important food flavor additive [41] (no product was obtained in batch condition due to the instability of the substrates to the prolonged exposure to light irradiation), and producing oxytocin (full conversion could be obtained in water with only 200 s residence time)).

A similar photoinduced oxidation of thiols to disulfides was presented by Wacharasindhu using Rose Bengal as photosensitizer and a commercial continuous microflow reactor, obtaining comparable results [42].

12.5.1.3 *Oxidation of amines to imines*

In 2014, Seeberger and co-workers reported the singlet oxygen-mediated oxidation of secondary (**21**) and primary (**22**) amines to the corresponding imines (**23** and **24**), taking advantage of a variable-temperature continuous-flow LED-photoreactor [43]. The generated imines were trapped by nucleophiles, such as trimethylsilyl cyanide, to synthesize α-aminonitriles **25** and **26**, useful precursors of α-amino acids (Scheme 12.8).

In order to suppress dimeric products formed by nucleophilic addition of amine **22** to the primary aldimine **24** formed *in situ* (Scheme 12.8, route *a*), they carried out the reaction by cooling the irradiation chamber at −50°C (Scheme 12.8, route *b*). Benzylamine was used to compare the efficiency of the reaction conducted under the optimized continuous-flow conditions and in a classical batch reactor (Scheme 12.9). A notable increment in the reaction rate

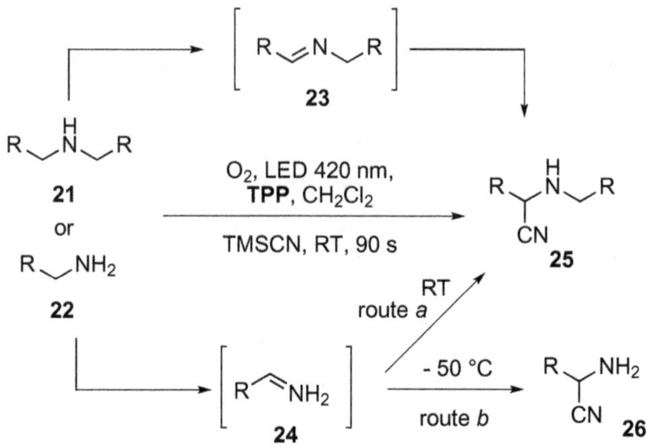

Scheme 12.8. Continuous flow synthesis of imines *via* oxidation.

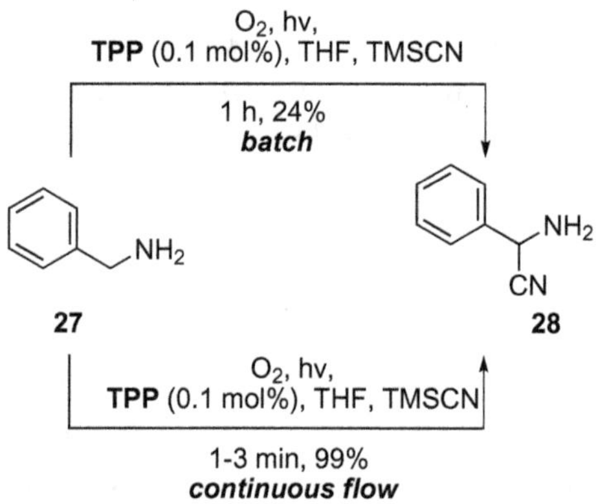

Scheme 12.9. Comparison between batch and continuous flow conditions for the oxidative cyanation of benzylamine.

and selectivity in the formation of the desired nitrile **28** was observed. The authors conveniently applied this strategy to the synthesis of unprotected amino acids [43] and fluorinated α-aminoacids starting from fluorinated amines [44].

By coupling the photooxidation with a second gas/liquid reaction with CO_2, hydantoins **29** have been prepared (Scheme 12.10) [45]. Using this approach, a variety of benzylic and aliphatic

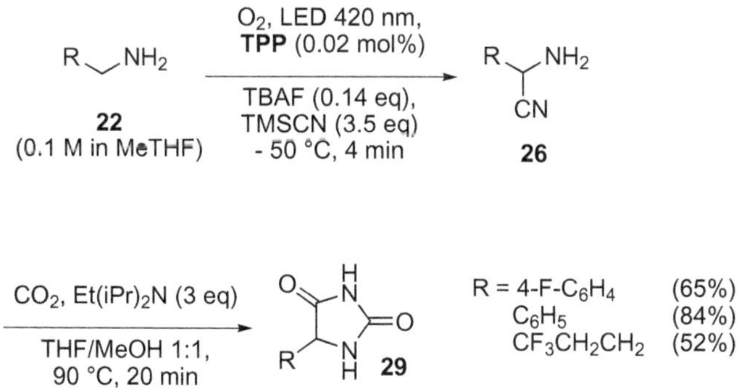

Scheme 12.10. The two-step semi-continuous flow synthesis of hydantoins from amines.

Scheme 12.11. Oxidative coupling reaction of *N*-aryl tetrahydroisoquinolines with nucleophiles.

unprotected hydantoins could be obtained in good yields without the need for chromatographic purifications.

12.5.2 *Photosubstitutions and additions*

The photoorganocatalyzed α-functionalization of tertiary amines through a dehydrogenative cross-coupling reaction [46] is one of the most studied reactions in continuous flow conditions.

In 2013, Rueping reported the functionalization of *N*-aryl tetrahydroisoquinolines catalyzed by Rose Bengal using a continuous flow reactor [19], trying to implement the procedure in the batch published by Tan few years before (Scheme 12.11) [47]. After optimization (low flow rate and increment of reactor volume using a longer tube), they were able to introduce different kind of groups such as nitroalkanes, TMSCN, dialkyl malonates, and dialkyl phosphites, obtaining the corresponding products in good to excellent yields (76%–92%) and showing higher performance than the reaction performed under batch conditions (3–5 hrs against 10–48 hrs).

Moreover, the authors applied their system to the synthesis of α-aminoamides by reacting *N,N*-dimethylanilines, isocyanides, and water in a Ugi-type reaction (Scheme 12.12). When this multicomponent reaction was carried out in batch [48], long reaction time and a quite expensive metal catalyst were required (2 days and iridium photoredox catalyst). With the employment of a recycling flow system (see Sec. 12.4.2), shorter reaction times (about 20 hrs) and the use of inexpensive, readily available Rose Bengal as catalyst were possible.

In 2015, Raston and Stubbs presented the reaction between *N*-phenyltetrahydroisoquinoline **34** and nitromethane using a rotating surface device (see Sec. 12.4.4) [26]. This thin film microfluidic vortex system (vortex fluidic device (VFD)) was initially employed in confined mode (tilt angle of 45°, rotational speed at 4000 rpm, and a final concentration of Rose Bengal 1 mol%), obtaining a high conversion and 80% of isolated yield in only 75 min (Scheme 12.13, route *a*). Notably, these optimized conditions were able to affect a 43-fold reduction in reaction time with comparable yields and a reduction in catalyst loading, compared to the Rueping's method. Moreover, the authors extended the strategy to an operational continuous mode, obtaining comparable results (Scheme 12.13,

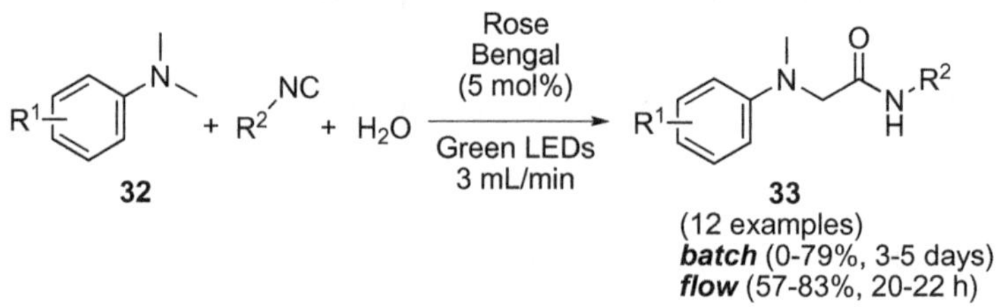

Scheme 12.12. Flow photoredox multicomponent reaction with Rose Bengal.

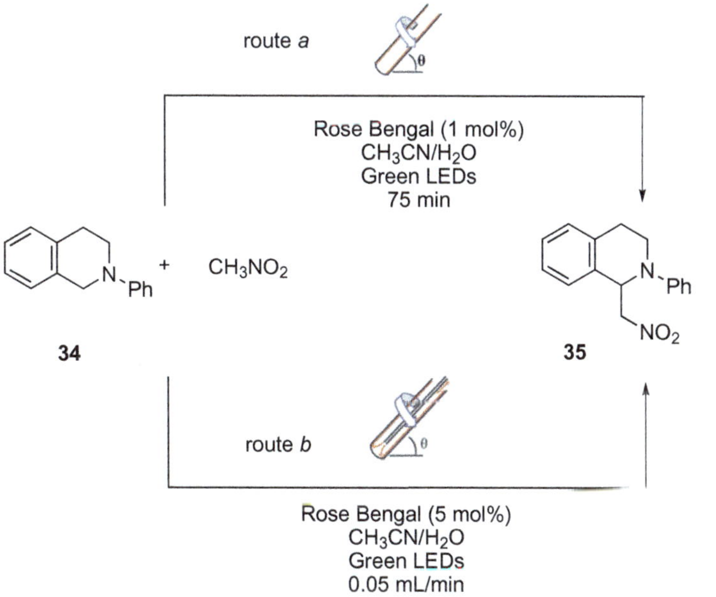

Scheme 12.13. Reaction between *N*-phenyltetrahydroisoquinoline **34** and nitromethane using the VFD in confined mode (route *a*) and flow mode (route *b*) (reprinted with permission from Ref. [26]. © The Royal Society of Chemistry).

route *b*). Interestingly, the oxidative multicomponent reaction published by Rueping was also carried out with the VFD in confined mode, showing a shorter reaction time (12 h) and greater yields, but no product was detected when the VFD system was operated in continuous mode.

Recently, Opatz published the oxidative α-cyanation of tertiary amines to give the corresponding α-aminonitriles **36–39** (Scheme 12.14) under the so called "sunflow" conditions [20]. The reaction was conducted under solar light irradiation with a microcapillary reactor. Among all the catalysts employed, Rose Bengal showed the best performances, producing very fast reaction times and high yield in comparison with the procedures given in the literature (Table 12.4).

Moreover, continuous flow methodologies can be very useful in photocatalyzed reactions involving unstable intermediates.

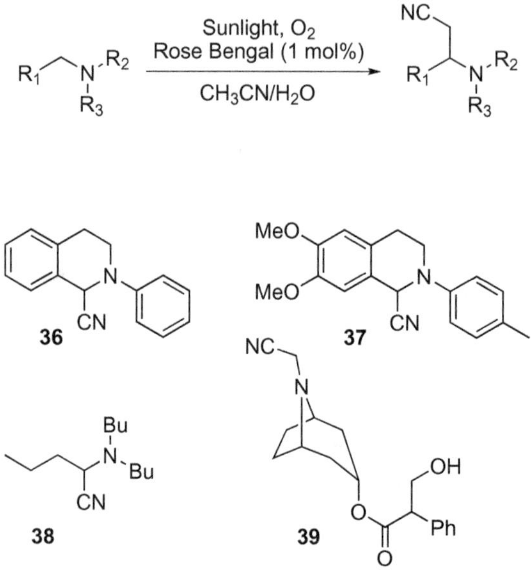

Scheme 12.14. Cyanation of various amines under "sunflow" conditions.

Table 12.4. Comparison of oxidative α-cyanation of tertiary amines under "sunflow" conditions and previous results.

	Sunflow		Literature [49]	
Compound	Yield (%)	Time (min)	Yield (%)	Time (min)
36	93	5	80	180
37	97	5	80	180
38	94	2	89	180
39	73	3	65	960

A remarkable example has been presented by Zeitler [18], who successfully applied organic dyes as photocatalysts for the cooperative organocatalytic asymmetric intermolecular alpha-alkylation of aldehydes (Scheme 12.15). After realizing the reaction in batch conditions [50], they tried to optimize the process in flow, using either a commercially available microreactor or an in-house constructed Booker-Milburn apparatus. The reaction time with the

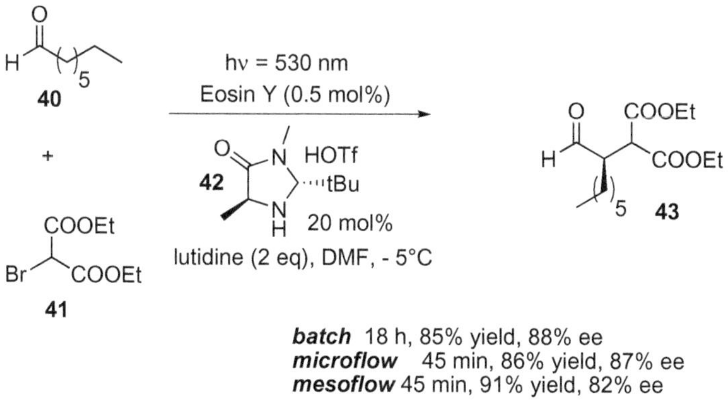

Scheme 12.15. Comparison of batch and flow for organocatalytic photoredox alpha-alkylation with catalyst **42** and Eosin Y.

microflow system was significantly shortened compared to what was observed under batch conditions, without erosion of yield and enantiomeric excess of **43**. Based on these promising results, they investigated the scalability of the process with a modified Booker-Milburn reactor (Sec. 12.4.2) with a bigger internal volume (about 10.5 mL), obtaining a 100-fold improvement of productivity and therefore feasibility for scale up. The reaction mechanism is quite complex due to the two catalytic cycles involved: the photoredox catalytic generation of the electron-deficient alkyl radical from **41** (with Eosin Y as catalyst) and the aminocatalyzed formation of the nucleophilic enamine intermediate from **40** (with MacMillan catalyst **42**).

Finally, Kappe's group presented a continuous-flow procedure for the α-trifluoromethylation of ketones **44**, where the substrate is initially converted into the corresponding silyl enol ether **46** and *in situ* subjected to a visible light-induced trifluoromethylation (Scheme 12.16) [51]. The flow setup consists of two different parts: in the first reactor, the reactive intermediate silyl enol ether **46** is formed, while in the second the photocatalyzed trifluoromethylation is carried out. It must be noted that the combination of flow conditions, Eosin Y as catalyst and CF_3SO_2Cl, seems more efficient and simpler compared to the conditions reported in the

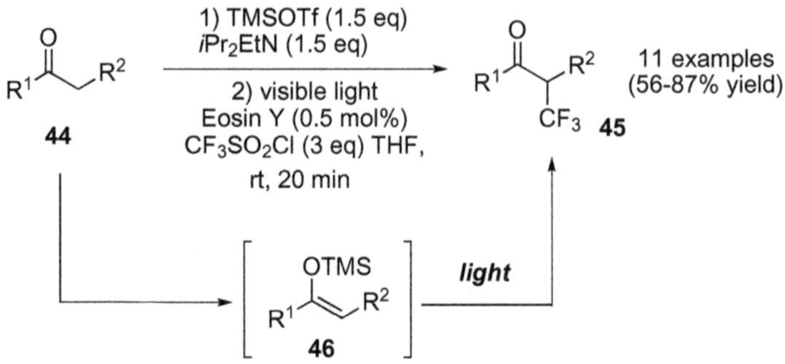

Scheme 12.16. Photocatalyzed continuous-flow preparation of alpha trifluoromethyl ketones **45**.

literature [52]. The two-step process requires only 20 min for completion, has a simple setup (no degassed solvents or inert atmosphere), and has a broad scope.

12.6 Conclusions

The field of flow photoorganocatalyzed reactions is in continuous expansion and it is difficult to foresee what future directions will be. However, one of the issues that, so far, has remained mostly unattended is the recovery of the catalyst. The extension of the existing methodologies to industrial applications cannot ignore this aspect. Attempts to support the organic photocatalyst on heterogenous materials such as PVC beads or aerogel monoliths have been recently reported [53], showing, however, problems of leaching and/or reduced activity. Reiser [54] recently reported the synthesis of a polyisobutylene-tagged *fac*-Ir(ppy)$_3$ complex and its application as recyclable visible-light photocatalyst in a continuous flow process, thus esulating from the contents of this chapter. But what the authors accomplished is indeed intriguing: they exploited thermomorphic solvent systems to separate the catalyst from the product solution. Thermomorphic couples of solvents have the peculiarity of being perfectly miscible with each other above a certain temperature and immiscible below it. The reaction takes place in a homogeneous phase above a certain temperature, then the solution is moved into a cooled phase separator unit. At this stage, the densest solvent, dissolving the product, is collected, while the less dense part, containing the catalyst, remains floating above and is consequently reconveyed in the heatable microreactor where it can continue to catalyze the reaction.

The genius of chemists is often surprising; therefore, it is reasonable to expect in the near future breakthrough solutions in the field of flow photoorganocatalyzed reactions as well [55].

References

[1] Wegner, J.; Ceylan, S.; Kirschning, A. *Chem. Commun.* **2011**, *47*, 4583–4592.

[2] Mizuno, K.; Nishiyama, Y.; Ogaki, T.; Terao, K.; Ikeda, H.; Kakiuchi, K. *J. Photochem. Photobiol. C*, **2016**, *29*, 107–147.

[3] Taylor, G. I. *J. Fluid. Mech.* **1961**, *10*, 161–165.

[4] Coyle, E. E.; Oelgemöller, M. *Photochem. Photobiol. Sci.* **2008**, *7*, 1313–1322.

[5] Su, Y.; Kuijpers, K.; Hessel, V.; Nöel, T. *React. Chem. Eng.* **2016**, *1*, 73–81.

[6] Ray, K. A.; Beenackers, A. C. M. *Catal. Today* **1998**, *40*, 73–83.

[7] Mukherjee, P. S.; Ray, K. A. *Chem. Eng. Technol.* **1999**, *22*, 253–260.

[8] Andrews, R. S.; Becker, J. J.; Gagné, M. R. *Angew. Chem. Int. Ed.* **2012**, *51*, 4140–4143.

[9] Tucker, J. W.; Zhang, Y.; Jamison, T. F.; Stephenson, C. R. J. *Angew. Chem. Int. Ed.* **2012**, *51*, 4144–4147.

[10] Turconi, J.; Griolet, F.; Guevel, R.; Oddon, G.; Villa, R.; Geatti, A.; Hvala, M.; Rossen, K.; Goller, R.; Burgard, A. *Org. Process Res. Dev.* **2014**, *18*, 417–422.

[11] Kirschning, A.; Solodenko, W.; Mennecke, K. *Chem. Eur. J.* **2006**, *12*, 5972–5990.

[12] Kimmel, S.; Tromberg, B. J.; Roberts, W. G.; Berns, M. W. *Photochem. Photobiol.* **1989**, *50*, 175–183.

[13] Ngen, E. J., Xiao, L.; Rajaputra, P.; Yan, X.; You, Y. *Photochem. Photobiol.* **2013**, *89*, 841–848.

[14] Lapkin, A. A.; Boddu, V. M.; Aliev, G. N.; Goller, B.; Polisski, S.; Kovalev, D. *Chem. Eng. J.* **2008**, *136*, 331–336.

[15] Wooton, R. C. R.; Fortt, R.; de Mello, J. R. *Org. Proc. Res. Dev.* **2002**, *60*, 187–189.

[16] Coley, C. W.; Abolhasani, M.; Lin, H.; Jensen, K. F. *Angew. Chem. Int. Ed.* **2017**, *56*, 9847–9850.

[17] Hook, B. D. A.; Dohle, W.; Hirst, P. R.; Pickworth, M.; Berry, M. B.; Booker-Milburn K. I. *J. Org. Chem.* **2005**, *70*, 7558–7564.

[18] Neumann, M.; Zeitler, K. *Org. Lett.* **2012**, *14*, 2658–2661.

[19] Rueping, M.; Vila, C.; Bootwicha, T. *ACS Catal.* **2013**, *3*, 1676–1680.

[20] Nauth A. M.; Lipp, A.; Lipp, B.; Opatz, T. *Eur. J. Org. Chem.* **2017**, 2099–2103.

[21] Park, C. Y.; Kim, Y. J.; Lim, H. J.; Park, J. H.; Kim, M. J.; Seo, S. W.; Park, C. P. *RSC Adv.* **2015**, *5*, 4233–4237.

[22] Jahnisch, K.; Dingedissen, U. *Chem. Eng. Technol.* **2005**, *28*, 426–427.

[23] Elvira, K. S.; Wootton, R. C. R.; Reis, N. M.; Mackley, M. R.; de Mello, A. J. *ACS Sustainable Chem. Eng.* **2013**, *1*, 209–213.

[24] Penders, I. G. T. M.; Amara, Z.; Horvath, R.; Rossen, K.; Poliakoff, M.; George, M. W. *RSC Adv.* **2015**, *5*, 6501–6504.

[25] Clark, C. A.; Lee, D. S.; Pickering, S. J.; Poliakoff, M.; George, M. W. *Org. Process Res. Dev.* **2016**, *20*, 1792–1798.

[26] Gandy, M. N.; Raston, C. L.; Stubbs, K. A. *Chem. Commun.* **2015**, *51*, 11041–11044.

[27] Yasmin, L.; Chen, X.; Stubbs, K. A.; Raston, C. L. *Sci. Rep.* **2013**, *3*, 2282.

[28] Ioannou, G. I.; Montagnon, T.; Kalaitzakis, D.; Pergantis, S. A.; Vassilikogiannakis, G. *ChemPhotoChem.* **2017**, *1*, 173–177.

[29] Schenck, O.; Ziegler, K. *Naturwissenschaften* **1944**, *32*, 157.

[30] Rudbäck, J.; Bergström, M. A.; Börje, A.; Nilsson, U.; Karlberg, A.-T. *Chem. Res. Toxicol.* **2012**, *25*, 713–721.

[31] Monnerie, N.; Ortner, J. *Sol. Energy Eng.* **2001**, *123*, 171–174.

[32] Mauryal, R. A.; Park, C. P.; Kim, D.-P. *Beilstein J. Org. Chem.* **2011**, *7*, 1158–1163.

[33] Oksana Shvydkiv, O.; Limburg, C.; Nolan, K.; Oelgemöller, M. *J. Flow Chem.* **2012**, *2*, 52–55.

[34] de Oliveira, K. T.; Millera, L. Z.; McQuade, D. T. *RSC Adv.* **2016**, *6*, 12717–12725.

[35] Levesque, F.; Seeberger, P. H. *Angew. Chem. Int. Ed.* **2012**, *51*, 1706–1709.

[36] Kopetzki, D.; Lvesque, F.; Seeberger, P. H. *Chem. Eur. J.* **2013**, *19*, 5450–5456.

[37] (a) Dhainaut, J.; Dlubala, A.; Guevel, R.; Medard, A.; Oddon, G.; Raymond, N.; Turconi, J. (Sanofi–Aventis, Fr.), WO 2011/026865, 2011; (b) Kraft, V.; Kretzschmar, G.; Rossen, K. (Sanofi –Aventis, Fr.), WO 2011/030223, 2011

[38] Heugebaert, T. S. A.; Stevens, C. V.; Kappe, C. O. *ChemSusChem* **2015**, *8*, 1648–1651.

[39] Canterino, M.; Di Somma, I.; Raffaele, M.; Andreozzi, R.; Caprio, V. *J. Photochem. Photobiol. A* **2010**, *210*, 69–76.

[40] Talla, A.; Driessen, B.; Straathof, N. J. W.; Milroy, L.-G.: Brunsveld, L.; Hessel, V.; Nöel, T. *Adv. Synth. Catal.* **2015**, *357*, 2180–2186.

[41] Blank, I.; Pascual, E. C.; Devaud, S.; Fay, L. B.; Stadler, R. H.; Yeretzian, C.; Goodman, B. A. *J. Agric. Food Chem.* **2002**, *50*, 2356–2364.

[42] Tankam, T.; Poochampa, K.; Vilaivan, T.; Sukwattanasinitt, M.; Wacharasindhu, S. *Tetrahedron* **2016**, *72*, 788–793.

[43] Ushakov, D. B.; Gilmore, K.; Kopetzki, D.; McQuade, D. T.; Seeberger P. H. *Angew. Chem. Int. Ed.* **2014**, *53*, 557–561.

[44] Vukelic, S.; Ushakov, D. B.; Gilmore, K.; Koksch, B.; Seeberger P. H. *Eur. J. Org. Chem.* **2015**, 3036–3039.

[45] Vukelic, S.; Koksch, B.; Seeberger, P. H.; Gilmore, K. *Chem. Eur. J.* **2016**, *22*, 13451–13454.

[46] Li, C.-J. *Acc. Chem. Res.* **2009**, *42*, 335–344.

[47] Pan, Y.; Kee, C. W.; Chen, L.; Tan C.-H. *Green Chem.* **2011**, *13*, 2682–2685.

[48] Rueping, M.; Vila, C. *Org. Lett.* **2013**, *15*, 2092–2095.

[49] Pacheco, J. C. O; Lipp, A.; Nauth, A. M.; Acke, F.; Dietz, J.-P.; Opatz, T. *Chem. Eur. J.* **2016**, *22*, 5409–5415.

[50] Neumann, M.; Fuldner, S.; Konig, B.; Zeitler, K. *Angew. Chem., Int. Ed.* **2011**, *50*, 951–954.

[51] Cantillo, D.; de Frutos, O.; Rincón, J. A.; Mateos, C.; Kappe, C. O. *Org. Lett.* **2014**, *16*, 896–899.

[52] Pham, P. V.; Nagib, D. A.; MacMillan, D. W. C. *Angew. Chem., Int. Ed.* **2011**, *50*, 6119–6122.

[53] Han, X.; Bourne, R. A.; Poliakoff, M.; George, M. W. *Chem. Sci.* **2011**, *2*, 1059–1067.

[54] Rackl, D.; Kreitmeier, P.; Reiser, O. *Green Chem.* **2016**, *18*, 214–219.

[55] Cambié, D.; Bottecchia, C.; Straathof, N. J. W.; Hessel, V.; Noël, T. *Chem. Rev.*, **2016**, *116*, 10276–10341.

Index

B

C

CATALYTIC SCIENCE SERIES

(Continued from page ii)

Vol. 8 *Petrochemical Economics: Technology Selection in a Carbon Constrained World*
by D. Seddon

Vol. 7 *Combinatorial Development of Solid Catalytic Materials: Design of High-Throughput Experiments, Data Analysis, Data Mining*
edited by M. Baerns and M. Holeňa

Vol. 6 *Catalysis by Gold*
edited by G. C. Bond, C. Louis and D. T. Thompson

Vol. 5 *Supported Metals in Catalysis*
edited by J. A. Anderson and M. F. García

Vol. 4 *Isotopes in Heterogeneous Catalysis*
edited by Justin S. J. Hargreaves, S. David Jackson and Geoff Webb

Vol. 3 *Zeolites for Cleaner Technologies*
edited by Michel Guisnet and Jean-Pierre Gilson

Vol. 2 *Catalysis by Ceria and Related Materials*
edited by Alessandro Trovarelli

Vol. 1 *Environmental Catalysis*
edited by F. J. J. G. Janssen and R. A. van Santen

www.ingramcontent.com/pod-product-compliance
Lightning Source LLC
LaVergne TN
LVHW021544021026
849161LV00005BA/947